AF790578

Springer Series in
OPTICAL SCIENCES 76

founded by H.K.V. Lotsch

Editor-in-Chief: W. T. Rhodes, Metz

Editorial Board: T. Asakura, Sapporo
K.-H. Brenner, Mannheim
T. W. Hänsch, Garching
F. Krausz, Wien
H. Weber, Berlin

Springer-Verlag Berlin Heidelberg GmbH

Physics and Astronomy ONLINE LIBRARY

http://www.springer.de/phys/

Springer Series in
OPTICAL SCIENCES

The Springer Series in Optical Sciences, under the leadership of Editor-in-Chief *William T. Rhodes*, Georgia Institute of Technology, USA, and Georgia Tech Lorraine, France, provides an expanding selection of research monographs in all major areas of optics: lasers and quantum optics, ultrafast phenomena, optical spectroscopy techniques, optoelectronics, information optics, applied laser technology, industrial applications, and other topics of contemporary interest.
With this broad coverage of topics, the series is of use to all research scientists and engineers who need up-to-date reference books.

The editors encourage prospective authors to correspond with them in advance of submitting a manuscript. Submission of manuscripts should be made to the Editor-in-Chief or one of the Editors. See also http://www.springer.de/phys/books/optical_science/os.htm

Editor-in-Chief

William T. Rhodes
Georgia Tech Lorraine
2-3, rue Marconi
57070 Metz, France
Phone: +33 387 20 3922
Fax: +33 387 20 3940
E-mail: wrhodes@georgiatech-metz.fr
URL: http://www.georgiatech-metz.fr
http://users.ece.gatech.edu/~wrhodes

Editorial Board

Toshimitsu Asakura
Faculty of Engineering
Hokkai-Gakuen University
1-1, Minami-26, Nishi 11, Chuo-ku
Sapporo, Hokkaido 064-0926, Japan
E-mail: asakura@eli.hokkai-s-u.ac.jp
(*Special Editor for Optics in the Pacific Rim*)

Karl-Heinz Brenner
Chair of Optoelectronics
University of Mannheim
B6, 26
68131 Mannheim, Germany
Phone: +49 (621) 292 3004
Fax: +49 (621) 292 1605
E-mail: brenner@rumms.uni-mannheim.de
URL: http://www.ti.uni-mannheim.de/~oe

Theodor W. Hänsch
Max-Planck-Institut für Quantenoptik
Hans-Kopfermann-Strasse 1
85748 Garching, Germany
Phone: +49 (89) 2180 3211 or +49 (89) 32905 702
Fax: +49 (89) 32905 200
E-mail: t.w.haensch@physik.uni-muenchen.de
URL: http://www.mpq.mpg.de/~haensch

Ferenc Krausz
Institut für Angewandte Elektronik
und Quantenelektronik
Technische Universität Wien
Gusshausstr. 27/359
1040 Wien, Austria
Phone: +43 (1) 58801 35937
Fax: +43 (1) 58801 35997
E-mail: krausz@iaee.tuwien.ac.at
URL: http://www.tuwien.ac.at

Horst Weber
Optisches Institut
Technische Universität Berlin
Strasse des 17. Juni 135
10623 Berlin, Germany
Phone: +49 (30) 314 23585
Fax: +49 (30) 314 27850
E-mail: weber@physik.tu-berlin.de
URL: http://www.physik.tu-berlin.de/institute/OI/Weber/Webhome.htm

Hans J. Coufal Demetri Psaltis
Glenn T. Sincerbox (Eds.)

Holographic Data Storage

With a Foreword
by Alstair M. Glass and Mark J. Cardillo

With 228 Figures, 64 in Color and 12 Tables

Springer-Verlag Berlin Heidelberg GmbH

Dr. Hans J. Coufal
IBM Corporation
Almaden Research Center
San Jose, CA 95120-6099, USA
E-mail: coufal@almaden.ibm.com

Professor Demetri Psaltis
California Institute of Technology
Department of Electrical Engineering
Pasadena, CA 91125, USA
E-mail: psaltis@caltech.edu

Professor Glenn T. Sincerbox
University of Arizona
Optical Sciences Center
Tucson, AZ 85721, USA
E-mail: sinbox@u.arizona.edu

Library of Congress Cataloging-in-Publication Data

Holographic data storage / H.J. Coufal, D. Psaltis, G. Sincerbox (Eds.)
p. cm. -- (Springer series in optical sciences, ISSN 0342-4111 ; 76)
Includes bibliographical references and index.
ISBN 9783540666912
1. Optical storage devices. 2. Holography. 3. Computer storage devices. 4. Optical data processing. I. Coufal, H. II. Psaltis, Demetri. III. Sincerbox, Glenn T., 1937- IV. Springer series in optical sciences ; v. 76.

TA1635 .H65 2000
621.39'767--dc21

00-023971

ISSN 0342-4111

ISBN 978-3-540-66691-2 Springer-Verlag Berlin Heidelberg New York

This work is subject to copyright. All rights are reserved, whether the whole or part of the material is concerned, specifically the rights of translation, reprinting, reuse of illustrations, recitation, broadcasting, reproduction on microfilm or in any other way, and storage in data banks. Duplication of this publication or parts thereof is permitted only under the provisions of the German Copyright Law of September 9, 1965, in its current version, and permission for use must always be obtained from Springer-Verlag. Violations are liable for prosecution under the German Copyright Law.

Springer-Verlag Berlin Heidelberg New York
a member of BertelsmannSpringer Science+Business Media GmbH

© Springer-Verlag Berlin Heidelberg 2000

The use of general descriptive names, registered names, trademarks, etc. in this publication does not imply, even in the absence of a specific statement, that such names are exempt from the relevant protective laws and regulations and therefore free for general use.

Data conversion by EDV-Beratung F. Herweg, Hirschberg
Cover concept by eStudio Calamar Steinen using a background picture from The Optics Project. Courtesy of John T. Foley, Professor, Department of Physics and Astronomy, Mississippi State University, USA
Cover production: *design & production* GmbH, Heidelberg

Printed on acid-free paper SPIN 10688525 56/3141/di 5 4 3 2 1 0

Foreword

In the late 1960s and early 1970s holographic storage was actively pursued by a number of development teams in major laboratories. The potential of high capacity information storage using optical techniques, exploiting the ability to address submicron dimensions ($\geq 1\,\mathrm{bit}/\mu^2$), had been realized many years previously, and led soon thereafter to the compact disc. However, the ability to use the thickness dimension with holographic techniques, thereby achieving storage densities in excess of 100 $\mathrm{Gb/cm^3}$, would indeed have been a breakthrough in storage technology! By the mid 1970s, however, no practical holographic recording system emerged from this work, even though experimental digital holographic systems had been demonstrated in the laboratory. There were fundamental reasons for this: The most important was that there was no satisfactory storage medium. Furthermore, holographic storage systems required many complex components that were still in their infancy and often extremely expensive with marginal performance, such as high power lasers, spatial light modulators, large detector arrays, and beam deflector optics. Despite extensive work on new materials including photochromics, photographic emulsions, thermomagnetic materials, amorphous semiconductors, thermoplastics, photopolymers, and photorefractive materials, none satisfied the stringent requirements for high recording sensitivity, dynamic range, signal/noise nor long storage time or degradation during readout. The focus of most of this early work was on in situ read–write–erase memory with stationary media and beam deflection for rapid random access. Eventually patience ran out, and with the emergence of magnetic recording the field of holographic storage was all but abandoned.

Out of this situation grew holographic optical image processing. Photorefractive materials developed for storage were found to be excellent candidates for doing high-speed optical correlations for pattern recognition and other special purpose applications, with the associated relaxed materials requirements. Research in this area has continued with many fascinating innovations, some of which are represented in this volume.

In the last few years we have seen a revival of activity in holographic storage. This has been driven by several factors. Firstly, the availability and performance of the requisite components for a system have changed dramatically. Laser technology now offers higher powers, many more wavelengths, and much lower cost. Spatial light modulators have been developed for high performance display applications such as liquid crystal devices and silicon micromirror arrays. The invention of CMOS cameras has brought with it the prospect of a high-performance, low-cost detector.

Secondly, although dramatic advances in storage technology have taken place, both with magnetic materials (disc and tape) and optical discs, we think we can see the limits of these technologies looming on the horizon: the paramagnetic limit in the case of magnetic storage and spot-size limits imposed by diffraction in the case of optical storage. The superior potential of holographic storage, both in storage capacity and data transfer rate, has stimulated renewed thinking.

Thirdly, a plethora of new applications and market opportunities requiring very large storage capacities and transfer rates seem to have stimulated a transition from systems based on exotic research concepts to more practical and realizable approaches. As an example, new emphasis is now being placed on write–once–read-many (WORM) memories. Low cost polymeric storage media have been developed which meet all the requirements for practical storage systems. New multiplexing techniques have been developed which permit very high density storage with simple optical system configurations. These approaches could satisfy the requirements of the rapidly expanding data warehousing and enterprise server industries, as well as the many applications associated with the explosion of bandwidth to meet the demands of internet traffic. They could also serve the emerging HDTV market, allowing consumers to record the huge video TV files or download over the internet.

As we think you will conclude from this text, over the last few years it appears that the major hurdles to achieving a practical high-performance holographic storage system have been overcome. As we enter the 21st century, holographic storage technology is poised for commercial acceptance.

Bell Laboratories, Lucent Technologies
July 2000

Alastair M. Glass
and
Mark J. Cardillo

Preface

"For nearly four decades, holographic memory has been the great white whale of technology research. Despite enormous expenditures, a complete, general-purpose system that could be sold commercially continues to elude industrial and academic researchers. Nevertheless, they continue to pursue the technology aggressively because of its staggering promise." John William Toigo, Scientific American, May 2000.

The time has finally come for Captain Ahab to confront his nemesis. Recent interest and progress in holographic data storage has been very substantial, yet the question arises whether holographic data storage will always be only an emerging technology or if it will finally emerge as a viable technology. While the jury is still out on the question of economic viability, the technical feasibility of holographic data storage has been demonstrated in laboratory experiments with impressive performance, close to the very high expectations.

A working holographic data storage system is a rather complex device. It obviously involves a data input device, such as a liquid crystal display, and optics to create a hologram of the input page in a suitable storage material. During read out of the data the hologram is used to reconstruct the stored data page onto a camera chip. To optimize the performance of the system the data have to be encoded before storage and then decoded during the readout process before they can be used. Such a device must store and retrieve user data reliably, without any noticeable error even after a long period of storage. This requires, therefore, the development of specialized hardware components for input and output, as well as suitable recording materials and complex algorithms. This book attempts to cover all of the above aspects, as well as the pertinent facts about competing and well established storage techniques.

This book is intended mainly for non-experts as an introduction to the field, but also serves as a synopsis of the current state of the art in the field for those currently involved in it. Individual chapters are contributed by groups working actively in the area, reviewing their work and that of others with extensive references to recent publications in the scientific literature. No one can do justice to the outstanding efforts of all the groups around the globe that have contributed to the field over the last forty years. We had to be selective, while attempting to be at the same time comprehensive, to fit the gist of holographic data storage into just one book.

Many of the research projects in industry and academia in the United States, that have brought holographic data storage close to becoming a technology, were sponsored by DARPA, AFRL and AFOSR and we would like to take this opportunity to thank Dr. L.N. Durvasula, Dr. A. Jambardino and Dr. A. Craig for their vision, leadership and support.

To those who taught us the basic optics and physics that underlie holographic storage, those who helped us first to appreciate and later to understand the issues that had to be solved, and to those who finally solved them with us and for all of us we owe a debt of gratitude.

Last, but by no means least, we would like to thank Dr. H. Lotsch of Springer-Verlag for the encouraging discussions that persuaded us to serve as editors of this book, and the team at Springer-Verlag who made it finally happen.

We would like to dedicate this book to the giants in this field on whose shoulders we are standing today with holographic data storage devices almost within reach. They laid the solid foundation that we have been privileged to build upon.

San Jose, *Hans J. Coufal*
Pasadena, *Demetri Psaltis*
Tucson, *Glenn T. Sincerbox*
July 2000

Contents

List of Contributors XXI

Part I Introduction

History and Physical Principles
G.T. Sincerbox 3

1 Holographic Storage Principles 4
1.1 Redundant Storage 6
1.2 Multiplexing 7
1.3 High Data Rate 9
1.4 Rapid Access 9
1.5 Novel Functions 9
2 Historical Development 10
2.1 Bell Labs and the Digital Page 11
2.2 IBM HOSP 11
2.3 RCA Holographic Memory 12
2.4 3M Holographic Data Storage System 12
2.5 Thompson-CSF Read–Write Memory Using Angular Multiplexing 13
2.6 NEC Holographic Coding Plate or Holotablet 13
2.7 Harris-Intertype Wide-Band Recorder 13
2.8 Hitachi Holographic Video Disk 14
2.9 Optical Data Systems Holoscan 14
2.10 Holographic Storage in the Soviet Union 14
2.11 NEC Holographic Disk 15
2.12 MEI Kanji Character Generation System 15
2.13 Tamarack Multistore 16
2.14 The PRISM Test Stand 16
2.15 Stanford University 16
2.16 Holoplex Memory Device for Fingerprint Verification 17
2.17 Rockwell Read-Only Demonstrator 17
2.18 IBM DEMON 18
3 Summary 18
References 19

Volume Holographic Multiplexing Methods
G. Barbastathis and D. Psaltis 21

1 Holographic Storage and Retrieval 21
1.1 Overview of Holographic Multiplexing Methods 25
1.2 Holographic Storage Geometries and Imaging Systems 28
2 Scattering from Volume Gratings 30
2.1 Volume Diffraction in the Born Approximation 31
2.2 Volume Diffraction of Scalar Fields 33
2.3 Volume Diffraction Calculations Using the k-Sphere Formulation 42
2.4 Visualization of the Multiplexing Methods on the Grating Space 47
2.5 Grating Manifold Motion and Fractal Multiplexing 52
3 Architectures for Holographic Memories 55
3.1 The Holographic 3-D Disk Geometry 55
3.2 The Holographic Random-Access Memory (HRAM) 57
3.3 The Phase Conjugate Geometry 57
4 Summary 59
References 59

Fundamental Noise Sources in Volume Holographic Storage
C. Gu, P. Yeh, X. Yi, and J. Hong 63

1 Cross-Talk Noise 63
1.1 Theoretical Formulation 64
1.2 Cross-Talk Noise and Signal-to-Noise Ratio 66
1.3 Storage Capacity 72
2 Intrinsic Scattering Noise 76
3 Noise Gratings 79
4 Conclusion 87
References 87

Part II Recording Media

Bit Error Rate for Holographic Data Storage
J.A. Hoffnagle and C.M. Jefferson 91

1 Definition of Bit Error Rate 92
2 BER in Terms of Pixel Distribution Functions 93
3 Experimental Distributions of CCD Pixel Values 94
4 Applications 99
References 100

Media Requirements for Digital Holographic Data Storage
R.M. Shelby 101

1 Ideal Media Parameters 101
1.1 Optical Quality 101
1.2 Sensitivity 103
1.3 Dynamic Range 104
1.4 Absorption 105
1.5 Volatility 105
2 Example Materials 105
3 Stability of Stored Data 107
3.1 Dark Decay 107
3.2 Decay During Readout: Fixing 107
3.3 Two-Color Recording 108
4 Hologram Fidelity and Bit Error Rate 108
5 Conclusions 109
References 109

Inorganic Photorefractive Materials
K. Buse and E. Krätzig 113

1 Charge Transport 114
2 Storage Properties: Dark Storage Time, Response Time, Capacity, Sensitivity 117
3 Theoretical Performance Limits 118
4 Various Crystals 119
5 Nondestructive Readout 121
6 Conclusions 122
References 123

Hologram Fixing and Nonvolatile Storage in Photorefractive Materials
S. S. Orlov and W. Phillips 127

1 Thermally Assisted Ionic Fixing 128
1.1 Hologram Fixing and Ionic Conduction in $LiNbO_3$ 130
1.2 Lifetime of Fixed Ionic Gratings 131
1.3 High-Low Fixing 134
2 Fixing by Spontaneous Polarization Modulation 136
3 Two-Photon Holographic Recording in Stoichiometric Lithium Niobate 138
3.1 Undoped Stoichiometric Lithium Niobate 139
3.2 Doped Stoichiometric Lithium Niobate 144
3.3 Summary on Two-Photon Recording in $LiNbO_3$ 146
References 146

Two-Color Holography in Lithium Niobate
R. Macfarlane, H. Guenther, Y. Furukawa, and L. Kitamura 149

1 Materials 151
2 Experimental 152
3 Spectroscopy and Sensitization 152
4 Photorefractive Properties 153
4.1 Sensitivity 154
4.2 Gating Ratio 155
4.3 Dynamic Range 155
4.4 Dark Decay 156
4.5 The Role of Iron 156
5 Conclusion 157
References 158

Overview of Photorefractive Polymers for Holographic Data Storage
B. Kippelen 159

1 Brief History of Photorefractive Polymers 159
2 Physics and Chemistry of Photorefractive Polymers 161
2.1 Photogeneration 161
2.2 Transport 161
2.3 Index Change: Electro-Optic and Orientational Effects 163
3 Performance of Current Photorefractive Polymers 164
3.1 Spectral Sensitivity 164
3.2 Dynamic Range 164
3.3 Material Stability 166
3.4 Speed 166
3.5 Applications 167
4 Trends and Outlook 167
References 168

Photopolymer Systems
R.T. Ingwall and D. Waldman 171

1 Introduction 171
2 Chemistry of Photopolymer Systems 171
2.1 Monomers 171
2.2 Photoinitiation Systems 173
2.3 Binders 174
3 Recording Characteristics of Photopolymers 175
4 Recording Mechanism 177
4.1 Refractive Index Changes 177
4.2 Component Segregation 180

5 Recording Thick Photopolymer Holograms 181
5.1 Light Absorption 181
5.2 Low Viscosity 184
6 Image Quality in Photopolymer Holograms 185
6.1 Shrinkage 185
7 Data Storage in Photopolymer Holograms 191
7.1 Multiplexing 191
7.2 Data Page Recording 194
8 Summary 195
References 195

Photopolymers for Digital Holographic Data Storage
L. Dhar, M.G. Schnoes, H.E. Katz, A. Hale, M.L. Schilling, and A.L. Harris 199

1 Hologram Formation in Photopolymer Systems 200
2 Photopolymer Materials 200
3 Formation of Thick, Optically Flat Media 201
4 Holographic Characterization of Photopolymer Media 202
4.1 Recording-Induced Bragg Detuning 202
4.2 Dynamic Range 203
5 Holographic Digital Data Storage in Photopolymer Media 205
6 Summary 207
References 207

Photoaddressable Polymers
T. Bieringer 209

1 Photoaddressable Polymers 209
1.1 Photochemistry of Azobenzene 210
1.2 Azobenzene Containing Polymers 211
1.3 Liquid Crystalline Side Chain Polymers 212
2 Materials Under Investigation 216
2.1 The Choice of the Main Chain 216
2.2 The Spacer 217
2.3 The Choice of the Azo Group 217
2.4 The Choice of the Mesogenic Group 218
2.5 The Azo Group Concentration 218
3 State of the Art in the Literature 219
4 Photoaddressable Polymers from Bayer 221
5 Photoaddressable Polymers Used in Holographic Data Storage 222
6 Open Questions and Outlook 223
References 223

Part III Components

Laser Sources
B. Pezeshki and S.S. Orlov 231

1 Laser Requirements 231
2 Diode-Pumped Solid-State Lasers 232
3 Semiconductor Lasers 234
References 239

Beam Deflectors and Spatial Light Modulators for Holographic Storage Application
G. Zhou, F. Mok, and D. Psaltis 241

1 Description of the Holographic Disk System 241
2 Recording Density 244
3 SLM Characteristics and System SNR 246
4 Recording Rate 250
5 Beam Deflector for Holographic Data Readout 253
References 256

Beam Conditioning Techniques for Holographic Recording Systems
R.K. Kostuk, M.P. Bernal Artajona, and Q. Gao 259

1 Defocusing 260
2 Random Phase Masks 261
3 Pseudo-Random Phase Masks 263
4 Axicons 265
5 Discussion and Summary 268
References 268

Detector Arrays for Digital Holographic Storage Applications
S. Campbell and E.R. Fossum 271

1 General Considerations for Detector Arrays 272
  1.1 Size, Power and Cost 272
  1.2 Number of Pixels, Readout Rate, and Pixel Size Considerations 273
  1.3 Noise, Dynamic Range, and Analog-to-Digital Converter Resolution 273
2 Detector Array Choices 274
  2.1 Quantum Efficiency 276
  2.2 Noise 276
3 Readout Rate 278

4 System Implementation ... 278
5 Conclusion ... 279
References ... 279

Part IV Channels

Modulation Codes for Holographic Recording
B. Marcus ... 283

1 Block Codes ... 283
1.1 Correlation Detection and Balanced Block Codes ... 284
1.2 Sparse Block Codes ... 285
1.3 Parity Thresholding ... 286
2 Strip Codes ... 286
2.1 Balanced and Pseudo-Balanced Strip Codes ... 287
2.2 Inter-Pixel Interference and Low-Pass Codes ... 288
2.3 Combined Constant-Weight Low-Pass Codes ... 289
References ... 291

Interleaving and Error Correction for Holographic Storage
M.A. Neifeld and W.-C. Chou ... 293

1 Capacity ... 294
2 Error Correction ... 297
3 Interleaving ... 300
4 Conclusions ... 305
References ... 305

Equalization for Volume Holographic Data Storage Systems
B. V. K. Vijaya Kumar, V. Vadde, and M. Keskinoz ... 309

1 Channel Modeling ... 310
2 Equalization Methods ... 311
2.1 Zero Forcing Equalization ... 312
2.2 LMMSE Equalization ... 313
2.3 Partial Response (PR) Equalization ... 314
3 Equalization Results ... 315
4 Implementation Issues ... 316
5 Summary ... 316
References ... 317

Gray-Scale Data Pages for Digital Holographic Data Storage
G.W. Burr and M.A. Neifeld ... 319

1 Motivation for Gray-Scale ... 319
2 Predistortion ... 321
3 Encoding Digital Data into Gray-Scale Pixels ... 323

4 Capacity Estimation ... 324
5 Optimizing the Error-Correction Coding to Obtain User Capacity . 326
6 Summary ... 327
References ... 328

Part V Demonstration Platforms

System Optimization for Holographic Data Storage Systems
G.W. Burr and M.P. Bernal Artajona ... 331

1 Noise ... 331
2 Camera Quantization ... 334
3 Choice of Fill-Factors and Apertures ... 335
4 Capacity-Estimation Procedure ... 337
5 Choice of ECC Design Point: Effect of Variations in Diffraction Efficiency ... 338
6 Summary ... 340
References ... 340

Tamarack Optical Head Holographic Storage
S. Redfield ... 343

1 Roots ... 343
2 Design Evolution ... 343
3 Final System Approach ... 346
3.1 Requirements ... 347
3.2 Optical Head ... 347
3.3 Media Disk ... 348
3.4 Data Format ... 349
4 Holographic Optical Head ... 349
4.1 Reference Path Optical Design ... 349
4.2 Object Path Optical Design ... 352
5 Mechanical Design ... 353
5.1 Page Motor Design ... 353
5.2 HOH-Media Positioning ... 353
5.3 Changer ... 354
6 Summary ... 356

High-Density, High-Performance Data Storage via Volume Holography: The Lucent Technologies Hardware Platform
K. Curtis, W.L. Wilson, M.C. Tackitt, A.J. Hill, and S. Campbell ... 359

1 Materials ... 360
2 Multiplexing Methods ... 361
3 Components ... 363

4 Holographic Demonstration System ... 365
5 System Evolution ... 367
6 Summary ... 367
References ... 368

IBM Holographic Digital Data Storage Test Platforms
C.M. Jefferson, G.W. Burr, and J.A. Hoffnagle ... 369

1 PRISM Photorefractive Materials Tester ... 369
2 DEMON I Holographic Data Storage Engine ... 374
3 DEMON II Advanced Holographic Digital Data Storage Engine ... 376
4 Innovative Optics ... 378
4.1 Axicons ... 378
4.2 Aspherical Apodizer ... 379
References ... 381

Digital Holographic Demonstration Systems by Stanford University and Siros Technologies
L. Hesselink ... 383

1 Optical Architectures ... 384
2 Capacity Versus Transfer Rate Tradeoff ... 384
3 Demonstration Platforms ... 386
4 The Stanford University all Digital System Demonstration (Science, 1994) ... 386
5 The Siros First Fully Automated Video Demonstration (1995) ... 391
6 The Siros Fully Automated System with Electronic Readout at Video Rates (PRISM, 1996) ... 392
7 The Stanford University and Siros Fully Electronic Data Readout System Achieving 1 Gbit/s (HDSS, 2000) ... 393
8 The Stanford University and Siros 100-Gbytes Capacity and 1 Gbit/s Readout System Demonstrator ... 395
References ... 396

Holographic Read-Only Memory
F. Mok, G. Zhou, and D. Psaltis ... 399

1 Specifications ... 400
2 Recorder ... 400
3 Reader ... 402
4 Replication ... 404

Digital Holographic Data Storage with Fast Access
J. Ma, T. Chang, S. Choi, and J. Hong ... 409

1 Introduction ... 409
2 System Architecture ... 410

3 System Operation 412
References 417

A Demonstration Platform for Phase-Coded Multiplexing
C. Denz, K.-O. Müller, F. Visinka, and T. Tschudi 419

1 Phase-Coded Multiplexing 419
1.1 Phase Code Generation 420
1.2 Arithmetic Image Operations 421
2 Design and Implementation of the Demonstrator 421
3 Experimental Results 423
3.1 Arithmetic Image Operations 425
3.2 Data Encryption 426
4 Summary 427
References 428

Volume Holographic Optical Correlators
P.A. Mitkas and G.W. Burr 429

1 Optical Correlation 429
2 Volume Holographic Correlators 431
3 Volume Holographic Database System Architecture 432
3.1 Associative Recall with Binary Data 433
3.2 Associative Recall with Image Data 434
4 Fuzzy Volume Holographic Search Engine 435
4.1 All-Optical Search-and-Retrieve Demonstration 436
5 Evaluation of Associative Recall 437
6 Conclusions 443
References 443

Part VI Competing Technologies

The Continuing Evolution of Magnetic Hard Disk Drives
E. Grochowski 447

1 Areal Density 448
2 Magnetic Recording Head Physics GMR 450
3 Magnetic Disk Design and Physical Spacing 452
4 The Mechanical HDD Design and Form Factor Evolution 453
5 Price 455
6 Performance and Coding 457
7 Super Paramagnetism and "Limits" for Magnetic Recording 459
8 Conclusion 462
References 462

Optical Disk Storage Roadmap
B.H. Schechtman ... 463

1 Product Categories ... 464
2 Technology Status and Outlook ... 467
3 Summary ... 472
References ... 472

Alternative Storage Techniques
G.R. Ashton and W.C. Mitchell ... 475

1 Three-Dimensional Optical Recording ... 475
1.1 Electron Trapping Optical Memory ... 475
1.2 Liquid Crystal Optical Disk ... 475
1.3 Surface-Enhanced Raman Optical Data Storage ... 476
1.4 Optical Tape Technology ... 476
2 New Storage Forms ... 477
2.1 Persistent Spectral Hole Burning ... 477
2.2 Two-Photon Three-Dimensional Recording ... 477
2.3 Charged Particle Beam Technology ... 477
2.4 Optical Storage Card ... 478
2.5 Scanning Probe Storage ... 478
3 Conclusions ... 479
References ... 479

Index ... 481

List of Contributors

Gary R. Ashton
Compaq Computer Corporation
Houston, TX 20555 SH 249, USA

George Barbastathis
Department
of Mechanical Engineering
Massachusetts Institute
of Technology
Room 3-461c
77 Massachusetts Ave.
Cambridge, MA 02139, USA
gbarb@mit.edu

María Pilar Bernal Artajona
Swiss Federal Institute
of Technology Lausanne
Institute of Applied Optics
BM 4.109
1015 Lausanne, Switzerland
maria-pilar.bernal@epfl.ch

Thomas Bieringer
BAYER AG
Central Research, Physics
Building E41
51368 Leverkusen, Germany
Thomas.Bieringer.TB@BAYER-AG.DE

Geoffrey W. Burr
IBM Almaden Research Center
650 Harry Road
San Jose, CA 95120, USA
burr@almaden.ibm.com

Karsten Buse
Universität Osnabrück
Fachbereich Physik
49069 Osnabrück, Germany
karsten.buse@uos.de

Scott Campbell
Photobit Corporation
135 N. Los Robles Ave.
7th Floor
Pasadena, CA 91101, USA
scottc@photobit.com

Tallis Chang
Rockwell Science Center
1049 Camino Dos Rios
Thousand Oaks, CA 91360, USA

Sung Choi
Rockwell Science Center
1049 Camino Dos Rios
Thousand Oaks, CA 91360, USA
sjchoi@rsc.rockwell.com

Wu-Chun Chou
Electrical and Computer
Engineering Department
The Optical Sciences Center
University of Arizona
Tucson, AZ 85721, USA
chou@ece.arizona.edu

Kevin Curtis
Bell Laboratories
Lucent Technologies
600 Mountain Avenue
Murray Hill, NJ 07974, USA

Cornelia Denz
Institut für Angewandte Physik
Technische Universität Darmstadt
Hochschulstr. 6
64289 Darmstadt, Germany
Cornelia.Denz@
physik.tu-darmstadt.de

Lisa Dhar
Bell Laboratories
Lucent Technologies
600 Mountain Avenue
Murray Hill, NJ 07974, USA
ldhar@lucent.com

Eric R. Fossum
Photobit Corporation
135 N. Los Robles Ave.
7th Floor
Pasadena, CA 91101, USA
fossum@photobit.com

Yasunori Furukawa
National Institute for Research
in Inorganic Materials
1-1 Namiki
Tsukuba, 305, Japan

Qiang Gao
Electrical and Computer
Engineering Department
University of Arizona
Tucson, AZ 85721, USA
gao@ece.arizona.edu

Edward Grochowski
IBM Almaden Research Center
650 Harry Road
San Jose, CA 95120-6099, USA
grchwski@almaden.ibm.com

Claire Gu
Department
of Electrical Engineering
University of California
Santa Cruz, CA 95064, USA
claire@cse.ucsc.edu

Harald Guenther
IBM Almaden Research Center
650 Harry Road
San Jose, CA 95120, USA

Arturo Hale
Bell Laboratories
Lucent Technologies
600 Mountain Avenue
Murray Hill, NJ 07974, USA
arturo@lucent.com

Alex L. Harris
Bell Laboratories
Lucent Technologies
600 Mountain Avenue
Murray Hill, NJ 07974, USA
alexh@lucent.com

Lambertus Hesselink
Stanford University
and Siros Technologies
Stanford, CA 94305-4035, USA
bert@kaos.stanford.edu

Adrian J. Hill
Bell Laboratories
Lucent Technologies
600 Mountain Avenuc
Murray Hill, NJ 07974, USA

John A. Hoffnagle
IBM Research Division
Almaden Research Center
650 Harry Road
San Jose, CA 95120-6099, USA
hoffnagl@almaden.ibm.com

John Hong
Rockwell Science Center
1049 Camino Dos Rios
Thousand Oaks, CA 91360, USA
jhhong@rsc.rockwell.com

Richard T. Ingwall
Aprilis, Inc.
750 Main Street
Cambridge, MA 02139, USA

C. Michael Jefferson
IBM Research Division
Almaden Research Center
650 Harry Road
San Jose, CA 95120-6099, USA
michael@almaden.ibm.com

Howard E. Katz
Bell Laboratories
Lucent Technologies
600 Mountain Avenue
Murray Hill, NJ 07974, USA
hek@lucent.com

Mehmet Keskinoz
Carnegie Mellon University
Department of Electrical
and Computer Engineering
Pittsburgh, PA 15213, USA
keskinoz@ece.cmu.edu

Bernard Kippelen
Optical Sciences Center
The University of Arizona
Tucson, AZ 85721, USA
kippelen@u.arizona.edu

Kenji Kitamura
National Institute for Research
in Inorganic Materials
1-1 Namiki
Tsukuba, 305, Japan

Raymond K. Kostuk
Electrical and Computer
Engineering Department
and The Optical Sciences Center
University of Arizona
Tucson, AZ 85721, USA
kostuk@ece.arizona.edu

Eckhard Krätzig
Universität Osnabrück
Fachbereich Physik
49069 Osnabrück, Germany
eckhard.kraetzig@uos.de

B.V.K. Vijaya Kumar
Carnegie Mellon University
Department of Electrical
and Computer Engineering
Pittsburgh, PA 15213, USA
kumat@ece.cmu.edu

Jian Ma
Rockwell Science Center
1049 Camino Dos Rios
Thousand Oaks, CA 91360, USA
jma@rsc.rockwell.com

Roger Macfarlane
IBM Almaden Research Center
650 Harry Road
San Jose, CA 95120, USA

Brian Marcus
IBM Almaden Research Center
650 Harry Road
San Jose, CA 95120-6099, USA
marcus@almaden.ibm.com

William C. Mitchell
Imation Corporation
1 Imation Place
Oakdale, MN, USA

Pericles A. Mitkas
Electrical and Computer
Engineering
Colorado State University
Fort Collins, CO 80523, USA

Fai Mok
Holoplex Inc.
600 South Lake Avenue. Suite 102
Pasadena, CA 91106, USA
fai@holoplex.com

Kai-O. Müller
Institut für Angewandte Physik
Technische Universität Darmstadt
Hochschulstr. 6
64289 Darmstadt, Germany
Kai-O.Mueller@
physik.tu-darmstadt.de

Mark A. Neifeld
Electrical and Computer
Engineering Department
The Optical Sciences Center
University of Arizona
Tucson, AZ 85721, USA
neifeld@ece.arizona.de

Sergei S. Orlov
Department of Electrical
Engineering and Information
Systems Laboratory
Stanford University, MC 4035
Stanford, CA 94305, USA
sergeio@kaos.stanford.edu

Bardia Pezeshki
SDL Inc.
80 Rose Orchard Way
San Jose, CA 95134, USA
bardiap@sdli.com

William Phillips
Department of Electrical
Engineering and Information
Systems Laboratory
Stanford University, MC 4035
Stanford, CA 94305, USA
bill@kaos.stanford.edu

Demetri Psaltis
Department
of Electrical Engineering
California Institute of Technology
Mail-Stop 136-93
Pasadena, CA 91125, USA
psaltis@caltech.edu

Demetri Psaltis
Holoplex Inc.
600 South Lake Avenue, Suite 102
Pasadena, California 91106, USA

Steve Redfield
Starzent, Inc.
557 Agua Fria
Santa Fe, NM 87501, USA
sjred@cybermesa.com

Barry H. Schechtman
National Storage
Industry Consortium
3655 Ruffin Road
Suite 335
San Diego, CA 92123-1833, USA
barry@nsic.org

Marcia L. Schilling
Bell Laboratories
Lucent Technologies
600 Mountain Avenue
Murray Hill, NJ 07974, USA

Melinda G. Schnoes
Bell Laboratories
Lucent Technologies
600 Mountain Avenue
Murray Hill, NJ 07974, USA
mgl@lucent.com

Robert M. Shelby
IBM Almaden Research Center
650 Harry Road
San Jose, CA 95120-6099, USA
shelby@almaden.ibm.com

Glenn T. Sincerbox
Optical Sciences Center
University of Arizona
Tucson, AZ 85721, USA
sinbox@u.arizona.edu

Michael C. Tackitt
Bell Laboratories
Lucent Technologies
600 Mountain Avenue
Murray Hill, NJ 07974, USA

Theo Tschudi
Institut für Angewandte Physik
Technische Universität Darmstadt
Hochschulstr. 6
64289 Darmstadt, Germany
Theo.Tschudi@
physik.tu-darmstadt.de

Venkatesh Vadde
Nokia Research Center
6000 Connection Drive
Mail Stop M6-440
Irving, TX 75039, USA
venkatesh.vadde@nokia.com

Franz Visinka
Institut für Angewandte Physik
Technische Universität Darmstadt
Hochschulstr. 6
64289 Darmstadt, Germany
Franz.Visinka@
physik.tu-darmstadt.de

David Waldman
Aprilis, Inc.
750 Main Street
Cambridge, MA 02139, USA
waldman_da@yahoo.com

William L. Wilson
Bell Laboratories
Lucent Technologies
600 Mountain Avenue
Murray Hill, NJ 07974, USA

Pochi Yeh
Department of Electrical
and Computer Engineering
University of California
Santa Barbara, CA 93106, USA
pochi@ece.ucsb.edu

Xianmin Yi
Department of Electrical
and Computer Engineering
University of California
Santa Barbara, CA 93106, USA
6500miny@engineering.ucsb.edu

Gan Zhou
Holoplex Inc.
600 South Lake Avenue. Suite 102
Pasadena, CA 91106, USA
gan@holoplex.com
Current address:
Chorum Technologies, Inc
1303E Arapaho Road
Richardson, TX 75081, USA
gzhou@chorumtech.com

Part I

Introduction

History and Physical Principles

G.T. Sincerbox

The physical principles of holography involve the recording of the interference pattern formed between two beams of light and the subsequent illumination of that recorded pattern by one of the beams to recreate the other beam. In the specific case of holographic data storage, one beam is an information beam containing a two-dimensional pattern of 1s and 0s representing digital data. Think of it as a special image. The other beam is a reference beam used to form the interference pattern that is recorded and subsequently used to reconstruct the information beam. Storage densities as high as 100 Gbits/inch^2 and data rates of 1.0 Gbits/s can be realized.

In 1891, Lippmann demonstrated the first method of color photography by interfering a beam of light with its own reflection [1]. Reflecting colored light from a mercury mirror, a node is created at the interface and a standing wave is established above the mercury surface. The extent of the standing wave depends on the purity (bandwidth) of the colored light. The antinodes of the standing wave expose a photographic emulsion floating on the mercury that, after development, forms a periodic structure of silver planes separated by $\lambda/2n$ (n is the index of refraction of the emulsion). When illuminated with white light, constructive interference occurs in the reflected light only at the wavelength that originally set up the layers producing a colored image. A shorter wavelength is generated if any shrinkage of the emulsion has occurred. To first order, the bandwidth of the reflected light is inversely proportional to the number of planes and hence to the emulsion thickness. The process was termed *interference heliography.* While some amazing color photographs were made on black and white film, very little was done with this process until the early 1960s, when it was applied to white light holography using counter-propagating laser beams. Data storage based on the Lippmann process was explored briefly at IBM, where the recording of 20 discrete wavelengths in one location was demonstrated [2].

In 1948 Dennis Gabor made the insightful observation that the phase information in a diffracted wavefront could be recorded by interfering it (the diffracted wave) with a coherent background. He went on to show by theory and experiment that the diffracted wave could be reproduced exactly by illumination of the recording with only the coherent background. Terming this process *holography* he postulated that it could be used in microscopy as a means of greatly magnifying x-ray images [3]. This was to be achieved by recording the Fresnel diffraction pattern of an object at one wavelength and reconstructing an image at another, longer, wavelength. The ratio of the wavelengths creates a tremendous magnification: in the case of x-ray recording and visible light reconstruction this could be 100 000×! The realization of a coherent x-ray source was problematic, and much of the early verification of the principles was done with visible light using arc lamps, filters and pinholes. In his 1949 paper, he states: " ... probably the most interesting feature of the new method for light-optical applications is the possibility of recording in one photograph the data of three-dimensional objects."

It was not until the early 1960s with the invention of the laser that the technology became practical for storing and retrieving images, and the concepts of holographic data storage were established by van Heerden [4]. In his seminal paper in 1963 on the theory of optical information storage in solids, van Heerden postulated that the recording of interference patterns in a three-dimensional medium could be used as a means of storing and retrieving information. In this paper, he also developed the use of three-dimensional storage as an associative memory. He went on to estimate that the storage density limit would be V/λ^3 (V is the recording volume): a number that is often cited and, more often, misused as an estimate of the density potential in holographic storage. The arrival of the laser provided the necessary coherent source and allowed the reference wavefront to be moved off-axis, thus eliminating the troublesome twin image problem that was inherent in Gabor's in-line method. Leith and Upatnieks applied communication theory to the holographic process and devised this off-axis use of a spatial carrier [5]. Although van Heerden discussed the multiplexing of numerous holograms in a common volume by changing either the angle of the reference beam or the wavelength of both beams, and Leith and his colleagues demonstrated the storage of multiple images by rotation of the photographic plate [6], it was not until 1973 that angular multiplexing was actually applied to the storage of information [7].

1 Holographic Storage Principles

In contrast to conventional holography, where light scattered from a three-dimensional scene or object is recorded and can be subsequently reconstructed for visual viewing, holographic storage uses the light coming from a very special 'electronic' binary object known as a *spatial light modulator* (SLM). As shown in Fig. 1, the SLM is used to display a two-dimensional binary pattern of 'ones' and 'zeros', which act very much like miniature open and closed shutters representing the information to be stored. A common example of a SLM is the liquid crystal displays commonly used for hand-held television and notebook computer displays.

A laser beam passes through the SLM (or is reflected by it in the case of a reflection SLM), becomes modulated in two dimensions by the displayed pattern, and is directed to a recording medium where it interferes with an-

Fig. 1. A section of a page of binary data, representing what would appear on the spatial light modulator. A typical page would contain 1024×1024 bits, and the size of each bit cell would be 15–20 μm

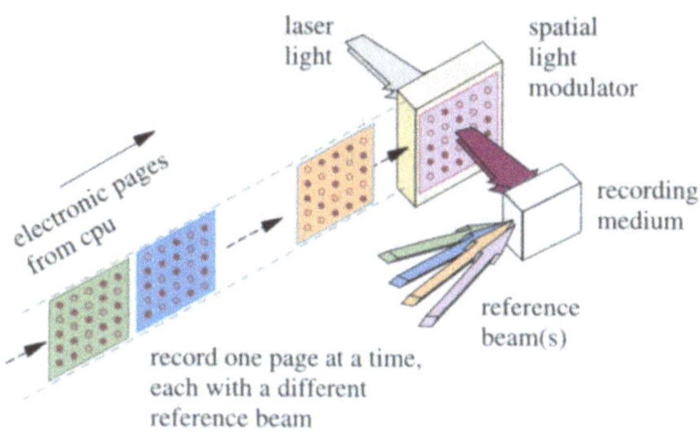

Fig. 2. Recording of digital data using angular multiplexing

other beam, called the *reference beam*, to form the hologram (Fig. 2). It is most common, and not shown in the figure, to interpose a lens between the SLM and the recording medium so that the Fourier transform of the SLM pattern is recorded rather than the diffraction from the pattern itself. As we will see later, this provides certain advantages in recording and retrieval. A new pattern, or page of information, is set up on the SLM and another hologram is recorded in the same volume of material with a different reference beam. As shown here, the reference beam angle is changed with each hologram recording, but some other attribute such as wavelength or a phase pattern could be used. A large number of different holograms are recorded in this manner, perhaps as many as 5 000, depending on the thickness and properties of the recording, medium. This process is termed *multiplexing*, and will be briefly described later in the context of the historical development of the technology. A later chapter will treat the subject in greater detail.

To retrieve a particular page of information that has been stored holographically, the recording medium (now a hologram) is illuminated with the reference beam that was used to record that page; in the case of angular multiplexing the corresponding reference beam direction is selected (see Fig. 3).

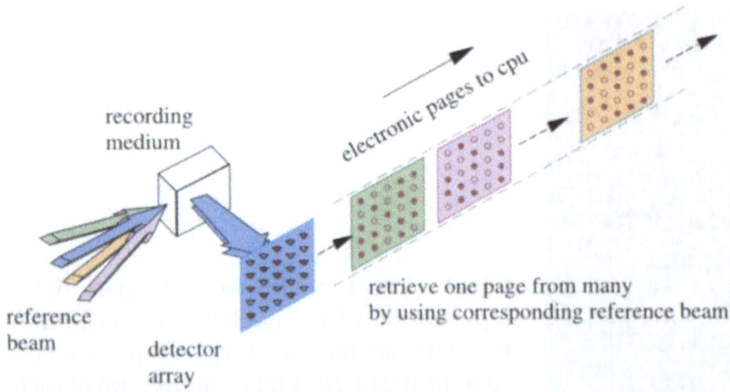

Fig. 3. Retrieval of the stored data

This light interacts with all of the recorded holograms but, because they are each highly tuned structures, only light is diffracted from the structure associated with that particular reference beam (direction). A wavefront is created that is identical to the light that was coming from the SLM when that page was displayed, illuminated and recorded. This light is imaged onto an array of photodetectors, converted to an electrical signal and transferred back to the cpu via the electronic channel. Each page can be retrieved independently by using the associated reference beam. Again, if a lens is used during recording, an identical lens must be placed between the medium and detector array to take the inverse transform and form the proper image.

Holographic storage and retrieval is an elegant process that invokes fundamental principles of the wave nature of light. Several inherent features, unique to this method of storage, are noteworthy. They will be mentioned briefly here and in much greater detail in subsequent chapters.

1.1 Redundant Storage

In contrast to magnetic and conventional optical recording, where an individual bit of information is represented by a highly localized change in some physical property, such as the ablation of a pit or the reversal of a magnetic domain, holographic recording of a single information bit is distributed throughout the entire recording volume. A one-to-one correspondence does not exist between an information bit and a microscopic element in the recording medium. The process by which this occurs is optical interference in which the light from an array (page) of bits, a complex signal beam, interferes with the light in a reference beam to create a three-dimensional pattern of high and low intensity. The recording material samples this 3-D pattern, producing a similar variation in some optical property such as absorption or refractive index. This recorded structure contains information about the amplitude and phase information in both beams of light at the time of recording and, upon illumination with a duplicate of only the reference beam, causes light to be diffracted to recreate the signal beam wavefront exactly. The wavefront, representing the page of information, is thus stored and can be retrieved at a later time and can be viewed directly or imaged onto a detector array as if the original object were still in place.

Because the interference pattern is three-dimensional, information about each data bit is distributed throughout the recording volume, thereby reducing the sensitivity to material defects that might otherwise obliterate one or more highly localized bits. The effect of a defect is to slightly reduce the signal-to-noise level of all the bits recorded in that volume. While the requirement for a defect-free recording material is less stringent for holography than for the more conventional recording technologies, scattering of light caused by defects and imperfections must be still controlled as this contributes to noise in the reconstructed image. This, in turn, reduces the signal-to-noise ratio and increases the bit-error rate – ultimately impacting on data reliability.

1.2 Multiplexing

The effective areal storage density (bits/unit area) can be significantly increased by using a thick recording layer to record multiple, independent pages of data. This should not be construed to mean that each page occupies a different depth in the recording volume, a common misconception, but rather that the holographic structure for one page is intermixed with the holographic structures of each of the other pages. This process is referred to as *multiplexing*. Retrieval of an individual page with minimum crosstalk from the other pages is a consequence of the volume nature of the recording and its behavior as a highly tuned structure. This effect, known as the Bragg effect, is depicted in Fig. 4, where the variation in diffraction efficiency from a holographic structure is shown as a function of a parameter that depends on mismatches in angle or wavelength between recording and playback. The specific use of angular multiplexing is depicted in Figs. 2 and 3 for recording and retrieving multiple pages or holograms. For our purposes it is sufficient to note that for a hologram in which the wavelength and angle used during recording and playback are identical, the efficiency of the reconstruction is a maximum. As the wavelength or readout angle is changed from this condition the efficiency decreases and eventually becomes zero. The point at which this happens depends on many parameters such as recording angles, initial wavelength and optical thickness of the recording material. For a given recording configuration, altering the thickness plays the central role. The inset table in the figure shows the corresponding angular or wavelength mismatches as a function of recording layer thickness for this point of zero diffraction efficiency. As thickness increases, the recorded structure becomes more highly tuned, much like an interference filter, such that smaller and smaller mismatches can be tolerated.

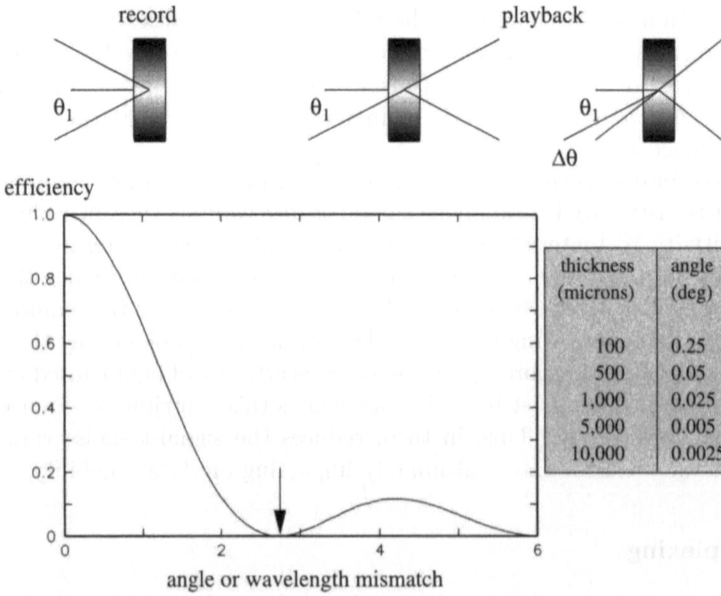

thickness (microns)	angle (deg)	wavelength (nm)
100	0.25	9.5
500	0.05	1.9
1,000	0.025	0.95
5,000	0.005	0.19
10,000	0.0025	0.095

Fig. 4. The Bragg effect, showing how the efficiency of reconstruction from a hologram decreases as the reference angle or wavelength deviates from the recording condition

Hence, for example, multiple holograms can be stored in a 1.0-mm-thick recording medium in increments of 0.025°, while a 10-mm-thick medium allows storage in 0.0025° increments. Using the angular technique, as many as 10 000 holograms have been stored in one location of a 6-mm-thick lithium niobate crystal [8]. In a similar 100 μm/1-mm-thick material, the wavelength sensitivity is 9.5 nm/0.95 nm. Wavelength multiplexing is simpler to implement but is highly dependent on the range over which lasers can be tuned, whereas angular multiplexing requires a more complex optical system. Today, wavelength tuning can be accomplished over a range of only about 20 nm, too small to multiplex many holograms. To overcome this, hybrid systems using a combination of angular and wavelength multiplexing have been explored. Several other techniques have also been developed for multiplexing that involve either rotation or translation of the recording medium. These include rotation of sample about axis normal to plane of incidence, rotation of sample about an axis in the plane of incidence (peristrophic multiplexing [9]), and small translation of the sample over a few microns (shift multiplexing [10]).

Another method of multiplexing has been demonstrated that uses the phase of a multitude of reference beams to distinguish one hologram from another [11]. Referred to as *deterministic phase encoding*, the individual beams used in angular encoding are incident on the recording medium simultaneously so that each hologram is recorded with the same set of beams. The sets differ from one another by an orthogonal code of phase retardations that have been impressed onto each 'beamlet'. Using a binary code of 0° and 180° retardation, unique codes can be constructed that allow the individual holograms to be reconstructed without crosstalk from the other holograms. This method is particularly attractive because access to a hologram can now be achieved by altering the phases of beams and not their direction – a faster and potentially simpler process. It is also possible to perform arithmetic operations between pages.

Increased storage density through multiplexing can provide a significant increase in overall storage capacity, and is a pathway by which holographic storage will be distinguished from conventional storage. These methods will be described in greater detail in subsequent chapters

1.3 High Data Rate

The two-dimensional page-oriented nature of holographic storage also utilizes the information capacity of an optical wavefront to allow data to be recorded and retrieved in parallel, a page at a time, rather than serially as in conventional storage. This affords a potential for extremely high data rates subject only to limitations imposed by the I/O devices (SLM, detector arrays) and electronic channels. For example, detector arrays are available today that can run at GHz speeds (using multiple output taps), and it is not difficult to conceive of massively parallel arrays running at effective data rates exceeding 1.0 Gbyte/s.

1.4 Rapid Access

A further distinguishing feature is the speed by which data can be accessed. Relying on movement of light beams and not mechanical mass with devices such as acousto-optic deflectors, or at least very minimal mass in the case of galvanometers, access times in the 10 μs to 1 ms range can be achieved. True random access can be provided, as it is not necessary to pass through other storage locations in accessing a target location. The absence of latency as present in rotating storage systems further enhances access.

1.5 Novel Functions

Finally, angular multiplexing provide an opportunity to perform novel functions such as associative retrieval. This is because a reference beam can reconstruct a signal beam and vice versa: a signal beam can reconstruct a reference beam. In its most basic form, all the holograms recorded at a given location are illuminated with a partial data page, rather than a particular reference beam, and each of the corresponding reference beams is reconstructed. The strongest reconstructed reference beam represents the page with the best correlation with the input pattern. For example, an index of images can be stored and rapidly searched for those images that match a particular search criterion. Analogous to pattern recognition, this promises to provide a very fast search mechanism that locates data by its content rather than its address. A detailed discussion of this subject will be presented in the chapter entitled "Volume Holographic Optical Correlations."

This abbreviated view of holographic storage does little justice to the breadth and complexity of this technology. The intent is not to describe the field in detail but to whet the appetite and establish a foundation for the discussion in the chapters to follow.

2 Historical Development

The development of holographic storage has not been easy. In fact, at the time of this writing, no commercial products are available – after almost 37 years! Actually, one can say the same for holography in general – after more than half a century we have only jewelry and credit card logos – very few products. Much of this can be attributed to the lack of what we now call the enabling technologies: the page composer, detector array, laser and recording material. These were not available in the 1960s and 1970s, and to a large extent required the emergence of a consumer electronics and entertainment industry. Watches, hand-held televisions and computer displays provided the economic driving forces for the development of liquid crystal displays, and the SLM became a reality. Camcorders, video cameras and digital image capture gave birth to affordable detector arrays. Laser development greatly benefited from optical storage products such as CD-audio, CD-ROM and magneto-optic recording. Holographic storage was a beneficiary of components developed for other applications. This, however, was not true of the materials. The unique requirements of holography established material requirements that were unlike any other application. Hence, a mass consumer-based application did not exist to drive the development of the recording material.

To list the companies and universities that have contributed to holographic storage is an impossible task. Almost four decades have provided a literature that is full of excellent papers on theoretical analysis and predictions, novel architectures, new multiplexing schemes, materials, systems designs, breadboard and prototype descriptions to name a few. Each of these papers has made an important contribution to the development of the field; many are technological milestones. To write a history of the development of holographic storage is the subject of a book, not a single chapter. Therefore, in what follows, I shall confine my discussion to the breadboards and prototypes that were developed in the early years. The more recent "systems" are covered in their respective chapters, and will be mentioned here very briefly only so we can see the progression and maturation of the technology.

2.1 Bell Labs and the Digital Page

Holography stimulated the imagination of researchers in the early 1960s as a means of recording complex wavefronts from three-dimensional objects in such a way that they could be reconstructed with amazing fidelity and detail. Researchers were so enamored with 3-D images that it was not until 1968 that the first published description of holographic *digital* storage occurred. In this report by L.K. Anderson of Bell Labs [12], he describes an experimental system consisting of an array of 1024 holograms (1.2 mm in diameter) recorded on a photographic emulsion. Each hologram contained 4096 bits of information, for a total capacity of 4 million bits. The system used an acousto-optic deflector that could address any of the 32×32 hologram locations in 6 μs. The reconstructed blocks of data were incident on a photodetector array constructed from 8×8 integrated circuit detector modules. An ultimate capacity of 100 million bits was predicted – far in excess of the 5.8 million bit twistor memory modules available at that time.

2.2 IBM HOSP

In the early 1960s, IBM, under contract to Wright Patterson Airforce Base, developed a read-only digital holographic storage system using bleached photographic plates (the famous 649F plates from Kodak) [13]. Called HOSP, for holographic optical storage prototype, it consisted of three subsystems, one for generating the bit mask, one for recording the holograms, and one for data retrieval. Using a 125-mW He-Ne laser, the recording system exposed 4096 (64×64), 1.0-mm-diameter holograms of a 32×32-bit mask that was slightly defocused from the Fourier transform plane. The bit mask consisted of a 35-mm film transport system that positioned frames of coded film against an all 1s aperture mask. Under computer control, the recording plate was moved in x and y to each new hologram position, a new mask was shuttled in place, and the exposure made. No multiplexing was used – it had not been invented yet. The capacity of a single plate was 0.5 Mbytes. Using a frequency-doubled Nd:YAG laser, readout was accomplished using a two-dimensional electro-optic digital light deflection technology that addressed 128×128 locations divided between four storage plates for a total system capacity of 2 Mbytes. This device consisted of 14 binary deflection stages, each comprising an electro-optic switch and a birefringent deflection element. The birefringent element of each deflection stage was twice the thickness of the preceding stage so that an n-stage device would produce $2n$ deflection locations with no jitter. True random access was achieved, with this deflector providing access to any hologram in 4 μs with a memory cycle time of 8 μs. The reconstructed image was inverse transformed, magnified and detected with a hybrid array of four silicon chips each containing 256 photodetector locations with associated transistor amplifying circuits. The active area of each detector was 800 μm in diameter. The sustained data rate of the reading system was 16 Mbytes/s with a 10:1 contrast ratio between the weakest 'one' and the strongest 'zero'. Prototypes of the three subsystems were built and successfully operated, but it never became a product.

2.3 RCA Holographic Memory

In 1973 researchers at RCA Laboratories described the design, development and implementation of an experimental 10^6 bit read–write holographic memory with no moving parts [14]. The memory employed an argon ion laser, acousto-optic beam deflection, a holographic beam splitter, a nematic liquid crystal page composer, a photoconductor-thermoplastic erasable storage medium, and a silicon photodiode array. All of the components were off-the-shelf except for the lenses, which were custom designed to reduce off-axis aberrations and enlarge the field of view. The memory was organized as 1024 holograms each containing 1024 bits. The individual bits were obtained from 2048 spots, as differential detection was used. The hololens was an array of holographically recorded lenses and served two functions: a beam splitter and a selective illumination system. Deflection of the light by an acousto-optic deflector to a particular hololens location caused the diffracted light to be directed through the page composer and the transform lens, and to a specific hologram memory location. The zero–order light through the hololens became the reference beam that when passed through a rooftop prism tracked the object beam. A pseudo-random phase plate was incorporated into the fabrication of the hololens. The page composer was a liquid crystal light valve with 2048 individually addressable apertures about 1 mm in size on 1.3 mm centers arranged in a quasi octagonal shape. It provided a contrast of 50:1 between 1s and 0s. Although all the pixels were recorded, readout was of a small subset. Twenty-pin photodiodes from five discrete clusters of four-element chips were used to detect 20 pixels representing 10 bits.

2.4 3M Holographic Data Storage System

A 7 Mbyte holographic store was described by Strehlow and coworkers at 3M in 1974 [15]. The array of 1024 × 1024 micro-holograms is arranged on a 10 × 10 cm area, with each hologram containing 56 bits. Access is with a two-dimensionally scanned electron-beam-pumped laser that provides random access in less than 10 μs. This laser source uses a cathode-ray tube with a semiconductor laser cavity of CdS to produce over 10^6 individually addressable sources. The 2.8 × 2.8 cm laser source array is expanded by a confocal arrangement to provide for readout from the 10 × 10 cm hologram array. A Fourier transform lens immediately behind the hologram array provides a spatially invariant image on 56 photodetectors arranged in a 4 × 14 array.

2.5 Thompson-CSF Read–Write Memory Using Angular Multiplexing

Although predicted by van Heerden in his 1962 paper, it was not until 1973 that Huignard and his colleagues at Thompson CSF first demonstrated the principles of angular multiplexing [7]. This enhancement is central to the achieving of high storage density as it makes better use of the full dynamic range of the recording medium. For many years it was the method of choice for multiplexing, and is still used extensively.

In their initial experimental configuration, a rather complex optical system was used involving cascaded acousto-optic deflectors, lens arrays and holographic gratings. The first deflector defined the spatial location of a hologram stack in the two-dimensional recording plane, and the second deflector established the angle between the reference and object beams. The use of a holographic grating in the reference path ensured that the beams tracked each other as they were deflected. Another configuration using electro-optic deflectors was also shown. Based upon components available in that era, superposition of 100 holograms was proposed for a total storage capacity of 10^{11} bits.

A few years later, they constructed an experimental breadboard to demonstrate and exercise all of the elements of a read–write memory [16]. A 5×5 array of iron-doped lithium niobate crystals was used as the recording medium. The system included many advanced features, such as a liquid crystal page composer with 8×8 points (remember, this is 1974), a silicon detector array, acousto-optic deflectors with 3 s access time, and Fourier transform optics. In one experiment, 10 pages of 10^4 bits were recorded in a larger $LiNbO_3$ crystal and recovered with a 1/0 contrast of 20.

2.6 NEC Holographic Coding Plate or Holotablet

In 1973, the Nippon Electric Company built a read-only holographic memory system for use as a graphical input device [17]. Based on a technique they developed called a *holographic coding plate*, a data pattern was stored in each individual hologram to represent the position of that hologram in a two-dimensional plane. In doing so, the position of a beam striking the hologram plane at an arbitrary location can be determined from the reconstructed bit pattern. An array of $2^m \times 2^n$ holograms each has a unique position code of $(m + n)$ bits. From this, they developed a Holotablet with a light pen linked to a stylus that could be used to 'write' on a 128×128 mm surface. Positions were resolved at 1 line/mm for a total capacity of 2.5×10^5 bits.

2.7 Harris-Intertype Wide-Band Recorder

A breadboard holographic recording system was built by the Electro-Optics Division of Harris-Intertype and reported in 1974 [18]. In order to demonstrate the feasibility of recording and reproducing digital data at rates up to

500 Mbits/s, a parallel recording technique was developed in which 128 data streams are created from a linear acousto-optic data composer and mixed with a common carrier (the reference beam). The signal and reference beam are deflected across the moving photographic film by a 40-sided polygonal mirror, where they are recorded in arrays of one-dimensional Fourier transform holograms. The recording rate is 4×10^6 holograms/s. Holographic data readout is detected by a photodiode array at 50 Mbits/s.

2.8 Hitachi Holographic Video Disk

In 1976, researchers at Hitachi reported on the development of a holographic videodisk that would contain 30 minutes of color video on a 300-mm-diameter disk [19]. Their choice of holography was to relieve the tight tolerances on the tracking and focus servo systems required by serially recorded video disks. Using Fourier transform recording and random phase sampling they coherently superimposed luminance and chrominance images into a 1-mm-diameter hologram. 54 000 holograms were recorded on a photoresist master to enable mass replication. They reported a signal-to-noise ratio in the reconstructed images of better than 35 dB. Techniques were developed to reduce cosmetic noise, phase coupling noise and moiré effects. Translational invariance of the Fourier transform was combined with stroboscopic illumination to achieve rotational immobility during readout.

2.9 Optical Data Systems Holoscan

One of the few systems to progress from the breadboard or prototype stage to an actual product was the Holoscan system made by Optical Data Systems in the early 1970s [20]. Intended for use in the credit and check verification field, this portable device scanned data stored in 0.5×0.75 mm Fourier transform holograms on 35 mm bleached photographic film. It was designed as a read-only memory with sequential rather than random access. Each hologram contained 56 bits arranged in a 14×4 array that could be read out from a photodetector array at 112 Kbits/s with a film transport speed of 60 in/s and an effective SNR of 10dB. The bit error rate was 2×10^{-5}. Total capacity was 1.5 Mbytes. Access was a combination of a lead-screw-driven mirror and film transport. An in-house system for the recording of holograms, called Holodac, was also built: this provided the requisite uniformity in bit luminance of better than $\pm 20\%$. Multiple copies were made from the master by a roll-to-roll contact printing process in which a hologram was made of a hologram: i.e., the diffracted wavefront from the master was interfered with the undiffracted zero order to produce the copied hologram.

2.10 Holographic Storage in the Soviet Union

In 1976, Andrei Mikaeliane described a rewritable holographic storage system developed in the (then) Soviet Union that was based on iron-doped lithium niobate [21]. A two-dimensional array of 1-mm-diameter holograms was recorded in a $25 \times 25 \times 1$ mm crystal of $LiNbO_3$: Fe. Recording was done with a HeNe laser for 2–3 min to obtain a diffraction efficiency of 1.5–2%. The longer HeNe wavelength permitted more readout cycles even though the speed of recording at 440 nm (HeCd) was 50 times faster. It was recognized that the read and write wavelengths had to be the same to permit full recovery of the image. HeCd was used for optical erasure. Over 10 write–erase cycles provided no visible change in the reconstructed image, after which deterioration was observed. Over 200 cycles was demonstrated with thermal erasure. Novel with this system was the use of a magnetic page composer consisting of a 60-μm-thick yttrium orthoferrite crystal. Sandwiched between a two-dimensional array of conducting loops, a pattern of 40×40 bits could be changed dynamically in less than 100 μs. Using polarized light and the Faraday effect, a contrast ratio of 300:1 was measured between 1s and 0s at 633 nm; 10% of the light was transmitted. Angular multiplexing was briefly explored with the recording of 5–6 holograms in one location. The effect of partial erasure was observed, and its influence on the system complexity noted.

2.11 NEC Holographic Disk

In 1980, Kubota et al. from Nippon Electric Company described a holographic disk containing tracks of one-dimensional Fourier transform holograms that can be used to synthesize speech for delivery to many users simultaneously [22]. An audio response system was developed with a 2000-word vocabulary and 256 Mbit/s transfer rate. Pulse code modulated speech signals are displayed on a 100-bit spatial light modulator and recorded as one-dimensional Fourier transforms on a photographic disk in the form of spiral tracks. The effect of motion-induced blur is minimized by orienting the transform along a disk radius and using short recording times. A conjugate reading beam is used to form a real image onto a detector array. The holograms are 0.7 mm in the radial direction and 10 μm wide tangentially.

2.12 MEI Kanji Character Generation System

In 1989, Matsushita Electric Industrial Company (MEI) reported on the analog storage of 16 000 Kanji characters as holograms in a character generation system for document editing [23]. 256 Fourier transform holograms were recorded of masks containing 63 Kanji characters in photographic emulsion for a recording density of 1.3×10^7 bits/cm^2. The reconstructed image of a page was incident on a camera tube from which the desired character could be scanned by an electron beam and digitized in resolutions ranging from 16×32 bits to 128×128 bits. Characters could be retrieved at 330 characters/s. Key to the quality of the reconstructed image was the use of a holographic beam splitter in the recording system. This beam splitter was the Fourier transform hologram of a four-level pseudo-random diffuser plate; the diffracted order provided an image of the diffuser and was used to illuminate the Kanji mask, and the zero order was used to form the reference beam. The system had been in operation for 8 years prior to publication.

2.13 Tamarack Multistore

Tamarack Storage Devices, a 1992 spin-off from an earlier MCC program known as Bobcat, had a program to develop a write-once holographic storage system using a photopolymer recording medium. Organized as a jukebox of 30 disk-shaped media cartridges, each holding 1 Gbyte of raw data, the system had a capacity of 30 Gbytes. Data was organized into stacks of 30 holograms with angular multiplexing over a $\pm 25°$ range. The stacks were arranged as concentric tracks around the circumference of the disk without overlap. Central to the design was a novel optical head in which the signal and reference optics, complete with all optoelectronic and optomechanical components except for the laser, moved along a radius of the disk. A given stack was accessed in 10 ms by a combination of linear radial motion of the head assembly and rotary stepping of the disk. The 512×512 data page was oversampled by 4 to 1 with a CCD array with sophisticated algorithms to compensate for any distortions in the image of the bit pattern. A complete description of this prototype can be found in the chapter entitled "Tamarack Optical Head Holographic Storage."

2.14 The PRISM Test Stand

In 1994, a consortium of seven companies and university research groups was formed under the sponsorship and partial support of DARPA. Named PRISM, for photorefractive information storage materials, it was the intent of this program to remove the materials bottleneck to successful implementation of holographic storage. The cornerstone of the PRISM program was the development of a sophisticated materials tester at IBM [24]. This complex system allowed the holographic recording and retrieval of large data pages in such a way that the bit error rate (BER) of the retrieved data could be measured. With very little contribution to the BER from the system, various recording materials could be analyzed and compared for their suitability for holographic storage. This storage systems approach to materials testing, pioneered by the PRISM team, has been key to the development and evaluation of suitable materials.

2.15 Stanford University

Stanford University reported the demonstration of a multiple page, fully digital holographic storage system in 1994 [25]. A $LiNbO_3$ crystal was mounted on a computer-controlled rotation stage that in turn was mounted on a computer-controlled translation stage. Vertical translation provided access to four different spatial locations, each containing 308 hologram pages of 1592 bits multiplexed by rotation of the crystal. A 480×440 SLM was used, and because of the mismatch between the dimensions of the off-the-shelf SLM and the detector array, a bit was represented by a block of 8×8 SLM pixels. It was recognized that anamorphic imaging could have been provided with a multiple lens arrangement to equalize the different horizontal and vertical pitches, but this would have added complexity to the optics. To improve the bit error rate over a system based on threshold detection, a differential encoding technique was used in which two adjacent pixels are used to record an individual bit; a "one" is represented by pixels that are on-off and a zero when the pixels are off-on. With the addition of simple error-correction codes, a bit error rate of 10^{-6} was obtained. Both digital color images and video data were retrieved.

2.16 Holoplex Memory Device for Fingerprint Verification

In 1995, Holoplex announced the availability of a holographic memory device for fingerprint identification application [26]. The Holoplex device is a read-only memory that stores 100 fingerprint images. Used in conjunction with an external optical joint-transform correlator, the system provides key entry to secure installations or equipment. The system compares the image of a fingerprint obtained from an input with the stored images and either grants or denies entry or use. The holograms were recorded in a polymeric recording medium on a disk substrate with peristrophic multiplexing. Rotation of the disk provided each fingerprint image to a joint transform correlator so that a correlation could be performed with the input fingerprint to see if it belonged to the set. The entire set of stored fingerprints can be compared in 1 s.

2.17 Rockwell Read-Only Demonstrator

As a participant in both the PRISM and HDSS consortia, Rockwell developed a compact demonstration platform to enable them to assess the practicality of the technology for use in avionics applications [27]. A read-only system was developed that combined the features of small size, high capacity and rapid access. The readout unit was constructed as a small module that could be removed from a larger recording system and hence did not carry with it the overhead associated with a larger recording laser and the spatial light modulator. High-speed access in tens of microseconds was obtained with acousto-optic deflectors that provided both spatial and angular multiplexing. Using $LiNbO_3$ as the recording medium, 20 000 holograms were recorded in 20 1-mm slices. Each hologram contained 26 Kbytes obtained from a 320×220

liquid crystal spatial light modulator and eight levels of gray scale. A raw bit-error rate of 1.6×10^{-3} was obtained using adaptive thresholding. Increasing the number of holograms per slice to 1200 and using 30 slices will provide the target capacity of 1 Gbyte.

2.18 IBM DEMON

In 1997, Burr and coworkers at IBM described a digital holographic storage system for the study of noise sources and the evaluation of modulation and error-correction codes [28]. Called DEMON (for demonstration), this system is a first demonstration of pixel-to-pixel matching between a real-time SLM and a video-rate CCD camera. All the optics except for the laser are contained within a 46 × 61 cm plate. Laser light from an argon ion laser is provided to the system via a fiber. A 640 × 480 liquid crystal SLM with 42 μm pixels is matched through the Fourier transform lens pair and sample to a 640 × 480 CCD chip with custom demagnification optics. Recording is on $LiNbO_3$, defocused 2–3 cm behind the transform plane to provide uniformity. Using an aperture, only the central nine orders of the Fourier spectrum are recorded (9 × 9 mm). Recording is in the 90° geometry with angular multiplexing.

3 Summary

Holography is such an elegant technology. How else can a wavefront be captured and later reconstructed in all its detail – reconstructed as if the original object that created the wavefront were still there? Quite remarkable when you stop and think about it. But the path to a holographic storage system has been long and, in some cases, painful and expensive. Looking back over the history of holographic storage one can see the enormous strides that have been made. To be sure, there have been false starts and many disappointments, but there has been unmistakable progress. It is evident in these brief descriptions that the technology is maturing. We have moved from a simple hologram of a fixed mask in photographic emulsion to the recording of thousands of holograms multiplexed in one volume from a dynamic spatial light modulator. We no longer need to demonstrate the technology; we know it works. The concerns now are with bit error rates, modulation and error-correcting codes, storage density, access times and data rate. Does it work? Yes. Does it compete? Much will depend on the recording material – we are very close to finding out.

Writing the history part of this chapter was more of a task than expected. At times it had some similarity to a genealogy; at other times it was like a scavenger hunt. Each paper provided a set of references that in turn pointed to another set and so on back into time. Some were easy to get; the old ones took some digging. I hope I have not overlooked an important breadboard or prototype in the process. The distinction between these systems and an experiment devised to demonstrate a new technique or recording material is often very fuzzy. The chapter would have been very different, and much longer, had I chosen to journal the events rather than the systems.

References

1. G. Lippmann, J. de Phys., **3**, 97 (1894).
2. H. Fleisher, P. Pengelly, J. Reynolds, R. Schools and G. Sincerbox, "An optically accessed memory using the Lippmann process for information storage," in Optical and Electro-optical Information Processing, J. Tippett et al., Eds, MIT Press, 1–30 (1965).
3. D. Gabor, Nature, **161**, 777 (1948).
4. P.J. van Heerden, "Theory of optical information storage in solids," Appl. Opt., **2**(4), 393–400 (1963).
5. E.N. Leith and J. Upatnieks, J. Opt. Soc. Am., **52**, 1123 (1962).
6. E.N. Leith, A. Kozma, J. Upatnieks, J. Marks and N. Massey, "Holographic data storage in three-dimensional media," Appl. Opt., **5**(8), 1303–1311 (1966).
7. L. d'Auria, J.P. Huignard and E. Spitz, "Holographic read-write memory and capacity enhancement by 3-D storage," IEEE Trans. Magn., MAG-**9**(2), 83–94 (1973).
8. G. Burr, F. Mok and D. Psaltis, "Storage of 10 000 holograms in $LiNbO_3$," CLEO1994.
9. K. Curtis, A. Pu and D. Psaltis, "Method for holographic storage using peristrophic multiplexing," Opt. Lett., **19**(13), 993–995 (1994).
10. D. Psaltis, M. Levene, A. Pu and G. Barbastathis, "Holographic storage using shift multiplexing," Opt. Lett., **20**(7), 782 (1995).
11. C. Denz, G. Pauliat, G. Roosen and T. Tschudi, "Volume hologram multiplexing using a deterministic phase encoding method," Opt. Commun., **85**, 171–176 (1991).
12. L.K. Anderson, "Holographic Optical Memory for bulk data storage," Bell Laboratories Record, **45**, 319–326 (1968).
13. J. Lipp and J. Reynolds, "A high capacity holographic storage system," in Applications of Holography, E.S. Barrakette et al., Eds. New York: Plenum Press, 377–388 (1970).
14. W.C. Stewart, R.S. Mezrich, L.S. Cosentino, E.M. Nagle, F.S. Wendt and R.D. Lohman, "An experimental read-write holographic memory," RCA Rev., **34**, 3–44 (1973).
15. W.H. Strehlow, R.L. Dennison and J.R. Packard, "Holographic data store," J. Opt. Soc. Am., **64**, 543–544 (1974).
16. L. d'Auria, J.P. Huignard, V.C. Slezak and E. Spitz, "Experimental holographic read-write memory using 3-D storage," Appl. Opt., **13**(4), 808–818 (1974).
17. N. Nishida, M. Sakaguchi, and F. Saito, "Holographic coding plate: a new application of holographic memory," Appl. Opt., **12**(7), 1663–1674 (1973).
18. A. Bardos, "Wideband holographic recorder," Appl. Opt., **13**(4), 832–840 (1974).
19. Y. Tsunoda, K. Tatsuno, K. Kataoka and Y. Takeda, "Holographic videodisk: an alternative approach to optical videodisks," Appl. Opt., **15**(6), 1398–1403 (1976).

20. K.K. Sutherlin, J.P. Lauer, and R.W. Olenick, "Holoscan: a commercial holographic ROM," Appl. Opt., **13**(6), 1345–1354 (1974).
21. A. Mikaeliane, "Holographic bulk memories using lithium niobate crystals for data recording," in *Optical Information Recording*, **2**, E.S. Barrekette et al., Eds, New York: Plenum Press, 217–233 (1978).
22. K. Kubota, Y. Ono, M. Kondo, S. Sugama, N. Nishida and M. Sakaguchi, "Holographic disk with high data transfer rate: its application to an audio response memory," Appl. Opt., **19**(6), 944–951 (1980).
23. I. Sato, M. Kato, K. Fujito and F. Tateishi, "Holographic memory system for Kanji character generation," Appl. Opt., **28**(13), 2634–2040 (1989).
24. M.-P. Bernal, H. Coufal, R.K. Grygier, J.A. Hoffnagle, C.M. Jefferson, R.M. Macfarlane, R.M. Shelby, G.T. Sincerbox and G. Wittmann, "A precision tester for studies of holographic optical storage materials and recording physics," Appl. Opt., **35**(14), 2360–2374 (1996).
25. J. Heanue, M. Bashaw and L. Hesselink, "Volume holographic storage and retrieval of digital data," Science, **265**, 749–752 (1994).
26. G. Zhou, Y. Qiao, F. Mok, and D. Psaltis, "A holographic memory product for fingerprint identification," Opt. Photonics News, March, 43 (1996).
27. I. Michael, W. Christian, D. Pletcher, T.Y. Chang, and J.H. Hong, "Compact holographic storage demonstrator with rapid access," Appl. Opt., **35**(14), 2375–2379 (1996).
28. G.W. Burr, J. Ashley, H. Coufal, R.K. Grygier, J.A. Hoffnagle, C.M. Jefferson and B. Marcus, "Modulation coding for pixel-matched holographic data storage," Opt. Lett., **22**(9), 639–641 (1997).

Volume Holographic Multiplexing Methods

G. Barbastathis and D. Psaltis

Data storage is the conversion of abstract information, representing the data, into physical changes in an appropriate medium. Data retrieval is the inference of the stored information from the storage-induced changes. The physical processes applied to different storage technologies differ widely, and, to a large extent, define their performance limits and applications. In this chapter we focus on the construction of holographic memories utilizing the effect of volume diffraction. Material physics related to holographic storage is the topic of later chapters.

As the name "holographic" suggests, the physical storage mechanism is the formation of *holograms* [1–4] by exposing photosensitive media to interfering light beams. Van Heerden [5] pointed out for the first time that multiple holograms can be superimposed, or *multiplexed*, in volume media. Individual holograms are accessed selectively in a way similar to the individual detection of multiple periodicities in crystal lattices using Bragg diffraction. The *Bragg selectivity* property of volume holograms, suggested and calculated by van Heerden [5], Leith and collaborators [6], and Kogelnik [7], is the basis of the enormous storage capacity of holographic memories.

1 Holographic Storage and Retrieval

A typical holographic storage system is shown in Fig. 1a. The laser produces a light beam of high spatial and temporal coherence, which is expanded, collimated, and split in two parts. The first, called the *signal arm*, is spatially modulated by a *page* of information after passing through a spatial light modulator (SLM).[1] By definition, a "page" is a two-dimensional spatial signal, analog or digital. The simultaneous, *parallel* storage (and retrieval) of entire pages, one at a time, is a distinguishing feature of holographic storage; most other storage technologies (magnetic, electronic, or optical disk – CD and DVD) are rather bit-oriented, and, therefore, *serial*.

The second optical arm of the storage system, called the *reference arm*, defines the address where the information is to be stored. In other words, the reference beam provides an identity for the page carried by the signal

[1] Typically, the SLM is implemented as a two-dimensional grid of liquid-crystal modulators followed by a polarizer, or an array of micro-cantilever-based deflectors.

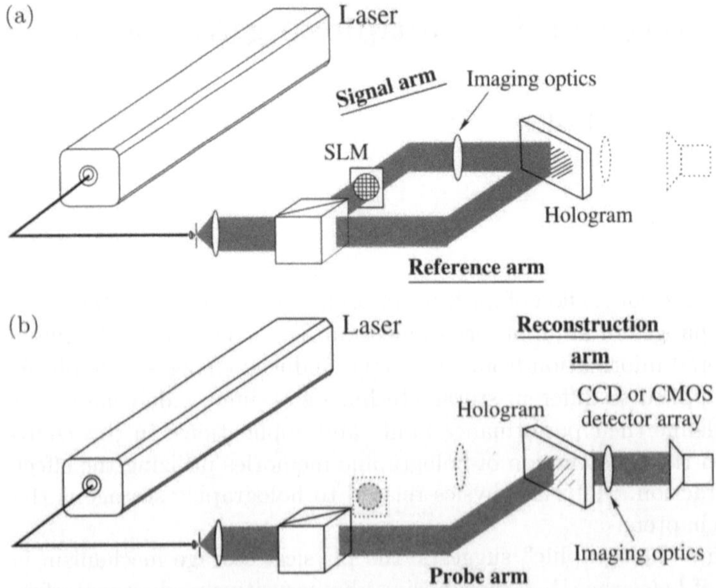

Fig. 1. Holographic memory architecture: (**a**) storage arrangement; (**b**) retrieval arrangement

beam, so that the information is distinguishable from other pages sharing the same volume inside the holographic medium. To this end, the reference arm is equipped with the appropriate addressing mechanism (not shown in Fig. 1a), mapping memory addresses to values of some physical property of the reference beam, e.g. angle of incidence [6,8–10] or wavelength [6,11,12]. The physical property used for addressing defines the holographic *multiplexing method.* We will be giving implementation details, performance data, and selection criteria for holographic multiplexing methods in this entire chapter.

Figure 1b is a diagram of the data retrieval system for the storage architecture of Fig. 1a. The holographic medium is probed, or *read out*, by a light beam[2] carrying the address of the page selected for retrieval. Physically, addressing during retrieval is similar to the storage phase: i.e. the probe beam replicates the reference beam used for storing the desired page.[3] The portion of probe light diffracted by the multiplexed volume holograms forms the *reconstruction*, which is detected by an appropriate detector array or camera, e.g. CCD or CMOS. The reconstructed beam is a replica of the desired page,

[2] Often, the probe beam need not be spatially coherent: therefore the light source used in the retrieval system need not be a laser.

[3] However, in some cases, the memory is read out at a wavelength longer than the wavelength used for recording. This method avoids erasure effects in materials where storage is non-permanent (e.g., photorefractives) but complicates the readout mechanism [13–16].

possibly with undesired superimposed *crosstalk* [17–22] from the remaining multiplexed pages, as well as other noise sources, e.g. scatter.

The capacity of a memory system is typically measured by the *surface storage density* [23], or simply density, defined as the number of bits of information[4] per unit area, or, equivalently,

$$\mathcal{D} = \frac{\text{total information in the memory}}{\text{area occupied}}. \tag{1}$$

The reason for normalizing capacity to area (as opposed to volume, for example) is to demonstrate the improvement over surface- or layer-type memories (CDs or DVDs, respectively) in overall capacity. For a single hologram, the density is determined by the page area and the minimum resolvable pixel size. We denote this number by $\mathcal{D}_{2D}^{\text{holographic}}$, measured in bits/$\mu$m^2. Compared with the density of a compact disk $\mathcal{D}^{\text{compact disk}}$, i.e. the number of pixels fitting on a compact disk area equal to that occupied by a holographic data page, $\mathcal{D}_{2D}^{\text{holographic}}$ is actually worse. This is because the imaging system of the compact disk is confocal (it images a single axial point only). Comparatively, an entire holographic page must be imaged with the same accuracy (i.e. with minimum distorion and blur) in the center as well as the edge (off-axis) pixels, which are affected worse by aberrations and vignetting. This forces the designer to increase the hologram pixel size, thereby decreasing the density so that

$$\alpha = \frac{\mathcal{D}_{2D}^{\text{holographic}}}{\mathcal{D}^{\text{compact disk}}} \leq 1. \tag{2}$$

An elegant solution to this problem involves the use of phase conjugation [24], which allows $\alpha \approx 1$ even for the page-based memory. Even if $\alpha < 1$, the overall density of the holographic memory is high because every location does not correspond to one but $M \gg 1$ superimposed holograms. Therefore, the overall storage density is

$$\mathcal{D}_{3D}^{\text{holographic}} = M \times \mathcal{D}_{2D}^{\text{holographic}}. \tag{3}$$

Typically, M is large enough that

$$\frac{\mathcal{D}_{3D}^{\text{holographic}}}{\mathcal{D}^{\text{compact disk}}} = \alpha \times M \gg 1, \tag{4}$$

even if $\alpha < 1$ as in non-phase-conjugate systems.

The gain in surface density is traded off with loss in dynamic range, i.e. detectable optical signal. The light efficiency performance of a holographic

[4] Here we are referring to *raw* information, i.e. without taking the error correction overhead into account.

memory is measured by a unitless quantity, the *diffraction efficiency* η, defined as

$$\eta = \frac{\text{Diffracted power}}{\text{Incident power}}. \tag{5}$$

The diffraction efficiency of individual holograms multiplexed within a memory with M holograms in total depends on M as [25]

$$\eta(M) = \frac{(M\#)^2}{M^2}. \tag{6}$$

The quantity $M\#$ (pronounced "M number") depends on the properties of the holographic material and the optical system. Typical values are in the range 0.01–10. The fundamental relationship (6) is true for any optical material if recording is accomplished by means of optical illumination [26]. A better $1/M$ dependence could be achieved if the hologram were constructed "voxel-by-voxel" or as a multilayered diffractive element. This would be completely impractical, since it would require several thousands of precisely aligned layers.

From this discussion, it follows that the density of holographic memory systems is determined by three design parameters:

(i) the page density $\mathcal{D}_{2D}^{\text{holographic}}$, which in turn is defined by the number of pixels N per page, the pixel size b and the $F\#$ of the imaging system;
(ii) the desirable dynamic range, or minimum acceptable diffraction efficiency $\eta_{\min}$, which, through (6), sets an upper limit $M_{\max;1}$ on the number M of the multiplexed holograms; and
(iii) the maximum number $M_{\max;2}$ of multiplexed holograms allowed by the particular properties of the multiplexing method used.

In several multiplexing methods, $M_{\max;2}$ is determined by the hologram thickness L and the geometry of the optical system delivering the reference beam to the hologram. We will refer to these as *Bragg* methods. In a second class of methods, called *fractal*, the limit $M_{\max;2}$ is set by the geometry and the page properties, and is therefore related to $\mathcal{D}_{2D}^{\text{holographic}}$. This peculiar property will become clear later on, after the corresponding multiplexing mechanisms are explained. Either way, our goal in this chapter is to provide the necessary tools for estimating how the selection of holographic multiplexing method affects both $\mathcal{D}_{2D}^{\text{holographic}}$ and $M_{\max;2}$. The complete design problem involves the optimization of $\mathcal{D}_{3D}^{\text{holographic}}$ by obtaining the optimal $\mathcal{D}_{2D}^{\text{holographic}}$ (parameter i) and M (parameters ii or iii). This is not attempted here.

The layout of the chapter is as follows: the next two sections (Sect. 1.1, 1.2) provide the definitions and brief descriptions of the most common holographic multiplexing techniques and recording–imaging geometries used in holographic storage systems. Volume diffraction, the natural mechanism that makes holographic multiplexing possible, is the topic of Sect. 2. The treatment begins with the general formal approach of scattering theory and is gradually simplified with Born's first approximation as well as the more intuitive wave-vector (k) and grating (K) space approaches. Specific multiplexing methods are considered individually as examples. Holographic memory architectures are the topic of Sect. 3.

1.1 Overview of Holographic Multiplexing Methods

The following is a list of known holographic multiplexing methods that we will be considering in this chapter:

Angle multiplexing, in-plane. [6,8] The reference beam is a plane wave, and the addressing mechanism is the angle of incidence of the reference. The reference beam angular positions are such that the reference beams and the optical axis of the signal beam are all on the same plane (Fig. 2a).

Angle multiplexing, out-of-plane. [9,27,28] Similar to (a) but the reference beams are on a plane normal to the plane defined by the optical axis of the signal beam and its projection on the surface of the holographic medium (Fig. 2b).

Peristrophic multiplexing. [10] The reference beam is a plane wave, and the addressing scheme is the relative rotational position of the storage medium. The axis of rotation is arbitrary, but typically it is chosen normal to the holographic medium (Fig. 2c).

Wavelength multiplexing. [6,11,12] The reference beam is a plane wave, and the addressing mechanism is the wavelength of incidence of the reference (Fig. 2d).

Phase-coded multiplexing. [29] The reference beam is a plane wave modulated in one dimension (in the plane defined by the optical axes of the reference and signal beams, Fig. 2e) by a phase SLM. The addressing mechanism is the pattern displayed on the SLM. The phase pattern must belong to a set of orthogonal codes. The easiest pattern sets to implement in practice are binary (phase π or $-\pi$), known as Walsh-Hadamard codes.

Shift multiplexing, in-plane. [30] The reference beam is a spherical wave, and the addressing mechanism is the location of the holographic medium relative to the reference beam. The relative displacements are small enough that multiplexed holograms in subsequent locations overlap significantly. The reference beam locations are such that the optical axes of the reference beams and the signal beam all belong to the same plane (Fig. 2f). Similar (shift) addressing can be implemented with a collection of plane waves at regular angular spacing, i.e. forming a one-dimensional fan [31] or a speckle pattern [32] as reference beam.

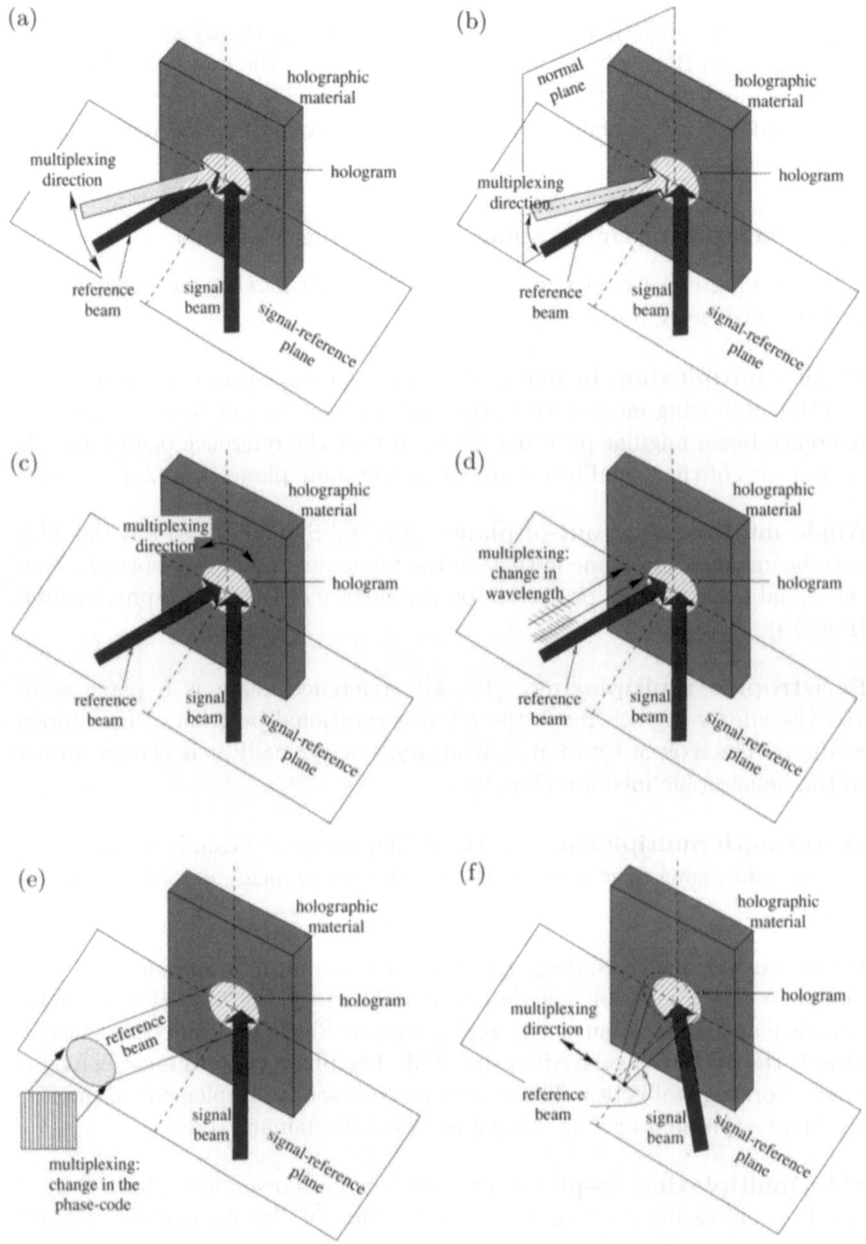

Fig. 2a–j. Holographic multiplexing methods: (**a**) angle multiplexing, in-plane; (**b**) angle multiplexing, out-of-plane; (**c**) peristrophic multiplexing; (**d**) wavelength multiplexing; (**e**) phase-code multiplexing; (**f**) shift multiplexing, in-plane

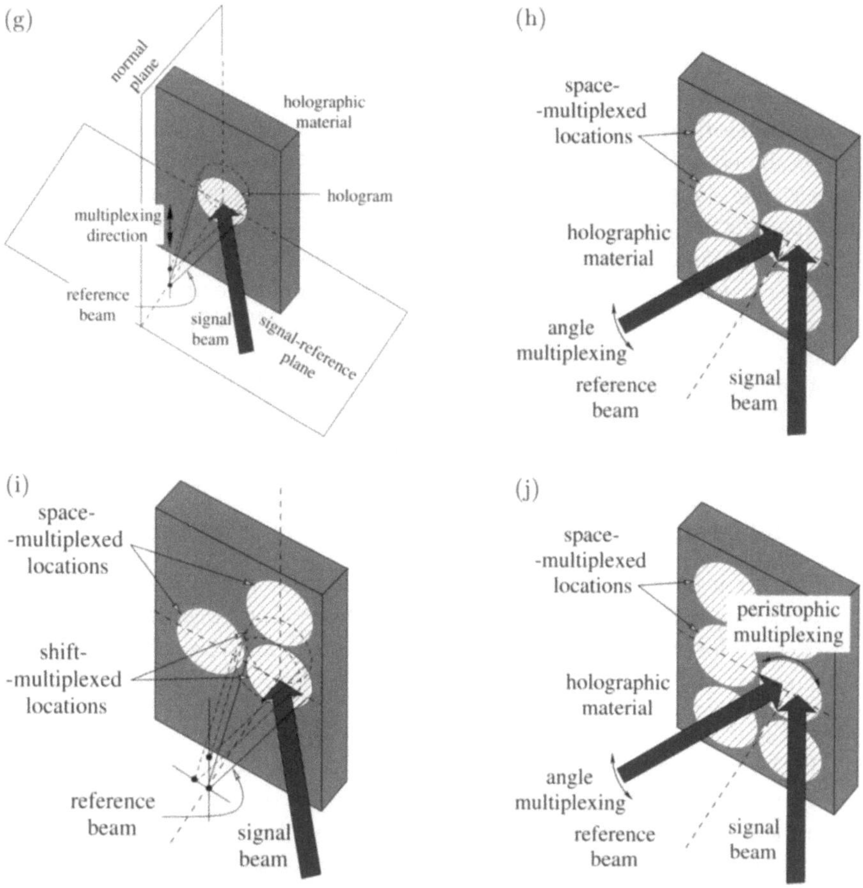

Fig. 2a–j. (*Cont.*) Holographic multiplexing methods: (**g**) shift multiplexing, out-of-plane; (**h**) spatial and angle multiplexing combined; (**i**) spatial and shift multiplexing combined; (**j**) spatial multiplexing combined with angle and peristrophic multiplexing

Shift multiplexing, out-of-plane. [33,34] Similar to (Sect. 1.1) with the spherical wave implementation, but the reference beams are on a plane normal to the plane defined by the optical axis of the signal beam and its projection on the surface of the holographic medium (Fig. 2g).

Spatial multiplexing. [23,35] Stacks of holograms multiplexed using one of the previous methods (a–g) are successively positioned in non-overlapping locations of the holographic medium. Each stack is well defined and separated from its neighbors by the lateral size of the holograms for all methods (e.g., the case of angle and spatial multiplexing is shown in Fig. 2h), except that in the case of shift multiplexing (Fig. 2i) the stacks themselves are overlapping. To avoid confusion when both shift and spatial multiplexing are present, we will say that two holograms are shift multiplexed if they overlap and spatially multiplexed if they do not overlap.

Combinations of multiplexing methods. Pairs of methods among (a–g) are combined with some restrictions (see Sect. 2.4) to achieve higher capacity, e.g. angle multiplexing in and out of plane [36], angle and peristrophic multiplexing [37], angle and wavelength multiplexing [38]. Spatial multiplexing with two multiplexing methods for each hologram stack has been used in high-capacity demonstrations involving angle multiplexing in and out of plane [28,39], angle and peristrophic multiplexing [40] (Fig. 2j), and shift multiplexing in and out of plane [41,42] (Fig. 2i).

According to the classification introduced earlier in Sect. 1, (a,d,e,f) are Bragg methods and (b,c,g) are fractal methods. We will defer the detailed explanation of this classification until Sect. 2.4.

1.2 Holographic Storage Geometries and Imaging Systems

The selection of a holographic storage geometry for the implementation of a particular multiplexing scheme is an important design decision. The three possibilities for arranging the reference, signal, probe and reconstructed beams with respect to a volume hologram are shown in Fig. 3.

- In the *transmission* geometry (Fig. 3a, also depicted in Fig. 1 and Fig. 2a–j) the two recording beams are incident on the same face of the holographic material. The reconstruction appears on the opposite face as an extension of the signal beam (i.e. as a replica of the signal beam if it were present and propagated beyond the holographic material). The transmission geometry has been used for the construction of holographic disks [23,35,40,41].
- The *90° geometry* (Fig. 3b) is similar to the transmission geometry, except that the reference and signal beams are incident on two normal faces

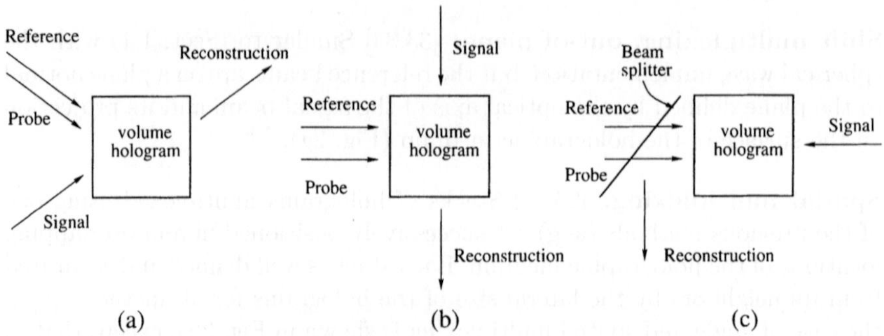

Fig. 3. Simplified holographic recording geometries: (**a**) transmission geometry; (**b**) 90° geometry; (**c**) reflection geometry

of a cube-like recording medium. It has been popular in high-capacity demonstrations [36,43,44] and compact implementations [45,46].

- In the *reflection* geometry (Fig. 3c) the recording beams are incident on two opposite faces of the holographic material and (approximately) counter propagating. In this case the reconstruction is counter propagating and on the same side of the medium as the probe beam (hence the name "reflection"). A beam splitter separates the reconstruction from the probe. The reflection geometry has been preferred in wavelength-multiplexed systems [12], because it provides the optimal wavelength Bragg selectivity (see Example 3, page 44). However, in the reflection geometry the reference beam is also reflected or scattered from the front face of the holographic medium into the detector, leading to a loss in dynamic range. Very often this fact makes other geometries preferable even for wavelength multiplexing, at the expense of a small loss in Bragg selectivity.

The geometry of the optical system used to image the reconstructed beam onto the detector is yet another design option. Four well-known possibilities, shown in Fig. 4, are

- the *image-plane* geometry (Fig. 4a), where the SLM is imaged onto the holographic material and the reconstruction in turn is imaged onto the CCD (d_i, d_o should satisfy the lens law, $1/d_i + 1/d_o = 1/f$ in this geom-

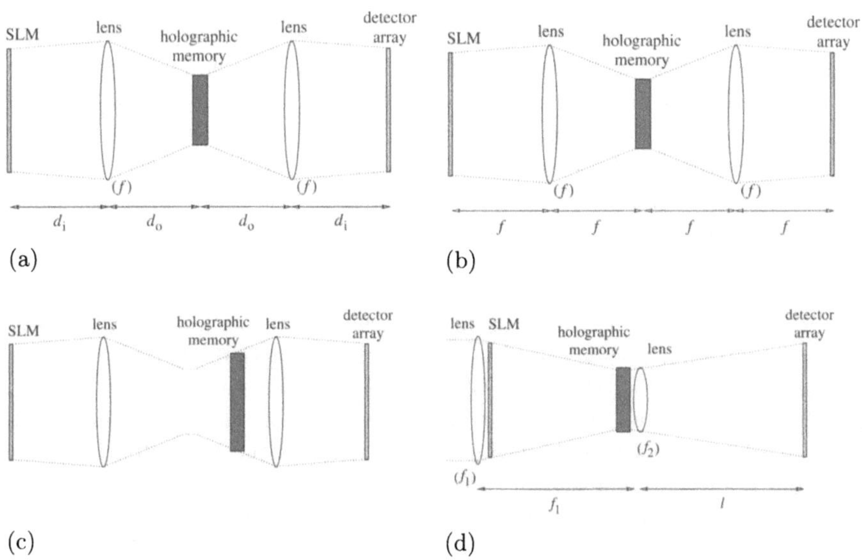

Fig. 4. Imaging architectures for holographic memories: (**a**) image plane, (**b**) Fourier plane, (**c**) Fresnel zone, (**d**) van der Lugt system. The symbol in parenthesis, e.g. (f), denotes the focal length of the corresponding lens

etry, as in Fig. 4a; another possibility – not shown – is to interpose 4–F systems between SLM, memory, and detector);

- the *Fourier-plane* geometry (Fig. 4b), where the SLM is Fourier-transformed onto the holographic material and the reconstruction in turn is Fourier-transformed onto the CCD;
- the *Fresnel-zone* geometry (Fig. 4c), where the material is placed in the Fresnel zone of the SLM;
- and the *van der Lugt* geometry (Fig. 4d), which is similar to the Fourier-plane geometry, except the SLM is adjacent to the Fourier-transforming lens. (The imaging condition is $1/f_1 + 1/l = 1/f_2$ and the magnification $|m| = f_1/l$ in this imaging geometry.)

Another option is to use a probe that is phase-conjugated with respect to the reference beam; the reconstruction itself is then phase-conjugated, and the need to use imaging elements is eliminated [24,46] (see Sect. 3.3).

This chapter does not contain an exhaustive analysis of the performance of all combinations of multiplexing schemes, recording geometries, and imaging systems. Our goal is to provide enough intuitive understanding of the physical mechanisms behind volume diffraction (Sect. 2) and practical experience with different systems (Sect. 3) for the reader to make informed design decisions in a variety of situations with minimum effort.

2 Scattering from Volume Gratings

In this section we develop the mathematical framework of three-dimensional scattering, which is the basis for understanding holographic multiplexing in volume media. In particular, we derive the field that is diffracted from a three-dimensional phase grating when the grating is illuminated by a probe field, first in the general case (Sect. 2.1) and then for the simplified case of scalar fields (Sect. 2.2). In Examples 1, 2 we use the theory of sections 2.1 and 2.2 to solve two simple problems of volume diffraction from slab-like gratings in the angle and shift multiplexing configurations, deriving the corresponding Bragg selectivities. The solution of Example 1 is generalized in Sect. 2.3 to the *k-sphere formulation*, a useful geometric construction for calculating volume diffraction without the need to use the diffraction integral.[5] Two examples follow showing the usage of the k-sphere: the wavelength selectivity is derived in Example 3, and the implementation of phase-code multiplexing is explained in Example 4. Section 2.4 extends the generalized k-space solution to another geometrical construction, the *grating space* (or K-space), useful for the visualization of volume holographic multiplexing techniques. This extension leads naturally to the classification of holographic multiplexing methods in two categories: *Bragg* and *fractal*. Peristrophic multiplexing as

[5] With the exception of shift multiplexing, which does require the diffraction integral.

either Bragg or fractal is analyzed with the help of the k-sphere and grating space in Example 5.

2.1 Volume Diffraction in the Born Approximation

In most holographic materials, volume holograms are recorded as modulation of the material refractive index, resulting after exposure to the interference pattern between the signal and reference beams. Consider such a dielectric material, as shown in Fig. 5, occupying volume $\mathcal{V}$ in space, with position-dependent dielectric constant $\epsilon(\boldsymbol{r})$. We assume that the modulation is weak and narrow-band, i.e.

$$\epsilon(\boldsymbol{r}) = \epsilon_0 + \tilde{\epsilon}(\boldsymbol{r})\mathrm{e}^{\mathrm{i}\boldsymbol{K}_{\mathrm{g}}\cdot\boldsymbol{r}}, \qquad \epsilon_0 \gg |\tilde{\epsilon}(\boldsymbol{r})| . \tag{7}$$

Here ϵ_0 is the dielectric constant before exposure, $\boldsymbol{K}_{\mathrm{g}}$ is the *carrier* wave-vector of the grating, and $\tilde{\epsilon}(\boldsymbol{r})$ is a low-bandwidth modulation (in the sense that the maximum spatial frequency contained in $\tilde{\epsilon}(\boldsymbol{r})$ is much less than $\boldsymbol{K}_{\mathrm{g}}$) representing the data stored in the volume hologram. The hologram is illuminated with an incident optical field $\boldsymbol{E}_{\mathrm{p}}$ at frequency ω. In this section we derive the diffracted field $\boldsymbol{E}_{\mathrm{d}}$, under some approximations, as the first-order solution to a wave scattering problem. The result is the *diffraction integral* (19).

The electromagnetic field satisfies Maxwell's equations in charge-free space:

$$\begin{aligned} \nabla \times \boldsymbol{E} &= \mathrm{i}\omega\boldsymbol{B} & \nabla \cdot \boldsymbol{D} &= 0 \\ \nabla \times \boldsymbol{H} &= -\mathrm{i}\omega\boldsymbol{D} & \nabla \cdot \boldsymbol{B} &= 0, \end{aligned} \tag{8}$$

where $\boldsymbol{E} = \boldsymbol{E}_{\mathrm{p}} + \boldsymbol{E}_{\mathrm{d}}$ is the total electric field, $\boldsymbol{D} = \epsilon\boldsymbol{E}$ is the dielectric displacement, $\boldsymbol{H}$ is the magnetic field, $\boldsymbol{B} = \mu\boldsymbol{H}$ is the magnetic induction,

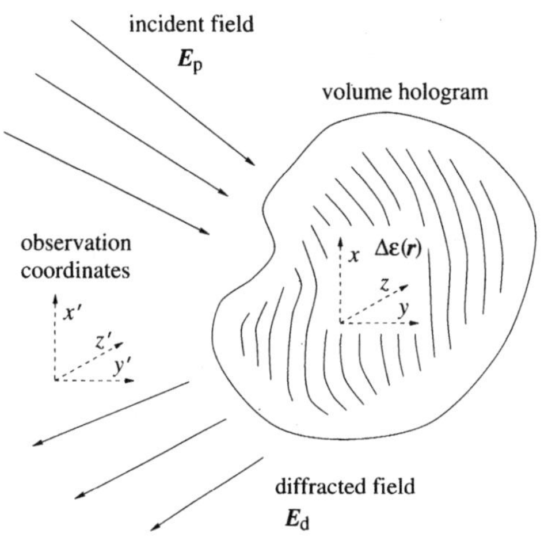

Fig. 5. Volume diffraction geometry

and μ is the magnetic permeability, hereafter assumed to be equal to μ_0, the magnetic permeability of vacuum. From the two curl equations we obtain

$$\begin{aligned} &\nabla \times \nabla \times \boldsymbol{E} = -\mu\epsilon\left(-\omega^2 \boldsymbol{E}\right) \\ \Rightarrow\ &\nabla\left(\nabla \cdot \boldsymbol{E}\right) - \nabla^2 \boldsymbol{E} = \omega^2 \mu\epsilon \boldsymbol{E}. \end{aligned} \tag{9}$$

To calculate $\nabla \cdot \boldsymbol{E}$, we use Gauss's law:

$$\nabla \cdot (\epsilon \boldsymbol{E}) = \nabla\epsilon \cdot \boldsymbol{E} + \epsilon \nabla \cdot \boldsymbol{E} = 0 \Rightarrow \nabla \cdot \boldsymbol{E} = -\frac{\nabla\epsilon \cdot \boldsymbol{E}}{\epsilon} \approx -\frac{\nabla\epsilon \cdot \boldsymbol{E}}{\epsilon_0}. \tag{10}$$

The modulation gradient is calculated explicitly as

$$\nabla\epsilon(\boldsymbol{r}) = \nabla\left(\tilde{\epsilon}(\boldsymbol{r})\mathrm{e}^{\mathrm{i}\boldsymbol{K}_\mathrm{g}\cdot\boldsymbol{r}}\right) = \mathrm{e}^{\mathrm{i}\boldsymbol{K}_\mathrm{g}\cdot\boldsymbol{r}}\nabla\tilde{\epsilon}(\boldsymbol{r}) + \mathrm{i}\,\tilde{\epsilon}(\boldsymbol{r})\mathrm{e}^{\mathrm{i}\boldsymbol{K}_\mathrm{g}\cdot\boldsymbol{r}}\boldsymbol{K}_\mathrm{g}. \tag{11}$$

The narrow-band grating assumption essentially means that the first term in the above expression is negligible, and we thus obtain

$$\frac{\nabla\epsilon \cdot \boldsymbol{E}}{\epsilon_0} = \mathrm{i}\frac{\left(\boldsymbol{K}_\mathrm{g} \cdot \boldsymbol{E}\right)\tilde{\epsilon}(\boldsymbol{r})\mathrm{e}^{\mathrm{i}\boldsymbol{K}_\mathrm{g}\cdot\boldsymbol{r}}}{\epsilon_0}. \tag{12}$$

The wave equation for the electric field in the presence of the volume hologram follows from (9) after substitution of (12):

$$\nabla^2 \boldsymbol{E} + \omega^2 \mu_0 \epsilon_0 \boldsymbol{E} = -\mathrm{i}\nabla\left(\frac{\left(\boldsymbol{K}_\mathrm{g} \cdot \boldsymbol{E}\right)\tilde{\epsilon}(\boldsymbol{r})\mathrm{e}^{\mathrm{i}\boldsymbol{K}_\mathrm{g}\cdot\boldsymbol{r}}}{\epsilon_0}\right) - \omega^2\mu_0\tilde{\epsilon}(\boldsymbol{r})\mathrm{e}^{\mathrm{i}\boldsymbol{K}_\mathrm{g}\cdot\boldsymbol{r}}\boldsymbol{E}. \tag{13}$$

Using the narrow-band grating approximation, the first term on the right-hand side of (13) simplifies to[6]

$$\approx -\mathrm{i}\left[\left(\boldsymbol{K}_\mathrm{g}\nabla\right)\boldsymbol{E} + \mathrm{i}\left(\boldsymbol{K}_\mathrm{g} \cdot \boldsymbol{E}\right)\boldsymbol{K}_\mathrm{g}\right]\tilde{\epsilon}(\boldsymbol{r})\mathrm{e}^{\mathrm{i}\boldsymbol{K}_\mathrm{g}\cdot\boldsymbol{r}}/\epsilon_0.$$

We now define the matrix operator

$$\begin{aligned} \mathrm{U}(\boldsymbol{r}) &= \tilde{\epsilon}(\boldsymbol{r})\mathrm{e}^{\mathrm{i}\boldsymbol{K}_\mathrm{g}\cdot\boldsymbol{r}}\left[\frac{-\mathrm{i}\left(\boldsymbol{K}_\mathrm{g}\nabla\right) + \boldsymbol{K}_\mathrm{g}\left(\boldsymbol{K}_\mathrm{g}\cdot\,\right)}{\epsilon_0} + \omega^2\mu_0\right] \\ &= \frac{\tilde{\epsilon}(\boldsymbol{r})}{\epsilon_0}\,\mathrm{e}^{\mathrm{i}\boldsymbol{K}_\mathrm{g}\cdot\boldsymbol{r}}\left[-\mathrm{i}\left(\boldsymbol{K}_\mathrm{g}\nabla\right) + \boldsymbol{K}_\mathrm{g}\left(\boldsymbol{K}_\mathrm{g}\cdot\,\right) + k^2\right], \end{aligned} \tag{14}$$

where we used $k = 2\pi/\lambda = \omega\sqrt{\mu_0\epsilon_0}$. The wave equation takes then the compact form

$$\left(\nabla^2 + k^2\right)\boldsymbol{E} = \mathrm{U}(\boldsymbol{r})\boldsymbol{E}; \tag{15}$$

i.e. the volume diffraction problem is formally equivalent to scattering of the vector field $\boldsymbol{E}$ from the potential $\mathrm{U}(\boldsymbol{r})$. Let $\mathrm{G}(\boldsymbol{r})$ denote Green's matrix for

[6] We denote by $\boldsymbol{K}_\mathrm{g}\nabla$ the *outer* product of the two operators, as opposed to $\boldsymbol{K}_\mathrm{g}\cdot\nabla$, which denotes the inner product as usual.

the homogeneous wave equation corresponding to (15). The solution to (15) is a series of diffracted orders [47]

$$\boldsymbol{E} = \boldsymbol{E}_\mathrm{p} + \sum_{k=1}^{\infty} \boldsymbol{E}_\mathrm{d}^{(k)}, \qquad \text{where} \tag{16}$$

$$\boldsymbol{E}_\mathrm{d}^{(k)}(\boldsymbol{r}) = \int_{\mathcal{V}} \mathrm{d}^3\boldsymbol{r}' \mathrm{G}(\boldsymbol{r}-\boldsymbol{r}')\mathrm{U}(\boldsymbol{r}')\boldsymbol{E}_\mathrm{d}^{(k-1)} \tag{17}$$

$$= \int_{\mathcal{V}} \mathrm{d}^3\boldsymbol{r}_1 \ldots \int_{\mathcal{V}} \mathrm{d}^3\boldsymbol{r}_k \mathrm{G}(\boldsymbol{r}-\boldsymbol{r}_1)\mathrm{U}(\boldsymbol{r}_1)\ldots\mathrm{G}(\boldsymbol{r}-\boldsymbol{r}_k)\mathrm{U}(\boldsymbol{r}_k)\boldsymbol{E}_\mathrm{p}(\boldsymbol{r}_k). \tag{18}$$

The iteration is initialized by $\boldsymbol{E}_\mathrm{d}^{(0)}(\boldsymbol{r}) = \boldsymbol{E}_\mathrm{p}(\boldsymbol{r})$. It is evident from (14) and (17) that the kth order term in the solution is of the same order of magnitude as $(\tilde{\epsilon}(\boldsymbol{r}')/\epsilon_0)^k$. Therefore, consistently with the weak modulation assumption, only the first term is non-negligible. This assumption is called *Born's approximation*; it is violated for strong gratings, when the coupled wave formulation [7] must be used. As we mentioned already (Sect. 1), when the number M of stored holograms is large, their strengths decrease as M^{-2} [25]; therefore the use of Born's approximation is valid in the context of holographic storage, and we are justified to express the diffracted field as

$$\boldsymbol{E}_\mathrm{d}(\boldsymbol{r}) = \int_{\mathcal{V}} \mathrm{d}^3\boldsymbol{r}' \mathrm{G}(\boldsymbol{r}-\boldsymbol{r}')\mathrm{U}(\boldsymbol{r}')\boldsymbol{E}_\mathrm{p}(\boldsymbol{r}'). \tag{19}$$

2.2 Volume Diffraction of Scalar Fields

The actual computation of the diffracted field, as given by the diffraction integral (19), is not always straightforward. In this section we restrict the calculation to scalar fields and develop an approach based on 3D Fourier analysis, which simplifies many calculations and leads to the development of the k-sphere formulation of Sect. 2.3. The key results of the present analysis are equation (27), which follows directly from (19) for a scalar incident field, and (32), which is (27) rewritten in the Fourier domain.

Let us assume that the incident field is a plane wave linearly polarized parallel to an arbitrary direction $\hat{\boldsymbol{e}}$, i.e.

$$\boldsymbol{E}_\mathrm{p} = \mathrm{e}^{\mathrm{i}\boldsymbol{k}_\mathrm{p}\cdot\boldsymbol{r}}\hat{\boldsymbol{e}} \qquad (\boldsymbol{k}_\mathrm{p} \perp \hat{\boldsymbol{e}})\,. \tag{20}$$

The diffracted field is then simply

$$\boldsymbol{E}_\mathrm{d}(\boldsymbol{r}) = \int_{\mathcal{V}} \mathrm{d}^3\boldsymbol{r}' \mathrm{G}(\boldsymbol{r}-\boldsymbol{r}')\tilde{\epsilon}(\boldsymbol{r}')\mathrm{e}^{\mathrm{i}\boldsymbol{K}_\mathrm{g}\cdot\boldsymbol{r}'} \frac{(\boldsymbol{K}_\mathrm{g}\cdot\hat{\boldsymbol{e}})\,(-\mathrm{i}\nabla+\boldsymbol{K}_\mathrm{g})+k^2}{\epsilon_0}\mathrm{e}^{\mathrm{i}\boldsymbol{k}_\mathrm{p}\cdot\boldsymbol{r}}. \tag{21}$$

Green's dyadic for homogeneous space is [48]

$$\mathrm{G}(\boldsymbol{r}) = \frac{\mathrm{e}^{ik_\mathrm{p}|\boldsymbol{r}|}}{4\pi\,|\boldsymbol{r}|} I_3, \qquad \text{(outgoing waves only)}, \tag{22}$$

where $k_{\mathrm{p}} = 2\pi/\lambda = |\boldsymbol{k}_{\mathrm{p}}|$, and I_3 is the 3×3 identity matrix. Substituting into (21) we obtain

$$\boldsymbol{E}_{\mathrm{d}}(\boldsymbol{r}) = \boldsymbol{s}(\boldsymbol{k}_{\mathrm{p}}) \int_{\mathcal{V}} \mathrm{d}^3\boldsymbol{r}' \frac{\mathrm{e}^{\mathrm{i}k_{\mathrm{p}}|\boldsymbol{r}-\boldsymbol{r}'|}}{4\pi|\boldsymbol{r}-\boldsymbol{r}'|} \tilde{\epsilon}(\boldsymbol{r}')\mathrm{e}^{\mathrm{i}(\boldsymbol{K}_{\mathrm{g}}+\boldsymbol{k}_{\mathrm{p}})\cdot\boldsymbol{r}'}, \tag{23}$$

where $$\boldsymbol{s}(\boldsymbol{k}_{\mathrm{p}}) = \frac{(\boldsymbol{K}_{\mathrm{g}} \cdot \hat{\boldsymbol{e}})(\boldsymbol{k}_{\mathrm{p}} + \boldsymbol{K}_{\mathrm{g}}) + k^2\hat{\boldsymbol{e}}}{\epsilon_0}. \tag{24}$$

Note that the diffracted field is also linearly polarized in the direction of $\hat{\boldsymbol{s}}$. The angle α formed between $\hat{\boldsymbol{s}}$ and the incident polarization $\hat{\boldsymbol{e}}$ is given by

$$\cos\alpha = \hat{\boldsymbol{s}} \cdot \hat{\boldsymbol{e}} = \frac{(\boldsymbol{K}_{\mathrm{g}} \cdot \hat{\boldsymbol{e}})^2 + k^2}{\sqrt{(\boldsymbol{K}_{\mathrm{g}} \cdot \hat{\boldsymbol{e}})^2 \left(|\boldsymbol{K}_{\mathrm{g}}|^2 + 2\boldsymbol{K}_{\mathrm{g}} \cdot \boldsymbol{k}_{\mathrm{p}} + 3k^2\right) + k^4}}. \tag{25}$$

For a more general scalar incident field

$$\boldsymbol{E}_{\mathrm{p}}(\boldsymbol{r}) = \tilde{E}_{\mathrm{p}}(\boldsymbol{r})\mathrm{e}^{\mathrm{i}\boldsymbol{k}_{\mathrm{p}}\cdot\boldsymbol{r}}\hat{\boldsymbol{e}}, \tag{26}$$

where $\tilde{E}_{\mathrm{p}}(\boldsymbol{r})$ is a low-bandwidth modulation, the diffracted field is

$$\boldsymbol{E}_{\mathrm{d}}(\boldsymbol{r}) = \boldsymbol{s}(\boldsymbol{k}_{\mathrm{p}}) \int_{\mathcal{V}} \mathrm{d}^3\boldsymbol{r}' \frac{\mathrm{e}^{ik_{\mathrm{p}}|\boldsymbol{r}-\boldsymbol{r}'|}}{4\pi|\boldsymbol{r}-\boldsymbol{r}'|} \tilde{\epsilon}(\boldsymbol{r}')\tilde{E}_{\mathrm{p}}(\boldsymbol{r}')\mathrm{e}^{i(\boldsymbol{K}_{\mathrm{g}}+\boldsymbol{k}_{\mathrm{p}})\cdot\boldsymbol{r}'}. \tag{27}$$

This expression has a simple intuitive intepretation: each infinitesimal region inside the hologram acts as a point source with amplitude proportional to the local amplitude of the incident field and the local value of the refractive index modulation; the diffracted field is obtained as coherent superposition of the fields emitted by all these infinitesimal point sources comprising the volume hologram.

It will be useful in many subsequent calculations to have the Fourier-space equivalent of (27). For this purpose, we will now calculate the spatial spectrum of the diffracted field. To keep the notation simple, we adopt the convention that the optical axis is $\hat{\boldsymbol{z}}$, the transverse (to the optical axis) component of an arbitrary vector $\boldsymbol{v}$ is $\boldsymbol{v}_{\perp} = \boldsymbol{v} \times \hat{\boldsymbol{z}}$, and the parallel component is $v_{\parallel} = \boldsymbol{v} \cdot \hat{\boldsymbol{z}}$, and denote the Cartesian components of $\boldsymbol{v}$ as $v_a = \boldsymbol{v} \cdot \hat{\boldsymbol{a}}$, where a is one of x, y, or z, and $\hat{\boldsymbol{a}}$ is the corresponding unit vector $\hat{\boldsymbol{x}}$, $\hat{\boldsymbol{y}}$, or $\hat{\boldsymbol{z}}$ respectively. We define the diffracted spectrum as

$$\mathcal{E}_{\mathrm{d}}(\boldsymbol{k}_{\mathrm{d}}, z) = \int_{\mathcal{V}_{\perp}} \boldsymbol{E}_{\mathrm{d}}(\boldsymbol{r})\mathrm{e}^{-\mathrm{i}\boldsymbol{k}_{\mathrm{d}}\cdot\boldsymbol{r}}\mathrm{d}^2\boldsymbol{r}_{\perp}. \tag{28}$$

Here $\mathcal{V}_{\perp}$ denotes the aperture of the hologram. Substituting (23) into (28) and using the Fourier expansion of Green's function,

$$\frac{\mathrm{e}^{\mathrm{i}k_{\mathrm{p}}|\boldsymbol{r}-\boldsymbol{r}'|}}{4\pi|\boldsymbol{r}-\boldsymbol{r}'|} = \frac{\mathrm{i}}{2} \int \frac{\mathrm{e}^{\mathrm{i}\boldsymbol{k}\cdot(\boldsymbol{r}-\boldsymbol{r}')}}{k_z} \mathrm{d}^2\boldsymbol{k}_{\perp}, \tag{29}$$

we obtain

$$\mathcal{E}_{\mathrm{d}}(\boldsymbol{k}_{\mathrm{d}}, z) = \frac{\mathrm{i}s(\boldsymbol{k}_{\mathrm{p}})}{2} \iint_{\mathcal{V}_{\perp}} \int_{\mathcal{V}} \mathrm{d}^3\boldsymbol{r}'\mathrm{d}^2\boldsymbol{r}_{\perp}\mathrm{d}^2\boldsymbol{k}_{\perp} \frac{\mathrm{e}^{\mathrm{i}(\boldsymbol{k}-\boldsymbol{k}_{\mathrm{d}})\cdot\boldsymbol{r}}}{k_{\mathrm{d}z}} \tilde{\epsilon}(\boldsymbol{r}')\mathrm{e}^{\mathrm{i}(\boldsymbol{K}_{\mathrm{g}}+\boldsymbol{k}_{\mathrm{p}}-\boldsymbol{k})\cdot\boldsymbol{r}'}. \tag{30}$$

Performing the integration with respect to $\boldsymbol{r}_{\perp}$ yields $\delta(\boldsymbol{k}-\boldsymbol{k}_{\mathrm{d}})$; we then integrate with respect to $\boldsymbol{k}_{\perp}$ and obtain

$$\mathcal{E}_{\mathrm{d}}(\boldsymbol{k}_{\mathrm{d}}, z) = \frac{\mathrm{i}\mathrm{e}^{\mathrm{i}k_{\mathrm{d}z}z}}{2}\boldsymbol{A}(\boldsymbol{k}_{\mathrm{p}}, \boldsymbol{k}_{\mathrm{d}}), \tag{31}$$

$$\boldsymbol{A}(\boldsymbol{k}_{\mathrm{p}}, \boldsymbol{k}_{\mathrm{d}}) = \frac{s(\boldsymbol{k}_{\mathrm{p}})}{k_{\mathrm{d}z}} \int_{\mathcal{V}} \mathrm{d}^3\boldsymbol{r}'\tilde{\epsilon}(\boldsymbol{r}')\mathrm{e}^{\mathrm{i}(\boldsymbol{K}_{\mathrm{g}}+\boldsymbol{k}_{\mathrm{p}}-\boldsymbol{k}_{\mathrm{d}})\cdot\boldsymbol{r}'}. \tag{32}$$

The quantity $\boldsymbol{A}(\boldsymbol{k}_{\mathrm{p}}, \boldsymbol{k}_{\mathrm{d}})$ is the three-dimensional Fourier transform of the index modulation $\tilde{\epsilon}(\boldsymbol{r})\mathrm{e}^{\mathrm{i}\boldsymbol{K}_{\mathrm{g}}\cdot\boldsymbol{r}}$, calculated at spatial frequency $\boldsymbol{k}_{\mathrm{p}} - \boldsymbol{k}_{\mathrm{d}}$. It represents the amplitude of the diffracted field propagating in the direction of $\hat{\boldsymbol{k}}_{\mathrm{d}}$ when the hologram is illuminated by a plane wave incident at direction $\hat{\boldsymbol{k}}_{\mathrm{p}}$.

Example 1: Changing the angle of the probe wave (angle multiplexing)

We will use the formulation developed so far to calculate the field diffracted from a very simple volume hologram, recorded by the interference of two plane waves in a slab-like medium with thickness L in the $\hat{\boldsymbol{z}}$ direction. The general geometry is shown in Fig. 6, and we simplify by setting $\theta_{\mathrm{f}} = \theta_{\mathrm{s}} = \theta$. During recording, the electric fields of the reference and signal beams are

$$\boldsymbol{E}_{\mathrm{f}} = \mathrm{e}^{\mathrm{i}k(z\cos\theta - x\sin\theta)}\hat{\boldsymbol{y}}, \qquad \boldsymbol{E}_{\mathrm{s}} = \mathrm{e}^{\mathrm{i}k(z\cos\theta + x\sin\theta)}\hat{\boldsymbol{y}}. \tag{33}$$

The dielectric modulation is proportional to $|\boldsymbol{E}_{\mathrm{f}} + \boldsymbol{E}_{\mathrm{s}}|^2$. From the four interference terms, we need only maintain[7] $\boldsymbol{E}_{\mathrm{f}}^{*}\boldsymbol{E}_{\mathrm{s}}$. Therefore,

$$\tilde{\epsilon}(\boldsymbol{r}) = \epsilon_1 \mathrm{e}^{\mathrm{i}\boldsymbol{K}_{\mathrm{g}}\cdot\boldsymbol{r}} \operatorname{rect}\frac{z}{L}, \qquad \boldsymbol{K}_{\mathrm{g}} = 2k\sin\theta\hat{\boldsymbol{x}}, \tag{34}$$

where ϵ_1 is a constant expressing the hologram strength ($\epsilon_1 \ll \epsilon_0$). The hologram is probed by a plane wave

$$\boldsymbol{E}_{\mathrm{p}} = \mathrm{e}^{\mathrm{i}k(z\cos\theta_{\mathrm{p}} - x\sin\theta_{\mathrm{p}})}\hat{\boldsymbol{y}}. \tag{35}$$

[7] By repeating the analysis presented below for the remaining three terms, it is easy to show that they do not contribute significantly to the diffracted field. This is the first significant difference with respect to thin holograms, where all four terms diffract.

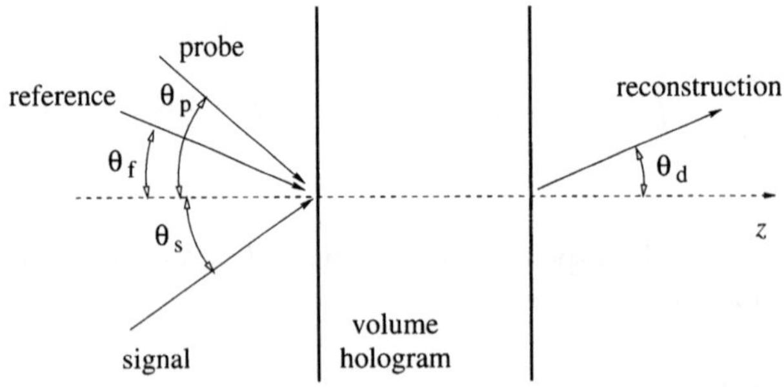

Fig. 6. Schematic of the geometry used in Examples 1, 3, 4. Examples 1, 4 also use $\theta_f = \theta_s = \theta$ for simplicity, and in Example 4 $\theta_p = \theta_f$ by definition

Substituting into (32) gives

$$\begin{aligned} \boldsymbol{A}(\boldsymbol{k}_p, \boldsymbol{k}_d) = & \frac{i\epsilon_1 k^2 L}{2\epsilon_0 k_{dz}} \delta\left[k(2\sin\theta - \sin\theta_p) - k_{dx}\right] \delta\left(k_{dy}\right) \\ & \times \operatorname{sinc}\left[L\left(\frac{\cos\theta_p}{\lambda} - \frac{k_{dz}}{2\pi}\right)\right] \hat{\boldsymbol{y}} \text{ where } \operatorname{sinc}(u) \equiv \frac{\sin(\pi u)}{(\pi u)}. \end{aligned} \tag{36}$$

Therefore, the diffracted field is non-zero only if the transverse components of the diffracted wave-vector satisfy

$$k_{dx} = k(2\sin\theta - \sin\theta_p), \qquad k_{dy} = 0. \tag{37}$$

The longitudinal component is constrained by the requirement

$$|\boldsymbol{k}_d| = \sqrt{k_{dx}^2 + k_{dy}^2 + k_{dz}^2} = k. \tag{38}$$

Suppose that $\theta_p = \theta + \Delta\theta$, where $|\Delta\theta| \ll \theta$. From (36), (37) and (38) we find

$$k_{dz} = k\left(\cos\theta + \Delta\theta \sin\theta\right), \qquad \text{and} \tag{39}$$

$$\begin{aligned} \boldsymbol{E}_d = & \frac{i\epsilon_1 kL}{2\epsilon_0 \cos\theta} e^{ik[(\sin\theta - \Delta\theta\cos\theta)x + (\cos\theta + \Delta\theta\sin\theta)z]} \\ & \times \operatorname{sinc} \frac{2L\,(\Delta\theta)\sin\theta}{\lambda}\hat{\boldsymbol{y}}. \end{aligned} \tag{40}$$

The diffraction efficiency, which expresses the proportion of incident intensity diffracted by the hologram, is, according to (5),

$$\eta = \frac{|\boldsymbol{E}_d|^2}{|\boldsymbol{E}_p|^2} = \left(\frac{\epsilon_1 kL}{2\epsilon_0 \cos\theta}\right)^2 \operatorname{sinc}^2\left(\frac{2L\,(\Delta\theta)\sin\theta}{\lambda}\right). \tag{41}$$

From (40) and (41) we observe that the diffracted field is also a plane wave propagating at angle $\theta - \Delta\theta$, and is attenuated by the *Bragg mismatch* factor given by the sinc function. The dependence of η on the angular detuning $\Delta\theta$

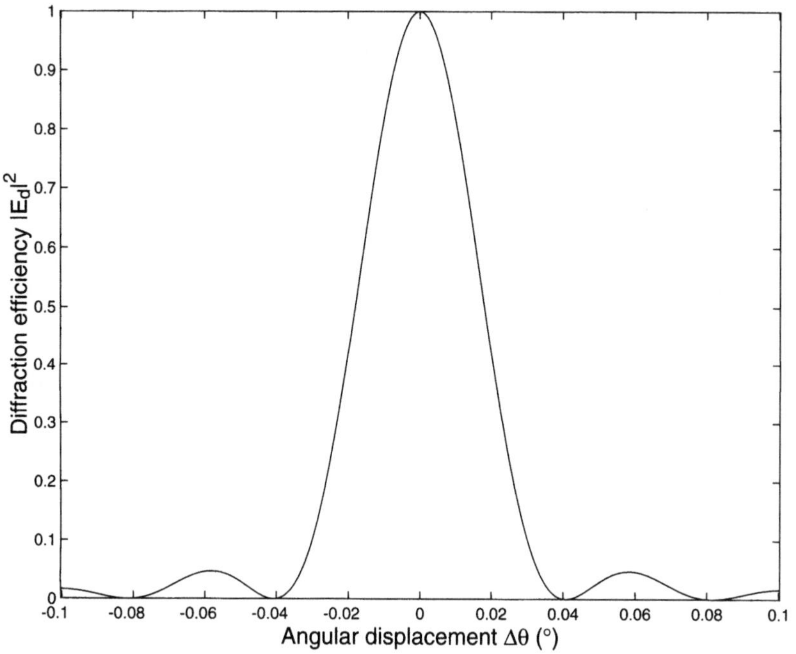

Fig. 7. Diffraction efficiency as function of the probe rotation $\Delta\theta$ in the $\hat{\boldsymbol{x}}$ direction relative to the reference

is shown in Fig. 7 for $\lambda = 488$ nm, $L = 1$ mm, and $\theta = 20°$. The diffraction efficiency is maximum when $\Delta\theta = 0$; this condition is called *Bragg matching.* On the other hand, when

$$\Delta\theta = (\Delta\theta)_{\mathrm{B}} = \frac{\lambda}{2L\sin\theta}, \tag{42}$$

the sinc reaches its first null, which means that the hologram does not diffract at all. The quantity $(\Delta\theta)_{\mathrm{B}}$ is called the *angle Bragg selectivity.*[8]

Now suppose that a second hologram is recorded with the same signal beam (a plane wave at direction $\hat{\boldsymbol{k}}_s$ and reference beam incident at angle $\theta + m\,(\Delta\theta)_{\mathrm{B}}$, where m is an integer). According to (40), when the superposition of the two holograms is illuminated at angle $\theta + m\,(\Delta\theta)_{\mathrm{B}}$, only the second hologram diffracts, while the first one is "silent." Provided $m\,(\Delta\theta)_{\mathrm{B}} \ll \theta$, it is also (approximately) true that when the superposition is illuminated at angle θ, the second hologram is "silent" and only the first hologram diffracts. This

[8] Note that so far we have implicitly assumed that the space surrounding the hologram is index matched to ϵ_0. In applications one must always correct the formulas for Snell refraction at the edges of the hologram. In the remainder of this chapter we will assume that the wavelength and angles of incidence are given for the index-matched case.

property forms the basis of *angle multiplexing* (see also Sect. 1.1, Fig. 2b). For the same numbers of the example in the previous paragraph, if $m = 1$ (multiplexing at the first null) and the total accessible angular range inside the medium is $\Delta\Theta = 5°$,[9] then it is possible to multiplex up to

$$M_{\max;2} = \frac{\Delta\Theta}{(\Delta\theta)_{\mathrm{B}}} = 122 \tag{43}$$

holograms. Note the inverse dependence of $(\Delta\theta)_{\mathrm{B}}$ on L (41) which indicates that in thicker media it is possible to multiplex more holograms within the accessible range $\Delta\Theta$. M also increases with the angle of incidence[10] θ.

Example 2: Changing the location of the probe wave (shift multiplexing)

Consider the simplified transmission geometry of Fig. 8, with a plane wave signal beam as in the previous example, but with a spherical reference beam. The reference emanates from a point source at $\boldsymbol{r}_{\mathrm{f}} = x_{\mathrm{f}}\hat{\boldsymbol{x}} + y_{\mathrm{f}}\hat{\boldsymbol{y}} + z_{\mathrm{f}}\hat{\boldsymbol{z}}$. We express this wave in the paraxial approximation as

$$E_{\mathrm{f}}(\boldsymbol{r}) = \exp\left(\mathrm{i}2\pi\frac{z - z_{\mathrm{f}}}{\lambda} + \mathrm{i}\pi\frac{(x - x_{\mathrm{f}})^2 + (y - y_{\mathrm{f}})^2}{\lambda(z - z_{\mathrm{f}})}\right). \tag{44}$$

Note that here and in the sequel we neglect a term of the form $1/\lambda(z - z_{\mathrm{f}})$ because it varies with z much slower than the exponential term. The signal beam is a plane wave propagating at angle $u \ll 1$ with respect to the $\hat{\boldsymbol{z}}$-axis.

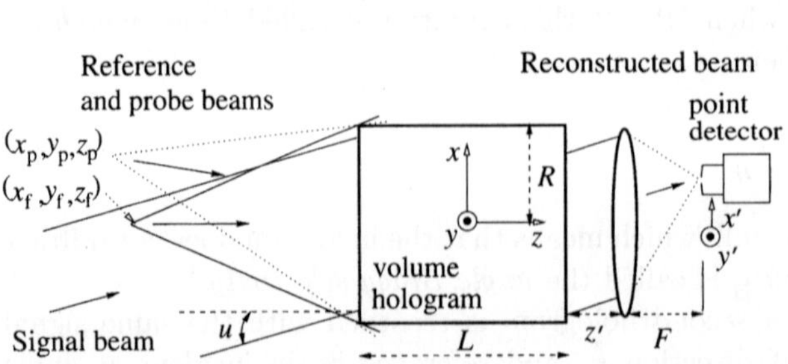

Fig. 8. Transmission geometry with spherical wave reference and plane wave signal beams

[9] The angular range that needs to be accessed outside the medium is $\Delta\Theta$ times the refractive index.

[10] This cannot be more than $1/\sqrt{\epsilon_0}$, though, because of Snell refraction at the entrance face – unless the hologram is immersed in index-matching liquid, or the signal and reference beams are incident at a different crystal face each, as in the 90° geometry (Fig. 3).

In the paraxial approximation, it is expressed as

$$E_\mathrm{s}(\boldsymbol{r}) = \exp\left[\mathrm{i}2\pi\left(1-\frac{u^2}{2}\right)\frac{z}{\lambda} + \mathrm{i}2\pi u\frac{x}{\lambda}\right] \tag{45}$$

After recording is complete, we probe the volume hologram with a spherical wave at the same wavelength λ emanating at $\boldsymbol{r}_\mathrm{p} = x_\mathrm{p}\hat{\boldsymbol{x}} + y_\mathrm{p}\hat{\boldsymbol{y}} + z_\mathrm{p}\hat{\boldsymbol{z}}$. The expression for the probe field is

$$E_\mathrm{p}(\boldsymbol{r}) = \exp\left(\mathrm{i}2\pi\frac{z-z_\mathrm{p}}{\lambda_\mathrm{p}} + \mathrm{i}\pi\frac{(x-x_\mathrm{p})^2+(y-y_\mathrm{p})^2}{\lambda_\mathrm{p}(z-z_\mathrm{p})}\right). \tag{46}$$

To find the diffracted field at detector coordinates $\boldsymbol{r}'$ (located near the focus $\boldsymbol{r}_\mathrm{s}$ of the signal beam), it is most convenient to use the volume diffraction formula (27). We simplify by assuming that the holographic medium is disk-shaped with radius R in the xy-plane, and thickness L along the z-direction, and making the paraxial approximation, i.e. we assume that R is much smaller than any longitudinal distance that the fields propagate. We then obtain

$$\begin{aligned}E_\mathrm{d}(\boldsymbol{r}'') = \iiint E_\mathrm{p}(\boldsymbol{r})\Delta\varepsilon(\boldsymbol{r})\,\mathrm{circ}\left(\frac{\sqrt{x^2+y^2}}{R}\right)\mathrm{rect}\,\frac{z}{L}\\ \times\exp\left(\mathrm{i}2\pi\frac{z''-z}{\lambda_\mathrm{p}} + \mathrm{i}\pi\frac{(x''-x)^2+(y''-y)^2}{\lambda_\mathrm{p}(z''-z)}\right)\mathrm{d}^3\boldsymbol{r}.\end{aligned} \tag{47}$$

Instead of the actual diffracted field, it is most informative to compute the Fourier transform of $E_\mathrm{d}(\boldsymbol{r}'')$ incident on the detector after going through a lens of focal length F, i.e.

$$\tilde{E}_\mathrm{d}(\boldsymbol{r}') = \iint_{-\infty}^{+\infty} E_\mathrm{d}(\boldsymbol{r}'')\exp\left(-\mathrm{i}2\pi\frac{x'x''+y'y''}{\lambda F}\right)\mathrm{d}x''\mathrm{d}y'', \tag{48}$$

where constant phase factors have been omitted. The limits of integration in (48) are taken to be infinite by assuming that the effective aperture of the system is set by the transverse size R of the hologram rather than by the lens. We can then substitute (47) into (48), and perform the x'', y'' integrations right away. The result is

$$\begin{aligned}\tilde{E}_\mathrm{d}(\boldsymbol{r}') = \iiint \left[\mathrm{i}\pi A(z)\left(x^2+y^2\right)\right]\exp\left\{\mathrm{i}2\pi\left[B_x(z)x + B_y(z)y\right]\right\}\\ \times\exp\left[\mathrm{i}\pi C(z)\right]\mathrm{circ}\left(\frac{\sqrt{x^2+y^2}}{R}\right)\mathrm{rect}\,\frac{z}{L}\mathrm{d}^3\boldsymbol{r}.\end{aligned} \tag{49}$$

The coefficients $A(z)$, $B_x(z)$, $B_y(z)$, $C(z)$ are expressed as

$$A(z) = \frac{1}{\lambda(z - z_{\mathrm{f}})} - \frac{1}{\lambda(z - z_{\mathrm{p}})}, \tag{50}$$

$$B_x(z) = -\frac{x_{\mathrm{p}}}{\lambda(z - z_{\mathrm{p}})} + \frac{x_{\mathrm{f}}}{\lambda(z - z_{\mathrm{f}})} - \frac{x'}{\lambda F} + \frac{u}{\lambda}, \tag{51}$$

$$B_y(z) = -\frac{y_{\mathrm{p}}}{\lambda(z - z_{\mathrm{p}})} + \frac{y_{\mathrm{f}}}{\lambda(z - z_{\mathrm{f}})} - \frac{y'}{\lambda F}, \tag{52}$$

$$C(z) = \frac{x_{\mathrm{p}}^2 + y_{\mathrm{p}}^2}{\lambda(z - z_{\mathrm{p}})} - \frac{x_{\mathrm{f}}^2 + y_{\mathrm{f}}^2}{\lambda(z - z_{\mathrm{f}})} + \left(\frac{x'^2 + y'^2}{\lambda F^2} - \frac{u^2}{\lambda}\right) z. \tag{53}$$

The integral (49) is further simplified by using cylindrical coordinates. We define the parameters ρ, ϕ, $B(z)$, $\alpha(z)$ by

$$\left\{\begin{array}{l} x = \rho\cos\phi,\ B_x(z) = B(z)\cos\alpha(z) \\ y = \rho\sin\phi,\ B_y(z) = B(z)\sin\alpha(z) \end{array}\right. \Leftrightarrow$$
$$\left\{\begin{array}{l} \rho = \sqrt{x^2 + y^2},\ B(z) = \sqrt{B_x(z) + B_y(z)} \\ \tan\phi = y/x, \quad \tan\alpha(z) = B_y(z)/B_x(z) \end{array}\right. \tag{54}$$

(the sign of the inverse tangent is taken to conform with the quadrant of x, y, and $B_x(z)$, $B_y(z)$, respectively). Equation (49) then becomes

$$\tilde{E}_{\mathrm{d}}(\boldsymbol{r}') = \int_{-L/2}^{L/2} \exp\left\{\mathrm{i}\pi C(z)\right\} \int_0^R \exp\left\{\mathrm{i}\pi A(z)\rho^2\right\} \times \int_{-\pi}^{\pi} \exp\left\{-\mathrm{i}2\pi B(z)\rho\cos\left(\phi - \alpha(z)\right)\right\} \mathrm{d}\phi\, \rho \mathrm{d}\rho\, \mathrm{d}z. \tag{55}$$

The innermost integral results in the zero-order Bessel function of the 1st kind,

$$\int_{-\pi}^{\pi} \exp\left\{-\mathrm{i}2\pi B(z)\rho\cos\left(\phi - \alpha(z)\right)\right\} \mathrm{d}\phi = 2\pi J_0\left(2\pi B(z)\rho\right). \tag{56}$$

The next-level integral occurs in the calculation of the 3D-PSF of a lens near focus [49] (pp. 435–449), and is written as

$$\int_0^1 \exp\left\{-\frac{\mathrm{i}}{2}u\rho^2\right\} J_0(v\rho)\rho\mathrm{d}\rho = \mathcal{L}(u, v), \tag{57}$$

where the real and imaginary parts of the function $\mathcal{L}(u, v)$ are expressed in terms of the *Lommel* functions. In terms of the $\mathcal{L}$ function, the diffracted field at the detector is expressed as

$$\tilde{E}_{\mathrm{d}}(\boldsymbol{r}') = 2\pi R^2 \int_{-L/2}^{L/2} \exp\left\{\mathrm{i}\pi C(z)\right\} \mathcal{L}\left(2\pi A(z)R^2, 2\pi B(z)R\right) \mathrm{d}z. \tag{58}$$

This result has an interesting interpretation. Since $\mathcal{L}(.,.)$ also describes the amplitude transmitted from a quadratic lens, (58) means that the diffracted light from the volume hologram is the coherent superposition of *several* "lenses" stacked in the $\hat{\boldsymbol{z}}$ direction. If the probe source is at the common front focus of all these virtual "lenses," then the "lenses" are all in phase and give a strong reconstruction in the back focal point. This is yet another interpretation of the Bragg matching condition. If the probe moves around, the "lens" contributions are in general out of phase, resulting in Bragg mismatch, and the reconstructed amplitude drops. Even richer behavior is obtained if the wavelength of the probe source is different from λ [16,50], but we will not consider these effects here.

Now suppose that the probe beam is displaced by δ_x in the $\hat{\boldsymbol{x}}$ direction with respect to the reference, i.e. in the plane defined by the signal beam direction and the optical axis. Specifically, we assume $\boldsymbol{r}_{\mathrm{f}} = (0, 0, -z_0)$ and $\boldsymbol{r}_{\mathrm{p}} = (\delta_x, 0, -z_0)$, with $z_0 > 0$. As the probe moves, the reconstruction moves also, to a location found by maximizing the slowly varying component of the integrand, i.e. setting the arguments to $\mathcal{L}(.,.)$ equal to zero. From (50) we see that $A(z) = 0$ always. From (51) and (52) it follows that we cannot satisfy $B(z) = 0$ simultaneously for all z. Instead, we satisfy

$$B_x(0) = 0 \Leftrightarrow \frac{x'}{F} = u - \frac{\delta_x}{z_0}, \tag{59}$$

$$B_y(0) = 0 \Leftrightarrow \frac{y'}{F} = 0. \tag{60}$$

If (59) and (60) are satisfied, then, ignoring terms of order z or higher in the Taylor expansion of $\mathcal{L}(.,.)$, we obtain

$$\tilde{E}_{\mathrm{d}}(\boldsymbol{r}') \approx 2\pi R^2 \int_{-L/2}^{L/2} \exp\left\{\mathrm{i}\pi C(z)\right\} \mathrm{d}z. \tag{61}$$

Next we approximate $C(z) \approx C_0 + C_1 z$, where, using (59) and (60),

$$C_0 = C(0) = -\frac{\delta_x^2}{\lambda z_0}, \qquad C_1 = \left.\frac{\mathrm{d}C}{\mathrm{d}z}\right|_{z=0} = -\frac{2\delta_x u}{z_0}. \tag{62}$$

Ignoring terms of higher order in z in the Taylor expansion of $C(z)$, we find the explicit (but approximate) expression

$$\tilde{E}_{\mathrm{d}}(u - \frac{\delta_x}{z_0}, 0, z') \approx \exp\left\{\mathrm{i}2\pi\frac{z'}{\lambda}\left[1 - \left(u - \frac{\delta_x}{z_0}\right)^2\right] - \mathrm{i}\pi\frac{\delta_x^2}{\lambda z_0}\right\} \times \operatorname{sinc}\frac{\delta_x u L}{\lambda z_{\mathrm{f}}}. \tag{63}$$

The *shift Bragg selectivity* is, by definition, the in-plane displacement of the probe beam required to reach the first null of the sinc function, i.e.

$$(\Delta\delta_x)_{\mathrm{B}} \approx \frac{\lambda z_{\mathrm{f}}}{uL}. \tag{64}$$

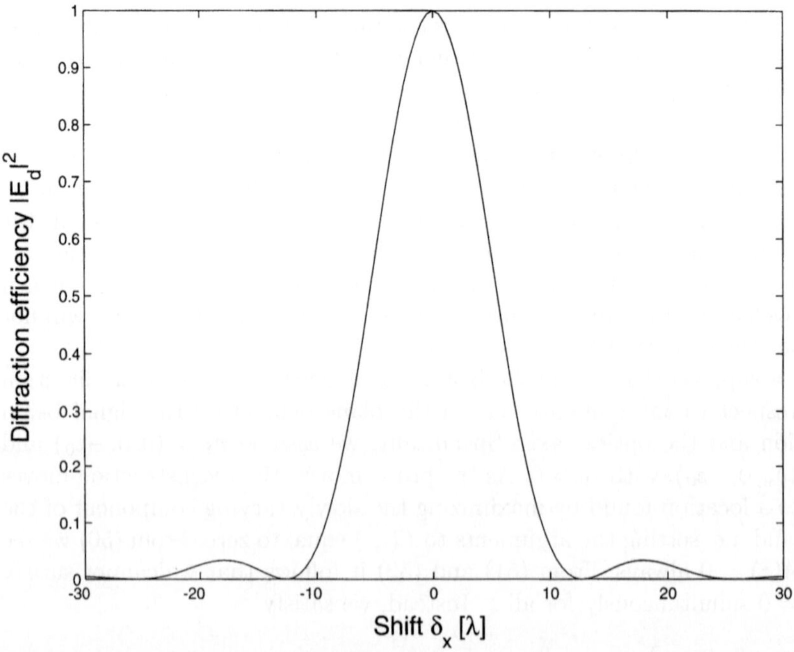

Fig. 9. Numerical calculation of the diffraction efficiency as a function of the probe displacement δ_x in the $\hat{x}$ direction

Note the analogy of (64) and (42) if we correspond $\Delta\theta \leftrightarrow \Delta\delta_x/z_{\mathrm{f}}$, the apparent relative angular motion resulting from shift δ_x.

The plot of the diffraction efficiency $\left|\tilde{E}_{\mathrm{d}}\right|^2$ calculated numerically using (58) with observation points given by (59) and (60) as function of δ_x is given in Fig. 9 for the geometry of Fig. 8 with $\boldsymbol{r}_{\mathrm{f}} = (0, 0, -10^4\lambda)$, $u = 0.2$, $R = 500\lambda$, $L = 4 \times 10^3\lambda$. From the plot we see that the location of the minima in diffraction efficiency is in good agreement with the estimate for the locations of the sinc nulls ($\Delta\delta_x = 12.5\lambda$) predicted by (64).

2.3 Volume Diffraction Calculations Using the k-Sphere Formulation

We now present a geometrical construction, the *k-sphere*, which often simplifies volume diffraction calculations. This construction is directly applicable to holograms recorded with a plane wave reference, but a simple correspondence can be made for spherical-reference holograms as well. The construction, also known as the *Ewald sphere*, is familiar from crystallography and solid-state physics [51]. Unlike x-ray diffraction, however, the assumption of infinitesimally small wavelength (equivalently, infinite diffraction volume) is not correct in the optical regime. Instead, we develop a methodology for using the k-sphere while still taking the finite hologram thickness into account.

For later convenience, we define the "Bragg mismatch vector"

$$\delta \boldsymbol{k}_{\mathrm{d}} = \boldsymbol{k}_{\mathrm{p}} + \boldsymbol{K}_{\mathrm{g}} - \boldsymbol{k}_{\mathrm{d}} \tag{65}$$

in terms of the probe, grating and diffracted wave-vectors respectively.

Consider again the problem of diffraction from a slab hologram,

$$\tilde{\epsilon}(\boldsymbol{r}) = \epsilon_1 \mathrm{e}^{\mathrm{i}\boldsymbol{K}_{\mathrm{g}} \cdot \boldsymbol{r}} \operatorname{rect} \frac{z}{L}, \tag{66}$$

except that now we do not place constraints (34) and (35) on $\boldsymbol{K}_{\mathrm{g}}$ and $\boldsymbol{k}_{\mathrm{p}}$. Generalizing the steps that led to (37) and (38) in the previous section, we obtain the more general relations

$$\boldsymbol{k}_{\mathrm{d}} \times \hat{\boldsymbol{z}} = (\boldsymbol{k}_{\mathrm{p}} + \boldsymbol{K}_{\mathrm{g}}) \times \hat{\boldsymbol{z}} \Leftrightarrow \delta \boldsymbol{k}_{\mathrm{d}} \times \hat{\boldsymbol{z}} = 0, \tag{67}$$

$$|\boldsymbol{k}_{\mathrm{d}}| = |\boldsymbol{k}_{\mathrm{p}}| \,. \tag{68}$$

These are interpreted as follows: *The diffracted wave vector is constrained to have magnitude equal to that of the incident wave vector, and transverse component equal to the transverse component of the vectorial sum of the incident wave vector and the grating vector.* In other words, *the tip of the diffracted wave vector lies at the intersection of a sphere having radius equal to the magnitude of the incident wave vector with a line parallel to the* z*-axis passing also from the tip of the vectorial sum of the incident wave vector and the grating vector.* Moreover, we obtain that

$$\eta \propto \operatorname{sinc}^2 \left(\frac{L\,(\boldsymbol{k}_{\mathrm{p}} + \boldsymbol{K}_{\mathrm{g}} - \boldsymbol{k}_{\mathrm{d}}) \cdot \hat{\boldsymbol{z}}}{2\pi} \right) = \operatorname{sinc}^2 \left(\frac{L\,\delta \boldsymbol{k}_{\mathrm{d}} \cdot \hat{\boldsymbol{z}}}{2\pi} \right), \tag{69}$$

i.e. the diffraction efficiency is proportional to the sinc squared of the projection onto the z*-axis of the vectorial sum of the incident wave vector and grating vector minus the diffracted wave vector, multiplied by the material thickness* L.

The geometric interpretation of the above statements is shown in Fig. 10a–c for the case of angle multiplexing. It is important to note that in Example 1, the basis of this treatment, the volume hologram was infinite in the transverse ($\hat{\boldsymbol{x}}$, $\hat{\boldsymbol{y}}$) dimensions;[11] this is not the case for realistic implementations. It is simpler to keep on ignoring this subtlety until we pick up on it again in Sect. 2.4 (see also [52], pp. 333–336). In the next example we show how the k-sphere formulation works in practice by deriving the reconstruction of a volume hologram at a wavelength different from the recording wavelength. The rederivation of (41) using the k-sphere formulation is left as an exercise for the reader.

[11] This approximation was not necessary in Example 2 (shift multiplexing) because that formulation allowed taking the finite lateral size into account.

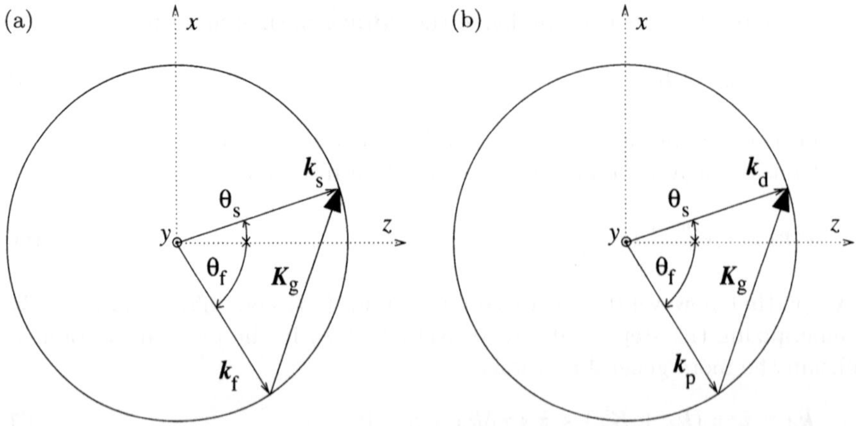

Fig. 10. Bragg diffraction on the k-sphere: (**a**) recording of grating $\boldsymbol{K}_{\mathrm{g}}$ by plane waves $\boldsymbol{k}_{\mathrm{f}}$, $\boldsymbol{k}_{\mathrm{s}}$; (**b**) Bragg-matched reconstruction, $\boldsymbol{k}_{\mathrm{p}} = \boldsymbol{k}_{\mathrm{f}}$; (**c**) probe beam rotated by $\Delta\theta$ (angle-multiplexing), and Bragg-mismatched by $\delta\boldsymbol{k}_{\mathrm{d}} = \boldsymbol{k}_{\mathrm{p}} + \boldsymbol{K}_{\mathrm{g}} - \boldsymbol{k}_{\mathrm{d}}$; (**d**) probe beam at wavelength detuned by $\Delta\lambda$ (wavelength multiplexing)

Example 3: Changing the wavelength of the probe wave (wavelength multiplexing)

Consider again the volume grating of Example 1 and Fig. 6, expressed by (34). The hologram is probed by a plane wave at wavelength $\lambda_{\mathrm{p}} = \lambda + \Delta\lambda$ ($|\Delta\lambda| \ll \lambda$) incident at the same angle θ_{f} of the recording reference beam. The k-sphere picture of the reconstruction is shown in Fig. 10d. Let Δk denote the wave-vector change resulting from wavelength detuning $\Delta\lambda$: i.e. $\Delta k \approx -2\pi\Delta\lambda/\lambda^2$. From (68) we obtain

$$\boldsymbol{k}_{\mathrm{d}} \cdot \hat{\boldsymbol{x}} = (\boldsymbol{k}_{\mathrm{p}} + \boldsymbol{k}_{\mathrm{s}} - \boldsymbol{k}_{\mathrm{f}}) \cdot \hat{\boldsymbol{x}} = k \sin\theta_{\mathrm{s}} - \Delta k \sin\theta_{\mathrm{f}}. \quad (70)$$

From (69) and using (70) it follows that

$$\boldsymbol{k}_{\mathrm{d}} \cdot \hat{\boldsymbol{z}} = \sqrt{(k+\Delta k)^2 - (\boldsymbol{k}_{\mathrm{d}} \cdot \hat{\boldsymbol{x}})} = k \cos\theta_{\mathrm{s}} + \Delta k \frac{1+\sin\theta_{\mathrm{s}}\sin\theta_{\mathrm{f}}}{\cos\theta_{\mathrm{s}}}. \tag{71}$$

Therefore, the amount of Bragg mismatch is

$$\delta \boldsymbol{k}_{\mathrm{d}} \cdot \hat{\boldsymbol{z}} = -\Delta k \frac{2\sin^2 \frac{1}{2}\left(\theta_{\mathrm{f}}+\theta_{\mathrm{s}}\right)}{\cos\theta_{\mathrm{s}}}. \tag{72}$$

From this and requiring that $\Delta\lambda$ be such that the diffraction efficiency reach the first null (69), we derive the *wavelength Bragg selectivity*

$$(\Delta\lambda)_{\mathrm{B}} = \frac{\lambda^2 \cos\theta_{\mathrm{s}}}{2L\sin^2 \frac{1}{2}\left(\theta_{\mathrm{f}}+\theta_{\mathrm{s}}\right)}. \tag{73}$$

For on-axis signal incidence ($\theta_{\mathrm{s}} = 0°$), $(\Delta\lambda)_{\mathrm{B}}$ is minimum when $\theta_{\mathrm{f}} = 180°$. Therefore, the reflection geometry (Fig. 3c) provides optimal hologram packing in wavelength-multiplexed systems [12].

Example 4: Phase-code multiplexing

Consider again the geometry of Fig. 6, except that the reference beam is not a single plane wave; it is phase modulated on one dimension and then Fourier-transformed as shown in Fig. 2e. Suppose that the SLM pixel spacing is b and the focal length of the Fourier-transforming lens is F. The SLM has J pixels in total, and each introduces phase delay α_j ($j = 1, \ldots, J$). We also assume for simplicity that the SLM pixels are well represented by point sources, and that $\theta_{\mathrm{f}} = \theta_{\mathrm{s}} = \theta$. Then the reference beam incident on the hologram is expressed, in the paraxial approximation ($Jb/F \ll \theta$), as

$$\boldsymbol{E}_{\mathrm{f}} = \mathrm{e}^{\mathrm{i}k(z\cos\theta - x\sin\theta)} \sum_{j=1}^{J} \alpha_j \mathrm{e}^{-\mathrm{i}jk\frac{b}{F}(z\sin\theta + x\cos\theta)} \hat{\boldsymbol{y}}. \tag{74}$$

The k-sphere representation of this reference beam has the fan-like appearance of Fig. 11a. Each plane wave component of the fan records a grating with the signal beam (33) as shown in the figure.

Suppose now that the composite hologram (incorporating all the individual gratings) is reconstructed by a probe beam identical to the reference except with a possibly different phase modulation α'_j ($j = 1, \ldots, J$). Therefore,

$$\boldsymbol{E}_{\mathrm{p}} = \mathrm{e}^{\mathrm{i}k(z\cos\theta - x\sin\theta)} \sum_{l=1}^{J} \alpha'_l \mathrm{e}^{-\mathrm{i}lk\frac{b}{F}(z\sin\theta + x\cos\theta)} \hat{\boldsymbol{y}}. \tag{75}$$

Each component of the probe fan reconstructs each component of the composite grating: thus, the diffracted beam is composed of J^2 total components.

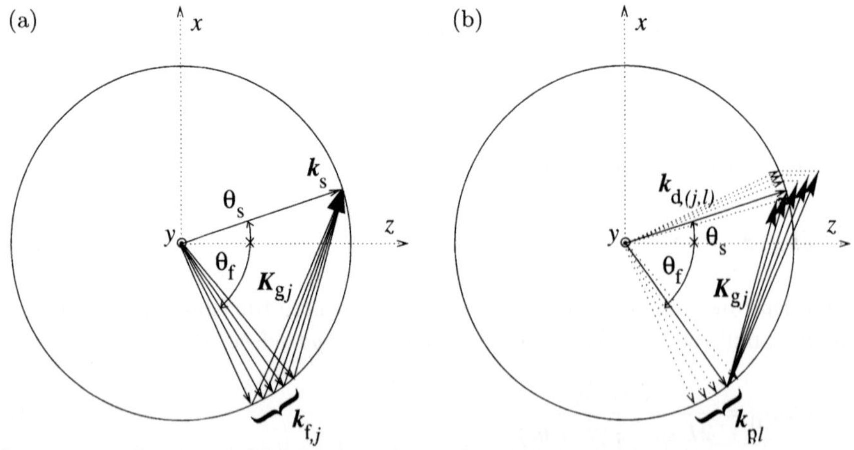

Fig. 11. Phase-code multiplexing on the k-sphere ($J = 5$): (**a**) recording of J grating vectors by the reference fan with the signal beam; (**b**) reconstruction of the composite grating by a single member of the reconstructing fan

Consider each probe beam component individually on the k-sphere as in Fig. 11b. We then notice that the lth probe component is Bragg-matched to the $j = l$th grating component, and angularly displaced with respect to the $J - 1$ remaining grating components that have $j \neq l$. Therefore, for each pair (j, l) the diffracted field is analogous to result (42) derived for the case of angular displacement between reference and probe beams, but with $\Delta\theta = (l - j)b/F$. According to the results of Example 1, the fields diffracted from pairs $j \neq l$ propagate at angles other than θ_s: therefore, they are undesirable.[12] Fortunately, these diffracted components are also Bragg mismatched, and can be eliminated completely from the reconstruction if the phase-SLM pixel spacing is selected so that the angular separation between the reference beam components equals the Bragg angle selectivity of the geometry: i.e. according to (42),

$$\frac{b}{F} = \frac{\lambda}{L \sin\theta}. \tag{76}$$

Provided this condition is satisfied, the diffracted field contains J components only (generated by the pairs $l = j$) copropagating in the direction of θ_s. It is left as exercise for the reader to show that

[12] It is easiest to estimate the effect of these terms in the Fourier-plane recording geometry, with reference to Fig. 11b: the $j \neq l$ terms diffract to replicas, or "ghosts," of the reconstruction displaced with respect to the original image by an amount proportional to $(l - j)b/F$.

$$\boldsymbol{E}_{\mathrm{d}} \approx \left(\sum_{j=1}^{J} \alpha_j^* \alpha_j' \right) \mathrm{e}^{\mathrm{i}k(z\cos\theta + x\sin\theta)} \hat{\boldsymbol{y}}. \tag{77}$$

Therefore, if the phase code modulated on the probe replicates the phase code of the reference ($\alpha_j = \alpha_j'$, $j = 1, \ldots, J$), then the original signal beam is reconstructed intact. On the other hand, the α_j's can be selected such that $\sum_{j=1}^{J} \alpha_j^* \alpha_j' = 0$; then the reconstruction is suppressed. The class of phase codes $(\alpha_j)_{j=1,\ldots,J}$ with $\alpha_j = \pm 1$ (i.e. binary phase modulation 0 or π) that satisfy this orthogonality criterion are known as Walsh–Hadamard codes.

Summarizing the operation of phase-code multiplexing:

- each multiplexed hologram is recorded with a reference beam modulated by a different member of a Walsh–Hadamard code;
- unlike the multiplexing methods of the previous Examples (1–3), the Bragg angular selectivity is utilized not for selective addressing but in order to avoid multiple reconstructions due to the fan nature of the reference/probe beams.[13]

Historically, phase-code multiplexing was conceived as described above [29]. However, it can be implemented with *any* phase code that has autocorrelation function similar to a δ-function, e.g. a random [53] or pseudo-random [32] speckle pattern. Another less obvious choice is to use constant phase (i.e. no phase modulation) and translate the medium relative to the reference fan between exposures. The autocorrelation of the phase code implemented by this shift operation is not a δ-function but a Hadamard function with periodic spikes (similar to the radiation patterns occuring in phased arrays with linear phase shift between elements). This version of phase coding was used in the first implementation of shift multiplexing [31]. Speckle multiplexing is different in that it relies on statistical decorrelation of the reference to multiplex holograms. For instance, shift selectivity is obtained in all three dimensions and not just in the Bragg direction (as is the case with the spherical reference). A detailed statistical analysis that predicts the crosstalk noise as function of packing density has not yet been carried out.

2.4 Visualization of the Multiplexing Methods on the Grating Space

In the geometries of Examples 1–4 the signal beam was invariably a single plane wave. This simplification eased the calculations; in reality, in order to store information in the hologram it is necessary to modulate the

[13] It would be possible to address multiplexed holograms individually using the phase-code method in a thin medium. However, each hologram would be reconstructed as the convolution of the corresponding signal beam with the fan-like reference rather than the signal beam itself. See [52], Chapters 8–9.

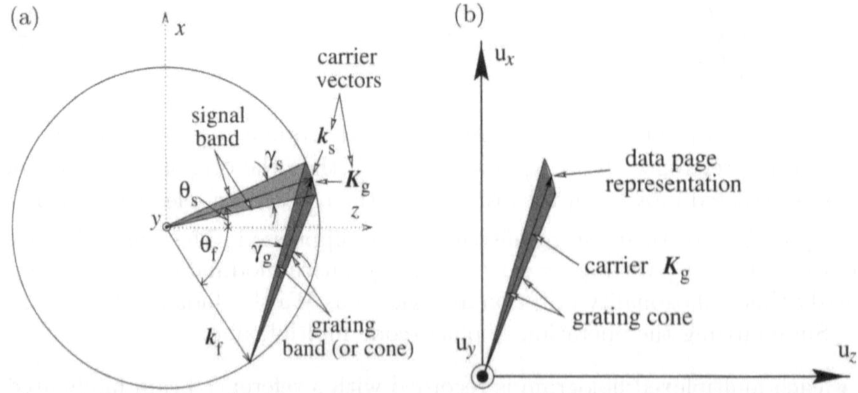

Fig. 12. From the k-sphere to the grating space representation: (**a**) recording of a single page as a hologram, represented in the k-sphere; (**b**) representation of the same data page in the grating space

signal beam with the data. The modulation manifests itself by broadening the recorded bandwidth by $\gamma_s/2$ around the signal beam carrier $\boldsymbol{k}_s$. Consequently, the recorded grating is also broadened and appears as a *grating cone* approximately centered at the location of the *carrier grating* $\boldsymbol{K}_g$, as shown in Fig. 12a. The solid angle γ_g subtended by the grating cone is finite if the recorded signal is bandlimited.[14]

It is useful to further abstract the k-sphere grating representation of Fig. 12a by constructing a space where the principal entities are the grating vectors. In this construction, known as the *grating space* (or K-space) [9,54], gratings are represented by the grating vectors originating at zero coordinates. The representation of the grating cone recorded according to the k-sphere representation of Fig. 12a, for example, in the grating space appears as in Fig. 12b. The grating space representation of the data page is the two-dimensional manifold $\mathcal{P}$, or "patch," formed by the tips of the grating vectors of the grating cone; it corresponds to the Fourier transform of the amplitude incident on the hologram, mapped onto the manifold.[15]

With the help of the grating space, it is possible to make an analogy that leads to useful intuition. Consider the multiplexing principle used for

[14] Strictly speaking, this is not true for finite-size pages and/or if the edges of the SLM pixels are sharp; however, this edge information is useless and can be discarded without loss in the function of the memory. The recorded pages can be considered to be bandlimited if the page data are bandlimited, which is always the case.

[15] For image plane holograms – Fig. 4a – the grating space representation corresponds to the Fourier transform of the pattern displayed on the SLM; for Fourier plane holograms, the correspondence is with the SLM pattern itself. Fresnel and van der Lugt holograms can be configured to correspond to either case with a quadratic phase modulation.

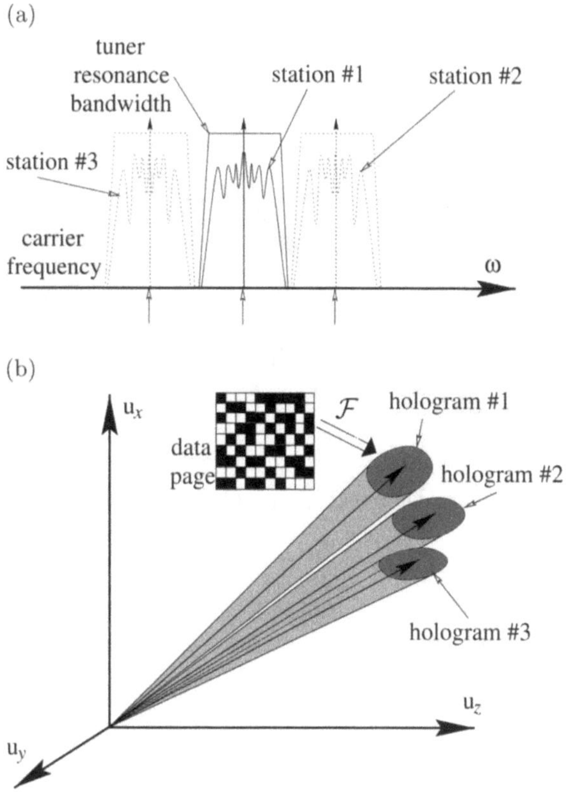

Fig. 13. Analogy of multiplexing: (**a**) one-dimensional temporal signals in the radio-waves, and (**b**) two-dimensional spatial signals as three-dimensional volume gratings

radio communications in the one-dimensional temporal frequency domain ω. As shown in Fig. 13a, radio stations transmit low-bandwidth modulations on their assigned carrier frequencies. Stations are selected by matching the receiver resonance frequency to the carrier frequency of the desired station. Ideally, the quality factor Q of the receiver circuit is chosen so that the resonance broadening matches the modulation bandwidth of the stations, and the carrier frequencies are separated by the same amount. Under these ideal conditions, individual stations are tuned to with minimal crosstalk from other stations, and are optimally packed to fit the maximum amount of stations within the range of frequencies available for transmission.

Holographic storage is the three-dimensional (3-D) space-domain generalization of the radio multiplexing principle. Indeed, the grating space is the 3-D spatial equivalent of the ω-space, in the sense that each sinusoidal grating $u_x x + u_y y + u_z z$ is represented by a triad of numbers, the grating vector $\boldsymbol{K} = (u_x, u_y, u_z)$. According to volume diffraction theory as we developed it so far for holograms of inifinite transverse size in Sects. 2.1–2.3, the multi-

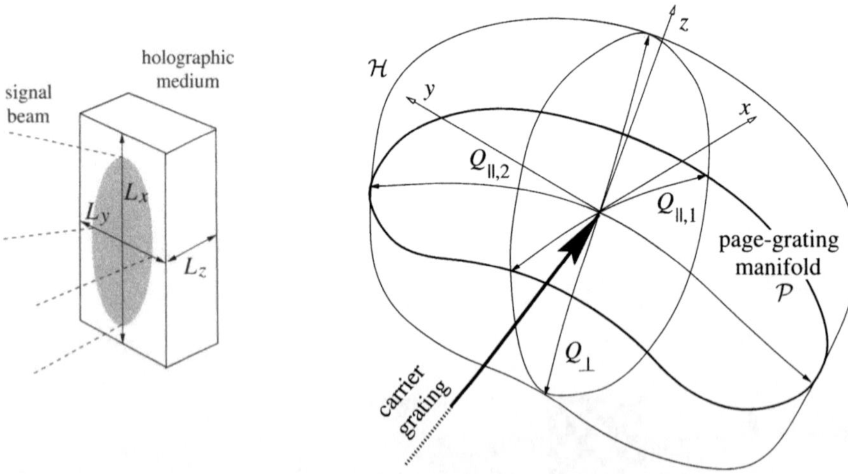

Fig. 14. Page-grating manifold $\mathcal{P}$, quality factors $Q_\perp$, $Q_{\parallel,1}$, $Q_{\parallel,2}$, and spatial bandwidth $\mathcal{H}$ of the volume hologram

plexed gratings are distinguishable (i.e. selectively addressable according to (69)) if they are separated by a multiple of the Bragg selectivity vector along the $\hat{\boldsymbol{z}}$ direction (mapped onto the grating space).

Let us now briefly consider the effect of the finite lateral hologram size. Suppose that the diffracting volume is described by

$$\mathcal{V} = \left[-\frac{L_x}{2}, \frac{L_x}{2}\right] \times \left[-\frac{L_y}{2}, \frac{L_y}{2}\right] \times \left[-\frac{L_z}{2}, \frac{L_z}{2}\right], \tag{78}$$

(see Fig. 14), and the configuration is otherwise similar to Example 1. From (32) we then obtain

$$\begin{aligned} |\boldsymbol{A}(\boldsymbol{k}_\mathrm{p}, \boldsymbol{k}_\mathrm{d})| \propto \operatorname{sinc} & \frac{K_{\mathrm{g}x} + k_{\mathrm{p}x} - k_{\mathrm{d}x}}{2\pi} L_x \\ & \times \operatorname{sinc} \frac{K_{\mathrm{g}y} + k_{\mathrm{p}y} - k_{\mathrm{d}y}}{2\pi} L_y \operatorname{sinc} \frac{K_{\mathrm{g}z} + k_{\mathrm{p}z} - k_{\mathrm{d}z}}{2\pi} L_z, \end{aligned} \tag{79}$$

instead of (36). Hence, the diffracted spectrum contains non-zero components for values of $\boldsymbol{k}_\mathrm{p}$ within a "cloud" surrounding the exact Bragg-matching condition (see also [52], pp. 333–336). The exact computational appreciation of this statement is non-trivial, but an approximate interpretation with the help of the grating space is the following: *individual gratings are distinguishable if they are separated by at least the lateral bandwidth of the volume hologram.* In simple terms, the lateral size of the hologram must be sufficient to accommodate the entire space-bandwidth product (SBP) of the stored data: i.e. larger than or equal to the data bandwidth for Fourier plane holograms, and larger than or equal to the SLM size for image plane holograms.[16] This requirement specifies how tightly grating vectors can be packed *within* the grating cone (Fig. 12b). On the other hand, the *angular width* γ_g of the grating cone is specified by the width of the Fourier-conjugate variable of the imaging geometry: i.e. the SLM size for Fourier transform holograms, and the data bandwidth for image plane holograms.

According to the above discussion, each data page (the 3-D equivalent of a radio station in our time–space analogy) is associated with three quality factors: the *perpendicular* factor $Q_\perp$, determined by the Bragg selectivity (and inversely proportional to L_z), and the two *lateral* factors $Q_{\|,1}$, $Q_{\|,2}$, determined by the grating cone size γ_g (see Fig. 14). For maximum storage efficiency, data page manifolds should be packed in the grating space as tightly as possible without overlapping, according to these quality factors (this is the 3-D space analog of radio station multiplexing).

The grating space page-manifold packing strategy is determined by the multiplexing method. The methods considered in Examples 1–3 (in-plane angle, in-plane shift, and wavelength) tend to pack pages along the perpendicular direction, as evidenced by the fact that hologram separation is determined by the Bragg selectivity of the respective geometries, and is inversely proportional to the thickness of the hologram L_z. Phase-code multiplexing (Example 4) is unique in that each page is distributed in J holograms, intended to be coherently reconstructed by the appropriate phase code; the separation of these holograms, however, is still set by the Bragg selectivity. Consequently, all the above methods are classified as *Bragg multiplexing methods*, which we now formally define as follows:

- Bragg multiplexing methods pack the page manifolds in grating space along the perpendicular direction, such that they are separated by $Q_\perp$.

If pages could be formed with $\gamma_g = 4\pi$, then *any* Bragg method would completely fill the grating space with them. Since this is practically impossible, however, the available storage bandwidth is effectively increased by *fractal multiplexing*, defined as follows:

- Fractal multiplexing methods pack the page manifolds in grating space along the lateral direction, such that they are separated by $Q_{\|,1}$ or $Q_{\|,2}$.

A typical grating space filling strategy by combining a Bragg and a fractal multiplexing method is illustrated in Fig. 15.

It is important to stress again that fractal methods do not rely on the Bragg selectivity of volume diffraction; hence, they are also applicable to infinitely thin holograms (with $Q_\perp = \infty$). The term "fractal" seems perhaps out of place for this application, but it makes sense in the context of volume holographic interconnects, where these concepts were first clarified [9,56].

[16] Fresnel and van der Lugt holograms have intermediate requirements; the details are left as an exercise for the reader. See also [55].

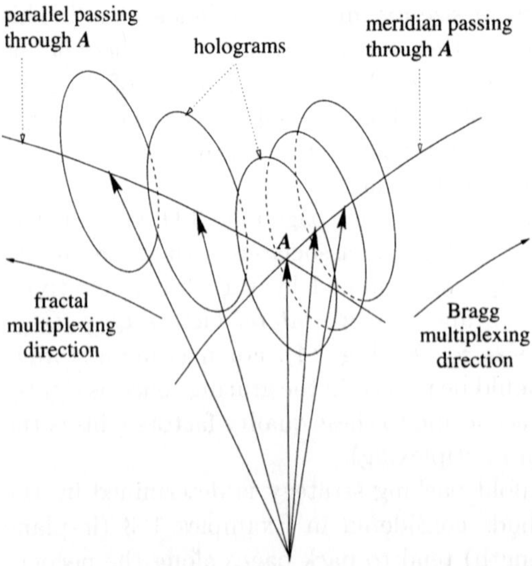

Fig. 15. Grating space picture of tight page manifold packing by combining Bragg and fractal multiplexing

That analysis is beyond the scope of this chapter, but we maintain the terminology for historical reasons. It is also amusing to note that only fractal multiplexing is *exactly* analogous to radio station multiplexing; Bragg methods cannot be implemented in the 1-D ω-space since the signal bandwidth is perpendicular to the multiplexing coordinate.

2.5 Grating Manifold Motion and Fractal Multiplexing

To understand why fractal and Bragg methods are fundamentally different, we compare in- and out-of-plane angle multiplexing in Fig. 16 for the particular case of the 90° geometry. Recall that Bragg multiplexing is possible because when the probe beam is rotated *in the plane defined by the reference and signal carrier beams* with respect to the reference beam, the tip of the vector $\boldsymbol{k}_\mathrm{p} + \boldsymbol{K}_\mathrm{g}$ moves rapidly out of the k-sphere, causing Bragg mismatch on the entire grating manifold. This is shown in Fig. 16a. The distance from the grating manifold to the k-sphere surface is not constant for all grating components, because the manifold motion with respect to the k-sphere is *not* along the manifold normal. This observation is the basis for explaining *crosstalk* in angle [17] and other multiplexing methods [18–22,30].

When the probe rotates *out of plane* (out-of-plane angle multiplexing, Fig. 2b), the reader may verify graphically from Fig. 16b or mathematically that the tip of $\boldsymbol{k}_\mathrm{p} + \boldsymbol{K}_\mathrm{g}$ rather moves almost tangentially to the k-sphere. As a result, Bragg mismatch occurs much slower as a function of out of plane probe rotation than in-plane rotation. A similar situation occurs for spherical-reference holograms if the probe is shifted out-of-plane relative to the reference (out-of-plane shift, Fig. 2g). We leave the analysis of these methods to the reader and consider peristrophic multiplexing in the next example.

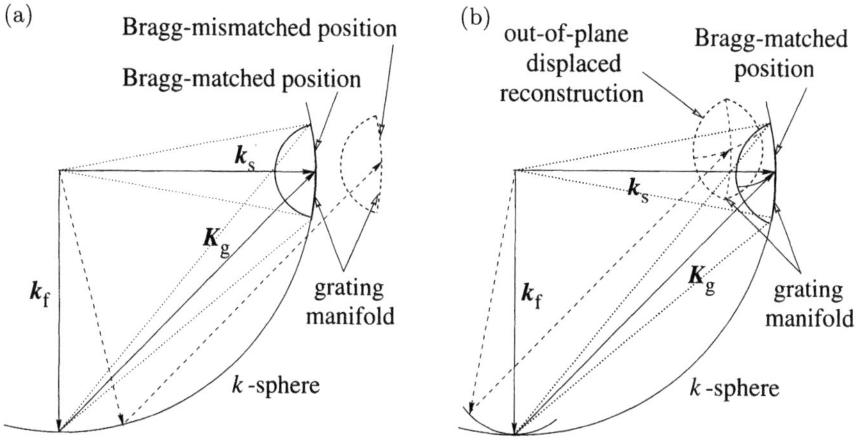

Fig. 16. k-sphere comparison of in- and out-of-plane rotation of the probe beam relative to the reference beam in the 90° geometry. (**a**) In-plane rotation causes the entire grating manifold to move off the k-sphere, thus leading to Bragg mismatch. (**b**) Out-of-plane rotation leads to tangential motion of the grating manifold with respect to the k-sphere

Example 5: Rotating the probe wave (peristrophic multiplexing)

The name of this method is derived from the greek word "peristrophee," which means "rotation." The name is used to refer to multiplexing holograms by rotating the probe beam relative to the recording medium *around the optical axis*, as shown in Fig. 2c (i.e. the axis normal to the recording medium). Consider again the geometry of Fig. 6, and suppose the probe is rotated by angle $\Delta\psi$ around $\hat{\boldsymbol{z}}$ with respect to the reference beam (equivalently, the probe might be coincident with the reference but the material rotates by $\Delta\psi$). From Fig. 16b it is obvious that the probe rotation around the axis, similar to out-of-plane rotation, results also in the tip of $\boldsymbol{k}_\mathrm{p} + \boldsymbol{K}_\mathrm{g}$ moving tangentially to the k-sphere.

Suppose now that an entire page has been recorded as a hologram in the Fourier plane geometry (Fig. 4c). Consider two holograms recorded with the same, on-axis signal beam but with relative rotation of the reference beams around the optical axis, as shown in Fig. 17. The two holograms are reconstructed simultaneously, but only one at a time makes it to the detector. The necessary rotation for the two reconstructions to not overlap is [10]

$$(\Delta\psi)_\mathrm{Fourier} = \frac{Nb}{f\,(\sin\theta_\mathrm{s} + \sin\theta_\mathrm{f})}, \tag{80}$$

where N and b denote the number of pixels and pixel size, respectively, of both the SLM and detector. In grating space terminology, selective reconstruction is possible because the holograms lie without the lateral quality factors of each

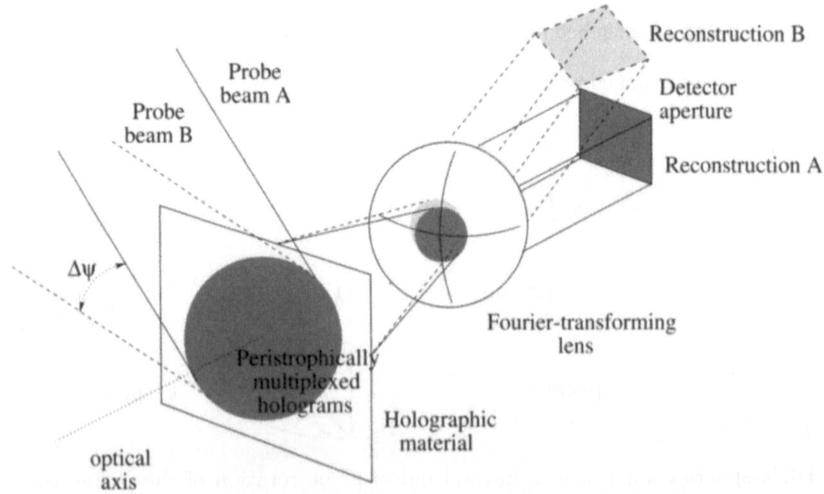

Fig. 17. Peristrophic multiplexing in the Fourier plane. The solid line denotes the hologram currently selected for reconstruction (hologram A). The dashed line (hologram B) is being reconstructed, but falls outside the detector aperture

other's page manifolds. The same idea works also in the image plane geometry, by placing a Fourier-plane stop to filter out the fractal reconstructions. The peristrophic selectivity in this case is [10]

$$(\Delta\psi)_{\text{image}} = \frac{2\lambda}{b\,(\sin\theta_{\text{s}} + \sin\theta_{\text{f}})}\,. \tag{81}$$

The fractal nature of peristrophic multiplexing is now clear. It has been combined [40] with angle multiplexing (a Bragg method) to fill the grating space in fashion similar to the illustration of Fig. 15.

Peristrophic multiplexing does have Bragg selectivity, given by [10]

$$(\Delta\psi)_{\text{B}} = \left\{ \frac{2\lambda}{L} \frac{\cos\theta_{\text{s}}}{\sin\theta_{\text{f}}\,(\sin\theta_{\text{s}} + \sin\theta_{\text{f}})} \right\}^{1/2}. \tag{82}$$

Returning to the k-sphere representation, $(\Delta\psi)_{\text{B}}$ denotes the probe rotation required to move the page manifold out of the k-sphere so that Bragg mismatch is caused. Because the manifold motion is tangential, however, Bragg mismatch requires more probe rotation than radial motion (occurring when the probe is rotated in plane). The slowness is quantified by the $L^{-1/2}$ dependence of the selectivity on thickness, in contrast with the L^{-1} found in Examples 1–3. Since $\lambda/L \ll 1$, it follows that the peristrophic selectivity is always much worse than the angular selectivity of (42) (all other parameters being comparable). The inverse square-root dependence is indicative of fractal geometries, where page juxtaposition is lateral rather than perpendicular (as defined in Sect. 2.4).

3 Architectures for Holographic Memories

3.1 The Holographic 3-D Disk Geometry

The holographic 3-D disk [35,23] is implemented in the familiar and convenient optical disk geometry. The general architecture is shown in Fig. 18. The disk head contains the hologram recording and/or readout optics. Each spot denotes a "stack" of holograms, i.e. a set of multiplexed holograms occupying the same location in the medium. Holograms belonging to different stacks are spatially multiplexed. The most appropriate recording geometries for the disk architecture are transmission and reflection; the 90° geometry is seldom used. Typically, one thinks of the holographic disk for write-once-read-many (WORM) or read-only (ROM) memory applications. The latter greatly simplifies the disk head design, since it requires no recording aparatus; moreover, a cheap monochromatic source (e.g. LED) can be used instead of the laser.

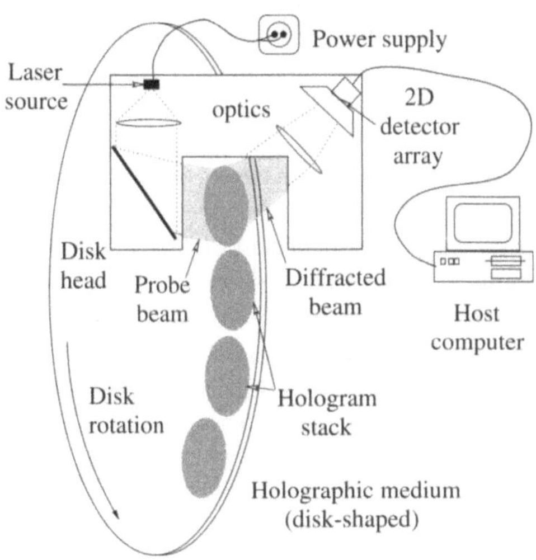

Fig. 18. The holographic three-dimensional (3-D) disk architecture

An angle-multiplexed disk in the transmission geometry is shown schematically in Fig. 19. It is left as an exercise for the reader to show that the Bragg (in-plane) angular selectivity of this configuration is

$$(\Delta\theta)_{\mathrm{B}} = \frac{\lambda \cos\theta_{\mathrm{s}}}{L \sin(\theta_{\mathrm{f}} + \theta_{\mathrm{s}})}. \tag{83}$$

(Example 1 was the special case $\theta_{\mathrm{f}} = \theta_{\mathrm{s}} = \theta$.) The basic requirement for an angle-multiplexed memory is an accurate, fast, and repeatable angular deflection mechanism with wide enough accessible angular range $\Delta\Theta$. The angular deflection mechanism can be implemented with a number of methods, e.g. motorized rotation stages, acousto-optic deflectors [57], micro-electromechanical

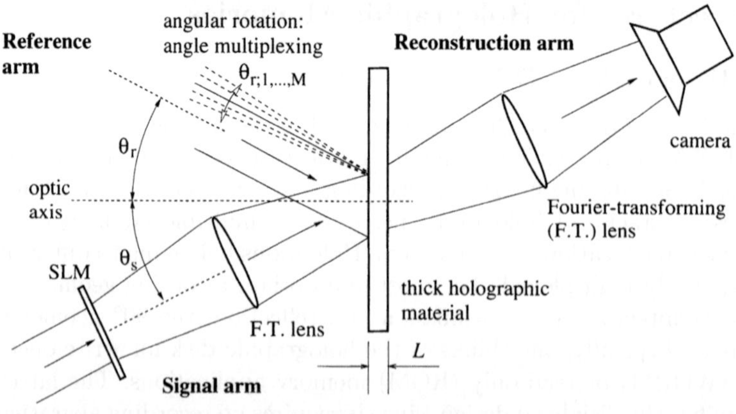

Fig. 19. Angle-multiplexed holographic memory system

(MEMS) deflectors [58–60], and liquid-crystal based deflectors [61]. The disk geometry, in combination with peristrophic multiplexing (implemented by adding a second degree of freedom to the deflector), has been used to demonstrate up to 1 000 holograms [37] and 10 bits/μm^2 [40] in thin photopolymer film.

The shift-multiplexed disk, shown in Fig. 20, is easier to implement mechanically. This is because the rotational motion of the disk provides in a natural way the translation necessary to access different holograms. Moreover, successive multiplexed holograms are accessed and read out using disk rotation only; it is not necessary to stop the disk during readout in order to implement the addressing mechanism. In the shift-multiplexed disk the stacks are not defined separately (cf. Fig. 18); instead, shift-multiplexed holograms are partially overlapping. Out-of-plane (fractal) multiplexed holograms are

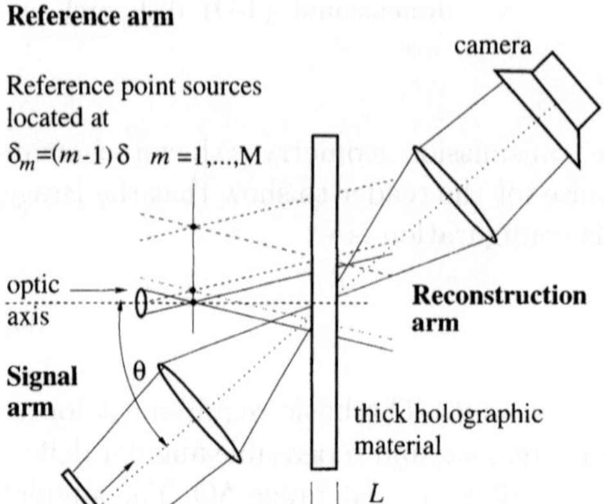

Fig. 20. Shift-multiplexed holographic memory system

accessed by radial head motion. This geometry has also been used in high-capacity demonstrations implemented with photopolymer (12 bits/μm^2 [41]) as well as lithium niobate (100 bits/μm^2 [42]). One practical problem in shift-multplexed disks is the partial mismatch or defocus of the reconstructions resulting from wobble in the motion of the disk relative to the readout head.

3.2 The Holographic Random-Access Memory (HRAM)

In the HRAM, the user is allowed to both record and reconstruct holograms in arbitrary addresses. This functionality makes the HRAM a candidate for several places in the computer memory hierarchy, e.g. as high-capacity buffer between a multimodal storage space and the processor, or as one mode of the processor's own random-access memory (RAM). In either case, the page-oriented nature of holographic storage provides high access speed because of parallelism, but it can be problematic when the application requires access to small records only (because of the overhead of recovering entire pages and searching them for the requested records).

The 90° geometry has been preferred for HRAM implementations [39,44], because it optimizes $M_{\mathrm{max};2}$ (the maximum number of holograms allowed by the numerical aperture of the reference arm optics and the Bragg selectivity of the multiplexing method – typically angle multiplexing). Unfortunately, the large number of pages is inconsistent with the small $M\#$ that this geometry allows. Other issues in HRAM are the hologram recording rate (which depends on the recording *sensitivity* of the holographic medium) and the volatility of the data (i.e. erasure of previously recorded holograms during the recording of a new page). Several refreshing schemes have been devised to deal with the latter difficulty, most of them based on phase-conjugate readout of the stored data [46,62–67](see the next section). Two-photon recording [68] provides a mechanism for controlling data recording and erasure through gating at a sensitizing wavelength.

3.3 The Phase Conjugate Geometry

In several holographic memory systems, phase conjugation offers relative advantages in the implementation. It is obtained as follows [69]: suppose that a hologram is recorded by a reference beam R and a signal (object) beam $Se^{i\phi}$ (Fig. 21a), where S is the actual signal and ϕ is the phase aberration introduced by the beam propagation[17] (including Fresnel diffraction). The interference pattern is expressed as $\left|R + Se^{i\phi}\right|^2$. When R is used for reconstruction (Fig. 21b), $Se^{i\phi}$ is obtained on the signal axis as a continuation of the signal beam, carrying over all the phase aberrations introduced in the signal path during recording. Therefore, the forward reconstruction is distorted

[17] It is usually safe to ignore the effects of absorption on the phase-conjugation process.

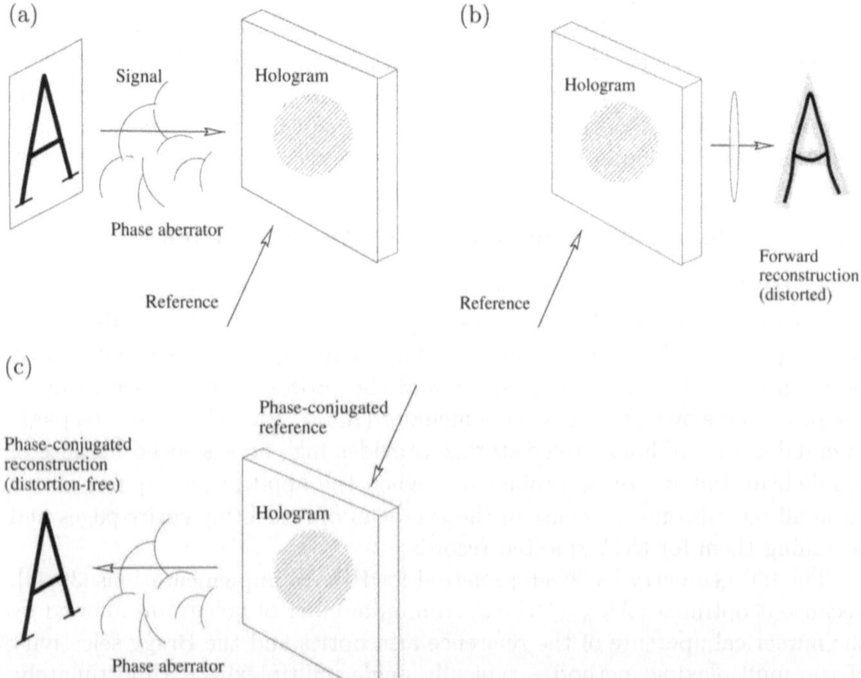

Fig. 21. Forward vs. phase-conjugated reference reconstruction: (**a**) recording with the reference beam and a phase-aberrated signal beam; (**b**) reconstructing with the same reference used for recording (lenses or other imaging optics are required); (**c**) reconstructing with the phase-conjugated (e.g. counter-propagating, in the case of a plane wave) reference without use of a lens

and needs correction by using the appropriate imaging optics. If, however, the phase-conjugated reference R^* is used for reconstruction (Fig. 21c), the on-axis reconstruction produced by the hologram contains the term $S^* \mathrm{e}^{-\mathrm{i}\phi}$. The reciprocal aberrator introduces an additional phase term $\mathrm{e}^{\mathrm{i}\phi}$ to the counter-propagating reconstruction, and, as a result, a distortion-free intensity image $|S|^2$ is obtained at the location of the original signal.

The main advantages of phase conjugate systems stem from the fact that reconstruction can be obtained without imaging elements. Therefore, the system is cheaper, lighter, and more compact. Moreover, the aberrations present in imaging systems are eliminated, and there is no need for custom-designed lenses in the reconstruction arm. The only remaining challenge in a system employing phase conjugation is to create the phase-conjugate reference. The simplest case is a plane wave reference, when the phase conjugate reference is obtained simply by a counter-propagating wave. A spherical reference is phase-conjugated by a mirror at the beam waist, but this method is very sensitive to the position and alignment of the mirror. In all cases, a phase-conjugate mirror (by use of four-wave mixing) can provide the phase-conjugate reference accurately, at the expense of some loss in dynamic range.

An example is the compact refreshable dynamic HRAM with liquid-crystal optoelectronic interface [46,70]. In this design, an optoelectronic circuit, the Dynamic Hologram Refresher (DHR), was performing detection, thresholding, and refreshing of the holograms periodically. The implementation of read-only memories is even simpler because the designer need only generate the phase-conjugated reference (the forward reference is only generated during recording, and is not needed by the memory user). An example to that effect is the compact memory module for uploading FPGA templates [71].

4 Summary

In summary, holographic multiplexing methods are classified as either *Bragg*, when the mechanism enabling selective recall is Bragg selectivity, or *fractal*, when holograms are Bragg degenerate, but they are reconstructed on distinct regions in the image or Fourier plane of the output space, and are selectively accessed with the help of a stop or pupil. Bragg-multiplexed holograms fill up the grating space in a direction perpendicular to the manifold surface, whereas fractal holograms are stacked along the axes of their manifold surfaces. Typically, one Bragg and one fractal method are combined to fill up the grating space.

The grating space packing efficiency of different multiplexing methods characterizes the capacity of a holographic memory, and it can be determined with geometrical calculations based on the analysis presented in this chapter. Subsequent chapters discuss other important parameters that need to be taken into account for the design of holographic memories.

References

1. D. Gabor, "A new microscopic principle," Nature, **161**:777, 1948.
2. D. Gabor, "Microscopy by reconstructed wavefronts," Proc. Roy. Soc., A **197**:454, 1949.
3. D. Gabor, "Microscopy by reconstructed wavefronts II," Proc. Phys. Soc., B **64**:449, 1949.
4. E. Leith and J. Upatnieks, "Wavefront reconstruction and communication theory," J. Opt. Soc. Am., **52**:1123, 1962.
5. P.J. van Heerden, "Theory of optical information storage in solids," Appl. Opt., **2**(4):393–400, 1963.
6. E.N. Leith, A. Kozma, J. Upatnieks, J. Marks, and N. Massey, "Holographic data storage in three-dimensional media," Appl. Opt., **5**(8):1303–1311, 1966.
7. H. Kogelnik, "Coupled wave theory for thick hologram gratings," Bell Syst. Tech. J., **48**(9):2909–2947, 1969.

8. D.L. Staebler, J.J. Amodei, and W. Philips, "Multiple storage of thick holograms in $LiNbO_3$," in *VII International Quantum Electronics Conference*, Montreal, 1972.
9. H. Lee, X.-G. Gu, and D. Psaltis, "Volume holographic interconnections with maximal capacity and minimal cross talk," J. Appl. Phys., **65**(6):2191–2194, 1989.
10. K. Curtis, A. Pu, and D. Psaltis, "Method for holographic storage using peristrophic multiplexing," Opt. Lett., **19**(13):993–994, 1994.
11. G. A. Rakuljic, V. Levya, and A. Yariv, "Optical data storage by using orthogonal wavelength–multiplexed volume holograms," Opt. Lett., **17**(20):1471–1473, 1992.
12. S. Yin, H. Zhou, F. Zhao, M. Wen, Y. Zang, J. Zhang, and F.T.S. Yu, "Wavelength-multiplexed holographic storage in a sensitive photorefractive crystal using a visible-light tunable diode-laser," Opt. Commun., **101**(5-6):317–321, 1993.
13. H.C. Külich, "A new approach to read volume holograms at different wavelengths," Opt. Commun., **64**(5):407–411, 1987.
14. H.C. Külich, "Reconstructing volume holograms without image field losses," Appl. Opt., **30**(20):2850–2857, 1991.
15. D. Psaltis, F. Mok, and H.Y.-S. Li, "Nonvolatile storage in photorefractive crystals," Opt. Lett., **19**(3):210–212, 1994.
16. G. Barbastathis and D. Psaltis, "Shift-multiplexed holographic memory using the two-lambda method," Opt. Lett., **21**(6):429–431, 1996.
17. C. Gu, J. Hong, I. McMichael, R. Saxena, and F. Mok, "Cross-talk-limited storage capacity of volume holographic memory," J. Opt. Soc. Am. A, **9**(11):1978–1983, 1992.
18. K. Curtis, C. Gu, and D. Psaltis, "Cross-talk in wavelength-multiplexed holographic memories," Opt. Lett., **18**(12):1001–1003, 1993.
19. A. Yariv, "Interpage and interpixel crosstalk in orthogonal (wavelength multiplexed) holograms," Opt. Lett., **18**(8):652–654, 1993.
20. K. Curtis and D. Psaltis, "Cross talk in phase-coded holographic memories," J. Opt. Soc. Am. A, **10**(12):2547–2550, 1993.
21. M.C. Bashaw, J.F. Heanue, A. Aharoni, J.F. Walkup, and L. Hesselink, "Cross-talk considerations for angular and phase-encoded multiplexing in volume holography," J. Opt. Soc. Am. B, **11**(9):1820–1836, 1994.
22. K. Curtis and D. Psaltis, "Cross–talk for angle-multiplexed and wavelength-multiplexed image plane holograms," Opt. Lett., **19**(21):1774–1776, 1994.
23. H.-Y.S. Li and D. Psaltis, "Three dimensional holographic disks," Appl. Opt., **33**(17):3764–3774, 1994.
24. W. Liu and D. Psaltis, "Pixel size limit in holographic memories," Opt. Lett., **24**(19):1340–1342, 1999.
25. F. Mok, G.W. Burr, and D. Psaltis, "A system metric for holographic memory systems," Opt. Lett., **21**(12):896–898, 1996.
26. D. Brady and D. Psaltis, "Control of volume holograms," J. Opt. Soc. Am. A, **9**(7):1167–1182, 1992.
27. F.H. Mok, G.W. Burr, and D. Psaltis, "Angle and space multiplexed random access memory (HRAM)," Optical Memory and Neural Networks, **3**(2):119–127, 1994.
28. G.W. Burr, F.H. Mok, and D. Psaltis, "Angle and space multiplexed storage using the 90° geometry," Opt. Commun., **117**(1-2):49–55, 1995.

29. C. Denz, G. Pauliat, and G. Roosen, "Volume hologram multiplexing using a deterministic phase encoding method," Opt. Commun., **85**:171–176, 1991.
30. G. Barbastathis, M. Levene, and D. Psaltis, "Shift multiplexing with spherical reference waves," Appl. Opt., **35**:2403–2417, 1996.
31. D. Psaltis, M. Levene, A. Pu, G. Barbastathis, and K. Curtis, "Holographic storage using shift multiplexing," Opt. Lett., **20**(7):782–784, 1995.
32. L. Dhar, K. Curtis, M. Tackitt, M. Schilling, S. Campbell, W. Wilson, A. Hill, C. Boyd, N. Levinos, and A. Harris, "Holographic storage of multiple high-capacity digital data pages in thick photopolymer systems," Opt. Lett., **23**(21):1710–1722, 1998.
33. G. Barbastathis, A. Pu, M. Levene, and D. Psaltis, "Shift-multiplexed holographic 3d disk," in *Optical Data Storage Proceedings*, San Diego, 1995, SPIE.
34. A. Pu and D. Psaltis, "Shift-multiplexed holographic 3-D disk system," in International Symposium on Optical Memory and Optical Data Storage, Maui, Hawaii, 1996.
35. D. Psaltis, "Parallel optical memories," Byte, **17**(9):179, 1992.
36. F.H. Mok, "Angle-multiplexed storage of 5000 holograms in lithium niobate," Opt. Lett., **18**(11):915–917, 1991.
37. A. Pu, K. Curtis, and D. Psaltis, "A new method for holographic data storage in photopolymer films," in *IEEE Nonlinear Optics: Materials, Fundamentals and Applications*, Waikoloa, Hawaii, 1994.
38. S. Campbell, X.M. Yi, and P. Yeh, "Hybrid sparse-wavelength angle multiplexed optical data storage system," Opt. Lett., **19**(24):2161–2163, 1994.
39. D. Psaltis and F. Mok, "Holographic memories," Sci. Am., **273**(5):70–76, 1995.
40. A. Pu and D. Psaltis, "High density recording in photopolymer–based holographic 3-D disks," Appl. Opt., **35**(14):2389–2398, 1996.
41. A. Pu and D. Psaltis, "Holographic 3-D disks using shift multiplexing," in *Summaries of Papers Presented at CLEO'96*, p. 165, Baltimore, MD, 1996.
42. A. Pu and D. Psaltis, "Holographic data storage with 100 bits/μm^2 density," in *Optical Data Storage Topical Meeting*, pp. 48–49, Tuscon, AZ, 1997.
43. F.H. Mok, M.C. Tackitt, and H.M. Stoll, "Storage of 500 high-resolution holograms in a $LiNbO_3$ crystal," Opt. Lett., **16**(8):605–607, 1991.
44. J.F. Heanue, M.C. Bashaw, and L. Hesselink, "Volume holographic storage and retrieval of digital data," Science, **265**(5173):749–752, 1994.
45. I. McMichael, W. Christian, D. Pletcher, T.Y. Chang, and J. Hong, "Compact holographic storage demonstrator with rapid access," Appl. Opt., **35**(14):2375–2379, 1996.
46. J.-J.P. Drolet, E. Chuang, G. Barbastathis, and D. Psaltis, "Compact, integrated dynamic holographic memory with refreshed holograms," Opt. Lett., **22**(8):552–554, 1997.
47. C. Cohen-Tannoudji, B. Diu, and F. Laloë, Quantum Mechanics, Hermann & Wiley-Interscience, Paris, France, 1977.
48. J.D. Jackson, Classical Electrodynamics, J. Wiley & Sons, 2nd edition, 1975.
49. M. Born and E. Wolf, Principles of Optics, Pergamon Presss, 6th edition, 1980.
50. G. Barbastathis and D.J. Brady, "Multidimensional tomographic imaging using volume holography," Proc. IEEE., **87**(12):2098–2120, 1999.
51. C. Kittel, *Introduction to Solid-State Physics*, J. Wiley, 6th edition, 1986.
52. J.W. Goodman, Introduction to Fourier Optics, McGraw-Hill, 2nd edition, 1996.

53. B. Markov, Yu.N. Denisyuk, and R. Ameskuita, "Three-dimensional displacement speckel hologram and its information capacity," Optika i Spektroskopiya, **84**(4):666–671, 1998.
54. C.X.-G. Gu, Optical neural networks using volume holograms, PhD thesis, California Institute of Technology, 1990.
55. G. Barbastathis, Intelligent holographic databases, PhD thesis, California Institute of Technology, 1998.
56. D. Psaltis, D. Brady, X.G. Gu, and S. Lin, "Holography in artificial neural networks," Nature, **343**(6256):325–330, 1990.
57. X. An and D. Psaltis, "Experimental characterization of an angle-multiplexed holographic memory," Opt. Lett., **20**(18):1913–1915, 1995.
58. R.A. Miller, G.W. Burr, Y.-C. Tai, D. Psaltis, C.-M. Ho, and R.R. Katti, "Electromagnetic MEMS scanning mirrors for holographic data storage," in *Solid-State Sensor and Actuator Workshop, Tranducer Res. Found.*, pp. 183–186, Cleveland Heights, OH, 1996.
59. R.A. Miller, G.W. Burr, Y.-C. Tai, and D. Psaltis, "Magnetically actuated micromirrors for use as optical deflectors," in *Proceedings of the Fourth International Symposium on Magnetic Materials, Processes, and Devices. Applications to Storage and Microelectromechanical Systems (MEMS). Eletrochem. Soc.*, pp. 474–481, Pennington, NJ, 1996.
60. R.A. Miller, G.W. Burr, Y.-C. Tai, and D. Psaltis, "A magnetically actuated mems scanning mirror," in *SPIE Proceedings: Miniaturized Systems with Micro-Optics and Micromechanics*, pp. 47–52, San José, CA, 1996.
61. X. Wang, D.W. Wilson, R.E. Muller, P.D. Maker, and D. Psaltis, "Liquid-crystal based grating beam deflector," SPIE Proc., **3468**:20–29, 1998.
62. D. Brady, K. Hsu, and D. Psaltis, "Periodically refreshed multiply exposed photorefractive holograms," Opt. Lett., **15**(14):817–819, 1990.
63. Y. Qiao, D. Psaltis, C. Gu, J. Hong, P. Yeh, and R.R. Neurgaonkar, "Phase-locked sustainment of photorefractive holograms using phase conjugation," J. Appl. Phys., **70**(8):4646–4648, 1991.
64. H. Sasaki, Y. Fainman, J.E. Ford, and S.H. Lee, "Dynamic photorefractive optical memory," Opt. Lett., **16**(23):1874–1876, 1991.
65. Y. Qiao and D. Psaltis, ""Sampled dynamic holographic memory," Opt. Lett., **17**(19):1376–1378, 1992.
66. S. Boj, G. Pauliat, and G. Roosen, "Dynamic holographic memory showing readout, refreshing, and updating capabilities," Opt. Lett., **17**(6):438–440, 1992.
67. T. Dellwig, C. Denz, T. Rauch, and T. Tschudi, "Coherent refreshment and updating for dynamic photorefractive optical memories using phase conjugation," Opt. Commun., **119**:333–340, 1995.
68. K. Büse, A. Adibi, and D. Psaltis, "Non-volatile holographic storage in doubly doped lithium niobate crystals," Nature, **393**(6686):665–668, 1998.
69. A. Yariv, Optical Electronics, Sounders College, 4th edition, 1991.
70. J.-J.P. Drolet, G. Barbastathis, and D. Psaltis, "Integrated optoelectronic interconnects using liquid-crystal-on silicon VLSI," in R.T. Chen and P.S. Guilfoyle, editors, Optoelectronic Interconnects and Packaging, SPIE Proceedings, volume CR-62, pp. 106–131, 1996.
71. J. Mumbru, G. Zhou, S. Ay, X.An, G. Panotopoulos, F. Mok, and D. Psaltis, "Optically reconfigurable processors," to appear in *SPIE Proceedings.*

Fundamental Noise Sources in Volume Holographic Storage

C. Gu, P. Yeh, X. Yi, and J. Hong

Volume holographic memory [1–8] is becoming a promising candidate for future storage technologies owing to its unique capability to offer large capacity and parallel access. Multiple holograms can be recorded in a photorefractive crystal [1,7,8] sequentially by using an exposure schedule [9] that equalizes the amplitudes. Different reference beam angles, or wavelengths, or phase distributions can be chosen for different exposures: these techniques are known as angle multiplexing, wavelength multiplexing, and phase multiplexing respectively. In this chapter, we consider some fundamental issues in volume holographic storage in photorefractive media. These include the cross-talk noise [10–15], scattering noise [16,17], and noise gratings formed during a multiple exposure schedule [18,19]. These fundamental issues are related to the potentials and limitations of volume holographic storage.

1 Cross-Talk Noise

In volume holographic storage, a large number of holograms are stored in the same volume of the recording medium. Therefore, cross-talk is an important issue for data retrieval. Here we first consider cross-talk noise in angle multiplexing, and then extend the same analysis to wavelength multiplexing. In an angle-multiplexed memory system, an object beam containing a page of information is recorded holographically in association with a uniquely oriented reference beam in the recording medium (e.g. a photorefractive crystal), as shown in Fig. 1. The procedure is repeated with different sets of object and reference beams to yield a volume memory in which any particular page of information can be uniquely accessed by reading the hologram with the corresponding reference beam. Even though the recorded data share the same volume, each page can be retrieved independently with reasonable fidelity, owing to the Bragg-selective nature of the readout. A unique benefit of such a system is the parallel nature of the readout, where by a given reference beam can retrieve large sectors of data simultaneously to be imaged onto a detector/memory array. In the readout process, an image is reconstructed by illuminating the volume with the corresponding reference beam. In an ideal case, no other stored images are read out by this beam. In reality, a small Bragg mismatch leads to a cross-readout of other images with lower efficiencies. The sum of such Bragg-mismatched components in the readout presents a form of cross-talk noise, and affects the storage capacity of the system.

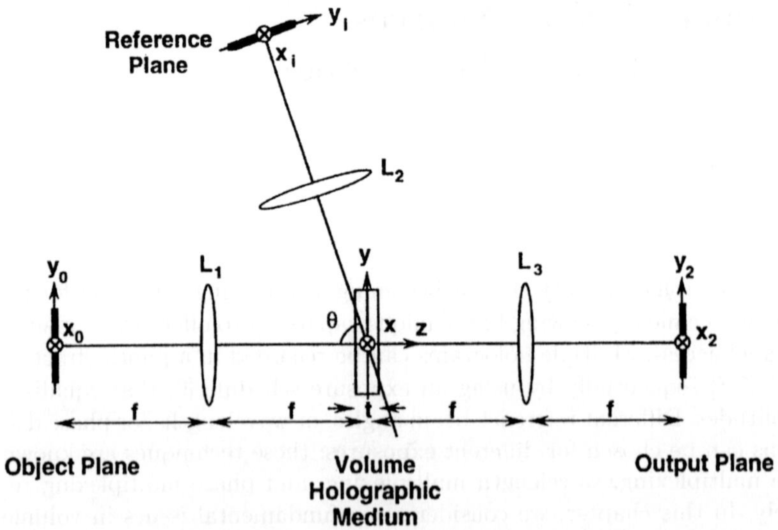

Fig. 1. An optical memory system where angle-multiplexed holograms are stored in a thick recording medium located in the Fourier domain

In the following, we will discuss cross-talk noise, assuming that the degeneracy noise is eliminated by distributing the reference points one-dimensionally along the y_i direction [20], and calculate the upper bound for the storage capacity given a required signal-to-noise ratio. The cross-talk noise will be shown to be a function of the position at the output plane, for a given readout beam. The storage capacity obtained here will be compared with that obtained by Ramberg. The comparison shows that, owing to the presence of the degeneracy noise, the storage capacity does not benefit from expanding the reference points from one-dimensional to two-dimensional (unless fractal spacing is used at the reference plane [21]). We will also give an optimum configuration that fully utilizes the storage capacity and completely eliminates the degeneracy noise.

1.1 Theoretical Formulation

Assume that $N = 2M + 1$ holograms are stored inside a volume holographic medium. We label these holograms as $m = -M, -(M-1), \cdots, 0, (M-1), M$. With the presence of these holograms, the modulation of the dielectric constant of the medium can be written as

$$\Delta\epsilon \propto \sum_{m=-M}^{M} R_m^* S_m + \text{c.c.}, \tag{1}$$

where R_m and S_m are the amplitudes of the reference beam and the object (signal) beam respectively, and c.c. represents complex conjugate. Such an index modulation can be created, for example, in a photorefractive crystal using a multiple exposure schedule [7], provided the total averaged intensity is almost uniform. In the configuration shown in Fig. 1, R_m is a plane wave corresponding to the mth point at the reference plane, and S_m is the transform of the mth object image near the Fourier plane (x, y). We will write R_m and S_m explicitly after examining the reading process.

During the readout, a plane wave with wave vector $\boldsymbol{k}_i$ is incident upon the volume holograms. The inhomogeneous dielectric constant (1) will cause scattering (diffraction) of the incident plane wave. According to scalar diffraction theory [22], the electric field can be written as

$$E(\boldsymbol{r}) \approx \mathrm{e}^{\mathrm{i}\boldsymbol{k}_{\mathrm{i}}\cdot\boldsymbol{r}} + \frac{k_0^2}{4\pi}\frac{\mathrm{e}^{\mathrm{i}kr}}{r}\int \mathrm{d}\boldsymbol{r}'\mathrm{e}^{-\mathrm{i}\boldsymbol{K}\cdot\boldsymbol{r}'}\Delta\epsilon(\boldsymbol{r}'), \tag{2}$$

where

$$\boldsymbol{K} = \boldsymbol{k}_{\mathrm{d}} - \boldsymbol{k}_i, \tag{3}$$

and $\boldsymbol{k}_{\mathrm{d}}$ is the wave vector of the diffracted plane wave. On the right-hand side of (2), the first term describes the incident plane wave, and the second term describes the diffracted light. The diffracted light consists of both the desired output image and the noise images due to cross-readout. Next, we evaluate the signal and noise terms for the configuration shown in Fig. 1.

The expression for the recorded holograms is given by (1). The mth reference beam R_m can be written as

$$R_m = \mathrm{e}^{\mathrm{i}\boldsymbol{k}_m\cdot\boldsymbol{r}}. \tag{4}$$

The mth object (signal) beam S_m is the transform of the mth image near the Fourier plane. Using the standard Fourier optics analysis [23], we obtain

$$S_m(\boldsymbol{r}) = \frac{\mathrm{e}^{2\mathrm{i}kf+\mathrm{i}kn\Delta_0}}{\mathrm{i}\lambda f}\mathrm{e}^{\mathrm{i}kz} \times \int \mathrm{d}x_0\mathrm{d}y_0 f_m(x_0, y_0)\mathrm{e}^{-\mathrm{i}\frac{2\pi}{\lambda f}(xx_0+yy_0)}\mathrm{e}^{-\mathrm{i}\frac{\pi}{\lambda f}\frac{z}{f}(x_0^2+y_0^2)}, \tag{5}$$

where $f_m(x_0, y_0)$ is the mth object image pattern, (x_0, y_0) are the coordinates of the object plane, (x, y, z) are the coordinates centered at the rear focal plane of lens L_1, f is the focal length, n is the index of refraction of the lens material, Δ_0 is the maximum thickness of the lens, and $k = 2\pi/\lambda$ with λ the wavelength of light in vacuum.

We can relate the diffracted wave vector in (2) to the coordinates of the output plane (x_2, y_2). A plane wave at the front focal plane of lens L_3 is converted to a point at the rear focal plane of lens L_3. Under the paraxial approximation, the relationship between the wave vector and the coordinates of the point is given by [23]

$$\boldsymbol{k}_{\mathrm{d}} = \left(\frac{2\pi}{\lambda}\frac{x_2}{f}, \frac{2\pi}{\lambda}\frac{y_2}{f}, \frac{2\pi}{\lambda}\left[1 - \frac{x_2^2}{2f^2} - \frac{y_2^2}{2f^2}\right]\right). \tag{6}$$

Using (4), (5), (1) and (6), we can evaluate the term in (2) that represents the diffracted light. The result is given by

$$g(x_2, y_2) \propto \sum_{m=-M}^{M} \int \mathrm{d}x_0 \mathrm{d}y_0 f_m(x_0, y_0)$$
$$\times V \mathrm{sinc}\left[\frac{a}{2\pi}(k_{mx} - k_{ix} + \frac{2\pi}{\lambda}\frac{x_0 + x_2}{f})\right]$$
$$\times \quad \mathrm{sinc}\left[\frac{b}{2\pi}(k_{my} - k_{iy} + \frac{2\pi}{\lambda}\frac{y_0 + y_2}{f})\right]$$
$$\times \quad \mathrm{sinc}\left[\frac{t}{2\pi}(k_{mz} - k_{iz} - \frac{\pi}{\lambda}\frac{x_2^2 - x_0^2 + y_2^2 - y_0^2}{f^2})\right], \tag{7}$$

where $\boldsymbol{k}_i$ is the wave vector of the reading plane wave, $\boldsymbol{k}_m$ is the wave vector of the mth reference beam during recording, the subscripts x, y and z indicate the corresponding components, a, b and t are the dimensions of the holographic medium in the x, y and z directions respectively, and $V = abt$ is the volume of the holographic medium.

1.2 Cross-Talk Noise and Signal-to-Noise Ratio

It is important to notice several qualitative results from (7). First, the output consists of all the stored images, represented by the summation over m. The amplitude of each image is determined by the difference between the input wave vector and that of the corresponding reference beam used for recording according to the sinc functions. Furthermore, the sinc functions that operate on the x and y components indicate the spatial frequency cut-off due to the finite size of the recording medium. If we assume that the transverse dimensions of the medium are much larger than the spatial bandwidth of the images, then the first two sinc functions in (7) can be approximated by δ functions, and (7) can be easily integrated. The result is given by

$$g(x_2, y_2) \propto \sum_{m=-M}^{M} f_m(-x_2 - \frac{\lambda f}{2\pi}\Delta K_{mix}, -y_2 - \frac{\lambda f}{2\pi}\Delta K_{miy})$$
$$\times t \mathrm{sinc}\left[\frac{t}{2\pi}(\Delta K_{miz} + \frac{\Delta K_{mix} x_2 + \Delta K_{miy} y_2}{f} + \frac{\Delta K_{mix}^2 + \Delta K_{miy}^2}{4\pi/\lambda})\right], \tag{8}$$

where

$$\Delta \boldsymbol{K}_{mi} = \boldsymbol{k}_m - \boldsymbol{k}_i \tag{9}$$

is the difference between the wave vector of the mth reference beam during recording and that of the input reading beam.

We can separate the signal term and noise terms in (8). Suppose the reading beam is the same as one of the reference beams during recording; the term with $\boldsymbol{k}_m = \boldsymbol{k}_i$ in (8) corresponds to the reconstructed image,

$$g_m(x_2, y_2) \propto f_m(-x_2, -y_2), \tag{10}$$

where the minus signs indicate a reversed image. Other terms in (8) with $\boldsymbol{k}_m \neq \boldsymbol{k}_i$ give rise to cross talk noise. Notice that the noise consists of all other images with $m \neq i$. Each noise image is a shifted pattern with reduced amplitude represented by the sinc functions. The amounts of shift and amplitude reduction depend on the momentum mismatch (Bragg mismatch) between the reading beam and the corresponding reference beam used for recording. A large momentum mismatch corresponds to a large spatial shift and a small amplitude.

In the configuration shown in Fig. 1, the wave vector of a plane wave is determined by the position of the reference point at the reference plane. As explained earlier, to eliminate the degeneracy noise, we distribute the reference points one-dimensionally along the y_i direction. Typically, we choose $x_i = 0$ for all reference points. According to Fourier optics [23], the wave vector of a plane wave resulting from a point (x_i, y_i) at the front focal plane of lens L_2 is

$$\begin{aligned}
k_x &= -\frac{2\pi}{\lambda}\frac{x_i}{f} = 0, \\
k_y &= -\frac{2\pi}{\lambda}\frac{y_i}{f}\cos\theta - \frac{2\pi}{\lambda}\left[1 - \frac{1}{2}\left(\frac{y_i}{f}\right)^2\right]\sin\theta, \\
k_z &= -\frac{2\pi}{\lambda}\frac{y_i}{f}\sin\theta + \frac{2\pi}{\lambda}\left[1 - \frac{1}{2}\left(\frac{y_i}{f}\right)^2\right]\cos\theta,
\end{aligned} \tag{11}$$

where θ is the angle between the optical axes of lenses L_1 and L_2, as shown in Fig. 1. Notice that the components of the wave vector are given in the (x, y, z) coordinate system, which has a relative rotation with respect to the local (x_i, y_i, z_i) coordinate system at the reference plane. We have also assumed that the reference points are distributed near the optical axis. Using (11), the momentum mismatch (9) can be written as

$$\begin{aligned}
\Delta K_{mix} &= 0, \\
\Delta K_{miy} &= \frac{2\pi}{\lambda}\frac{y_m - y_i}{f}\left(-\cos\theta + \frac{1}{2}\sin\theta\frac{y_m + y_i}{f}\right), \\
\Delta K_{miz} &= \frac{2\pi}{\lambda}\frac{y_m - y_i}{f}\left(-\sin\theta - \frac{1}{2}\cos\theta\frac{y_m + y_i}{f}\right).
\end{aligned} \tag{12}$$

The noise terms in (8) can be written as

$$g_{\text{noise}}(x_2, y_2) \propto \sum_{m \neq i} f_m(-x_2 - \frac{\lambda f}{2\pi}\Delta K_{mix}, -y_2 - \frac{\lambda f}{2\pi}\Delta K_{miy})$$
$$\times\, t\text{sinc}\left[\frac{t}{2\pi}(\Delta K_{miz} + \frac{\Delta K_{mix}x_2 + \Delta K_{miy}y_2}{f} + \frac{\Delta K_{mix}^2 + \Delta K_{miy}^2}{4\pi/\lambda})\right]. \tag{13}$$

To estimate the maximum noise intensity $I_{\text{noise}} = |g_{\text{noise}}|^2$, we further assume that the object images are randomly distributed light patterns with $|f_m| = 1$ for the "on" pixels. In this case, the averaged noise intensity can be written as

$$\text{Noise} = \sum_{m \neq i} t^2 \text{sinc}^2\left[\frac{t}{2\pi}(\Delta K_{miz} + \frac{\Delta K_{mix}x_2 + \Delta K_{miy}y_2}{f} + \frac{\Delta K_{mix}^2 + \Delta K_{miy}^2}{4\pi/\lambda})\right]. \tag{14}$$

Remember that the intensity of the signal image is

$$\text{Signal} = t^2 |f_i(-x_2, -y_2)|^2. \tag{15}$$

Therefore, the intensity ratio between the noise and signal ("on" pixels) is

$$\frac{\text{Noise}}{\text{Signal}} = \sum_{m \neq i} \text{sinc}^2\left[\frac{t}{2\pi}(\Delta K_{miz} + \frac{\Delta K_{mix}x_2 + \Delta K_{miy}y_2}{f} + \frac{\Delta K_{mix}^2 + \Delta K_{miy}^2}{4\pi/\lambda})\right]. \tag{16}$$

Using (12), (16) can be written as

$$\frac{\text{Noise}}{\text{Signal}} = \sum_{m \neq i} \text{sinc}^2\left\{\frac{t}{\lambda}\frac{y_m - y_i}{f}\sin\theta \times \left[1 + \cot\theta\left(\frac{y_2}{f} + \frac{y_m + y_i}{2f} - \cos\theta\frac{y_m - y_i}{2f}\right) + \cos\theta\frac{y_m - y_i}{f}\frac{y_m + y_i}{2f} - \frac{y_2}{f}\frac{y_m + y_i}{2f} - \sin\theta\frac{y_m - y_i}{2f}\frac{(y_m + y_i)^2}{4f^2}\right]\right\}. \tag{17}$$

Now, we examine (17) by expanding the argument of the sinc function in terms of $\xi = y_\alpha / f$, where $\alpha = i, m, 2$. To the order of $O(\xi)$,

$$\frac{\text{Noise}}{\text{Signal}} = \sum_{m \neq i} \text{sinc}^2\left(\frac{t}{\lambda}\frac{y_m - y_i}{f}\sin\theta\right). \tag{18}$$

This expression is zero if we properly choose the separation between reference points so that $[t(y_m - y_i)/\lambda f]\sin\theta$ is an integer for all m. Therefore, to minimize the cross talk-noise, we choose the separation between two adjacent reference points as

$$\Delta = \frac{\lambda f}{t\sin\theta}. \tag{19}$$

Note that Δ is minimum when $\theta = 90°$. This result indicates that the maximum number of object images can be stored by using the $\theta = 90°$ configuration and positioning the reference points as

$$y_m = m\Delta = m\frac{\lambda f}{t\sin\theta}. \tag{20}$$

With this optimum configuration, we can calculate (17)

$$\frac{\text{Noise}}{\text{Signal}} \approx \sum_{m\neq i} \text{sinc}^2\left[\frac{t}{\lambda}\frac{y_m - y_i}{f}\left(1 - \frac{y_2}{f}\frac{y_m + y_i}{2f}\right)\right], \tag{21}$$

where we have kept terms up to the order of $O(\xi^3)$.

Figure 2 plots Noise/Signal as a function of both the location of the output plane y_2/f and the location of the reading point i, according to (21). In Fig. 2, the number of stored patterns is $N = 2001$, i.e. $M = 1000$; and $t/\lambda = 10^4$. Notice from Fig. 2 that the noise is minimum at the center of the output plane, and increases as we move to the edge of the output plane. This is because we have chosen the separation between reference points (19) such that the centers of output images are at the zeros of the sinc functions. By

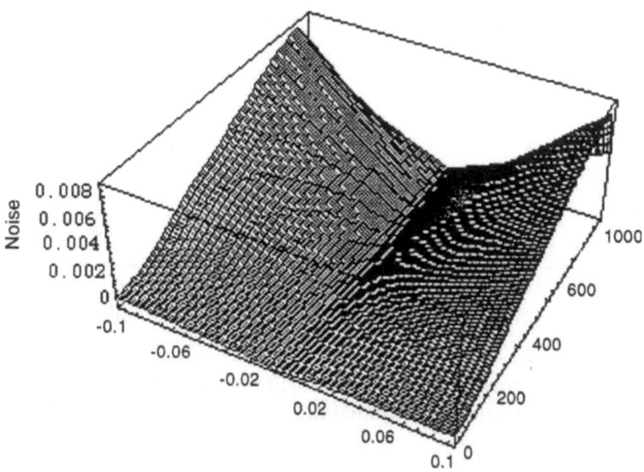

Fig. 2. The relative cross-talk noise Noise/Signal as a function of both the location of the output plane y_2/f and the location of the reading point i

so doing, it is also guaranteed that the overall noise for each output image is minimized, since the sinc function is almost symmetric near its zeros. For the same reason, as in Fig. 2, the noise increases as the reading point i moves away from the center, $i = 0$. However, we can see from Fig. 2 that the maximum noise occurs at a reading point close to the edge reference point $i = M$. This can be explained as follows. The major contribution of noise comes from the cross readout of images whose corresponding reference points are adjacent to the reading point. For the outmost reading point, $i = M$, there is only one adjacent reference point, $i = M - 1$. As a result, the cross-talk noise is smaller than that for the reading points located slightly away from the edge (with smaller i).

In Fig. 3, we plot Noise/Signal as a function of M according to (21) (solid line). The noise is evaluated at $y_2/f = 0.1$ and $i = 0.9M$, where the cross-talk noise is a maximum. Figure 3 shows that the noise increases as the total number of stored images increases. The dotted line in Fig. 3, which almost coincides with the solid line, is a fitting curve using polynomial expansions. We notice that the curve is almost linear with a slope of 10^{-5}. It is interesting to note that $10^{-5} = \lambda(y_{2max})/tf$. We will show in the following that the noise is determined by the thickness of the crystal, the wavelength and the numerical aperture of the images.

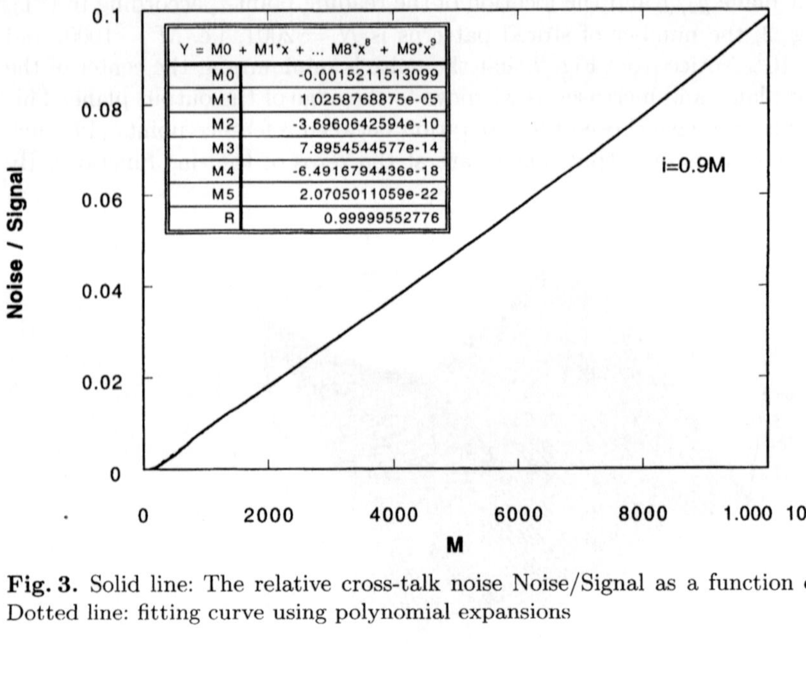

Fig. 3. Solid line: The relative cross-talk noise Noise/Signal as a function of M. Dotted line: fitting curve using polynomial expansions

We can estimate the maximum cross-talk noise using (21) and (20). Rewrite (21) as

$$\frac{\text{Noise}}{\text{Signal}} \approx \sum_{m \neq i} \text{sinc}^2\left[(m-i)\left(1-\beta\frac{m+i}{2}\right)\right], \tag{22}$$

where

$$\beta = \frac{\lambda y_{2\max}}{tf}. \tag{23}$$

Taking $i = M$ and calculating the noise, we obtain

$$\begin{aligned}\frac{\text{Noise}}{\text{Signal}} &\approx \sum_{m=-M}^{M-1} \text{sinc}^2\left[(m-M)\left(1-\beta\frac{m+M}{2}\right)\right],\\ &\approx \sum_{n=1}^{n_0} \text{sinc}^2\left[n\left(1-\beta\frac{2M-n}{2}\right)\right],\end{aligned} \tag{24}$$

where n_0 is the value for the first zero of the sinc function, i.e.

$$n_0\beta\frac{2M-n_0}{2} \approx 1. \tag{25}$$

This leads to

$$n_0 \approx \frac{1}{\beta M} \tag{26}$$

for $1 \ll M \ll 1/\beta$. Following (24), we have

$$\frac{\text{Noise}}{\text{Signal}} < n_0 \text{sinc}^2\left(1-\beta\frac{2M-1}{2}\right) \approx n_0(\beta M)^2 = (\beta M). \tag{27}$$

Notice that the above noise is estimated for $i = M$. As shown in Fig. 2, the maximum cross-talk noise occurs for i close to but not equal to M. We estimate the maximum cross-talk noise as

$$\left.\frac{\text{Noise}}{\text{Signal}}\right|_{\max} < 2(\beta M). \tag{28}$$

This is valid since the maximum noise is always less than twice the noise at $i = M$ for the reason discussed in explaining Fig. 2.

Using (28), we can estimate the signal-to-noise ratio (intensity) for the storage of $N = 2M + 1$ images. The result is

$$\text{SNR} = \frac{\text{Signal}}{\text{Noise}} > \frac{tf}{\lambda d M} \approx \frac{2tf}{\lambda d N}, \tag{29}$$

where $d = 2y_{2\max}$ is the linear dimension of the output plane.

1.3 Storage Capacity

The maximum cross-talk noise gives an upper bound for the storage capacity of the volume holographic memory. Given a required signal-to-noise ratio, the maximum number of images that can be stored in a volume holographic medium is

$$N_{\max} \approx \frac{2tf}{\lambda d(\mathrm{SNR})_{\mathrm{re}}}, \tag{30}$$

where $N_{\max}$ is the upper bound of the storage capacity, t is the thickness of the volume holographic medium, f is the focal length, λ is the wavelength of light, d is the linear dimension of the output plane (object image), and $(\mathrm{SNR})_{\mathrm{re}}$ is the required signal-to-noise ratio in terms of intensity.

There are some fundamental differences between the result obtained here and that obtained by Ramberg. Ramberg's is based on the assumption that the reference points are randomly distributed on the two-dimensional plane. Even in such a case involving a random distribution of reference points, there is a finite probability of the occurrence of degenerate gratings [24] where the sharing of gratings by different object/reference pairs gives rise to cross-talk. Fluctuations in the reconstruction fidelity caused by degenerate gratings can be quantified by noting that the probability of their occurrence is proportional to the angular bandwidth of the Bragg selectivity. The noise calculated by Ramberg is in terms of an ensemble average over all possible reference point locations, and consideration for such fluctuations is absent. In reality, if $N_{\max}$ holograms are recorded with random reference point positions, the reconstructed images will have randomly varying SNR. For some reconstruction where degeneracy occurs, the SNR may even be less than 1. The average SNR over all reconstructions is the value given by his calculation. In practice, it will be desirable to have regularly arranged reference points and similar SNR values for all reconstructions. To effectively use volume holograms as optical memory medium, it is important to eliminate the degeneracy noise completely. In our calculation, the noise level can be zero as for the center of the output images. This indicates that the degeneracy noise is systematically eliminated. Systematic approaches, to eliminate degeneracy noise for regularly arranged reference points can be found in [24]. The arrangement presented here (20) is one of the approaches and gives an optimum configuration. The similarity between the two results also shows that the storage capacity does not benefit from expanding the reference points from one-dimensional to two-dimensional.

It is interesting to extend the above analysis to the case of wavelength multiplexing [11,12]. We found that in the reflection holograms configuration, as shown in Fig. 4, the cross-talk noise does not significantly limit the storage capacity. The basic limiting factors for wavelength-multiplexed reflection holograms will be the laser tuning range and the bandwidth of material response.

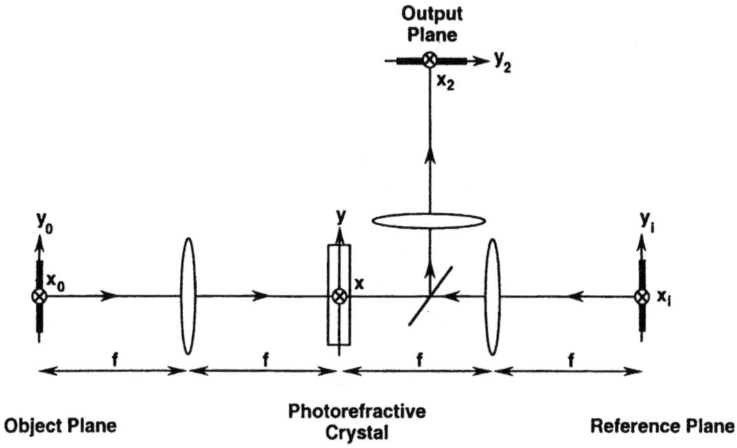

Fig. 4. An optical memory system in which wavelength-multiplexed holograms are stored in a thick recording medium located in the Fourier domain

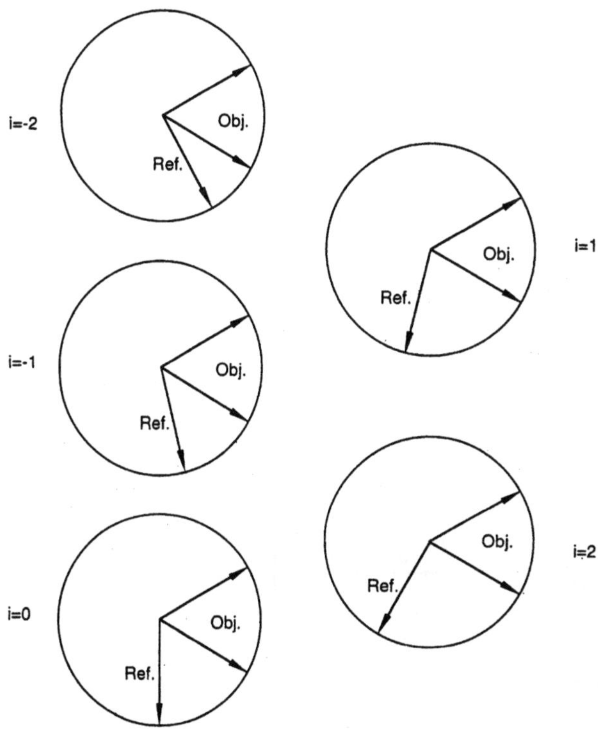

Fig. 5. Five exposures of angle-multiplexed holograms

The difference between the cross-talk behavior in angle multiplexing and in wavelength multiplexing [13,11,15] can be explained with the help of the so-called K-space (or grating space) representation. Figure 5 shows the momentum space (normal surface) representation of the recording of five angle-multiplexed holograms. Each of the reference wave vectors represents the reference beam angle during each exposure. The range of object beams represents the Fourier components (spatial frequency) of the corresponding image. Figure 6 shows the momentum space (normal surface) representation of the recording of five wavelength-multiplexed holograms. In the case of wavelength multiplexing, all the reference beams have the same direction of propagation. The wavelength is indicated by the size of circles. Figure 7 shows the K-space representation of the five angle-multiplexed holograms. In this figure, the five circles in Fig. 5 are stacked together so that all reference wave vectors point

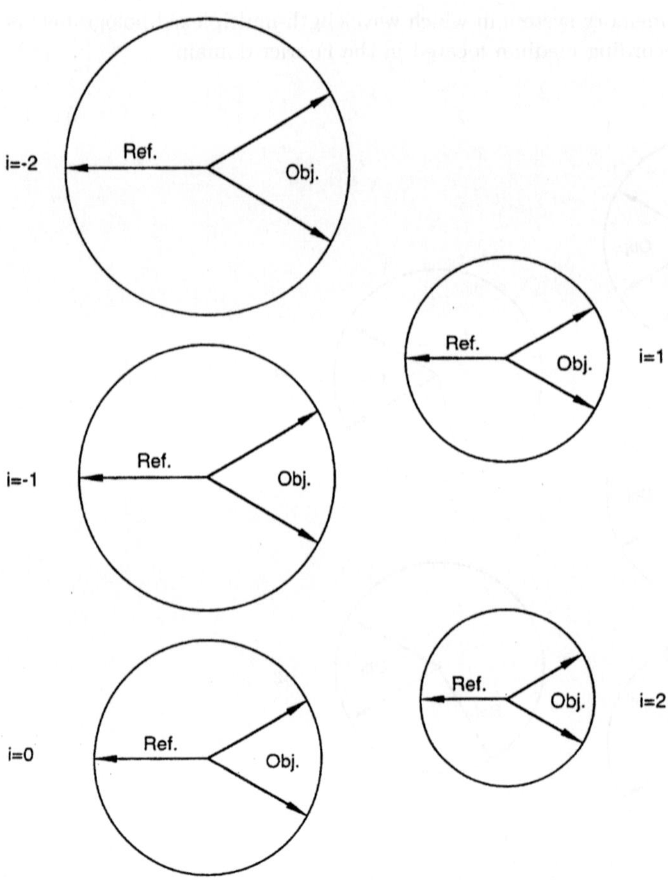

Fig. 6. Five exposures of wavelength-multiplexed holograms

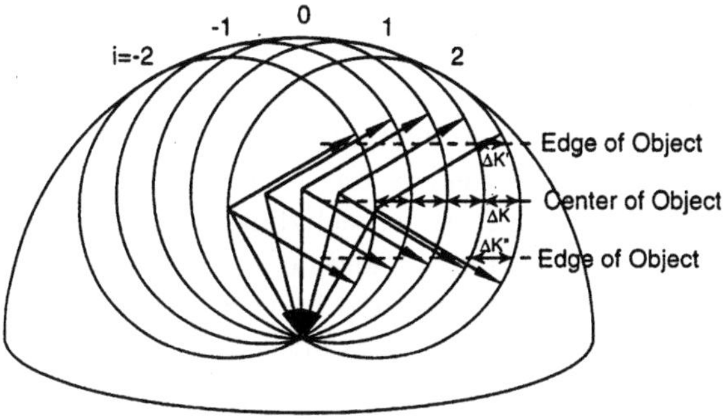

Fig. 7. K-space representation of the five angle-multiplexed holograms

to the same origin of the K-space. The five arcs describing object Fourier components represent the range of grating wave vectors. The reference angle separations are chosen so that the Bragg mismatch ΔK near the center of all objects is the same for all holograms where the cross-talk is minimum. Notice that the Bragg mismatch $\Delta K'$ and $\Delta K''$ at the edges of objects are different from that at the center ΔK. Therefore, the cross-talk noise for edge pixels is larger than that for the center pixels. This effect becomes strong as the reference beam angle deviates from the reference center ($i = 0$). This is the reason why the cross-talk noise increases as the number of holograms increases. In the case of wavelength multiplexing with reflection holograms, Fig. 8 shows that the Bragg mismatch ΔK is about the same everywhere. Therefore, the cross-talk noise saturates as the number of holograms increases. The difference in these two cases arises from the different packing schemes in the K-space. By separating the reference beam angles non-uniformly, in the case of angle multiplexing, it is possible that cross-talk noise also saturates [15] so that it does not limit the storage capacity.

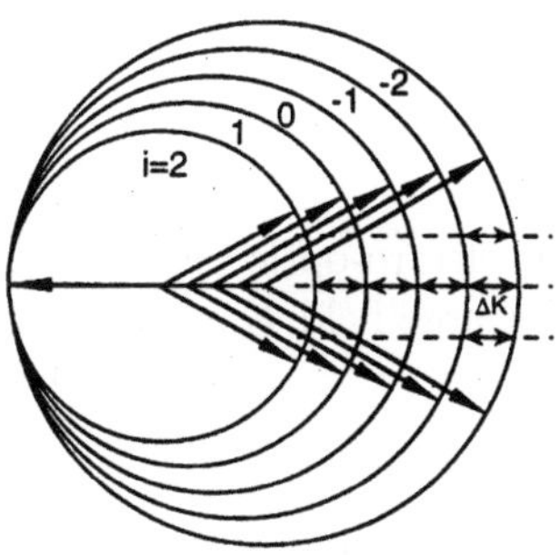

Fig. 8. K-space representation of the five wavelength-multiplexed holograms

2 Intrinsic Scattering Noise

Another important issue in photorefractive storage is limitation of the dynamic range by scattering noise [16,17]. The fluctuation of refractive index introduced by the random space charge field is one of the noise sources in photorefractive media. In electro-optic crystals, including photorefractive crystals, charge particles, such as ionized dopants and defects, are randomly distributed. The space charge electric field generated by these charge particles induces a fluctuation in dielectric constant (or equivalently the index of refraction) due to the Pockels effect. Such a fluctuation exists even in the presence of hologram recording and two-wave mixing. When the crystal is illuminated by light, scattering occurs in a manner similar to that of Rayleigh scattering. This kind of scattering exists in all photorefractive crystals, where ionized dopants and/or defects are responsible for the formation of index gratings, and leads to statistical noise in photorefractive devices. The noise due to scattering from these noisy index gratings ultimately limits the minimum modulation depth for a signal index grating.

Consider the scattering in a medium with a fluctuation in dielectric constant, $\Delta\epsilon$. The differential scattering cross-section is written as [22]

$$\frac{\mathrm{d}\sigma}{\mathrm{d}\Omega} = |S|^2, \tag{31}$$

where

$$S = \frac{k^2}{4\pi}\int \mathrm{d}\boldsymbol{r}'\boldsymbol{p}'^* \cdot [(\Delta\epsilon)\epsilon^{-1}\boldsymbol{p}]\exp[\mathrm{i}(\boldsymbol{k}'-\boldsymbol{k})\cdot\boldsymbol{r}'], \tag{32}$$

where k is the magnitude of the wave vector, $\boldsymbol{p}$ and $\boldsymbol{p}'$ are unit vectors representing the polarization states of the incident and scattered light respectively, $\boldsymbol{k}$ and $\boldsymbol{k}'$ are their corresponding wave vectors, and ϵ is the dielectric tensor.

In an electro-optic medium, a random space charge field can produce a fluctuation in the dielectric tensor via the Pockels effect. The fluctuation in ϵ is written [25]

$$\Delta\epsilon = -\frac{\epsilon(r\boldsymbol{E}_{\mathrm{sc}})\epsilon}{\epsilon_0}, \tag{33}$$

where $\boldsymbol{E}_{\mathrm{sc}}$ is the random space charge field, r is the electro-optic tensor, and ϵ_0 is the dielectric constant in vacuum. The space charge field $\boldsymbol{E}_{\mathrm{sc}}$ is related to the charge density ρ by Poisson's equation

$$\nabla\cdot\epsilon_{\mathrm{dc}}\boldsymbol{E}_{\mathrm{sc}} = \rho, \tag{34}$$

where ϵ_{dc} is the dielectric tensor for a dc field. Substituting (33) and (34) into (32) and using the properties of the Fourier transform, the integral in (32) can be written in terms of the charge density

$$S = \frac{-ek^2}{4\pi\epsilon_0}\frac{\boldsymbol{p}'^*\cdot[\epsilon(r\boldsymbol{K}\boldsymbol{p})]}{\boldsymbol{K}\cdot\epsilon_{dc}\cdot\boldsymbol{K}}\mathcal{F}, \tag{35}$$

where

$$\mathcal{F} = \frac{1}{e} \int \rho(\boldsymbol{r}') \exp[\mathrm{i}\boldsymbol{K} \cdot \boldsymbol{r}'] \mathrm{d}\boldsymbol{r}', \tag{36}$$

where e is the electronic charge and $\boldsymbol{K} = \boldsymbol{k}' - \boldsymbol{k}$. In deriving the above equation, we have assumed that the space charge field is parallel to the vector $\boldsymbol{K}$.

Consider the charge particles as point charges. Suppose that at a given time there are N positive and N negative charge particles. The charge density function is written as

$$\rho(\boldsymbol{r}') = e \sum_{i=1}^{N} [\delta(\boldsymbol{r}' - \boldsymbol{r}_i) - \delta(\boldsymbol{r}' - \boldsymbol{x}_i)], \tag{37}$$

where $\boldsymbol{r}_i$ and $\boldsymbol{x}_i$ represent the positions of the ith positive and negative charge particles respectively. Here we assume that $\boldsymbol{r}_i$ and $\boldsymbol{x}_i$ are random variables. We introduce probability density functions $p(\boldsymbol{r}_i, \boldsymbol{x}_i)$ to describe the probability distribution of the ith positive and negative charge particles. Suppose that the charge particles are uniformly distributed inside a crystal with linear dimensions L_x, L_y and L_z along the x, y and z-directions respectively. If we further assume that the mobility of these charge particles is not zero (either the thermal and/or photoconductivity of the material is not zero), then the electric field due to these charge particles inside the crystal is electrostatically screened [26]. The screening leads to a probability density function given by

$$p(\boldsymbol{r}, \boldsymbol{x}) = \frac{1}{V} \frac{k_{\mathrm{D}}^2}{4\pi} \frac{\exp(-k_{\mathrm{D}}|\boldsymbol{r} - \boldsymbol{x}|)}{|\boldsymbol{r} - \boldsymbol{x}|} \quad \text{inside the crystal}, \tag{38}$$

where $V = L_x L_y L_z$ is the volume of the crystal, k_{D} is the Debye wave number given by

$$k_{\mathrm{D}}^2 = \frac{e^2 (N/V)}{k_{\mathrm{B}} T \epsilon_{dc}}, \tag{39}$$

k_{B} is the Boltzmann constant and T is temperature. Equation (38) is known as the Debye distribution and is a result of electrostatic screening. This distribution can be obtained from the screened point charge potential given by [26].

The differential scattering cross-section, (31), can now be evaluated according to the ensemble average of $|\mathcal{F}|^2$. Notice, from (36) and (37), that $\mathcal{F}$ is a function of $\boldsymbol{r}_i$ and $\boldsymbol{x}_i$ ($i = 1, ..., N$). Here we assume that the probability density function $p(\boldsymbol{r}_i, \boldsymbol{x}_i)$ is the same for $i = 1, ..., N$. Using (38), the ensemble average is obtained as

$$\langle |\mathcal{F}|^2 \rangle = 2N \frac{K^2}{K^2 + k_{\mathrm{D}}^2}, \tag{40}$$

where $K = |\boldsymbol{K}|$. The expression for the differential scattering cross-section is given by

$$\frac{\mathrm{d}\sigma}{\mathrm{d}\Omega} = 2e^2 N \frac{k^4}{(4\pi\epsilon_0)^2} \left| \frac{\boldsymbol{p}'^* \cdot [\epsilon(r\boldsymbol{K}\boldsymbol{p})]}{\boldsymbol{K} \cdot \epsilon_{\mathrm{dc}} \cdot \boldsymbol{K}} \right|^2 \frac{K^2}{K^2 + k_{\mathrm{D}}^2}. \tag{41}$$

In the case of an isotropic crystal, (41) can be simplified to

$$\frac{\mathrm{d}\sigma}{\mathrm{d}\Omega} = 2e^2 N \frac{k^4}{(4\pi)^2} \left| \frac{n^2 \boldsymbol{p}'^* \cdot [(r\boldsymbol{K})\boldsymbol{p}]}{K \epsilon_{\mathrm{dc}}} \right|^2 \frac{1}{K^2 + k_{\mathrm{D}}^2}. \tag{42}$$

Notice that the differential cross-section is proportional to the total number of charge particles, and it depends on the angle between the incident and the scattered beams. In addition, the polarization state of the scattered light in a photorefractive crystal may be different from that of Rayleigh scattering. This difference arises from the tensor properties of r.

Consider a photorefractive grating written by two coherent beams with modulation depth m. The scattering is one of the noise sources. With the presence of an index grating, the probability density function can be written as

$$p(\boldsymbol{r}, \boldsymbol{x}) = \frac{1}{V}(1 + m \cos \boldsymbol{K}_{\mathrm{g}} \cdot \boldsymbol{r}) \frac{k_{\mathrm{D}}^2}{4\pi} \frac{\exp(-k_{\mathrm{D}}|\boldsymbol{r} - \boldsymbol{x}|)}{|\boldsymbol{r} - \boldsymbol{x}|} \quad \text{inside the crystal,} \tag{43}$$

where $\boldsymbol{K}_{\mathrm{g}}$ is the wave vector of the grating. When the linear dimension of the crystal is much larger than the grating spacing, and the number of charge particles is much greater than 1, the differential scattering cross-section can be written as

$$\begin{aligned} \frac{\mathrm{d}\sigma}{\mathrm{d}\Omega} = e^2 \frac{k^4}{(4\pi\epsilon_0)^2} & \left| \frac{\boldsymbol{p}'^* \cdot [\epsilon(r\boldsymbol{K}\boldsymbol{p})]}{\boldsymbol{K} \cdot \epsilon_{dc} \cdot \boldsymbol{K}} \right|^2 \\ & \times \left\{ \left| \frac{mN}{2} \frac{K^2}{K^2 + k_{\mathrm{D}}^2} \mathrm{sinc}[(u - u_{\mathrm{g}})L_x] \mathrm{sinc}[(v - v_{\mathrm{g}})L_y] \mathrm{sinc}[(w - w_{\mathrm{g}})L_z] \right|^2 \right. \\ & \left. + 2N \frac{K^2}{K^2 + k_{\mathrm{D}}^2} \right\}, \end{aligned} \tag{44}$$

and u, v, w are related to $\boldsymbol{K}$ by $\boldsymbol{K} = 2\pi(u\hat{x} + v\hat{y} + w\hat{z})$. We note that the last term is the same as that in (41), and the term with sinc functions represents the Bragg scattering of the sinusoidal grating. During grating detection, the Bragg scattering term gives rise to a signal, while the background scattering term results in a noise. The product of sinc functions has a sharp peak centered at $\boldsymbol{K} = \boldsymbol{K}_{\mathrm{g}}$. The signal-to-noise ratio can be characterized by the ratio between these two terms, evaluated at $\boldsymbol{K} = \boldsymbol{K}_{\mathrm{g}}$, i.e.

$$\mathrm{SNR} = \frac{[mNK^2/2(K^2 + k_{\mathrm{D}}^2)]^2}{2NK^2/(K^2 + k_{\mathrm{D}}^2)}, \tag{45}$$

where $K = |\boldsymbol{K}_{\mathrm{g}}|$.

Equation (45) determines the minimum modulation depth that is required to write a detectable index grating. In the limiting case of $K \gg k_D$, the signal to noise ratio is SNR $= m^2N/8$. If the charge density is of the order of 10^{15} cm^{-3}, the crystal has a thickness 1 cm, and the area of the beam cross-section is 1 mm^2, the minimum modulation depth that leads to a SNR $= 1$ is $m = \sqrt{8 \times 10^{-13}}$. This is equivalent to an intensity ratio between the two writing beams $I_1/I_2 = 2 \times 10^{-13}$, which gives a limitation to the dynamic range.

3 Noise Gratings

In the above discussion, the cross-talk noise applies to any volume holographic medium, and the dynamic range limit applies to any electro-optic medium. There are other noise sources, such as the noise gratings formed during a multiple exposure sequence [18,19]. These noise gratings are induced by the diffracted beams from previously recorded holograms and the writing beams. It is important to note that energy coupling occurs not only between the writing beams, but also between the diffracted and writing beams (especially in photorefractive media). During the recording process of the pth exposure, the writing beams generate new gratings while erasing the previously stored gratings. In addition, diffraction occurs due to the readout of the previously stored gratings, even if they are Bragg-mismatched for the writing beams (e.g. cross-talk readout). The additional diffracted beams will interact with the writing beams and form noise gratings. This may be especially problematic in optical memory applications where a large number of gratings are superimposed in a photorefractive crystal, and the creation of photo-induced noise gratings during the exposure sequence may significantly degrade the signal-to-noise ratio of the optical memory. Previously, the diffraction properties of index gratings in photorefractive media were analyzed assuming an arbitrary momentum mismatch (Bragg mismatch) [19]. Here, we use a similar method to calculate the amplitude of noise gratings and discuss the resulting signal-to-noise ratio in optical memory.

Referring to Fig. 9, consider a sequence of exposures. In the first exposure, reference beam R_1 and signal beam S_1 are used to record the grating with wave vector $\boldsymbol{K}_1$. In the second exposure, reference beam R_2 and signal beam S_2 are used to record the grating with wave vector $\boldsymbol{K}_2$. During the second recording process, the first grating $\boldsymbol{K}_1$ is also read out by the reference beam R_2 with a momentum mismatch (Bragg mismatch). The diffracted beam A_1 will interact with the reference beam R_2 and write a noise grating with wave vector $\boldsymbol{K}_{21}$. Here, we assume that the reference beam is much stronger than the signal beam (which is usually a Fourier component of an image) and neglect the Bragg-mismatched diffraction attributed to the signal beam S_2.

Conservation of momentum requires that

$$K_{21x} = K_{1x}, \qquad K_{21y} = K_{1y}, \tag{46}$$

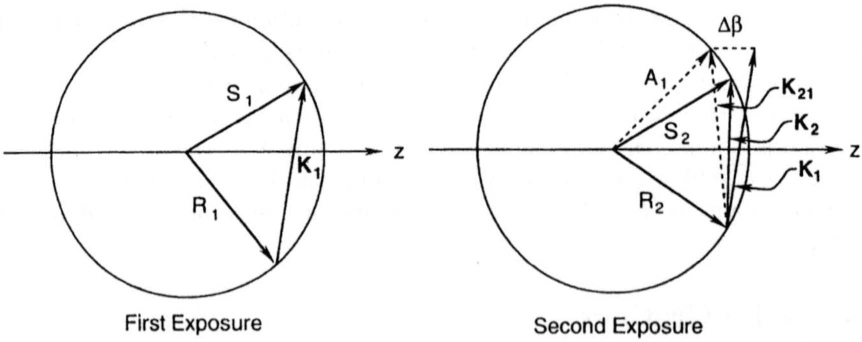

Fig. 9. A sequence of two exposures

where the x and y directions are parallel to the surface of the crystal, and the dimension along these two directions is assumed to be infinite. In the z direction, the crystal has a finite thickness L, and the wave vectors can be written

$$K_{21z} - K_{1z} = \Delta\beta, \tag{47}$$

where $\Delta\beta \neq 0$. In other words, the momentum mismatch $\boldsymbol{\Delta k}$ has only a z component.

The spatio-temporal equations [19,27,28] that describe such an interaction in photorefractive media can be derived from the wave equation [29,30] and the Kukhtarev equations [31]. In the following discussion, we assume that the photorefractive time constant is much longer than the time of flight for a light beam to propagate through the thickness of the crystal, i.e. $\tau \gg nL/c$ where τ is the photorefractive time constant, L is the thickness of the crystal, n is the index of refraction of the medium, and c is the speed of light in vacuum. We also assume that the slowly varying envelope approximation is valid. Under these assumptions, the coupled mode equations [19,27,28] can be written

$$\begin{aligned}
\frac{\partial R_2}{\partial z} &= -\sigma_2 S_2 - \sigma_{21} A_1, \\
\frac{\partial S_2}{\partial z} &= \sigma_2^* R_2, \\
\frac{\partial A_1}{\partial z} &= \sigma_{21}^* R_2,
\end{aligned} \tag{48}$$

where we have neglected the coupling between A_1 and S_2 since its modulation depth $A_1 S_2^*/I_0$ is much smaller than that for grating $\boldsymbol{K}_{21}$, which is $A_1 R_2^*/I_0$, and its grating spacing is much larger. The grating amplitudes σ_2 and σ_{21}

satisfy the following equations [19,27,28]:

$$\begin{aligned} \frac{\partial \sigma_2}{\partial t} &= -\frac{1}{\tau}\left(\sigma_2 - \frac{\Gamma}{2}\frac{R_2 S_2^*}{I_0}\right), \\ \frac{\partial \sigma_{21}}{\partial t} &= -\frac{1}{\tau}\left(\sigma_{21} - \frac{\Gamma}{2}\frac{R_2 A_1^*}{I_0}\right), \end{aligned} \tag{49}$$

where τ is the photorefractive time constant, I_0 is the total averaged intensity, and Γ is the photorefractive coupling constant, which is generally complex. Notice that in the above equations σ_{21} represents the superposition of all gratings (including grating $\boldsymbol{K}_1$) that couple A_1 and R_2: i.e.

$$\sigma_{21} = \sum_{K_x = K_{1x}, K_y = K_{1y}} \sigma_K \exp[\mathrm{i}(K_{21z} - K_z)z], \tag{50}$$

where σ_K is the amplitude of the grating $\boldsymbol{K}$ and $\exp[\mathrm{i}(K_{21z} - K_z)z]$ represents its momentum mismatch. For simplicity, we have assumed that the photorefractive coupling constant Γ is the same for all gratings. The boundary conditions are given by

$$\begin{aligned} R_2(t, z = 0) &= R_{20}, \\ S_2(t, z = 0) &= S_{20}, \\ A_1(t, z = 0) &= 0, \end{aligned} \tag{51}$$

and the initial conditions are given by

$$\begin{aligned} \sigma_2(t = 0, z) &= 0, \\ \sigma_{21}(t = 0, z) &= \sigma_1 \mathrm{e}^{\mathrm{i}\Delta\beta z}, \end{aligned} \tag{52}$$

where σ_1 is the grating amplitude of the first grating $\boldsymbol{K}_1$, and $\Delta\beta$ is the momentum mismatch. These conditions imply that the second exposure starts at $t = 0$. At $t = 0$, only the previously recorded grating $\boldsymbol{K}_1$ is inside the photorefractive crystal. At $t > 0$, the first grating $\boldsymbol{K}_1$ will be erased and the other two gratings $\boldsymbol{K}_2$ and $\boldsymbol{K}_{21}$ will be recorded.

In the following discussion, we will derive analytical solutions under the undepleted pump approximation and the short writing time assumption. We assume that during the second exposure, the writing beams R_2 and S_2 are constants. This assumption is valid when the two-wave mixing (TWM) effect is small and/or the exposure time is short. At $t = 0$, (48) can be solved with the initial conditions given by (52) and the boundary conditions given by (51). The solutions are

$$\begin{aligned} R_2(t = 0, z) &= R_{20}, \\ S_2(t = 0, z) &= S_{20}, \\ A_1(t = 0, z) &= -\sigma_1^* R_{20} \frac{1}{i\Delta\beta}(\mathrm{e}^{-\mathrm{i}\Delta\beta z} - 1). \end{aligned} \tag{53}$$

Substituting (53) into (49), we obtain, at $t = 0$,

$$\frac{\partial \sigma_2}{\partial t} = \frac{1}{\tau}\frac{\Gamma}{2}\frac{R_{20}S_{20}^*}{I_0},$$

$$\frac{\partial \sigma_{21}}{\partial t} = -\frac{1}{\tau}\left[\sigma_1 \mathrm{e}^{\mathrm{i}\Delta\beta z} - \frac{\Gamma}{2}\frac{I_{2R}}{I_0}\sigma_1 \frac{1}{\mathrm{i}\Delta\beta}(\mathrm{e}^{\mathrm{i}\Delta\beta z} - 1)\right], \tag{54}$$

where $I_{2R} = |R_{20}|^2$ is the intensity of the second reference beam. If we further assume that the exposure time Δt is short, we can obtain the approximate solutions

$$\sigma_2 = (1 - \mathrm{e}^{-\frac{\Delta t}{\tau}})\frac{\Gamma}{2}\frac{R_{20}S_{20}^*}{I_0}, \tag{55}$$

and

$$\sigma_{21} = \mathrm{e}^{-\frac{\Delta t}{\tau}}\sigma_1 \mathrm{e}^{\mathrm{i}\Delta\beta z} + (1 - \mathrm{e}^{-\frac{\Delta t}{\tau}})\frac{I_{2R}}{I_0}\frac{\Gamma}{\Delta\beta}\sigma_1 \sin(\Delta\beta z/2)\mathrm{e}^{\mathrm{i}\Delta\beta z/2}. \tag{56}$$

Equation (55) represents the recording of the signal grating $\boldsymbol{K}_2$. The first term in (56) represents the erasure of previous grating $\boldsymbol{K}_1$. The second term in (56) represents the growth of the noise grating $\boldsymbol{K}_{21}$. Notice that the noise grating is proportional to the amplitude of the Bragg-mismatched grating σ_1 and inversely proportional to the momentum mismatch $\Delta\beta$. In addition, the amplitude of the noise grating is a periodic function with period $4\pi/\Delta\beta$, and the noise grating has a momentum mismatch $\Delta\beta/2$.

If the exposure time is comparable to the photorefractive time constant, we need to consider the growth of both signal and noise gratings, taking into account the time variation of beam amplitudes. In practice, the first exposure that does not generate any noise gratings is the longest. Consequently, the most significant noise gratings are generated in the second exposure. Suppose that steady state is reached during the first exposure: then the second exposure time is given by $\Delta t = \tau \ln 2$, where τ is the photorefractive time constant. To show the effect of noise gratings formed during the second exposure, we solve (48) and (49) numerically. Figure 10a shows the intensity distributions as functions of the position inside the crystal for the signal and the noise beams, at exposure time $t = 0$, $t = (\tau \ln 2)/3$, and $t = \tau \ln 2$. The parameters are chosen as $\Delta\beta L = 5$, $R_2(z = 0) = 1$, $S_2(z = 0) = 0.1$, $A_1(z = 0) = 0$, $\Gamma L = 5\mathrm{i}$ (purely imaginary so that the two-wave mixing effect is minimum), and $\sigma_1 = (0.1/1.01)\Gamma/2$. Figure 10b shows the corresponding grating amplitudes $|\sigma_2/(\Gamma/2)|$ and $|\sigma_{21}/(\Gamma/2)|$. At $t = 0$, the signal grating is 0 and the only existing grating is the previously recorded one with $|\sigma_1/(\Gamma/2)| \approx 0.1$. As the exposure proceeds, the signal grating grows while the previously recorded grating is partially erased, almost uniformly throughout the crystal. At the same time, the noise grating also grows and interferes with the previously recorded grating. At the front face of the crystal ($z = 0$), there is no noise

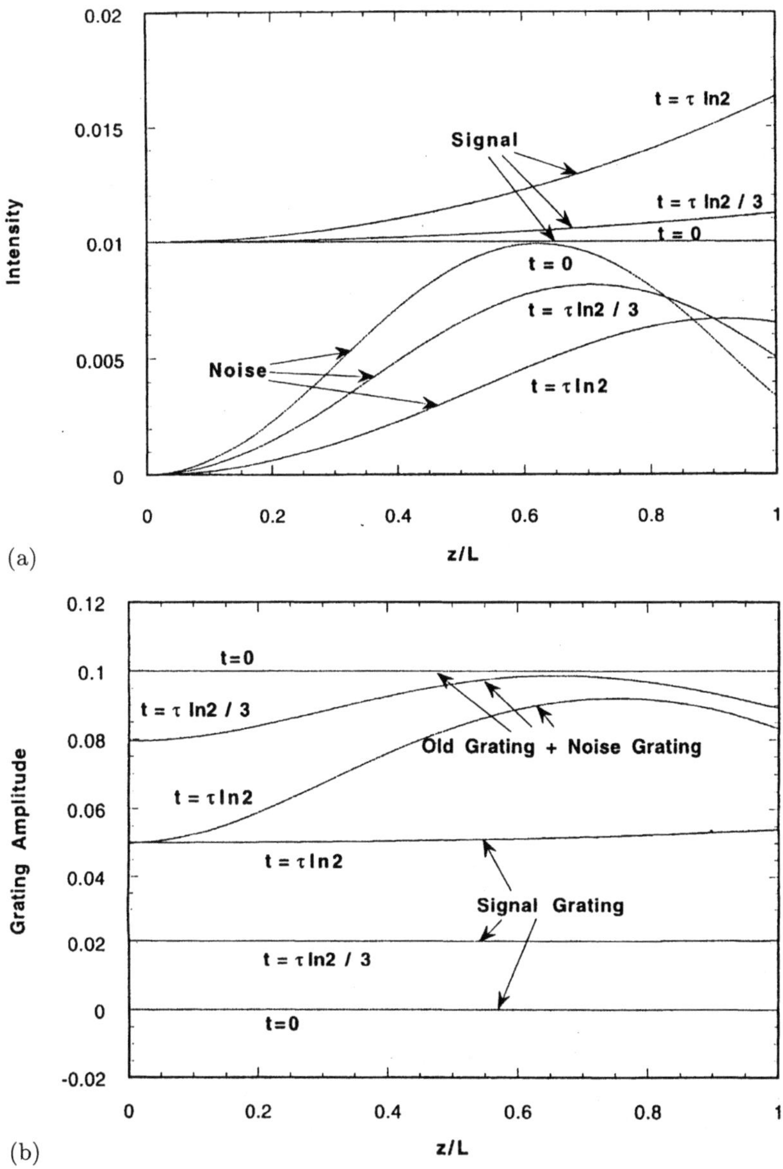

Fig. 10. Intensity distributions and grating amplitudes as functions of the position inside the crystal during the second exposure with exposure time $t = 0$, $t = (\tau \ln 2)/3$ and $t = \tau \ln 2$: (**a**) intensities for the signal ($|S_2|^2$) and the noise ($|A_1|^2$) beams; (**b**) grating amplitudes $|\sigma_2/(\Gamma/2)|$ and $|\sigma_{21}/(\Gamma/2)|$. The Bragg-mismatch is chosen as $\Delta\beta L = 5$

grating: therefore, at the end of the exposure, $t = \tau \ln 2$, the newly recorded grating and the previously recorded grating have the same amplitude, as shown in Fig. 10b. At $0 < z < L$, the grating amplitude for $|\sigma_{21}/(\Gamma/2)|$ (denoted as Old Grating + Noise Grating in Fig. 10b) deviates from its value at $z = 0$, resulting from the interference between the previously recorded grating and the noise grating. The corresponding intensity distribution for the noise beam (Fig. 10a) also deviates from the Bragg-mismatched readout of the old grating. Note that during the second exposure, the noise intensity at the output ($z = L$) increases with time, owing both to the formation of the noise grating and to the partial erasure of the previously recorded grating. Since the noise grating originates from the Bragg mismatched readout of previously recorded gratings, it is sensitive to the amount of Bragg-mismatch. Figures 11a and 11b show the intensity distributions and the grating amplitudes with the parameters $\Delta\beta L = 30$ and the rest the same as those in Fig. 10. Notice that the noise effect is significantly reduced with the increased Bragg mismatch.

In the case of optical memory, a large number of gratings are stored in a single photorefractive crystal. The above analysis can be extended to the general situation. During the pth exposure, suppose N gratings (including noise gratings) are already recorded in the previous exposures, R_2 is the reference beam, and S_2 is one of the Fourier components of the signal image. σ_1 now indicates the grating amplitude of any of the N gratings, and $\Delta\beta$ represents the corresponding momentum mismatch. A_1 is the diffracted beam due to the Bragg-mismatched reading of grating σ_1 by the reference beam R_2. σ_{21} describes the superposition of gratings that couple beams A_1 and R_2. The equations (there will be a sum over all possible A_1s in the equation for $\partial R_2/\partial z$) and the results are the same as those above if we neglect the depletion of the writing beams and the coupling between noise beams.

Consider the simultaneous growth of both the signal and the noise gratings in the pth exposure. Notice that the noise beam consists of the diffractions from both the previously recorded Bragg-mismatched gratings and the newly formed noise gratings. Here we discuss the effect of the newly formed noise gratings. Although an accurate discussion of the noise intensity requires numerical solutions of the spatio-temporal equations (e.g. results shown in Figs. 10 and 11), we can obtain analytical results under the approximations of weak pump depletion and short exposure times. The results will demonstrate the basic properties of the noise gratings. Since the diffraction efficiency is related to the grating amplitude integrated over the entire crystal, we compare the grating amplitudes between the signal and the noise gratings by defining a signal-to-noise ratio as the absolute value of the ratio between the amplitude of the signal grating and the peak amplitude of the noise grating, i.e.

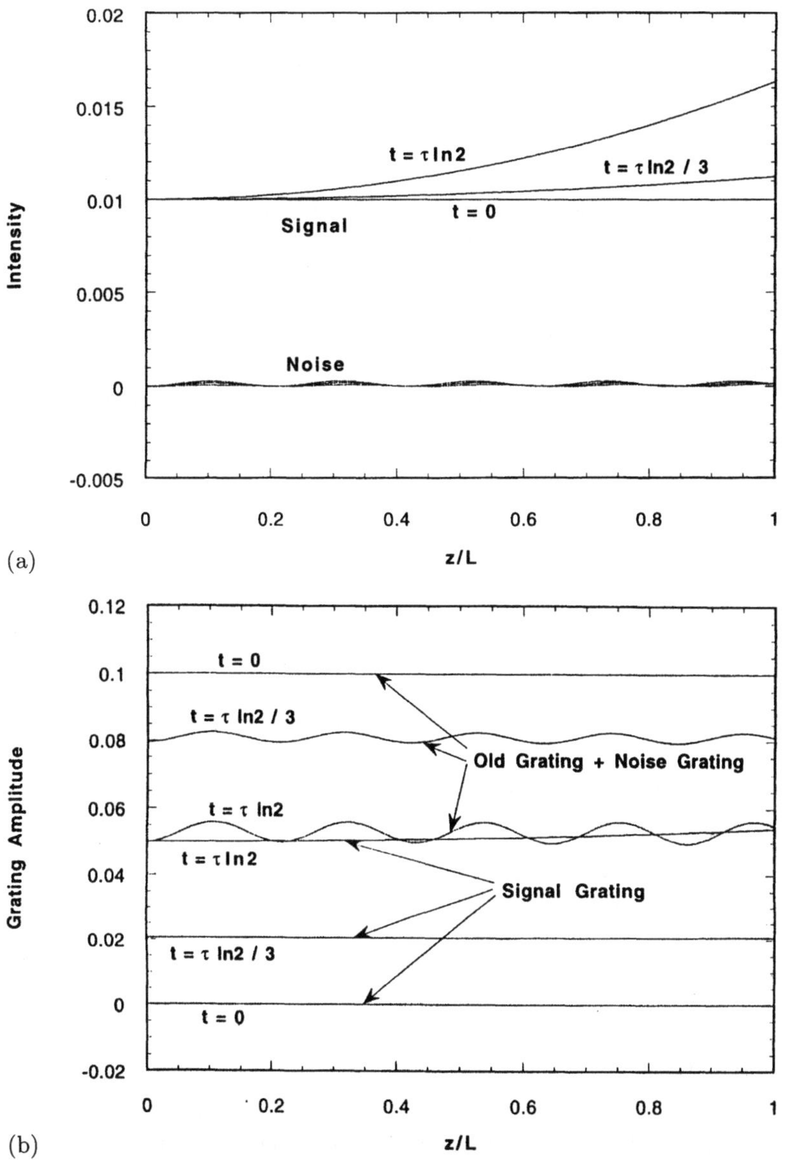

Fig. 11. Intensity distributions and grating amplitudes as functions of the position inside the crystal during the second exposure with exposure time $t = 0$, $t = (\tau \ln 2)/3$ and $t = \tau \ln 2$: (**a**) intensities for the signal ($IS_2|^2$) and the noise ($IA_1|^2$) beams; (**b**) grating amplitudes $|\sigma_2/(\Gamma/2)|$ and $|\sigma_{21}/(\Gamma/2)|$. The Bragg mismatch is chosen as $\Delta\beta L = 30$

$$
\begin{aligned}
\frac{S}{N} &= \left| \frac{(1 - \mathrm{e}^{-\Delta t/\tau}) \Gamma R_{20} S_{20}^{*} / 2I_0}{(1 - \mathrm{e}^{-\Delta t/\tau})(I_{2R}\Gamma / I_0 \Delta\beta) \sigma_1 \mathrm{e}^{\mathrm{i}\Delta\beta z/2}} \right|, \\
&= \left| \frac{\Delta\beta R_{20} S_{20}^{*}}{2 I_{2R} \sigma_1} \right|. \qquad (57)
\end{aligned}
$$

This signal-to-noise ratio indicates the relative strength of the noise grating with respect to that of the signal gratings, which is independent of the incident angle of the reading beam. Notice that in the above equation, only the second term in (56) is considered as noise. The first term in (56) describes the erasure of previously recorded gratings. It is interesting to note that the N previously recorded gratings will be erased with the same rate, according to (56). Since the N previously recorded gratings include both signal gratings and noise gratings, therefore, the signal-to-noise ratio determined in one exposure will not change in the later exposures. Furthermore, the exposure time Δt is chosen so that the signal gratings recorded in all exposures have equal strength. On the other hand, the amplitude of the noise gratings is proportional to the amplitude of the previously recorded grating σ_1, which becomes smaller and smaller in later exposures. Therefore, the signal-to-noise ratio determined in the pth exposure is larger than that determined in previous exposures. In other words, the most significant noise gratings are created in the early exposures (except the first one, which does not create any noise gratings).

The minimum signal-to-noise ratio occurs in the second exposure. We can estimate the minimum signal-to-noise ratio assuming that during the first exposure the recording time is so long that the steady state is reached. We further assume that the modulation depth and the total averaged intensity (equivalently the time constant τ) are the same for all exposures. Therefore,

$$
|\sigma_1| = \left| \frac{\Gamma}{2} \frac{R_{20} S_{20}^{*}}{I_0} \right|, \qquad (58)
$$

and the signal-to-noise ratio in the second exposure is obtained as

$$
\frac{S}{N} = \left| \frac{I_0}{I_{2R}} \frac{\Delta\beta}{\Gamma} \right|. \qquad (59)
$$

Equation (59) indicates that we can select our reference beam in the second exposure such that the momentum mismatch $\Delta\beta$ is large and the signal-to-noise ratio is high. Suppose we choose $\Delta\beta = 5\Gamma$ and $I_{2R} = I_0/2$ then the minimum signal-to-noise ratio is larger than 10. Remember that the diffraction efficiency is approximately proportional to the square of the grating amplitude. Therefore, the signal-to-noise ratio in terms of the diffraction efficiency can be greater than 100. This result indicates that the noise gratings can be minimized by increasing the momentum difference between the gratings (i.e. $\Delta\beta$) with large grating amplitude.

In the case of multiple exposures in holographic recording, the holograms that are recorded during the first few exposures are the strongest. Therefore, an exposure schedule that takes into account the noise gratings suggests selecting larger momentum mismatch for the first few exposures. For example, for the five exposures shown in Fig. 5, the exposure sequence should be $i = -2$, $i = 2$, $i = 0$, $i = -1$ and $i = 1$.

There are other important issues in volume holographic storage that are not covered in this chapter, such as materials, fixing technique, beam scanning technique, etc. With the development of technology, we hope that volume holographic storage will become practical in the near future.

4 Conclusion

In conclusion, we have considered some fundamental noise sources in volume holographic storage, including cross-talk noise, intrinsic scattering noise, and noise gratings formed during a multiple exposure schedule. These fundamental noise sources are important in understanding the potentials and limitations of volume holographic storage systems. The cross-talk noise gives an upper bound for the storage capacity. The scattering noise ultimately limits the dynamic range of the recording material. To avoid the noise gratings, we have suggested an exposure schedule that takes into account the separation between various holograms.

Acknowledgments

The authors acknowledge technical discussions with Demetri Psaltis, Kevin Curtis, Scott Campbell, and Foster Dai. Claire Gu also acknowledges support from the National Science Foundation. This work at UC Santa Barbara has been supported by the U.S. Air Force Office of Scientific Research and Office of Naval Research.

References

1. D.L. Staebler, J.J. Amodei, and W. Phillips, "Multiple Storage of Thick Phase Holograms in $LiNbO_3$," IEEE J. Quantum Electron. **QE-8**, 611 (1972).
2. P.J.V. Heerden, "Theory of Optical Information Storage in Solids," Appl. Opt., **2**, 393–400 (1963).
3. E.G. Ramberg, RCA Rev., **33**, 5 (1972).
4. D. Psaltis, Byte, **17**, 179 (1992).
5. L. Hesselink and M.C. Bashaw, Photonics Spectra, **30**, 44 (1996).
6. J. Ashley, M.P. Bernal, M. Blaum, G.W. Burr, and others, Laser Focus World, **32**, 81 (1996).
7. F. Mok, M.C. Tackitt, and H.M. Stoll, Opt. Lett., **16**, 605 (1991).
8. F. Mok, Opt. Lett., **18**, 915 (1993).
9. D. Psaltis, D. Brady, and K. Wagner, "Adaptive Optical Networks Using Photorefractive Crystals," Appl. Opt., **27**, 1752–1759 (1988).

10. C. Gu, J. Hong, I. McMichael, R. Saxena, and F. Mok, J. Opt. Soc. Am. A., **9**, 1978 (1992).
11. C. Gu and J. Hong, paper WT2 in OSA Annual Meeting, Albuquerque, New Mexico, 1992.
12. K. Curtis, C. Gu, and D. Psaltis, Opt. Lett., **18**, 1001 (1993).
13. G.A. Rakuljic, V. Leyva, and A. Yariv, Opt. Lett., **17**, 1471 (1992).
14. A. Yariv, Opt. Lett., **18**, 652 (1993).
15. G.W. Burr, D. Psaltis, and K. Curtis, paper OWB4 in OSA Topical Meeting on Optical Computing, Palm Springs, March 16–19, 1993.
16. C. Gu and P. Yeh, Opt. Lett., **16**, 1572 (1991).
17. F. Vachss, C. Gu, J. Hong, and T. Chang, Technical Digest on Photorefractive Materials, Effects, and Devices, 1991, **14**, 92 (Optical Society of America, Washington, D.C., 1991).
18. C. Gu and J. Hong, Opt. Commun., **93**, 213 (1992).
19. C. Gu, J. Hong, and P. Yeh, J. Opt. Soc. Am. B, **9**, 1473 (1992).
20. H. Lee, X.-G. Gu and D. Psaltis, J. Appl. Phys., **65**, 2191 (1989).
21. G.W. Burr, F.H. Mok, and D. Psaltis, CLEO 1994, page 9, paper CMB7 (1994).
22. J.D. Jackson, *Classical Electrodynamics*, John Wiley & Sons, pp. 427–432 (New York) 1975.
23. J.W. Goodman, *Introduction to Fourier Optics*, McGraw-Hill, pp. 57–96 (San Francisco) 1968.
24. C.X.-G. Gu, Optical Neural Networks using Volume Holograms, Ph.D. Thesis, California Institute of Technology, 1990, pp. 39–116.
25. See, for example, A. Yariv and P. Yeh, *Optical Waves in Crystals* (Wiley, New York 1984.
26. See, for example, C. Kittel, *Introduction to Solid State Physics* (Wiley, New York) 1976.
27. D.M. Lininger, D.D. Crouch, P.J. Martin, and D.Z. Anderson, Opt. Commun., **76**, 89 (1990).
28. D.Z. Anderson and R. Saxena, J. Opt. Soc. Am. B **4**, 164 (1987).
29. See, for example, P. Yeh, IEEE J. Quantum Electron., **25**, 484 (1989).
30. See, for example, H. Kogelnik, Bell Syst. Tech. J., **48**, 2909 (1969).
31. N.V. Kukhtarev, V.B. Markov, S.G. Odulov, M.S. Soskin, and V.L. Vinetskii, Ferroelectrics, **22**, 949 (1979).

Part II

Recording Media

Bit Error Rate for Holographic Data Storage

J.A. Hoffnagle and C.M. Jefferson

The HOST (Holographic Optical Storage Team) tester is an apparatus specifically designed and constructed to test holographic data storage media under conditions similar to those of a product, and to allow virtually any desired experimental parameters, such as sample orientation or reference beam angle, to be varied with complete freedom and control. It was designed with custom optics that provide diffraction-limited imaging superior to what the media allow, thereby ensuring that if inadequate image quality was observed when a piece of medium was tested, one could be sure that this was due to the material under test, and was not an artifact of the tester. The construction of the tester has been desribed in detail in [1]. Here we wish to discuss the quantitative presentation of media test results in terms of bit error rate (BER). It is evidently essential to have a numerical measure of the quality of the data that is received after transmission through the recording channel in order to characterize the performance of the tester itself, as well as the suitability of the recording medium being tested.

From the beginning of the PRISM project, it was intended to use the BER of the reconstructed data for this purpose, although at the onset of the program a satisfactory procedure for computing BER had yet to be devised. What was required was a figure of merit for what could loosely be called the "image quality" or "fidelity" of the data pages that were read out after having been stored in the holographic medium. This, in turn, depends on many considerations, including the performance of components such as the SLM, optics, and CCD camera that constitute the tester, the mechanical stability and accurate alignment of the tester components, optical properties such as homogeneity and surface polish of the recording medium, and the physics of the holographic reconstruction. The BER provides a such a quantitative measure of image quality, describing the performance of the test stand and samples, which is directly applicable to the ultimate goal of data storage.

It should be emphasized that the BER is a property of the *entire recording channel.* It depends on the hardware of the tester, the recording material, the amount of data written per page, and whatever modulation or error correction coding may be used. This makes it a very versatile diagnostic tool – by comparing the BER observed under differing experimental conditions one can identify specific contributions to image quality – but it also means that a single BER measurement says nothing about the performance of an isolated piece of the apparatus; it only describes the recording channel as a whole. The following discussion pertains to the performance of the optical components of the test stand, including the storage medium itself. Consequently, we exclude considerations of modulation and error-correction coding, in order to focus on the error rate for the raw data, which depends sensitively on the optics of the test stand. This does not mean that we are unaware of the important inprovements in BER that can be achieved through coding techniques, merely that we wish to study the basic optical performance of the tester.

1 Definition of Bit Error Rate

The basic notion of BER is independent of the way in which the data is formatted or recorded: one takes some input bitstream, passes it through some or all of the recording channel to obtain a corresponding output bitstream, and compares the input and output to deduce the number of errors (discrepancies between input and output). The BER is the ratio of the number of errors to the total number of bits in the bitstream, in the limit where the latter goes to infinity. In practice, of course, one deals with bitstreams of finite size: consequently this simple approach of *error counting* is only useful if the BER is large enough for one to actually observe a statistically significant number of errors. For a well-designed system, the BER may be small enough that this is not true for bitstreams of a manageable size. We therefore wish to find a way to estimate the BER, as defined in terms of error counting, but with a smaller bitstream than the error-counting method would require.

Such an estimate is possible because implicit in the concept of BER is the assumption that there is a probabilistic component to the observation of bit errors. The transformation of logical zeros and ones in the input bitstream into measured physical values at the output of the CCD camera is assumed to include some processes, e.g. detector noise or image degradation due to imperfections in the optics or sample, that give rise to a result that is described in terms of probability distributions rather than purely deterministic expressions. If one can determine these underlying probability distributions it is then possible to derive the BER without recourse to error counting.

Recall the basic recording scheme: binary data bits are transformed by the SLM into an array of bright and dark regions in the object beam. Ideally, the SLM and associated optics should modulate the object beam with the best possible contrast and uniformity thus logical 0 could correspond to regions of equal and very low light intensity, and logical 1 to regions of equal and much greater light intensity. During readout the object beam is holographically reconstructed and imaged onto a CCD array in such a way that there is a one-to-one mapping of data bits and CCD pixels. Within each pixel the photoelectric effect generates a charge proportional to the incident light intensity; this charge is converted to a voltage, amplified, and digitized. The output of the CCD camera is an array of integers proportional to the light intensity at each of the pixels in the sensor. If the holographic reconstruction were perfect, the CCD would record a uniformly large number of photoelectrons in the pixels corresponding to 1s in the original data and a uniformly small number of photoelectrons in the pixels corresponding to 0s. In this case the original binary data could be recontructed by identifying the pixel values below some threshold with 0s and the pixel values above the threshold with 1s. In fact the camera output is distributed over a range of values, which may result in ambiguity as to whether a particular pixel represents a 0 or a 1. In the next section we shall show how the distribution of pixel values can be used to derive the BER as the probability of having an erroneous bit in the reconstructed binary data. As explained in the introduction, the use of a global threshold is generally not the best way to decode the data: local thresholding [2], differential encoding [3], or more sophisticated modulation coding techniques [4] can greatly reduce the measured BER. However, since we are interested in studying the *fidelity* of the reconstruction of an image that ideally would consist of identical 0s and 1s, the global threshold is appropriate.

2 BER in Terms of Pixel Distribution Functions

The utility of the BER as a diagnostic of storage channel performance depends in part on the implicit assumption that for a large enough data set, errors can be regarded as random events, described by a rate constant equal to the fraction of erroneous bits in the entire transmitted bitstream. Generalizing this probabilistic approach to the output of the CCD camera, before binarization, allows the derivation of an expression for the BER that does not depend on error counting. Let us divide the CCD pixels into two populations, those corresponding to bits that were 0s in the input bitstream and those corresponding to 1s, and suppose that for each of these populations the probability that the output of a pixel has the value x is given by a probability distribution

$$W_j(x),\ j = 0 \text{ or } 1. \tag{1}$$

For the moment let x be a real number; physically it could correspond to the output of the camera preamplifier just before digitization. The discretization introduced by the camera electronics will be taken into account below. The probability distributions W are defined for $0 \leq x < \infty$ and are normalized,

$$\int_0^\infty W_j(x)\,\mathrm{d}x = 1. \tag{2}$$

Ideally, the Ws would be delta functions; in practice they are peaked, with some nonvanishing width. Let us assume that there is a unique $x \equiv x_\mathrm{c}$ where the Ws cross, i.e. $W_0(x) \gtrless W_1(x)$ for $x \lessgtr x_\mathrm{c}$, and that 0s and 1s occur with equal frequency in the input data. These assumptions are not essential, but they are generally at least approximately fulfilled, and they simplify the expression for the BER. Then, chosing x_c to be the threshold between 0s and 1s, and identifying the BER with the probability of a misinterpreted bit, we have simply

$$\mathrm{BER} = \frac{1}{2}\left[\int_0^{x_c} W_1(x)\,\mathrm{d}x + \int_{x_c}^{\infty} W_0(x)\,\mathrm{d}x\right]. \tag{3}$$

This expression generalizes the error-counting method in the sense that for a sufficiently large datastream the expectation value for the number of errors is the BER given by (3) times the number of bits in the datastream.

3 Experimental Distributions of CCD Pixel Values

The CCD camera does not directly provide the continuous probability distributions (1). Rather, when each pixel is read out, the accumulated thermoelectric and photoelectric charge is converted to a voltage and digitized, yielding an integer value, I, between 0 and some maximum value I_{max}. Since I is a discrete approximation to the continuous variable x, the likelihood of a pixel having the value I reflects the underlying probability distributions $W_j(x)$. A single CCD image contains a large number of pixel values, from which one can construct histograms that approximate the W_j, as follows: The range of possible pixel values is divided into m "bins" of equal width, ΔI, centered on the values

$$I_k = \left(k - \frac{1}{2}\right)\Delta I, \qquad k = 1 \dots m, \tag{4}$$

and for each bin the number of pixels in the CCD output having

$$I_k - \frac{1}{2}\,\Delta I \le I < I_k + \frac{1}{2}\,\Delta I \tag{5}$$

is counted. The histograms are the sets

$$N^{(j)}(I_k), \qquad j = 0 \text{ or } 1, \tag{6}$$

of observed pixel counts, for those pixels corresponding to values of 0 and 1 in the input bitstream. Denoting the total number of 0s and 1s in the input bitstream by

$$N_{\mathrm{tot}}^{(j)} = \sum_{k=1}^{m} N^{(j)}(I_k), \tag{7}$$

and assuming that ΔI is small, then the expectation value for the number of pixels in the bin centered on I_k is approximately equal to

$$N^{(j)}(I_k) = N_{\mathrm{tot}}^{(j)}\, W_j(I_k)\,\Delta I. \tag{8}$$

This provides the connection between experimental histogram data and the underlying probability distributions. However, in order to carry out the integrals in (3) one still needs a model of the functional form of the Ws. This is not a trivial requirement, because it involves extrapolating the Ws from regions in which there is adequate experimental data but little overlap of W_0 and W_1 into regions where the overlap is important but there is little experimental data.

Under some conditions, it may be possible to derive *a priori* expressions for the functional form of the Ws. For instance, if random fluctuations in the optical field are the dominant source of experimental noise, then the amplitude of the optical field should follow a Rician distribution [5],

$$W_{\mathrm{R}}(p, s; x) = \frac{x}{s^2} \exp\left(-\frac{p^2 + x^2}{2s^2}\right) \mathrm{I}_0\left(\frac{px}{s^2}\right), \quad (9)$$

for both W_0 and W_1 [6]. The parameters p and s that appear in (9) describe the average field amplitude and its variance respectively, I_0 denotes the modified Bessel function of the first kind of order 0, and the independent variable x is proportional to the square root of the intensity recorded by the camera. Such distributions have been observed experimentally under conditions of low data density and high signal strength, so that image degradation due to electronic noise and pixel misalignment are negligible, as illustrated in Fig. 1. If incoherently scattered photons, stray light, dark current, or Johnson noise in the CCD preamplifier are the dominant noise sources, then the CCD output I is expected to follow a Gaussian distribution,

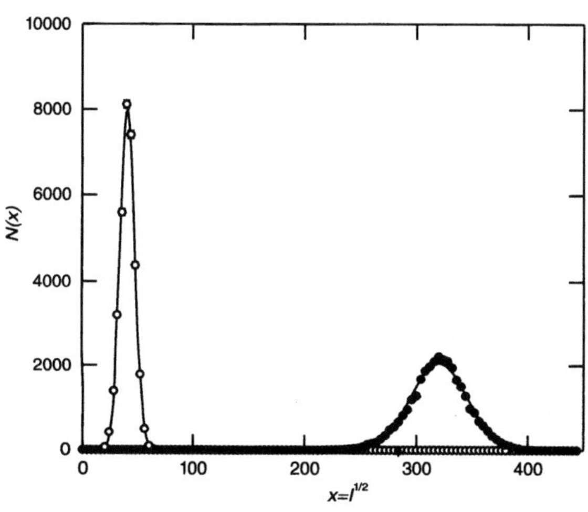

Fig. 1. Histograms for a well-aligned 64-kbyte image. The independent variable is the square root of the CCD output. Open and closed circles denote the measured $N_0(x)$ and $N_1(x)$ respectively. Solid lines are best fits to Rician distributions

$$W_{\mathrm{G}}(x_0, \sigma; x) = \frac{1}{\sqrt{2\pi}\,\sigma} \exp\left[-\frac{(x - x_0)^2}{2\sigma^2}\right]. \tag{10}$$

Generally, however, there is no reason to expect the *W*s to have any simple analytical form. Both coherent ("Rician") and incoherent ("Gaussian") noise processes will normally be present, resulting in a complicated pixel distribution. Moreover, not all contributions to the observed pixel distribution come from simple stochastic processes.

When interpreting the experimental histograms, it is important to realize that several different processes act to smear out the ideal pixel distributions. Some stochastic processes were considered in the previous paragraph; they give rise to pixel distributions that are Rician, Gaussian, or a convolution of these. Other processes are in principle deterministic, e.g. fixed-pattern noise in the CCD, background illumination, interpixel cross-talk, and misregistration of the optical image to the CCD. The last of these is of great practical importance, especially when one attempts to achieve the greatest possible data density. One consequence of misregistration is that the pixel distribution becomes *environment-dependent.* This is illustrated in Fig. 2 for the simple case of an image that is shifted parallel to the rows of the CCD. With

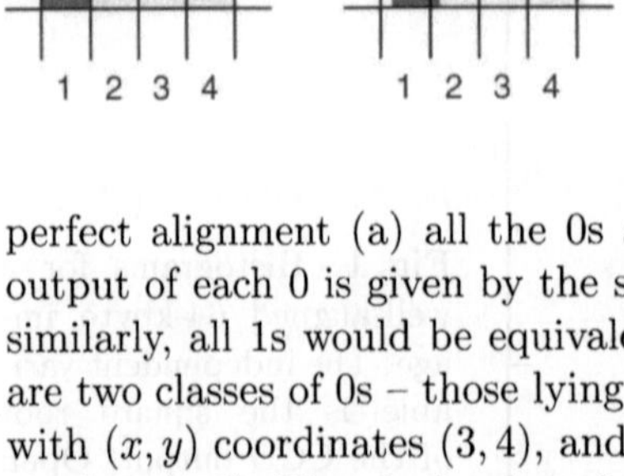

Fig. 2a,b. Schematic illustration of pixel misregistration. The heavy black lines delineate the CCD pixels, and the light and dark rectangles represent the optical binary data pattern. Ideally the optical data and CCD pixels exactly coincide, as in (**a**), whereas the misregistered data pattern (**b**) is shifted horizontally by 0.2 pixels in this example. The numbers label the x- and y-coordinates of the CCD pixels

perfect alignment (a) all the 0s are equally illuminated, and therefore the output of each 0 is given by the same distribution as that of every other 0 ; similarly, all 1s would be equivalent. As a result of misalignment (b), there are two classes of 0s – those lying to the right of another 0, such as the pixel with (x, y) coordinates $(3, 4)$, and those lying to the right of a 1, such as the pixel at $(4, 3)$. The pixels whose left-hand neighbors are 0s receive less illumination than those whose neighbors are 1s. Similarly, there are two classes of 1s – exemplified by the pixels at coordinates $(3, 1)$ and $(3, 2)$. In this simple case, we can think of the environment of a pixel as referring to the magnitude and direction of the misregistration and the value of the neighboring pixel, 0 or 1. The expected output of any given pixel depends on the value of the input bit *and* the environment, and the functions W depend on the environments of all the pixels.

The situation illustrated in Fig. 2 is an artificially simplistic one. At the highest possible data density, 1024×1024 bits imaged onto a CCD sensor with 9 μm square pixels, several kinds of misregistration are inevitably present: shifts in both x- and y-directions of the data array with respect to the CCD array, rotation of the data array with respect to the CCD array, and magnification error and distortion in the optics. The result is that there may be a very large number of different pixel environments, leading to a complicated pixel intensity distribution. To deal with this difficult situation, we have adopted a very simple physical picture in which the full probability distribution is regarded as a convolution of a simple analytical function describing the stochastic distribution that one would have for perfect registration with a function that takes into acount the many possible pixel environments. This is analogous to the inhomogeneously broadened lineshape that arises in spectroscopy as the sum of the natural line profile over many inequivalent environments. When the pixel distribution is used to calculate the BER, it is clear from (3) that the only parts of the Ws that matter are $W_0(x)$ for $x > x_c$ and $W_1(x)$ for $x < x_c$. We are therefore led to consider only the relatively simple functions that describe the flanks of the full pixel distributions, as shown schematically in Fig. 3. Unlike the full probability distributions, W_j,

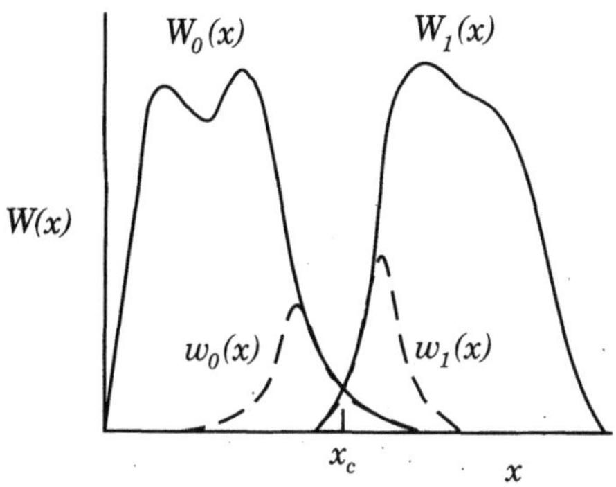

Fig. 3. Schematic representation of the distributions of measured pixel intensities. The full lines denote the complete, normalized probability distributions W_0 and W_1 for pixels corresponding to zeros and ones respectively. The broken lines show simpler, analytic functions w_0 and w_1 that approximately equal the full distributions near x_c

the approximating functions, w_j, are not normalized to unity; rather their amplitudes are chosen to best approximate the Ws near the threshold, x_c. Provided that w_0 is a good approximation to W_0 for $x > x_c$ and w_1 is a good approximation to W_1 for $x < x_c$, then the exact distributions in (3) can be replaced by their approximate expressions, to yield

$$\mathrm{BER} \approx \frac{1}{2}\left[\int_0^{x_c} w_1(x)\,\mathrm{d}x + \int_{x_c}^{\infty} w_0(x)\,\mathrm{d}x\right]. \tag{11}$$

Chosing the w_j to have a simple form, the integrals can be performed analytically. In particular, Gaussian functions were empirically found to give a good fit to the tails of the experimental histograms. Using the notation of (10), we postulate

$$w_j(x) = A_j \, W_{\mathrm{G}}(x_j, \sigma_j; x), \qquad j = 0 \text{ or } 1. \tag{12}$$

Then (11) can be evaluated in terms of the complementary error function, erfc, as

$$\mathrm{BER} \approx \frac{\sqrt{\pi/2}}{N \, \Delta I} \left[A_0 \, \sigma_0 \, \mathrm{erfc}\left(\frac{x_{\mathrm{c}} - x_0}{\sqrt{2}\sigma_0} \right) + A_1 \, \sigma_1 \, \mathrm{erfc}\left(\frac{x_1 - x_{\mathrm{c}}}{\sqrt{2}\sigma_1} \right) \right]. \tag{13}$$

The BER can be computed simply by fitting Gaussian functions to the flanks of the measured histograms. The principle is illustrated in Fig. 4, which shows histograms of a 256-kbit page of data that has been deliberately misaligned to increase the error rate. The solid curves show least-squares fits to the flanks of the histograms, and by using these fitted probability distributions in (3) one obtains a calculated BER of 2.83×10^{-3}. This can be compared with the BER derived by simply counting errors: with the optimal threshold there are 739 erroneous bits out of a total of 262 144 bits on the page, for a BER of 2.82×10^{-3}, in good agreement with the prediction based on (3).

The data of Fig. 4 illustrate the correspondence of the BER computed from the probability distributions W with the straightforward approach of counting errors, when the CCD images are sufficiently poor to give a high error rate. As image quality improves, W_0 and W_1 become narrower and further separated, while the number of errors decreases. Eventually, the number of actual errors goes to zero, at which point the error-counting approach gives no more information. However, (3) still gives a well-defined result, even when the error rate is small enough that the average number of errors per page is

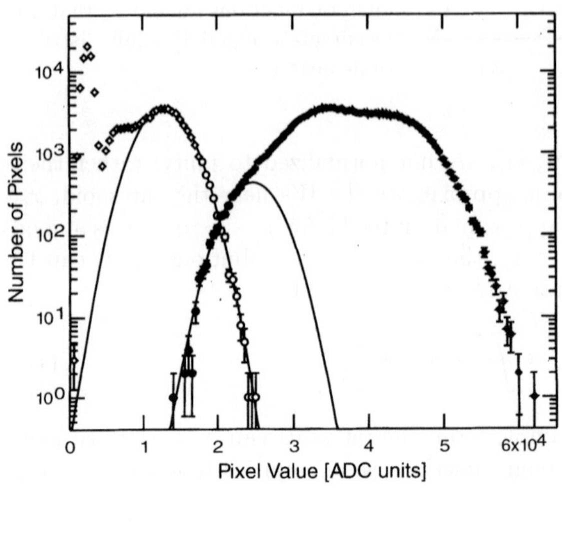

Fig. 4. Histograms of the CCD response for a poorly aligned image. Open and closed symbols denote the observed distributions for pixels corresponding to input bits of 0 and 1 respectively. The error bars are purely statistical. Solid curves are Gaussian functions fitted to the flanks of the experimental distributions

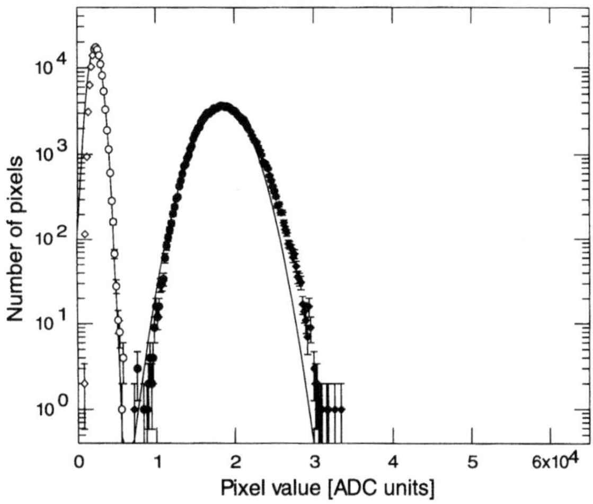

Fig. 5. Similar to Fig. 4 but with better image alignment

less than unity. This is illustrated in Fig. 5, which shows the histograms for a hologram of a 256-kbit page of data, this time with the image more accurately aligned on the CCD. Here the distributions W_0 and W_1 are separated well enough that there are no errors on the data page; the computed BER of 8.7×10^{-7} gives a quantitative evaluation of the expected error rate.

4 Applications

The computation of BER from measured pixel distribution functions has been used extensively both in the development of the tester and for the characterization of holographic storage media. The performance of the tester itself is measured by inserting plates of high-quality fused silica in place of the storage medium. The most exacting measurements, of course, are carried out at the ultimate data density of one bit per CCD pixel. This requires that the SLM be pixel-matched to the CCD sensor across the entire 1024×1024 pixel field of view, to an accuracy much better than the size of a single pixel. Under these conditions the tester is capable of a BER of better than 10^{-8}, a level of performance that would have been impractical to quantify through error counting. This limit is presumably set by residual distortion in the imaging optics and nonuniformity of the object beam illumination. It exceeds by far the performance that can be obtained when reading out recorded holograms. For instance, we have presented data on the first recovery of holographically recorded megapixel data pages [7]. Using iron-doped $LiNbO_3$, which has the best optical quality of any currently available holographic storage material, a BER of 2.4×10^{-6} was achieved. The large difference between the BER of the holographically stored image and the imaging performance of the tester alone is an indication that we are indeed measuring the limitations of the storage medium, as the tester is designed to do.

References

1. M.-P. Bernal, H. Coufal, R.K. Grygier, J.A. Hoffnagle, C.M. Jefferson, R.M. Macfarlane, R.M. Shelby, G.T. Sincerbox, P. Wimmer, and G. Wittmann (1996): A precision tester for studies of holographic optical storage materials and recording physics. Appl. Opt., **35**, 2360–2374.
2. X.A. Shen, A.-D. Nguyen, J.W. Perry, D.L. Huestis, and R. Kachru (1997): Time-domain holographic digital memory. Science, **278**, 96–100.
3. J.F. Heanue, M.C. Bashaw, and L. Hesselink (1994): Volume holographic storage and retrieval of digital data. Science **265**, 749–752.
4. G.W. Burr, J. Ashley, H. Coufal, R.K. Grygier, J.A. Hoffnagle, C.M. Jefferson, and B. Marcus (1997): Modulation coding for pixel-matched holographic data storage. Opt. Lett., **22**, 639–641.
5. J.W. Goodman (1985): *Statistical Optics*. Wiley, New York.
6. J.E. Weaver and T.K. Gaylord (1981): Evaluation experiments on holographic storage of binary data in electro-optic crystals. Opt. Eng., **20**, 404–411.
7. R.M. Shelby, J.A. Hoffnagle, G.W. Burr, C.M. Jefferson, H. Coufal, R.K. Grygier, H. Günther, R.M. Macfarlane, and G.T. Sincerbox (1997): Pixel-matched holographic data storage with megabit pages. Opt. Lett., **22**, 1509–1511.

Media Requirements for Digital Holographic Data Storage

R.M. Shelby

The end user of a holographic data storage system is concerned with its performance as described by storage capacity, data input and output rates, stability of stored data and device compactness, all of which must be delivered at a specified (very low) bit error rate (BER). To a large extent, the possibility of delivering such a system is limited by the properties of the materials available as storage media [1]. The connections between materials properties and system performance are complex, and many trade-offs are possible in adapting a given material to give the best results. Though these issues will be explored more deeply in other chapters, we will attempt to outline in a general way the properties that are desirable for a holographic storage medium, and to assess the difficulties that arise in achieving them with real materials.

Some media-related parameters are straightforward to understand and specify (for example measures of optical quality such as wavefront distortion), and many of these can likely be achieved through sufficient engineering effort. Other properties depend in a deeper way on the very physical processes that make optical recording in the materials possible, and optimization of these properties has proven resistant to the efforts of researchers, especially when a detailed understanding of these processes is lacking.

1 Ideal Media Parameters

The following is a brief discussion of the more quantifiable properties a successful holographic storage material should possess.

1.1 Optical Quality

The high data rate and storage density of holographic storage depend critically on the distortion-free imaging of the input data to the detector in such a way that each pixel of a 1000×1000 array on a spatial light modulator is precisely registered on the corresponding detector pixel. Doing so requires very high-performance optics, of which the storage medium is one component. Furthermore, if the medium is moved to access different areas, this motion must not compromise this performance. For example, if one imagines a disk configuration in the usual Fourier optics arrangement with typical values for lens focal length and pixel size, very stringent parallelism requirements result, of a fraction of an arc-second.

These very high requirements on material homogeneity and precision fabrication tolerances may be relaxed if a viable technique for phase-conjugate readout of stored holograms can be developed [2–4]. This method has been shown to compensate the distortions introduced by imperfect optics and storage media, but a scheme to implement it in a practical storage device remains to be demonstrated.

Another more microscopic aspect of optical quality is the background noise floor at the detector produced by the intrinsic light scattering of the material. This scattering level imposes a fundamental minimum on the efficiency of a stored data hologram, and thus on the storage density [5].

A number of different materials have been tested using the IBM Holographic Optical Storage Tester to quantitatively compare their optical quality. Imaging quality is determined based on the BER (as described in the previous chapter) of data pages of various densities, imaged through the medium, while scattered light noise floor is represented as the efficiency needed for a hologram to have the mean intensity of the "on" pixels equal to the mean scattered light intensity. Data are plotted for several materials in Fig. 1. Each point on this plot represents a measured BER for a transmitted image or a

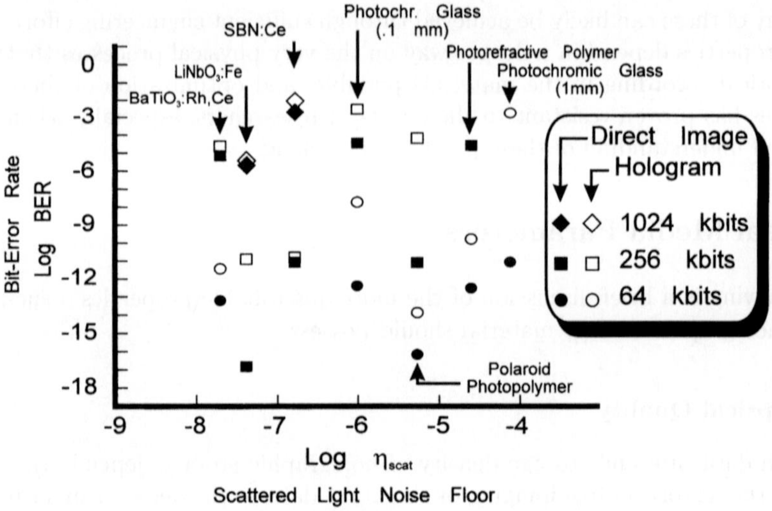

Fig. 1. Optical quality comparison of several holographic storage materials. Bit error rates at several data page densities (see legend) of transmitted images (*filled symbols*) and holograms (*open symbols*) are plotted vs. scattered light level for each material. The photochromic glass and photorefractive polymer materials are described in [6,7]

reconstructed hologram (to be discussed below in more detail), for a sample exhibiting a given scattered light noise floor.

The data shown are for typical available samples, and as such do not necessarily represent intrinsic materials properties nor ultimately achievable performance for a given material. However, this plot serves as a useful comparison between different materials. For example, one sees that organic materials as a class can have good imaging properties, but tend to have higher scattered light level. It is notable that only one material, lithium niobate, is available in sufficient optical quality and size, and displays recording properties sufficient to record and replay a hologram of the highest density, i.e. 1024 kbits per page. This material also has very low scattered light and – as will be discussed below – reasonable dynamic range, and it has been the "workhorse" material for holographic storage systems demonstrations [8,9].

1.2 Sensitivity

Several definitions are in common use to describe holographic recording sensitivity [10,11]. Though sensitivity definitions based on refractive index modulation Δn and aborbed fluence are the most physically appealing, it is more convenient in the context of storage to define sensitivity in terms of the square root of efficiency η obtained in writing plane-wave, unity modulation depth gratings with a given fluence It, where I is the total intensity in W/cm^2 and t is the exposure time. In these terms, the sensitivity is

$$S_{\eta^{1/2}} = \frac{\eta^{1/2}}{It\ell} \ (\mathrm{cm/J}), \tag{1}$$

where ℓ is the effective interaction length.

For purposes of materials comparison, it is useful to recognize that different materials are available in very different thicknesses, depending on the mechanism of the absorption that sensitizes the photoresponse, the magnitude of index change available, and issues of material fabrication. Thus, we define a modified "sensitivity" that gives the response of the material in terms of the square root of diffraction efficiency for a given exposure fluence, for a sample *of the thickness to be used in the actual storage system*:

$$S'_{\eta^{1/2}} = S_{\eta^{1/2}} \times \ell \ (\mathrm{cm^2/J}). \tag{2}$$

In a storage application, one must further assume that the efficiency for data holograms written with an information-bearing object beam is related in some simple way to the sensitivity as defined here: for example, the response is approximately proportional to the average modulation depth, $m = 2(I_\mathrm{o}I_\mathrm{r})^{1/2}/(I_\mathrm{o} + I_\mathrm{r})$, where I_o and I_r are the average object beam and reference beam intensities. If the efficiency of a hologram is given by $\eta = (mS'_{\eta^{1/2}}It)^2$, one can give a reasonable minimum value for the sensitivity. For a mega-pixel hologram of sufficient efficiency that it can be read with reasonable energy (e.g. about $\eta = 3 \times 10^{-5}$ for 100 μJ in 1 ms), and assuming $m \approx 0.2$ and writing fluence of 1 mJ/cm^2 , sensitivity should be at least $S'_{\eta^{1/2}} \approx 20$ cm^2/J. With this sensitivity, and with optics that allow the media area occupied by a hologram to be reasonably small (e.g ~ 0.1 cm^2 or less), in principle each hologram can be written and read in less than 1 ms with ~100 mW of laser power.

1.3 Dynamic Range

Dynamic range refers to the total response of the medium when divided up among many holograms multiplexed in a common volume of material. If media response is quantified as the square root of efficiency, dynamic range can be defined for M multiplexed holograms as [12]:

$$M\# = \sum_{i=1}^{M} \eta_i^{1/2} \, . \tag{3}$$

This quantity is called the "M number" of the material, a quantity that also describes the scaling of diffraction efficiency with number of holograms:

$$\eta = \left(\frac{M\#}{M}\right)^2 . \tag{4}$$

Equations (3) and (4) are generally applicable across the full range of potential holographic media, and allow for useful comparisons.

Experimentally, $M\#$ can be estimated by direct application of (3) to a series of exposures that exhaust the material's response, or by modeling its response. The first method is useful with irreversible materials, which by definition give useful sensitivity over a limited range of total exposure fluence. With sensitivity defined above ((1) and (2)), we have

$$M\# = \int_0^{E_{\max}} S'_{\eta^{1/2}}(E) \mathrm{d}E \, , \tag{5}$$

where E represents exposure fluence and $E_{\max}$ the maximum useful fluence.

For photorefractive crystals, as holograms are written, previously written holograms are partially erased, and recording times must be adjusted such that the early holograms in the series are written closer to saturation, since they will experience the greatest erasure. When the response curve can be approximated as exponential, it has been demonstrated [12] that $M\#$ is given by the ratio of writing rate (build-up of $\eta^{1/2}$ per unit time) to erasure rate (decay of $\eta^{1/2}$ with subsequent exposure) for a given set of exposure conditions.

The quantity $M\#$ is most useful under a specific set of exposure conditions. Physically, it depends on the total refractive index modulation that can be achieved for multiplexed holograms and is approximately proportional to thickness, and it depends on modulation depth and sample geometry. Dynamic range has a strong impact on the data storage density that can be achieved. To reach a density of 100 bits/μm^2 with megabit data pages of target diffraction efficiency of 3×10^{-5} and area on the medium of 0.1 cm^2 requires $M\# = 5$, which is barely achievable with known recording materials given exposure conditions consistent with high-fidelity holograms.

1.4 Absorption

Sensitivity is linked to the absorption coefficient of the material: in general one expects sensitivity to be optimized for a given material when the absorption length and the useful thickness of the material are similar [13,14]. If the absorption depth is too short, the hologram will not be recorded uniformly throughout the volume, and Bragg selectivity will be reduced, resulting in decreased storage density. If the absorbing centers are reduced in concentration to modify the absorption to allow greater thickness, sensitivity will likely be reduced; this reduction could be more than in proportion to the reduction in concentration, depending on the details of the recording mechanisms and the relative concentrations of other species (chemical species, traps, donors, etc.) that might be involved.

The concentration of sensitizing species usually affects dynamic range as well. These issues often prevent the scaling up of the thickness of a material to achieve the desired Bragg selectivity for multiplexing purposes, and require non-selective multiplexing techniques such as peristrophic multiplexing to be used [15].

1.5 Volatility

Data stored as holograms is subject to decay in the dark if the microscopic recording mechanism does not produce a stable refractive index modulation, and to decay during readout of the data if the index modulation is not stable during repeated irradiation by the readout laser beam. The second problem in particular has received a lot of attention recently, as it presents a fundamental problem in the implementation of reversible holographic storage media [16–18]. For a holographic storage system, stability on the timescale of years is desirable.

2 Example Materials

Many kinds of materials have been investigated as holographic storage media. Table 1 is a comparison between several recording materials from a data storage perspective. This table is not an exhaustive list of materials, but is

Table 1. Materials Comparison

	Image quality	Scat.	Fidelity	$S'_{\eta^{1/2}}$ (cm^2/J)	$M\#$	Stability	Thickness (mm)
$LiNbO_3$:Fe	+++	+++	+	0.02	1	0	10
$LiNbO_3$ (stoichio.)	++	++	++	0.07	0.3	–	10
$LiNbO_3$[b] (2-color)	++	++	+	~0.02[c]	~1[c]	++	10
Polaroid[d] photopol.	+++	–	–	20	1.4	++	0.5
PQ/PMMA[e]	+	–	+	0.2–0.5	1	++	~2
Bayer photoaddr. polymer	+++	0	++	0.002–0.02			~0.1

[a] See following chapters for more details on these materials.
[b] [18]
[c] Values depend on writing and gating intensity.
[d] [14]
[e] [21]

intended to compare a few materials that are among the best available as data storage media.

One of the earliest and most productive from a research standpoint is the class of photorefractive crystals [19]. Of these, iron-doped lithium niobate has been the workhorse for demonstrations of holographic data storage. Its sensitivity is sufficient for demonstration purposes, but it lacks a factor of 100× for practical applications and is subject to erasure during readout. Two-color recording in lithium niobate [16,18,20] eliminates the data volatility problem, but rather high intensities are required to have sensitivity and $M\#$ that approach suitable values.

Organic photopolymers are promising as irreversible write-once storage media [22,23]. They can be very sensitive, allowing data to be written rapidly, and have good dynamic range, with $M\#$ for plane wave recording greater than 10. Some impressive demonstrations of the recording of data holograms have been made [24]. Some of the problems that remain include a higher scattered light level than for inorganic materials, difficulty in making media thicker than a fraction of a millimeter, and distortions due to shrinkage of the material upon polymerization or to index modulations or noise recorded by non-information-bearing parts of the beams.

Also shown in the table are entries for phenanthrenequinone-doped PMMA polymer [21] and a "photoadressable" polymer system developed at Bayer. The PMMA system offers excellent optical quality. It is based on a photoreaction between the dopant and polymer followed by diffusion of unreacted chromophore: this requires a long thermal treatment, which is a disadvantage from a system viewpoint. The Bayer system, based on photoreorientation of azo-dyes, is still at an early stage of development.

3 Stability of Stored Data

Development of materials with decreased data volatility has been a focus of recent materials research. Here we highlight the main issues in this area.

3.1 Dark Decay

In the case of photorefractive crystals, dark decay may be caused by residual dark conductivity, which allows the space-charge gratings recorded in the material to relax, or in some cases by ionic mobility which allows a compensating ionic space charge to build up [25]. In either case, the decay is intimately connected to the microscopic defect structure of the material. Dark decay in some photorefractives is sufficiently long for data storage. Unfortunately, it is often the case that attempts to improve other properties by manipulating the defect concentration, Fermi level etc. result in shortening of dark decay.

For irreversible recording materials such as photopolymers, the dark stability issue is more one of aging processes, mechanical relaxation of strains created during polymerization, effects due to residual unreacted components of the original photoactive mixture, or unwanted chemistry of by-products of the polymerization process such as bleached photosensitizer. These materials are relatively new and are undergoing rapid development, and their long-term properties are often not well understood.

3.2 Decay During Readout: Fixing

Reversible materials such as photorefractives suffer intrinsically from the problem of erasure during data readout. The data cannot simply be read out at a wavelength where the material is insensitive, owing to Bragg mismatch in reconstruction of an object beam with large spatial frequency bandwidth. Various schemes have been proposed to circumvent this problem, and more details are found in subsequent chapters. Thermal fixing uses the high-temperature ionic mobility of light ions (often protons) present in the crystal to form a complementary grating that is stable during illumination at room temperature. This procedure has been shown to produce high-quality data holograms that are resistant to erasure during readout [17], but is probably too unwieldy to be used in a practical storage system.

A second technique for complementary grating formation involves ferroelectric domain reversal by applied electric fields, as demonstrated in strontium barium niobate [26]. The technique is limited, however, by the minimum stable size that domains can have to gratings with rather large spatial frequencies, and thus is difficult to apply for data storage.

Some irreversible media suffer from data erasure during readout due to saturation. If a material responds by producing a net refractive index change proportional to exposure, and it eventually saturates at some fluence, the recorded holograms will be washed out. Thus, such media need a means to "turn off" their sensitivity after recording to prevent erasure by the readout beam.

3.3 Two-Color Recording

Two-color or "gated" recording conceptually can solve the problem of data volatility during readout. The idea is to use a light beam at one wavelength of light to sensitize the material. In the presence of this gating light, holograms are recorded by reference and object beams of another wavelength, usually to the red of the gating light. In the absence of gating light, the medium is insensitive, allowing nondestructive readout of data at the same wavelength at which it is written.

This scheme has been implemented with some success in lithium niobate [16,20,18] through the use of both instrinsic and intentionally introduced dopants at appropriate energies within the bandgap. The details of this work are reviewed elsewhere in this volume. For the purposes of this discussion we note that it is possible to achieve a ratio of 10^3–10^4 or more between the rates of erasure during reading and gated writing.

4 Hologram Fidelity and Bit Error Rate

To produce a faithful reconstruction of a light beam carrying the image of a data page, the storage medium must accurately record the interference pattern produced by the superposition of reference and object beams. Ideally the medium response should be linear, and independent of spatial frequency over the bandwidth of the interference pattern. In most storage demonstrations and materials tests that have been done, the holograms have relatively small bandwidth, and the relatively weak variation of response with spatial frequency of most materials has not caused significant distortions. To maximize storage density, optics must be used that produce both object and reference beams covering a wide range of angles, and this may prove more demanding.

The information-carrying interference pattern rides on a background intensity; the fringes between the reference and a complex object beam have non-unity visibility. The reference beam and especially the object beam have spatial intensity variations, typically of lower spatial frequency than the interference pattern, and the storage medium will record these variations in intensity as a refractive index pattern, which then distorts subsequent hologram recordings as well as reconstruction of earlier ones. This is a particular problem with $LiNbO_3(Fe)$, which exhibits a strong photovoltaic effect that enhances its response to low spatial frequencies [27]. It will also be a concern for any recording material with a very high dynamic range that is therefore capable of large refractive index modulation. Ideally one would prefer to find a way to suppress the low spatial frequency response of the material if possible.

Distortions caused by such noise mechanisms and by others such as non-linear material response are reflected by the higher bit error rate for holograms than would be expected from the BER of images in a given material (see Fig. 1). In some cases, an increase in BER is observed as increasing number of holograms is recorded.

5 Conclusions

In this chapter we have given a brief outline of some of the properties that define the performance of a holographic data storage medium, and have given a few examples of promising materials. In each case, the properties of available materials fall short of requirements for a practical storage system. Considerable current research effort is under way to address this problem and to engineer a material that approaches the ideal for holographic storage.

References

1. M.-P. Bernal, G.W. Burr, H. Coufal, R.K. Grygier, J.A. Hoffnagle, C.M. Jefferson, R.M. Macfarlane, R.M. Shelby, G.T. Sincerbox, and G. Wittmann. Holographic data storage materials. Mat. Res. Bull, **21**, 51–60, September (1996).
2. F. Ito, K.-I. Kitayama, and H. Oguri. Compensation of fiber holographic image distortion caused by intrasignal photorefractive coupling by using a phase-conjugate mirror. Opt. Lett., **17**, 215–217 (1992).
3. M.C. Bashaw, A. Aharoni, and L. Hesselink. Alleviation of image distortion due to striations in a photorefractive medium using a phase-conjugated reference wave. Opt. Lett., **17**, 1149–1151 (1992).
4. F. Zhao and K. Sayano. Compact read-only memory with lensless phase-conjugate holograms. Opt. Lett., **21**, 1295–1297 (1996).
5. G.W. Burr, W.C. Chou, M.A. Neifeld, H. Coufal, J.A. Hoffnagle, and C.M. Jefferson. Experimental evaluation of user capacity in hologrpahic data storage systems. Appl. Opt., **37**, 5431–5443 (1998).
6. R. Wortmann, P.M. Lundquist, R. Twieg, C. Geletneky, C.R. Moylan, Y. Jia, R.G. DeVoe, D.M. Burland, M.-P. Bernal, H. Coufal, R.K. Grygier, J.A. Hoffnagle, C.M. Jefferson, R.M. Macfarlane, R.M. Shelby, and G.T. Sincerbox. A novel sensitized photochromic organic glass for holographic optical storage. Appl. Phys. Lett., **69**, 1657–1659 (1996).

7. P.M. Lundquist, C. Poga, R.G. DeVoe, Y. Jia, W.E. Moerner, M.-P. Bernal, H. Coufal, R.K. Grygier, J.A. Hoffnagle, C.M. Jefferson, R.M. Macfarlane, and G.T. Sincerbox. Holographic digital data storage in a photorefractive polymer. Opt. Lett., **21**, 890–892 (1996).
8. J. Heanue, M. Bashaw, and L. Hesselink. Volume holographic storage and retrieval of digital data. Science, **265**, 749–752 (1994).
9. G.W. Burr, J. Ashley, H. Coufal, R.K. Grygier, J.A. Joffnagle, C.M. Jefferson, and B. Marcus. Modulation coding for pixel-matched holographic data storage. Opt. Lett., **22**, 639–641 (1997).
10. G.C. Valley and M.B. Klein. Optimal properties of photorefractive materials for optical data processing. Opt. Eng., **22**, 704–711 (1983).
11. A. Glass, P. Günter, J. Huignard, M. Klein, E. Krätzig, N. Kukhtarev, J. Lam, R. Mullen, M. Petrov, O. Schimer, S. Stepanov, J. Strait, and G. Valley. *Photorefractive Materials and Their Applications I.* Springer-Verlag (1988).
12. F.H. Mok, G.W. Burr, and D. Psaltis. A system metric for holographic memory systems. Opt. Lett., **21**, 896–899 (1996).
13. G.W. Burr and D. Psaltis. Effect of the oxidation state of $LiNbO_3$:Fe on the diffraction efficiency of multiple holograms. Opt. Lett., **21**, 893–895, 1996.
14. D.A. Waldman, H.-Y.S. Li, and E.A. Cetin. Holographic recording properties in thick films of ulsh-500 photopolymer. Proc. SPIE, **3291**, 89–103 (1998).
15. K. Curtis, A. Pu, and D. Psaltis. Method for holographic storage using peristrophic multiplexing. Opt. Lett., **19**, 993–995 (1994).
16. Y.S. Bai, R.R. Neurgoankar, and R. Kachru. High efficiency nonvolatile holographic storage with two step recording in praseodymium doped lithium niobate by use of continuous wave lasers. Opt. Lett., **22**, 334 (1997).
17. J.F. Heanue, M.C. Bashaw, A.J. Daiber, R. Snyder, and L. Hesselink. Digital holographic storage system incorporating thermal fixing in lithium niobate. Opt. Lett., **21**, 1615–1617 (1996).
18. H. Günther, R. Macfarlane, Y. Furukawa, K. Kitamura, and R. Neurgaonkar. Two-color holography in reduced near-stoichiometric lithium niobate. Appl. Opt., **37**, 7611–7623 (1998).
19. L. Hesselink and M. Bashaw. Optical memories implemented with photorefractive media. Opt. Quantum Electron., **25**, 611–651 (1993).
20. L. Hesselink, S.S. Orlov, A. Liu, A. Akella, D. Lande, and R.R. Neurgaonkar. Photorefractive materials for nonvolatile volume holographic data storage. Science, **282**, 1089–1094 (1998).
21. G.J. Stechman, I. Solomantine, G. Zhou, and D. Psaltis. Characterization of phenanthrenequinone-doped poly(methyl methacrylate) for holographic memory. Opt. Lett., **23**, 1310–1312 (1998).
22. D.A. Waldman, H.-Y.S. Li, and M.G. Horner. Volume shrinkage in slant fringe gratings of a cationic ring-opening holographic recording material. J. Imaging Sci. Technol., **41**, 497–514 (1997).
23. V. Colvin, R. Larson, A.L. Harris, and M. Schilling. Quantitative model of volume hologram formation in photopolymers. J. Appl. Phys., **81**, 5913–5923 (1997).
24. A. Pu and D. Psaltis. High-density recording in photopolymer-based holographic three-dimensional disks. Appl. Opt., **35**, 2389–2398 (1996).

25. A. Yariv, S. Orlov, G. Rakuljic, and V. Leyva. Holographic fixing, readout, and storage dynamics in photorefractive materials. Opt. Lett., **20**(11), 1334 1336 (1995).
26. Y. Qiao, S. Orlov, D. Psaltis, and R.R. Neurgaonkar. Electrical fixing of photorefractive holograms in $Sr_{0.75}Ba_{0.25}Nb_2O_6$. Opt. Lett., **18**, 1004–1006 (1993).
27. A.M. Glass, D. von der Linde, and T.J. Negran. High voltage bulk photovoltaic effect and the photorefractive process in $linbo_3$. Appl. Phys. Lett., **25**, 233–235 (1974).

Inorganic Photorefractive Materials

K. Buse and E. Krätzig

Light-induced refractive index changes – so-called photorefractive effects – in inorganic electro-optic crystals were discovered by Ashkin *et al.* in 1966 [1]. Though in the beginning these effects seemed to be very undesirable ("optical damage"), Chen *et al.* recognized only two years later the significance for holographic data storage [2]. In 1975, Staebler *et al.* reported the recording of 500 thermally fixed volume phase holograms in $LiNbO_3$:Fe, each hologram with more than 2.5% readout efficiency [3]. The method is based on the Bragg condition allowing the superposition of many volume ("thick") holograms at the same site under different angles.

During hologram recording a light pattern has to be transposed into a refractive index pattern. Interfering light beams generate bright and dark regions in an electro-optic crystal. When light of suitable wavelength is chosen, charge carriers – usually electrons [4] – are excited in the bright regions and become mobile. The charge carriers migrate in the crystal and are subsequently trapped at new sites. By these means electronic space-charge fields are set up that give rise to a modulation of refractive index via the electro-optic effect. Index changes up to 10^{-3} are obtained. The trapped charge can be released and the field pattern erased by uniform illumination or by heating. On the one hand, this reversibility is highly desired for erasable memories; on the other hand, the problem of destructive readout arises.

Photorefractive effects have been observed in many electro-optic crystals, among them $LiNbO_3$, $LiTaO_3$, the ferroelectric perovskites $BaTiO_3$, $Ba_{1-x}Ca_xTiO_3$, $KNbO_3$ and $KTa_{1-x}Nb_xO_3$ (KTN), the tungsten-bronze-type crystals $Ba_2NaNb_5O_{12}$ and $Sr_{1-x}Ba_xNb_2O_6$ (SBN), the nonferroelectric sillenites $Bi_{12}TiO_{20}$ (BTO), $Bi_{12}SiO_{20}$ (BSO), and $Bi_{12}GeO_{20}$ (BGO), the semiconductors GaAs and InP, and others. Very early crucial influences of dopants and thermal treatments were discovered [5,6]. Though we discuss here mainly aspects of holographic storage, photorefractive crystals are also of particular interest for many unique devices, such as self-pumped phase-conjugating mirrors, parametric amplifiers and oscillators, and novelty filters [7].

In this chapter we describe the charge transport processes that critically determine the performance of the material. Figures of merit for the storage properties will be introduced, the limits will be evaluated, and the crystals mentioned above will be discussed in terms of their applicability for holographic data storage. Finally, fixing techniques will be briefly reviewed.

1 Charge Transport

The charge driving forces are well known: diffusion, bulk photovoltaic currents [8], and drift of charge carriers in external or pyroelectric fields [9] generate the space-charge field. Compensating drift currents arise, and saturation is achieved if they are as large as the driving currents. Application of large external electric fields is often inconvenient, and pyroelectric fields occur only in the case of high light intensities. Thus diffusion and the bulk photovoltaic effect are of major importance.

Extrinsic or intrinsic defects occuring in more than one valence state are sources and traps of the charge carriers. Figure 1 shows different charge transport situations in a band diagram. The energetic positions of the levels shown correspond to the thermal energy required to excite a charge carrier from the filled level. Photon energies necessary to release electrons are typically larger because of the Franck–Condon principle. The simplest and for many applications the desired mechanism is provided by a one-center system with monopolar conductivity (shown in Fig. 1 for electron conductivity). One impurity type occurs in two different valence states, and charge is redistributed via either conduction or valence band [10]. An example is iron-doped $LiNbO_3$. Iron ions occur only in the valence states 2+ and 3+ [11]. Electrons are excited from Fe^{2+} to the conduction band and trapped by Fe^{3+} elsewhere.

Electron–hole competition complicates the situation [4,12,13]. Electrons and holes can be created at the same center, as shown in Fig. 1, but it is also possible that an additional independent center is present that generates holes. In any case, diffusion currents of electrons and holes compensate each other, which is disadvantageous for storage.

Many materials show more complicated transport mechanisms, because two or more photorefractive levels participate in the charge transport simul-

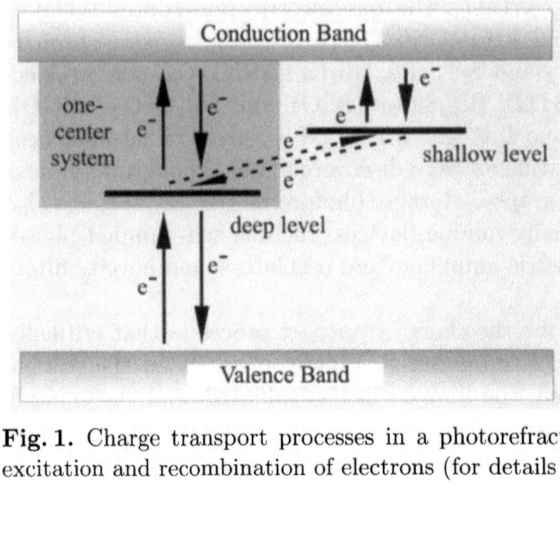

Fig. 1. Charge transport processes in a photorefractive crystal. Arrows indicate excitation and recombination of electrons (for details see text)

taneously. Additional independent centers [14,15] or impurities occuring in three or more valence states [16] may be the origin of these additional levels. We have to distinguish between so-called deep and shallow levels. "Shallow" indicates here that charge can be released thermally from this level, which is practically impossible for a deep site. A system with the simultaneous presence of a deep and a shallow energy level (shown in Fig. 1 for electron conduction) has detrimental consequences for storage. The shallow levels are empty in the dark because all charge carriers have been thermally excited and trapped by deep sites. Upon illumination, the shallow levels trap charge carriers from either the conduction or the valence band. As a consequence, the response time grows (smaller photoconductivity [17]) and, moreover, often the absorption increases, too, because the filled shallow traps absorb light effectively [15].

Deep and shallow levels exchange charge via conduction or valence band. Direct exchange of charge is possible, if the concentrations are large enough that different centers are located close together. This may occur in the case of intrinsic defects [18]. For very high defect concentrations additional bands are created, because centers of the same type couple and exchange charge directly. More details about the charge transport may be found in a recent review [19].

The charge transport situation can be influenced by doping, thermal annealing and selection of experimental parameters (light intensity and wavelength, crystal temperature). Thus it is possible to establish in almost all materials the desired one-center charge transport situation. We will concentrate in the following on a one-center system, because we are interested in the performance limits of different materials. Iron–doped lithium niobate ($LiNbO_3$:Fe) will be used as an example to describe in more detail the ongoing processes and the resulting dependences. However, most of the theoretical consideration is valid for any photorefractive crystal.

Quantitative description of the formation of the space-charge pattern under nonuniform illumination requires solution of the Maxwell equations, together with the current and the rate equations. Generation and recombination of conduction band electrons are described by $(\mathrm{d}N_\mathrm{e})/(\mathrm{d}t) = +SIc_{\mathrm{Fe}^{2+}} - \gamma c_{\mathrm{Fe}^{3+}} N_\mathrm{e}$, where N_e is the concentration of electrons in the conduction band, S is the photon absorption cross-section, I is the light intensity, $c_{\mathrm{Fe}^{2+}}$ and $c_{\mathrm{Fe}^{3+}}$ are the concentrations of Fe^{2+} and Fe^{3+} ions, and γ is the recombination coefficient. The general problem leads to a system of coupled nonlinear differential equations that cannot be solved analytically [10]. Reasonable approximations reveal that simple exponential evolutions govern build-up and decay of the space-charge field. Analysis yields for a fully modulated interference pattern (visibility one) the steady-state amplitude E_sc of the space-charge field [10]:

$$E_{sc} = (E_D^2 + E_{phv}^2)^{1/2} \qquad E_D = \frac{k_B T}{e} K \qquad E_{phv} = \frac{\beta\gamma}{e\mu S} c_{Fe^{3+}} , \tag{1}$$

where E_D is the diffusion field, E_{phv} is the photovoltaic field, k_B is the Boltzmann constant, T is the temperature, e is the elementary charge, K is a typical spatial frequency of the hologram (2π/period length of the interference pattern), β is the bulk photovoltaic coefficient (photovoltaic current density/$(c_{Fe^{2+}} I)$), and μ is the charge carrier mobility. The field E_{sc} might be reduced by space-charge limiting effects ($E_q < E_D$ or $E_q < E_{phv}$ with the space-charge limiting field $E_q = [e/(\epsilon\epsilon_0)](1/K)N_{eff}$, dielectric constant ϵ, permittivity of free space ϵ_0, and effective trap density N_{eff}).

The photovoltaic effect dominates in $LiNbO_3$ and $LiTaO_3$ crystals containing 0.01 wt.% or more Fe^{3+}, but diffusion is the major transport mechanism in all other materials. We have to distinguish between different geometries, e.g. transmission and 90° geometry (see Fig. 2). Typical spatial frequencies in these geometries are $2\pi/0.4\ \mu m^{-1}$ and $2\pi/0.15\ \mu m^{-1}$, respectively. These values will be used in the following if we refer to transmission or 90° geometry. The corresponding diffusion fields at room temperature are 4 kV/cm and 10 kV/cm. Photovoltaic fields in $LiNbO_3$ can reach values up to 100 kV/cm [8]. Larger fields cannot be created because electric breakdowns take place [20]. We will treat in the following the situation where the light is polarized perpendicular to the plane of incidence (ordinary polarization), if nothing else is mentioned. In-plane polarization yields recording beams of crossed polarization in the case of 90° geometry, and no intensity pattern is present. Extraordinary polarization can be used in the transmission geometry, and electro-optic coefficients, dynamic range and sensitivity might be higher, but extraordinarily polarized light tends to create holographic scattering [21], which is unacceptable for storage.

The refractive index is modulated via the linear electro-optic effect (see e.g. [22]) and the refractive index changes are

$$\Delta n = -\frac{1}{2} n^3 r E_{sc} , \tag{2}$$

where n is the refractive index and r is the electro-optic coefficient. Proper n and r values must be selected considering the material symmetry, the crystal

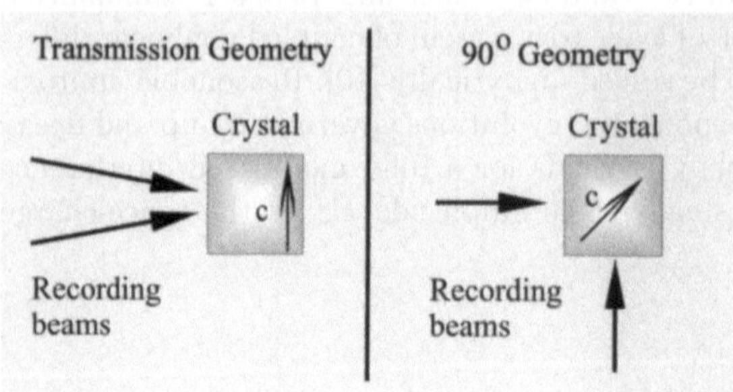

Fig. 2. Transmission and 90° recording geometries. The small arrows show the direction of the crystallographic **c** axis

cut, the polarization of light, and the direction of the space-charge field E_{sc}. For example, $LiNbO_3$ with ordinarily (o) or extraordinarily (e) polarized light and a field along the optical axis requires $n = n_{o,e}$ and $r = r_{113\,333}$.

2 Storage Properties: Dark Storage Time, Response Time, Capacity, Sensitivity

Useful figures of merit are required in order to allow comparison and evaluation of the materials in terms of storage performance.

The *dark decay* of the refractive index modulation follows typically a monoexponential function $\Delta n(t) = \Delta n(t=0)\exp(-t/\tau_{\text{dark}})$. Large dark storage times τ_{dark} are desired. The requirement depends on the application. Storage times of hours might be sufficient for RAM systems, a couple of years are fine for storage of files on a PC, but more than 100 years are required for archival storage to compete with microfilm.

Monoexponential growth of the space-charge fields is characterized by the *response time* τ, which depends on light intensity according to $\tau = \epsilon\epsilon_0/\sigma_{\text{ph}}(I)$, where σ_{ph} is the photoconductivity.

Subsequent superposition of several holograms requires a special exposure schedule, if holograms of equal efficiency are the target [23]. Suppose that Δn is the saturation value of the amplitude of a single hologram: the diffraction efficiency (ratio between the intensities of diffracted and incident light) of an individual of M multiplexed holograms will be $\eta = [(M\#)/M]^2$, if M is large enough to ensure that η is much smaller than 1. The $M\#$ is a useful measure for the *storage density* and *capacity* [24]. It is given by

$$M\# = \frac{\pi\,\Delta n\,d}{\lambda\cos\Theta}\exp(-1) \quad \text{for} \quad \alpha d = 2\,, \tag{3}$$

if the time constants for recording and erasure are equal. Here d is the thickness of the crystal, λ is the vacuum wavelength of the light, Θ is half the angle between the recording beams in the medium, and α is the intensity absorption coefficient. Build-up of Δn requires the presence of excitable electrons, which cause absorption. Considering this trade-off, the largest $M\#$ is achieved for $\alpha d = 2$. The photon budget determines the minimum efficiency η necessary to operate a storage system. Assuming, e.g., $\eta = 10^{-6}$, an $M\#$ of 10 indicates that 10 000 holograms can be multiplexed. However, the number of useful data pages can be reduced significantly owing to hologram cross-talk, light scattering, or homogeneous photovoltaic fields. The spatial resolution of the stored data pages is not limited by the photorefractive crystals. In the case of digital storage, the number of bits per page depends on the spatial light modulator and on the camera. For a 1000×1000 pixel-matched system and the parameters mentioned above we get a raw capacity of 10 Gbit. A crystal of about 1 cm thickness is required to make possible multiplexing of 10 000 holograms with low cross-talk, and beam diameters of about 1 mm are necessary to resolve 1000 × 1000 individual pixels. Using these numbers, we end up with a storage density of 1 Tbit/cm^3.

There are various ways of defining the *sensitivity.* A useful measure is $W_{\eta=1\%}$, the exposure density (laser-light intensity multiplied by the recording time) required to record a hologram of 1% diffraction efficiency. Recording and erasure of the refractive index changes Δn are governed by monoexponential functions [10]. Using Kogelnik's equation [25] for small diffraction efficiencies ($\eta \approx [(\pi \, \Delta n \, d)/(\lambda \cos \Theta)]^2 \exp(-\alpha d)$) we end up with

$$W_{\eta=1\%} = \frac{1}{10} \frac{\tau(I) \, I}{M\#} \quad \text{for} \quad \alpha d = 2 \, . \tag{4}$$

Small response times τ and large values of $M\#$ are required to achieve good sensitivities (small $W_{\eta=1\%}$ values).

3 Theoretical Performance Limits

There is no obvious theoretical limit to the *dark storage time*; it can reach very high values.

Crystals with thicknesses d up to 1 cm are available, and are supposed to have the highest *storage density.* Absorption is limited to $\alpha = 2/d = 2$ cm^{-1}, which determines the concentration $Fe^{2+} = 4.3 \times 10^{17}$ cm^{-3} [11]. Thus the largest possible E_{sc} (all electrons are redistributed) is $E_{sc} = E_q = [e/(\epsilon\epsilon_0)] \, (1/K) \, N_{eff} = 180$ kV/cm and 66 kV/cm for transmission and 90° geometry, respectively (ϵ, Table 1). A field of 180 kV/cm is above the breakdown threshold, but the $M\#$ describes the multiplexing properties, and the amplitude of each single hologram is well below the break-down field if many holograms are superimposed. We get $\Delta n_o = 13 \times 10^{-4}$ and $\Delta n_o = 5 \times 10^{-4}$ ((2), n_o and r_{113} from Table 1), which yields according to (3) the $M\#$ values 30 and 16 for transmission and 90° geometry respectively ($\lambda = 514$ nm, $\Theta = 45°$, $\alpha d = 2$). A similar estimate holds for $LiTaO_3$. However, in diffusion-controlled media the space-charge fields are lower, and limited by $E_{sc} = E_D$. The $M\#$ can be estimated using (1)–(3). $M\#$s range from 0.3 to 10 for different materials (Table 2).

Regarding *response time and sensitivity,* we have to consider the quantum limit, i.e. each incident photon moves in the optimum situation one electron to the desired position. Considering green light (photon energy 2.4 eV) and an energy difference of about 1 eV between the impurity level and the band, about 1.4 eV is left to move the electron against the space-charge field. Supposing a typical period length of 0.4 µm (transmission geometry), we have to move the electron about 0.2 µm against the field. Thus the energy of one photon is sufficient to move the electron against fields of up to 70 kV/cm. The required transport lengths are shorter in the 90° geometry, and the fields are even higher. From this perspective it is reasonable to treat the situation where each photon redistributes one electron. In this case we can replace $\tau(I)$

by $(\epsilon\epsilon_0/e)K(hc/\lambda)\,d\,E_{sc}(1/I)$, where h is the Planck constant and c is the vacuum speed of light, and get for $I = 1\ \mathrm{W/cm^2}$ and parameters of $LiNbO_3$ the result $\tau \approx 100$ ms (ϵ from Table 1, $K = 2\pi/0.4\ \mu\mathrm{m}^{-1}$, $\lambda = 514$ nm). This yields $W_{\eta=1\%} \approx 0.3\ \mathrm{mJ/cm^2}$.

4 Various Crystals

Table 1 summarizes important parameters, and Table 2 presents typical storage characteristics of various inorganic photorefractive crystals.

Storage performances of $LiNbO_3$ and $LiTaO_3$, of the perovskites $BaTiO_3$, $Ba_{0.77}Ca_{0.23}TiO_3$ (BCT), $KTa_{0.52}Nb_{0.48}O_3$ (KTN) and $KNbO_3$, of the tungsten–bronze structure $Sr_{0.61}Ba_{0.39}Nb_2O_6$ (SBN), of the sillenite $Bi_{12}TiO_{20}$ (BTO), and of the semiconductor GaAs are summarized. *Dark storage* and *response time* (Table 2) may vary by orders of magnitude for the same material depending on doping (elements, concentrations), annealing (oxidation, reduction), and experimental conditions (light intensity, crystal temperature, etc.). Thus the values shown are only typical numbers. It is possible to improve one value at the expense of another. For example, the response time of $KNbO_3$ can be decreased to a few ms by electrochemical reduction [55], but simultaneously the dark storage time is decreased by orders of magnitude.

The largest possible *storage densities* are calculated ((1)–(3), Table 1) considering photovoltaic charge transport for $LiNbO_3$ and $LiTaO_3$, and diffusion for the other materials. $LiNbO_3$ and $LiTaO_3$ have larger storage densities

Table 1. Refractive index n_o for light polarized perpendicular to the optical axis, if there is any, electro-optic coefficient r and dielectric coefficient ϵ for fields along the optical axis, if there is any, for different photorefractive crystals (light wavelength 514 nm; room temperature values)

Material	n_o			r (pm/V)			ϵ_{33}	
$LiNbO_3$	2.33		[26]	11	*1	[26]	28	[26]
$LiTaO_3$	2.21		[26]	8	*1	[26]	43	[26]
$BaTiO_3$	2.42		[27]	22	*	[28]	83	[28]
BCT [a]	2.47		[29]	36	*	[30]	230	[31]
KTN [b]	2.31		[32]	21	*	[33]	485	[32]
$KNbO_3$	2.33		[26]	40	*	[28]	35	[28]
SBN [c]	2.37		[34]	47	*	[35]	880	[26]
BTO [d]	2.65		[36]	5.75	+	[37]	52	[36]
GaAs	3.45	2	[38]	1.4	+2	[26]	12	[26]

[a] $Ba_{0.77}Ca_{0.23}TiO_3$,
[b] $KTa_{0.52}Nb_{0.48}O_3$,
[c] $Sr_{0.61}Ba_{0.39}Nb_2O_6$,
[d] $Bi_{12}TiO_{20}$
* r_{113}, + r_{231}, [1] 633 nm, [2] 1150 nm

Table 2. Various storage parameters for different photorefractive crystals (light wavelength 514 nm; room temperature values; Trans., transmission geometry; 90°, 90° geometry). The values for $M\#$ and $W_{\eta=1\%}$ are calculated from (1)–(4)

Material	Dark storage time		Response time (ms) @1 W/cm^2		$M\#$ Storage capacity Trans.	90°	$W_{\eta=1\%}$ Sensitivity (mJ/cm^2) Trans.	90°
$LiNbO_3$	1 a	[39]	3000	[40]	30	16	10	19
$LiTaO_3$	10 a	[39]	250 [1]	[39]	30	16	0.8	1.6
$BaTiO_3$	1 h – 1 a	[41]	500	[42]	1.5	5	33	10
BCT [a]	10 s	[43]	400	[43]	2.5	9	16	4
KTN [b]	0.6 a	[44]	200	[45]	1.2	4	16	5
$KNbO_3$	1 h – 30 d	[46,47]	100	[48]	2.4	8	4	1.3
SBN [c]	1 h – 30 d	[49,50]	100	[49]	3.0	10	3	1
BTO [d]	10 s	[51]	2 [2]	[52]	0.5	1.7	1.2	0.35
GaAs	1 ms	[53]	0.2 [3]	[54]	0.12	0.41	0.02	0.006

[a] $Ba_{0.77}Ca_{0.23}TiO_3$,
[b] $KTa_{0.52}Nb_{0.48}O_3$,
[c] $Sr_{0.61}Ba_{0.39}Nb_2O_6$,
[d] $Bi_{12}TiO_{20}$
[1] 351 nm, [2] 633 nm, 3 W/cm^2 and $Bi_{12}SiO_{20}$, [3] 1160 nm, 0.13 W/cm^2

in the transmission geometry because more charge per interference fringe is available to build-up the space-charge field. However, the 90° geometry is advantageous for diffusion recording, because higher spatial frequencies yield larger diffusion fields. Using the response time τ and the storage capacity already determined, the *sensitivity* is estimated based on (4).

The available storage capacity may be reduced significantly in many samples by several undesired effects. For example, light scattering reduces the visibility of the interference pattern, and therefore the fields and the dynamic range are also lowered. The experimentally required exposures $W_{\eta=1\%}$ [39,56] for recording in $LiNbO_3$ and $LiTaO_3$ are one order of magnitude larger than the values presented in Table 2. It is also known that the measured refractive index changes in, e.g., $BaTiO_3$ and KTN crystals are often about 4 to 5 times below the calculated values [33,57]. The numbers for storage capacity and sensitivity in Table 2 are optimum values, which may be achieved in carefully selected and optimized samples.

Table 2 shows clearly which material is the best for which application. Crystals of the lithium-niobate family ($LiNbO_3$ and $LiTaO_3$) are the favorites for long-term high-capacity storage. The performance of the rarely used $LiTaO_3$ exceeds that of $LiNbO_3$, but the drawback in this case is that ultraviolet light is required. Good availability, excellent homogeneity, and high robustness are further advantages of $LiNbO_3$ and $LiTaO_3$. Furthermore, the processes occuring in $LiNbO_3$ are well understood, e.g. the proportionality between the space-charge field and the concentration of Fe^{3+} ions (1) is verified [40]. The crystals can be tailored for special storage systems by optimization of the Fe concentration, by annealing and by additional doping. For example, doping with two deep independent levels which can be addressed with light of different wavelength allows optical fixing of the information [58]. Doping with large amounts of Mg increases the conductivity and improves the sensitivity [40]. More details about the specific properties of $LiNbO_3$ and other photorefractives can be found in a recent review [59].

The perovskites do not look very promising for storage, but the tungsten-bronze structure SBN and the sillenite BTO are useful for sensitive short-term low-capacity storage. Thus these materials are the best for, e.g., holographic double-exposure interferometry [60]. The properties of all sillenite-type crystals $Bi_{12}TiO_{20}$ (BTO), $Bi_{12}SiO_{20}$ (BSO), and $Bi_{12}GeO_{20}$ (BGO) are very similar with one exception: The optical activity of BTO is much smaller than that of BSO and BGO [36], which is of importance especially for thick samples. Parameters of the semiconductor GaAs are shown for completeness. Small dark storage times and small storage capacities make GaAs unattractive for holographic storage applications.

The sensitivities of the materials are orders of magnitude away from the quantum performance limits, and the low sensitivity is indeed the major drawback of inorganic photorefractives. A detailed study of the properties of cerium-doped SBN revealed the reasons [61], which apply for many photorefractive crystals. Only a few percent of the photons release electrons; most of the light energy is immediately transferred to heat. Trapping of electrons is very efficient because of Coulomb attraction, which limits the life-time and the transport lengths. It is unlikely that these problems can be completely overcome by a new inorganic photorefractive crystal.

Laser technology may solve the remaining problems of speed. The available output power of continuous-wave lasers increases rapidly. Pulsed systems become smaller and cheaper. Recording and reconstruction of holograms in photorefractives by, e.g., femtosecond light pulses has been demonstrated [62]. Another approach to solve the problem of insufficient sensitivity is the usage of multiple-quantum wells [63,64]. Electric space-charge fields yield large absorption changes, which modulate the refractive index due to the Kramers-Kronig relations. Multiple-quantum wells are fast, and there seems to be a lot of room for further improvements.

5 Nondestructive Readout

The retrieval of information stored in electro-optic crystals requires homogeneous illumination and thus leads to erasure effects. In many cases asymmetries in the recording and read-out processes (intensity, time, etc.) are not sufficient. For this reason several methods have been proposed to stabilize volume phase holograms versus the readout light.

By 1971 thermal fixing had already been discovered in $LiNbO_3$ [65]. By heating, a hologram resulting from electronic and ionic charges is formed, which is stable against homogeneous illumination. Protons play the major role as mobile ions in the thermal fixing process [20,66]. Electrical fixing has also been demonstrated [67,68]. An external electric field slightly smaller than the coercitive field has been applied to convert the space–charge pattern into a domain pattern.

To preserve the possibility of desired optical erasure, the use of two-photon excitation has been proposed for hologram recording [69]. Then readout without erasure is possible using reduced light intensity [70]. Similar methods that avoid large recording intensities are based on the population of shallow levels [71] or of photochromic centers [58].

There seems to exist a simple possibility for reading volume holograms nondestructively: the use of readout light of low photon energy, insufficient to excite charge carriers. However, information losses then arise because of the Bragg condition. A method has been proposed to improve the situation with the help of anisotropic diffraction [72]. Another solution to increase the bandwidth of the readout light has been developed [73,74], based on a spectrum of spatially adapted wave vectors. Finally, the use of frequency-difference holograms has been suggested and demonstrated for nondestructive readout [75,76]. The spatial frequency shift is achieved by nonlinear mixing of a hologram with a carrier frequency grating, but recording with two different wavelengths is required.

6 Conclusions

Good availability, excellent homogeneity, robustness, large dark storage times, large storage capacities, reversibility and the availability of efficient and convenient fixing techniques make $LiNbO_3$ and $LiTaO_3$ the favorites among the inorganic photorefractive materials for the application of holographic data storage. This is the reason why many demonstrators are based on this material. Tungsten–bronze- and sillenite-type crystals are more sensitive, but small dark storage times and low storage capacities limit their applicability to short-term, high-resolution storage of images, e.g. for the purpose of holographic interferometry. Knowledge about the charge transport nowadays enables tailoring of the crystals considering system issues such as the desired Bragg selectivity, the light wavelength and the light intensity. Concentrations and valence states of impurities may be adjusted by doping or thermal annealing. An obstacle of crystals might be the price. However, production of doped $LiNbO_3$ on a large scale has the potential to reduce the price per sample to a few dollars. Then inorganic photorefractive crystals are a good choice for the realization of holographic read–write memory sytems.

References

1. A. Ashkin, G.D. Boyd, J.M. Dziedzic, R.G. Smith, A.A. Ballman, J.J. Levinstein, and K. Nassau, Appl. Phys. Lett. **9**, 72 (1966).
2. F.S. Chen, J.T. LaMacchia, and D.B. Fraser, Appl. Phys. Lett. **13**, 223 (1968).
3. D.L. Staebler, W.J. Burke, W. Phillips, and J.J. Amodei, Appl. Phys. Lett. **26**, 182 (1975).
4. R. Orlowski and E. Krätzig, Solid State Commun. **27**, 1351 (1978).
5. G.E. Peterson, A.M. Glass, and T.J. Negran, Appl. Phys. Lett. **19**, 130 (1971).
6. J.J. Amodei, W. Phillips, and D.L. Staebler, Appl. Opt. **11**, 390 (1972).
7. P. Günter and J.-P. Huignard, eds., *Topics in Applied Physics: Photorefractive Materials and Their Applications I and II,* Topics Appl. Phys., Vols. 61 and 62 (Springer, Berlin, Heidelberg 1988, 1989).
8. A.M. Glass, D. von der Linde, and T.J. Negran, Appl. Phys. Lett. **25**, 233 (1974).
9. K. Buse, J. Opt. Soc. Am. B **10**, 1266 (1993).
10. N.V. Kukhtarev, V.B. Markov, S.G. Odoulov, M.S. Soskin, and V.L. Vinetskii, Ferroelectrics **22**, 949, 961 (1979).
11. H. Kurz, E. Krätzig, W. Keune, H. Engelmann, U. Gonser, B. Dischler, and A. Räuber, Appl. Phys. **12**, 355 (1977).
12. G.C. Valley, J. Appl. Phys. **59**, 3363 (1986).
13. F. P. Strohkendl, J.M.C. Jonathan, and R.W. Hellwarth, Opt. Lett. **11**, 312 (1986).
14. D.L. Staebler and W. Phillips, Appl. Phys. Lett. **24**, 268 (1974).
15. G.A. Brost, R.A. Motes, and J.R. Rotgé, J. Opt. Soc. Am. B **5**, 1879 (1988).
16. K. Buse and E. Krätzig, Appl. Phys. B **61**, 27 (1995).
17. L. Holtmann, Phys. Status Solidi A **113**, K89 (1989).
18. F. Jermann and J. Otten, J. Opt. Soc. Am. B **10**, 2085 (1993).
19. K. Buse, Appl. Phys. B **64**, 273 (1997).
20. K. Buse, S. Breer, K. Peithmann, S. Kapphan, M. Gao, and E. Krätzig, Phys. Rev. B **56**, 1225 (1997).
21. R. Magnusson and T.K. Gaylord, Appl. Opt. **13**, 1545 (1974).
22. J.F. Nye, *Physical Properties of Crystals.* Oxford University Press (London), 1979.
23. D. Psaltis, D. Brady, and K. Wagner, Appl. Opt. **27**, 1752 (1988).
24. F.H. Mok, G.W. Burr, and D. Psaltis, Opt. Lett. **21**, 896 (1996).
25. H. Kogelnik, Bell Syst. Tech. J. **48**, 2909 (1969).
26. *Landolt Börnstein — Numerical Data and Functional Relationships in Science and Technology, New Series,* Editor in Chief: O. Madelung, **III/11, III/16, III/18, III/28**. Springer Verlag (Berlin, Heidelberg, New York, London, Paris, Tokyo, HongKong).
27. K. Buse, S. Riehemann, S. Loheide, H. Hesse, F. Mersch, and E. Krätzig, Phys. Status Solidi A **135**, K87 (1993).
28. M. Zgonik, K. Nakagawa, and P. Günter, J. Opt. Soc. Am. B **12**, 1416 (1995).
29. M. Simon, F. Mersch, C. Kuper, S. Mendricks, S. Wevering, J. Imbrock, and E. Krätzig, Phys. Status Solidi A **159**, 559 (1997).
30. Ch. Kuper, K. Buse, U. van Stevendaal, M. Weber, T. Leidlo, H. Hesse, and E. Krätzig, Ferroelectrics **208**, 213 (1998).
31. Ch. Kuper, R. Pankrath, and H. Hesse, Appl. Phys. A **65**, 301 (1997).

32. S. Loheide, S. Riehemann, F. Mersch, R. Pankrath, and E. Krätzig, Phys. Status Solidi A **137**, 257 (1993).
33. S. Loheide, S. Riehemann, R. Pankrath, and E. Krätzig, Ferroelectrics, **160**, 213 (1994).
34. D. Kip, S. Aulkemeyer, K. Buse, F. Mersch, R. Pankrath, and E. Krätzig, Phys. Status Solidi A **154**, K5 (1996).
35. S. Ducharme, J. Feinberg, and R.R. Neurgaonkar, IEEE J. Quantum Electron. **23**, 2116 (1987).
36. F. Mersch, K. Buse, W. Sauf, H. Hesse, and E. Krätzig, Phys. Status Solidi A **140**, 273 (1993).
37. J.P. Wilde and L. Hesselink, J. Appl. Phys. **67**, 2245 (1990).
38. D.T.F. Marple, J. Appl. Phys. **35**, 1241 (1964).
39. E. Krätzig and R. Orlowski, Appl. Phys. **15**, 133 (1978).
40. R. Sommerfeldt, L. Holtmann, E. Krätzig, and B.C. Grabmaier, Phys. Status Solidi A **106**, 89 (1988).
41. G.D. Bacher, M.P. Chiao, G.J. Dunning, M.B. Klein, C.C. Nelson, and B.A. Wechsler, Opt. Lett. **21**, 18 (1996).
42. D. Rytz, M.B. Klein, R.A. Mullen, R.N. Schwartz, G.C. Valley, and B.A. Wechsler, Appl. Phys. Lett. **52**, 1759 (1988).
43. N. Korneev, D. Mayorga, S. Stepanov, H. Veenhuis, K. Buse, C. Kuper, H. Hesse, and E. Krätzig, Opt. Commun. **160**, 98 (1999).
44. L.A. Boatner, E. Krätzig, and R. Orlowski, Ferroelectrics **27**, 247 (1980).
45. S. Loheide, D. Sabbert, F. Mersch, H. Hesse, and E. Krätzig, Ferroelectrics **166**, 99 (1995).
46. R.J. Reeves, M.G. Jani, B. Jassemnejad, R.C. Powell, G.J. Mizell, and W. Fay, Phys. Rev. B **43**, 71 (1991).
47. M. Ewart, M. Ryf, C. Medrano, H. Wüest, M. Zgonik, and P. Günter, Opt. Lett. **22**, 781 (1997).
48. C. Medrano, M. Zgonik, N. Sonderer, Ch. Beyeler, S. Krucker, J. Seglins, H. Wüest, and P. Günter, J. Appl. Phys. **76**, 5640 (1994).
49. K. Buse, U. van Stevendaal, R. Pankrath, and E. Krätzig, J. Opt. Soc. Am. B **13**, 1461 (1996).
50. K. Megumi, H. Kozuka, M. Kobayashi, and Y. Furuhata, Appl. Phys. Lett. **30**, 631 (1977).
51. P. Tayebati and D. Mahgerefteh, J. Opt. Soc. Am. B **8**, 1053 (1991).
52. M.P. Petrov, I.A. Sokolov, S.I. Stepanov, and G.S. Trofimov, J. Appl. Phys. **68**, 2216 (1990).
53. Y. Fainman, J. Ma, and S.H. Lee, Materials Science Reports **9**, 53 (1993).
54. S. Bian and J. Frejlich, Opt. Lett. **19**, 1702 (1994).
55. D. Fluck, P. Amrhein, and P. Günter, J. Opt. Soc. Am. B **8**, 2196 (1991).
56. E. Krätzig and R.A. Rupp, SPIE **673**, 483 (1986).
57. U. van Stevendaal, K. Buse, H. Malz, H. Veenhuis, and E. Krätzig, J. Opt. Soc. Am. B **15**, 2868 (1998).
58. K. Buse, A. Adibi, and D. Psaltis, Nature **393**, 665 (1998).
59. K. Buse, Appl. Phys. B **64**, 391 (1997).
60. D. Dirksen, F. Matthes, S. Riehemann, and G. von Bally, Opt. Commun. **134**, 310 (1997).
61. K. Buse, A. Gerwens, S. Wevering, and E. Krätzig, J. Opt. Soc. Am. B **15**, 1674 (1998).

62. L.H. Acioli, M. Ulman, E.P. Ippen, J.G. Fujimoto, H. Kong, B.S. Chen, and M. Cronin-Golomb, Opt. Lett. **16**, 1984 (1991).
63. Q.N. Wang, D.D. Nolte, and M.R. Melloch, Appl. Phys. Lett. **59**, 256 (1991).
64. Q.N. Wang, R.M. Brubaker, and D.D. Nolte, J. Opt. Soc. Am. B **11**, 1773 (1994).
65. J.J. Amodei and D.L. Staebler, Appl. Phys. Lett. **18**, 540 (1971).
66. H. Vormann, G. Weber, S. Kapphan, and E. Krätzig, Solid State Commun. **40**, 543 (1981).
67. F. Micheron and G. Bismuth, Appl. Phys. Lett. **20**, 79 (1972).
68. J. Ma, T. Chang, J. Hong, R. Neurgaonkar, G. Barbastathis, and D. Psaltis, Opt. Lett. **22**, 1116 (1997).
69. D. von der Linde, A.M. Glass, and K.F. Rodgers, Appl. Phys. Lett. **25**, 155 (1974).
70. H. Vormann and E. Krätzig, Solid State Commun. **49**, 843 (1984).
71. K. Buse, L. Holtmann, and E. Krätzig, Opt. Commun. **85**, 183 (1991).
72. M.P. Petrov, S.I. Stepanov, and A.A. Kamshilin, Optics and Laser Technology 149 (1979).
73. H.C. Külich, Appl. Opt. **30**, 2850 (1991).
74. E. Chuang and D. Psaltis, Appl. Opt. **36**, 8445 (1997).
75. S. Fries, S. Bauschulte, E. Krätzig, K. Ringhofer, and Y. Yacoby, Opt. Commun. **84**, 251 (1991).
76. S. Fries, Appl. Phys. A **55**, 104 (1992).

Hologram Fixing and Nonvolatile Storage in Photorefractive Materials

S. S. Orlov and W. Phillips

Virtually all volume holographic recording materials require a fixing process to reduce or eliminate their sensitivity to light of the recording wavelength as they are being read out. This is true because illuminating the medium with light of the readout wavelength will ultimately either erase the recorded information, as in the case of photorefractive crystals, or deplete the dynamic range and cause spurious "noise gratings" that reduce the SNR to unacceptable levels, as in the case of photopolymers. This problem appears to be intrinsic to holographic media because the materials have to have a linear response in order to record superimposed (multiplexed) volume holograms with a minimum of scatter. Fixing is not an issue in established two-dimensional optical data storage materials (e.g. magneto-optic, phase change) because they use highly nonlinear thermal recording mechanisms, so that the information can be read nondestructively at lower powers.

In this chapter we will examine the most-often-used strategies for fixing in holographic storage media, as well as the effect that fixing has upon other properties of the media. Some of the issues addressed are: lifetime of the fixed holograms (optical, thermal), fixed/unfixed diffraction efficiency (fixing ratio), and increase in noise upon fixing.

Some other holographic materials, such as photopolymerizable material systems, e.g. photopolymers, are irreversible write-once materials. Recording of successive multiplexed holograms depletes the concentration of photoinitiator, dye and monomer, thereby reducing the remaining sensitivity. Nevertheless, reading out the recorded holograms leads to a reduction of diffraction efficiency and the build-up of scatter, leading to lower signal-to-noise ratios at the detector. Fixing is therefore necessary to eliminate remaining sensitivity and prevent degradation of data upon readout. Fixing approaches for photopolymerizable materials are based on using up any remaining photoactivity by heating or flooding with uniform incoherent light.

Index gratings in photorefractive crystals decay upon continuous readout (or under uniform illumination) because of the same mechanism that is responsible for their creation – electronic excitation and transport. Hologram fixing techniques for these materials involve reproduction of the original electronic pattern into the form in which the species responsible for the formation of the hologram are not optically active. Thermal ionic fixing has been demonstrated in a variety of photorefractive materials including $LiNbO_3$, $Bi_{12}SiO_{20}$, $KNbO_3$, $BaTiO_3$, and KLTN [1–4]. Fixing by thermally assisted ionic compensation and spontaneous polarization modulation will be discussed here. In addition we will discuss two-photon methods of nonvolatile recording.

1 Thermally Assisted Ionic Fixing

Fixing of holograms can be achieved by transforming the electronic space charge pattern into an optically inactive quasi-permanent ionic charge pattern. This was first demonstrated by Staebler *et al.* in $Fe:LiNbO_3$ in the early 1970s [1]. The identity of the ions primarily responsible for ionic conduction and fixing was, however, not determined until 1981 [5].

The ionic and electronic conductivity each obey the Arrhenius-type dependence on the temperature T:

$$\sigma_i = en_i\mu_0 \exp(-E_a/k_bT). \tag{1}$$

Efficient ionic fixing is based on the great disparity between the dark electronic conductivity at elevated temperatures and the ionic conductivity. At elevated temperature the ionic conductivity is dominant, and ions readily compensate for the holographic electric field pattern created by photoexcited electrons by mimicking their spatial distribution. At low temperature the ionic conductivity is very low, enabling the quasi-permanent storage of the ionic replica of the initial electronic hologram. The residual ionic conductivity at low temperature thus determines the lifetime of the *fixed* hologram. At elevated temperature ($T >\sim 70°C$ in lithium niobate) the ionic conductivity, prevails over the dark electronic conductivity and the ionic motion relaxes the electronic space charge. When the crystal is cooled back to room temperature and illuminated, the electronic grating is partially erased, leaving behind the ionic charge field, which now represents a fixed ionic hologram (Fig. 1).

The ionic hologram is quasi-stabilized owing to partial compensation by electronic grating [6,7]. This means that thermal redistribution of the ionic grating is opposed by the electronic space-charge field. The shelf lifetime of the fixed grating depends largely on the ionic impurity concentration and the degree of electronic compensation. The lifetimes of uncompensated ionic gratings can range from a few months (in as-grown $LiNbO_3$) to 2–3 years in dehydrated lithium niobate, while strongly compensated gratings with low reconstruction efficiency may possess lifetime of 10 years and more [7].

A typical history of a hologram is sketched in Fig. 2 [6,7]. In phase I an electronic grating, previously recorded, is heated up to cause ionic transport. This leads to a compensated grating that is represented by the net space-charge field amplitude $E_1^{(1)}$. In phase II the grating is left in the dark or is exposed to infrared light that is not photoactive. This stage corresponds to a slow dark decay of both the electronic grating and the ionic compensating grating that adiabatically follows the former. In many applications, however, the hologram is exposed to a "reading" light as in phase III. This light causes

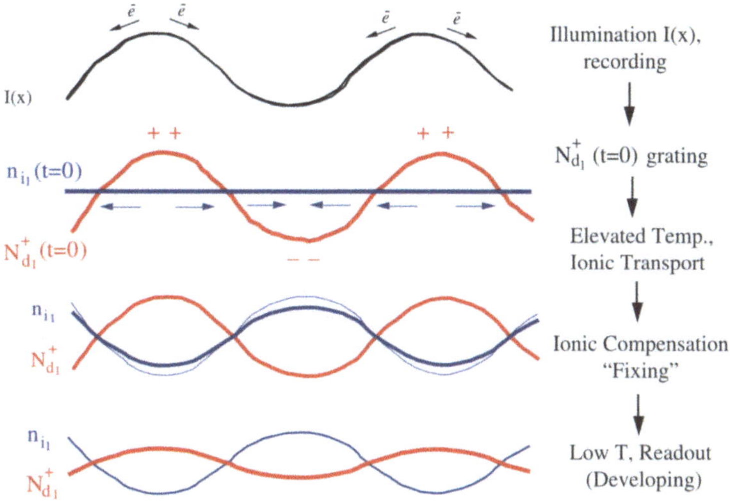

Fig. 1. Mechanism of hologram fixing via thermally assisted ionic drift

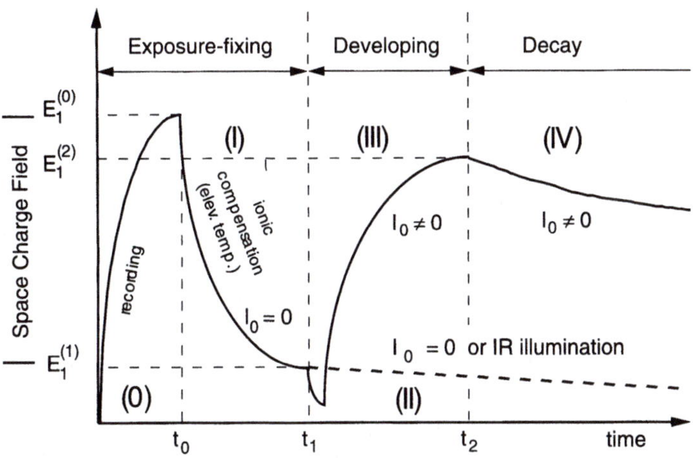

Fig. 2. A typical life history of a hologram in photorefractive materials

a partial redistribution of the trapped electrons, culminating in a quasi-stable field $E_1^{(2)}$. This field will proceed to decay to zero under illumination due to ionic transport (phase IV).

The problem of grating dynamics has been considered extensively [6–13]. Experimental data reproduced in Fig. 3 show the dark decay of holographic gratings recorded in Fe:$LiNbO_3$ at elevated temperature [7]. Two distinct stages of the process (fast and slow) can be identified as fast ionic compensation (phase I) and the much slower thermal decay of the electronic grating screened by the mobile ions (phase II) [6].

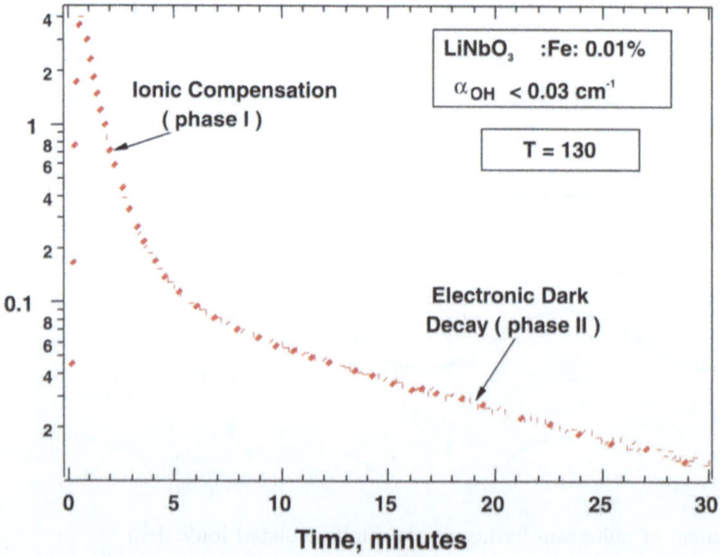

Fig. 3. Diffraction efficiency (in percent) vs. time for a hologram recorded and stored in Fe-doped $LiNbO_3$ (partially reduced) at 130°C. The grating spacing is $2\pi/K = 1.15$ µm. Two stages of the dark decay can be identified as ionic compensation (fast stage of the decay, phase I) and a much slower decay due to conduction by the thermally excited electrons (phase II)

Phase II consists of the developing (revealing) of the compensated grating by continuous illumination ($I_0 \neq 0$), typically at or near room temperature. This phase also describes what happens when we read the hologram. This causes an increase in the electronic conductivity, so that it greatly exceeds that of the ions ($\sigma_e \gg \sigma_i$). The mobile electrons diffuse and drift in the electric field induced by semi-permanent ionic charge distribution. Therefore, even under spatially homogeneous illumination I_0 the equilibrium trapped electrons distribution N_{d1}^+ is not uniform, but rather screens the "fixed" ionic pattern [6,7,9]. The degree of this electronic screening is greatly dependent on the amount of available traps N_e, the grating spatial frequency K, and the strength of the photovoltaic effect [6,7]. The treatment of the photovoltaic effect in this stage makes it necessary to distinguish two different types of boundary condition imposed on the electron current density and the total electric space-charge field [7] (i.e. open-circuit and short-circuit boundary conditions). Also, since the electronic conductivity is dominant in this phase the ions can be considered as stationary.

1.1 Hologram Fixing and Ionic Conduction in $LiNbO_3$

The hologram thermal ionic fixing in photorefractive lithium niobate has been studied extensively [1,5–7,10–14,25,16,17]. In general, it is agreed that above 70–80°C the ionic conductivity in this crystal largely prevails over the dark electronic conduction due to Fe^{2+} electron detrapping. At temperatures somewhat below 60°C, however, the conductivity is predominantly due to electrons, and is characterized by a small activation energy (0.1–0.4 eV) arising from small polaron electronic conduction associated with the Nb_{Li}^{4+} defect center [18]. This defect (i.e. Nb^{5+} ion occupying a Li^{+} crystal lattice cite) is due to the inherent Li-deficiency in congruent lithium niobate, and it plays an important role in considerations of the material dark conductivity at and near room temperature. In the presence of light, however, the photoconductivity is governed by Fe impurities in Fe-doped lithium niobate [19].

It is well established [5,20] that the ionic conductivity in as-grown and hydrogen-doped lithium niobate crystals is predominantly due to the H^{+} ions. Hydrogen is normally located in the oxygen planes along the O-O bond, and its relative content can be evaluated, therefore, as the strength (peak or integral) of the OH^{-} stretching vibration absorption line near 2.87 μm [21]. Vormann *et al.* [5] have shown unambiguously that in as-grown and H^{+}-doped crystals the ionic conductivity is due to the hydrogen, and deduced the activation energy of $E_a = 1.2$ eV for its migration. It was recognized, though, that in as-grown lithium niobate the residual ionic conductivity gives extrapolated lifetime of an ionic hologram of only 50–70 days, at room temperature, and, therefore, a substantial reduction of the ion density is essential to increase the available storage time [6,22]. The latter can be achieved via substantial dehydration of as-grown crystals [7].

1.2 Lifetime of Fixed Ionic Gratings

A hologram recorded at elevated temperature and left in the dark experiences fast ionic compensation (Fig. 2, phase (1)), which of course also takes place during recording if the ionic conductivity is larger than or comparable to the electronic photoconductivity. Fast ionic compensation is further followed by a much slower electronic grating decay arising from electron detrapping from Fe^{2+} sites (Fig. 2, phase (2)). The ionic compensation is very strong because of the relatively high density ($>10^{18}$ cm^{-3}) of compensating ions, and therefore the residual weak index grating may have both electro-optic (that due to the remaining uncompensated electric field) and photochromic contributions [23]. The dark electronic grating lifetime strongly depends on the grating spacing since it is primarily determined by diffusion rather than by conduction of the thermally induced electrons [6,7]. At the same time, ionic decay time (fast stage of the decay) is almost independent of the grating spacing. The slow decay can be enhanced by light, and therefore is indeed electronic. It is worth emphasizing that above ~90°C, even in heavily Fe-doped oxidized lithium niobate crystals, the ionic conductivity is higher than the dark electronic conduction (due to both thermal Fe-detrapping and the small polaron Nb_{Li}^{4+} shallow trap defect).

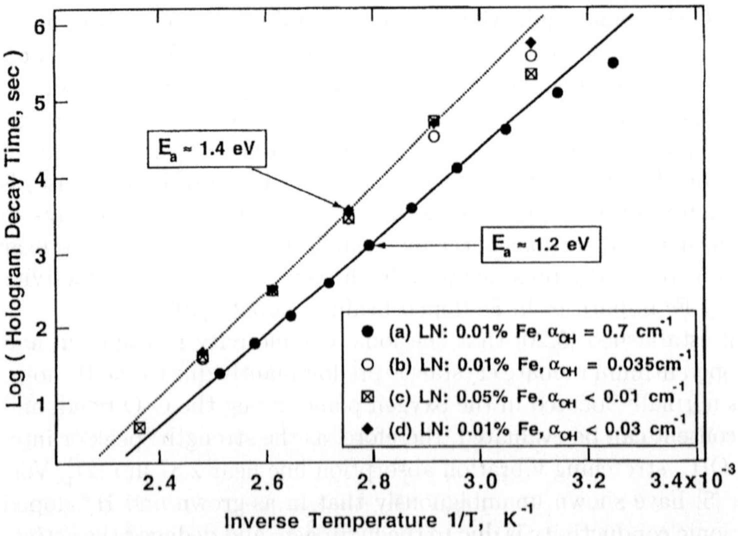

Fig. 4. Arrhenius plot of the ionic hologram lifetime in (**a**) as-grown ($E_a = 1.2$ eV) crystal, (**b**,**c**) samples with low hydrogen impurity content and different Fe doping ($E_a = 1.4$ eV), (**d**) Li_2O equilibrated (VTE processed) crystal. The fall-off in the hologram lifetime in the low temperature range (i.e. when $T < 70°C$) is due to the shallow trap (small polaron defect) electronic decay in nonstoichiometric lithium niobate

The lifetime of a fixed ionic grating is determined mainly by the conductivity σ_i of ionic species at given temperature T, i.e. where $\tau \propto \varepsilon\varepsilon_0/\sigma_i$ is given by (1). In order to determine the ionic conductivity and the corresponding activation energies we measure the dark decay time (Fig. 2, phase (1)) of the holograms recorded and stored at different (although constant throughout each experiment) temperatures. We find that in as-grown congruent lithium niobate crystals with fairly high hydrogen content (measured as the strength of the OH^- absorption near 2.87 μm) the ionic conductivity exhibits a ~1.2 eV activation energy and is relatively high (Fig. 4, sample (a)). This activation energy is characteristic for hydrogen conductivity [5,20]. The lifetime extrapolated to room temperature (24°C) is about 50 days, which is too short for the majority of applications [22]. Partial electronic screening under illumination slows down the ionic decay but at the same time decreases the grating strength [6,7]. Therefore, the most feasible mechanism for increasing the fixed hologram lifetime is the decrease in the ionic conductivity, which can be achieved by reducing the total density of conducting ions. High temperature (~950°C) post-growth processing in dry oxygen atmosphere results in a substantial reduction, up to 100 times compared with as-grown, of OH^- density in the crystal, and leads to a significant increase in the ionic hologram lifetime. However, we find (Fig. 4b,c) that after substantial hydrogen removal, the ionic conductivity is determined by a species other than hydrogen. The ionic conductivity in dehydrated crystals does not depend on the hydrogen density, whereas as-grown and hydrogen-doped samples [5,20] exhibit a linear dependence of σ_i on $[OH^-]$. The activation energy also differs from that typical for the H^+ impurity transport, and is equal to ~1.4 eV. The fixed ionic hologram lifetime in dehydrated samples at room temperature (24°C) can be extrapolated to be about 2 years [7].

In order to investigate the influence of lithium vacancies on the ionic conductivity [24], a nearly stoichiometric sample was prepared using the vapor transport equilibration (VTE) technique [25,26]. The processing temperature was 1080°C, and the processing time was 400 hours for a 1-mm-thick *a*-cut lithium niobate crystal. The Li_2O content in this sample was estimated to be at least 49.5% (congruent lithium niobate contains 48.6% Li_2O [27]) from the phase-matching angle measurements for the second harmonic generation at 1.064 μm [28]. In the temperature range studied in our experiments the ionic conductivity in the VTE sample also exhibits a ~1.4 eV activation energy (Fig. 4, sample (d)) and is nearly the same as that of congruent dehydrated lithium niobate. This may rule out mechanisms of ionic conduction that involve a substantial role for lithium vacancies/lithium deficiency. Migration of lithium self-interstitials [26] may be a possible mechanism responsible for the ionic conductivity with activation energy of 1.4 eV.

It should be noted that the ionic conductivity with the 1.4 eV activation energy *and* the substantially increased ionic hologram lifetimes at room temperature can be found only in oxidized lithium niobate crystals with very low hydrogen impurity content ($[OH^-] < 5 \times 10^{17}$ cm^{-3}, that is for OH^- absorption at 2.87 μm less than 0.05 cm^{-1}). In samples with higher hydrogen contents the dark decay time constant may exhibit activation energies ranging widely from 0.9 eV to 1.4 eV, depending on the density of H^+ impurity, the range of the temperatures used in the experiments, and the oxidation state and density of the iron dopant [5,15–17,29].

To summarize, holographic gratings based on charge redistribution inevitably decay because of ionic and electronic conduction. Ionic gratings are partially screened by trapped electrons upon readout. The lifetimes of *fixed* ionic holograms are limited by the finite ionic conductivity at low (i.e. room) temperature. A significant increase in fixed ionic hologram lifetime is realized in lithium niobate with low hydrogen impurity content. The residual ionic conductivity (decay time constant) in these samples exhibits ~1.4 eV activation energy and is *not* due to protonic conduction. Fixed hologram lifetimes of about 2 years at room temperature are projected in dehydrated lithium niobate crystals. A few main conclusions are worth emphasizing:

(a) At elevated temperature ($T >$ ~60°C for lithium niobate) the initial decay of the recorded grating is due to ionic compensation (phase I). Under dark storage conditions, the (phase II) hologram decay is due to the thermal electronic transport slowed down by ionic screening.

(b) A near-perfect development and fixing (i.e. $E_1(t_2) \approx E_1(t_0)$ in Fig. 2) can be obtained provided that the screening of the residual ionic grating by electronic grating is weak upon readout. This happens in weakly iron-doped oxidized lithium niobate crystals with strong photovoltaic response. The fixing efficiency can be close to 100% in such crystals.

(c) Under illumination (phase IV) the hologram is quasi-stabilized. The residual decay is due to the transport of the ionic charge backbone. Major reduction of the ionic conductivity at the operating temperature is necessary to bring this decay rate to the acceptable range of lifetimes = 10 years.

(d) Substantial increase in the fixed (ionic hologram) lifetime can be realized in lithium niobate crystals with low hydrogen content. The residual ionic conductivity (and associated hologram decay time constant) exhibits a ~1.4 eV activation energy, and is not due to protonic conduction. This conductivity mechanism can be attributed to migration of lithium interstitials. Ionic hologram decay time is independent of material stoichiometry, and can be projected to be about 2 years in dehydrated crystals. Further increase of the fixed hologram lifetime can be achieved using strong electronic compensation in strongly Fe-doped crystals, but only at the expense of the reconstruction efficiency.

1.3 High-Low Fixing

It is also feasible to write holograms into a $LiNbO_3$ crystal while it is held at an elevated temperature, allowing ionic compensation to occur simultaneously with hologram recording. The advantages of real-time ionic compensation (the so-called "high-low" fixing technique) as compared with low-high-low fixing are:

- high asymmetry in read/write time constants, which increases the effective dynamic range of the medium;
- supression of photoinduced scattering and other self-diffraction effects;
- no photovoltaic damage during the recording stage.

We performed a number of experiments on holographic recording of digital data in Fe-doped $LiNbO_3$ crystal (0.03 wt.%) in transmission geometry using high-low fixing. Holograms were stored at high temperature ($T = 190°C$) in order to provide a real-time charge compensation of photoinduced electronic grating by mobile ions for subsequent nonvolatile readout of nonphotoactive ionic holograms at room temperature. The stored holograms were read out in a lens-free phase-conjugate geometry. The experiments were carried out with a 400 mW frequency-doubled YAG laser (532 nm). The light in the signal arm is spatially filtered and expanded to form a collimated beam with a diameter of 1.5 cm. The beam is reflected by a 480 × 480 pixel portion of the TI DMD micromirror spatial light modulator (SLM), and focused by a lens placed right in front of the SLM ($1/f = 2 \times 1/6$ cm $= 1/3$ cm). The Fe-doped LiNbO3 crystal (0.03 wt.%) with dimensions $10 \times 20 \times 2.5$ mm^3 (the last dimension corresponds to the crystal thickness) is placed 9 mm behind a Fourier plane (spot diameter 3 mm) in order to avoid a large-intensity spike in the center of the image-bearing spot. The collimated reference beam has a diameter of 4 mm. The average angle between the reference and the signal beam is 45° in air.

The crystal is heated up to 190°C, and holograms are sequentially recorded using two types of angular multiplexing techniques: rotation of the crystal (50 steps with 0.23° separation between subsequent holograms) and tilt of the reference beam (10 steps with 0.12° separation). The exposure time is 2 s for each hologram. After recording, the crystal is cooled down to room temperature, and the compensated holograms are revealed by illuminating with intense white light (2.5 W/cm^2) for 25 min. The crystal is then flipped around (180° rotation) to provide a phase-conjugate readout of the revealed holograms. The readout is performed with a Kodak CCD camera with 1000×1000 pixels of 9×9 μm^2.

In summary, the experiment results on high-low recording are as follows:

- All holograms show similar diffraction efficiency ($\eta = 2 \times 10^{-4}$) without any scheduling of the recording time. This means that the effective dynamic range of the crystal is much larger than that for the "low-high-low" fixing technique in the 90° geometry. This high diffraction efficiency permits the simultaneously reduction of readout power (40 mW) and camera integration time (1 ms).
- No substantial photovoltaic damage is observed.
- There is no photoinduced scattering (although intrinsic surface and volume scattering is higher than for the 90° recording geometry).
- Resolution of the holograms is high enough to resolve 1×1 features, but because of the pixel size mismatch between SLM and CCD the image quality is not very good.
- Holograms do not show any decay or damage after 14 hours of exposure to an intense white light (2.5 W/cm^2).

A critical limitation of storage density in holographic storage media arises from the effective dynamic range of the material. When M pages are stored, the diffraction efficiency of each page often obeys the relationship [30]

$$\eta = (M\#/M)^2, \tag{2}$$

where $M\#$ is a proportionality constant characteristic of the material and optical configuration. A high $M\#$ allows the diffraction efficiency and the signal to remain high. Typical numbers realized with $LiNbO_3$ are $M\# = 5$, $M = 1000$, and $\eta = 5 \times 10^{-6}$. In the high-low fixing method, the situation is dramatically improved. First, (2) is not directly applicable in the case of recording at elevated temperatures. Indeed, using the high-low fixing procedure for $LiNbO_3$ in the transmission geometry, the $M\#$ is found to increase in proportion to M. (Stated another way, the exponent of M is closer to 1 than 2.) In early work at RCA Laboratories [23], an $M\#$ of 100 was observed when 100 holograms were stored, and an $M\#$ of about 500 was observed when 500 holograms were stored. These holograms were recorded to extremely high diffraction efficiency (>1%).

The $M\#$ is huge because reillumination erases previously recorded holograms much more slowly at elevated temperatures. The simplified physical reason is that the ionic compensation removes the huge coulombic repulsion between electrons in the stored grating. The dominant decay mechanism is then the relatively slow process of diffusion.

We have stored 500 holograms in the transmission geometry. The erasure of holograms was negligible. Uniform exposure time results in uniform diffraction efficiency. Because the gratings do not decay when additional holograms are recorded on top of prior holograms, the effective dynamic range is so large that dynamic range is no longer a factor that limits storage density.

In the experiments described above, using a 400 mW 532 nm laser, typical times to write a single hologram that could be read with a BER of $\sim 10^{-3}$ or better were 1–2 s. Corresponding recording transfer rates for a 1000×1000 page size are 0.5 to 1.0 mpel/s. Actual data rates are reduced by the modulation and error correction code efficiencies. Net transfer rates for recording are therefore in the range of 128 Kbytes/s or slower.

The low write rates in the transmission geometry create an additional problem. Writing in the crystal at elevated temperature is accompanied by slow thermal erasure of previously recorded holograms over timescales of roughly 1 h. This limits the length of a recording session for a heated block of recording medium. Using the write rate projection of the previous paragraph, each individually heated block would have a storage capacity of 450 Mbytes. Improved record rates could be achieved by reducing the amount of scattered light reaching the detector, leading to reduction of the diffraction efficiency required for low BER, and by increasing the laser power.

2 Fixing by Spontaneous Polarization Modulation

Fixing of photoinduced gratings can also be achieved by means of partial polarization reversal in regions of high depoling space charge fields. This was accomplished both in ferroelectrics with "diffused" phase transition, such as $Sr_{0.75}Ba_{0.25}Nb_2O_6$ [31], and in "classical" ferroelectrics with a sharp phase transition point, such as $BaTiO_3$ [32]. Fixing typically involves application of an external depoling field below the coercive field for bulk ferroelectric switching (Fig. 5). The microscopic mechanisms responsible for polarization switching [33,34] include antiparallel microdomain nucleation and their subsequent lateral and sidewise expansion. The domain nucleation probability $p(x)$, which is given by [33]

$$p(x) \propto \exp[-\alpha/E(x)], \tag{3}$$

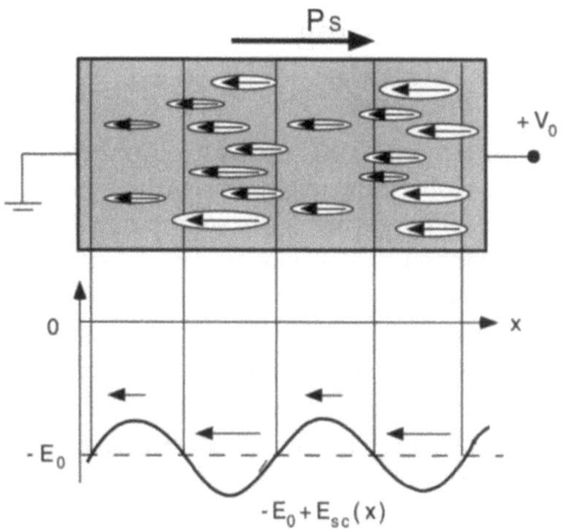

Fig. 5. Ferroelectric domain fixing schematic. Antiparallel microdomains are nucleated most intensely in the regions of most negative electric field. They continue to grow, leading both to a partial compensation of the hologram space charge and to depoling of the crystal (reduction of effective electro-optic coefficient). An external depoling electric field can be applied (during or after the recording) in order to enhance microdomain nucleation and growth. Microdomains distribution spatially correlated with space charge field constitutes the fixed polarization hologram

is maximal in the regions of large depoling fields ($E(x) = E_{\mathrm{sc}}(x) + E_0$; α in (3) is a temperature-dependent parameter). The antiparallel domain nucleation can be largely enhanced by an externally applied voltage E_0 that negatively biases the internal space charge (Fig. 4). The nucleated microdomains typically have a spike-shape form oriented along the c-axis, because such geometry decreases the depolarization energy [34]. The nucleation is followed by domain growth. The latter leads both to stronger compensation of the electronic space charge and to partial depoling of the crystal as a whole. The microdomains formation is correlated with the spatial profile of the internal space charge field through both the nucleation probability (3) and the growth rate, which also depends on the local field amplitude [33]. Spatial variation of the spontaneous polarization P_{s} is, therefore, also correlated with the space charge field $E_{\mathrm{sc}}(x)$. The first fundamental Fourier harmonic P_{s_1} of the perturbed spontaneous polarization represents the fixed polarization hologram. The magnitude of the space charge variation, ρ_1, of the polarization grating is given by $\rho_1 = -\nabla P_{\mathrm{s}_1}$. The correlated portion of the polarization variation is usually quite small, typically of the order of 10^{-3} to 10^{-4} compared with the total P_{s} [9]. Similarly to the case of the fixed ionic grating, upon illumination the P_{s_1} grating is compensated by dynamic photorefractive space charge [9].

3 Two-Photon Holographic Recording in Stoichiometric Lithium Niobate

An alternative procedure to thermal or domain fixing, optically gated recording focuses on using an additional light source during the recording process that is not present during readout [35]. The gating light source can be incandescent, and does not have to be coherent. The material is sensitive to the writing wavelength only in the presence of gating or bias light, while it is insensitive to readout light by itself. This optically induced light sensitivity is potentially a very attractive mechanism for data recording and readout; it allows long-term storage and virtually nondestructive readout. Additional advantages of the two-photon recording scheme include low absorption of the reading light (which produces stronger signals for the same index modulation) and virtual absence of optical damage during hologram readout.

In early work, pulsed lasers were used to achieve large intensities to take advantage of optical nonlinearities. Transition metals were doped into a variety of host materials, including $LiNbO_3$, KTN, and $LiTaO_3$. Experiments showed that by using a fundamental harmonic (1.064 μm) of a YAG laser for writing the data and its second harmonic (532 μm) for gating, long erasure times could be obtained, but the required near-IR intensities were on the order of 10^8 W/cm^2 [35]. Alternatively, Mn- and Fe-codoped photochromic congruent $LiNbO_3$ and rare-earth-doped (e.g. Pr) $LiNbO_3$ and SBN have been used [36–38]. In recent years, substantial advances in two-photon recording with near-infrared CW lasers have been made using near-stoichiometric doped and undoped lithium niobate materials [39–41]. The key idea for optimization was to modify the host composition and the reduction state of the crystalline material [39,40]. It has been demonstrated that even for moderate cw intensities of the recording light (5 W/cm^2 or less) the two-photon sensitivities of near-stoichiometric-reduced lithium niobate can be as high as the nongated recording sensitivity of Fe-doped $LiNbO_3$ materials with green light [39]. Digital hologram storage has also been successfully demonstrated by using these improved materials [41].

To characterize the sensitivity of a medium, several measures have been developed that generally relate the measured diffraction efficiency or the index of refraction change under illumination to the incident light intensity. The recording photosensitivity is typically defined as follows:

$$S(I_{\mathrm{g}}) = \frac{\partial\sqrt{\eta}}{\partial t}\,\frac{1}{I_{\mathrm{w}}L}, \tag{4}$$

where η is the diffraction efficiency, L is the thickness of the medium, I_{w} is the recording intensity, and I_{g} is the gating intensity. Typical sensitivity of Fe-doped $LiNbO_3$ in the green is from 0.02 to 0.10 cm/J, depending on the total doping and the reduction state of Fe-dopant.

3.1 Undoped Stoichiometric Lithium Niobate

In order to achieve optically gated recording, there must be present: (i) deep traps that are partially filled with electrons to generate excited carriers during gating and recording of the hologram; and (ii) shallow levels to trap photogenerated electrons, with sufficiently long lifetimes (hundreds of milliseconds) to absorb light at the writing wavelength. Efficient two-photon gated recording can be realized using intrinsic point defects of lithium niobate. Lithium niobate crystals are in general not stoichiometric, as the chemical formula is given by $Li_{1-x}Nb_{1+x/5}O_3$ (and $x = 0.028$ for congruently melting material). Intrinsic point defects include Nb_{Li} (that is, a niobium ion on the Li-site) antisite defects: $Nb_{Li}V_{Nb}$ and $Nb_{Li}Nb_{Nb}$ [18]. The latter represents a two-electron trap, or, when filled with a pair of electrons with opposite spins, a diamagnetic bipolaron $Nb^{4+}Nb^{4+}$. Illumination with blue-green light dissociates the bipolarons, and these electrons enter the conduction band and become self-trapped on the Nb lattice sites, forming Nb^{4+} small polarons (Fig. 6) [42]. Small polaron levels can be viewed as shallow traps with a limited electron lifetime. For small polarons, the absorption due to optical excitation (hopping between different sites) is very broadband, and peaks in the near-infrared region of the spectrum at about 780 nm, which is twice the binding energy of polarons (0.8 eV). Photoexcitation by recording with near-infrared light and subsequent carrier redistribution and migration creates the index hologram. The lifetime of the excited Nb^{4+} states depends on the density, the capture cross-section, and the reduction state of the deep defects, as well as the temperature, and the intensity of the recording IR light. When the gating light is turned off, the carriers eventually become trapped in the deep traps, and the material becomes essentially transparent to near-IR readout radiation.

Optimization of the undoped crystalline material can be achieved through changing its stoichiometry and the reduction state [39,40] (Fig. 7). The gated photosensitivity of the material to near-IR light is proportional to the density

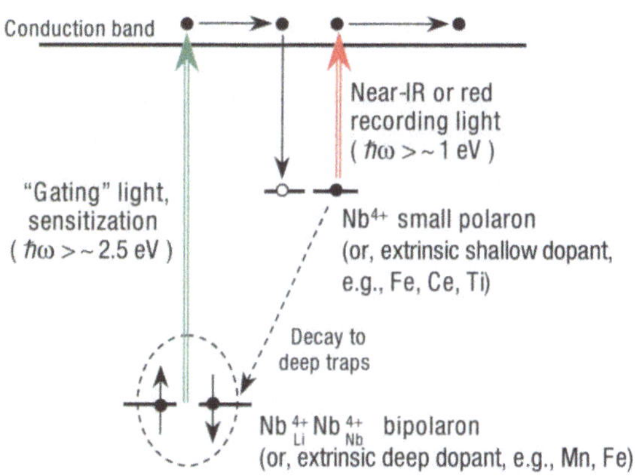

Fig. 6. Mechanism of optically gated recording in lithium niobate

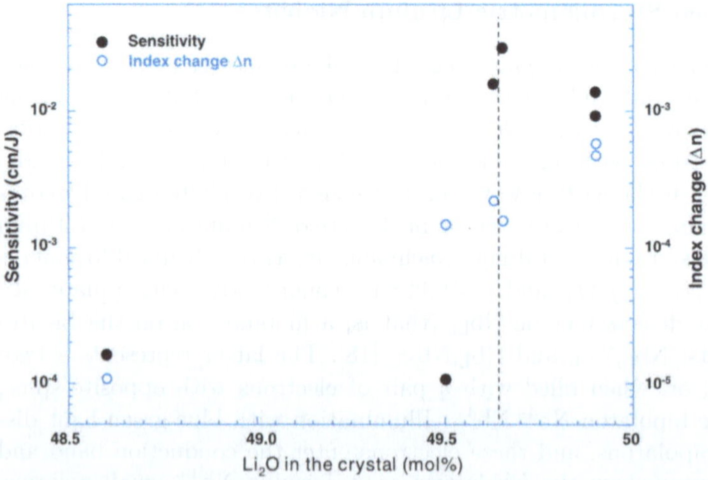

Fig. 7. Gated photosensitivity and index change versus material composition (Li content). All crystals are partially reduced. Gating intensity is 0.25 W/cm^2 at 457.9 nm. Recording intensity 10 W/cm^2 at 800 nm

of excited small polarons, and therefore is proportional to their recombination lifetime. High defect densities, as occur in congruent lithium niobate (48.6 mol% Li_2O), lead to substantial lattice disorder and fast non-radiative, phonon-assisted decay of Nb^{4+} polaron levels into deep traps and their recombination to bound bipolarons [18,43]. As the material becomes closer to stoichiometric and the defect density decreases, the two-photon sensitivity S exhibits a sharp, two order of magnitude increase with increased Li/Nb ratio up to 49.6 mol% and then S saturates (Fig. 7). The lifetime of the upper Nb^{4+} levels in lithium niobate at room temperature changes from <40 ms in as-grown congruent composition [36] to several seconds in lightly reduced near-stoichiometric (with $[Li_2O] = 49.65$ mol%) lithium niobate (Fig. 8).

Post-growth reduction of near-stoichiometric crystals produces more filled bipolaron sites, which provide efficient absorption for the gating light of the blue-green region of the spectrum. Also, reduction converts the contaminant acceptor ions into donors (for example, Fe^{3+} into Fe^{2+}), which decreases the small polaron capture cross-section and increases their lifetime and, therefore, the material sensitivity under similar gating conditions. The lifetime of small polarons in optimally reduced crystals can be as long as several seconds, and depends on the degree of reduction. More excessive reduction, however, leads to very long polaron lifetimes (tens of seconds) that adversely affect recording, in that the holograms exhibit a substantial dark decay right after writing.

Reduction of lithium niobate at high temperatures (900–1000°C) involves the migration of lithium vacancies from the bulk to the surface of the material (with corresponding back-transport of free electrons for charge compen-

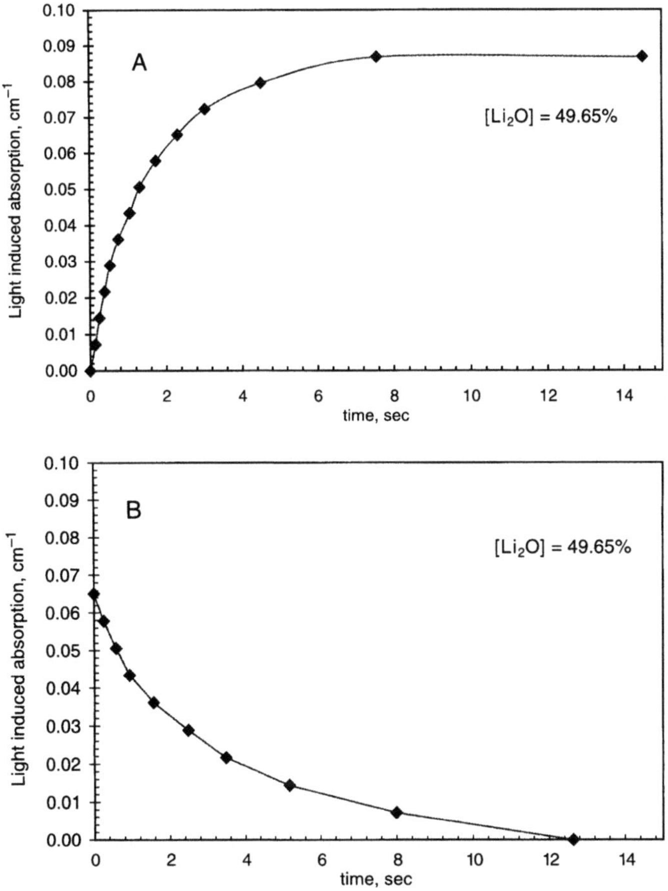

Fig. 8. Build-up (A) and dark decay (B) of light-induced absorption (at $\lambda_{\mathrm{probe}} =$ 670 nm) in stoichiometric lightly reduced lithium niobate sample. Gating intensity is 0.5 W/cm^2 at 532 nm

sation). Surface reaction at processing temperature involves a consumption of nonstoichiometric cells and generation of the free electrons [24]. At lower temperatures free electrons are trapped on the defect sites, forming bipolarons, while local charge compensation is still maintained by a change of the density of lithium vacancies:

$$\left(\mathrm{Nb}_{\mathrm{Li}}^{5+}\mathrm{Nb}_{\mathrm{Nb}}^{5+})^{4+}\right) + 2\bar{e} \rightarrow \left(\mathrm{Nb}_{\mathrm{Li}}^{4+}\mathrm{Nb}_{\mathrm{Nb}}^{4+}\right)^{2+} . \tag{5}$$

Reduction of crystal samples was performed in a high-temperature tube furnace in a flow of dry ultra-high-purity argon gas. The oxygen content during processing was under 10 ppm and was limited by the purity of the argon gas used. Optimum reduction steps for undoped near-stoichiometric material usually require a processing temperature of about 900°C for duration of 15 to

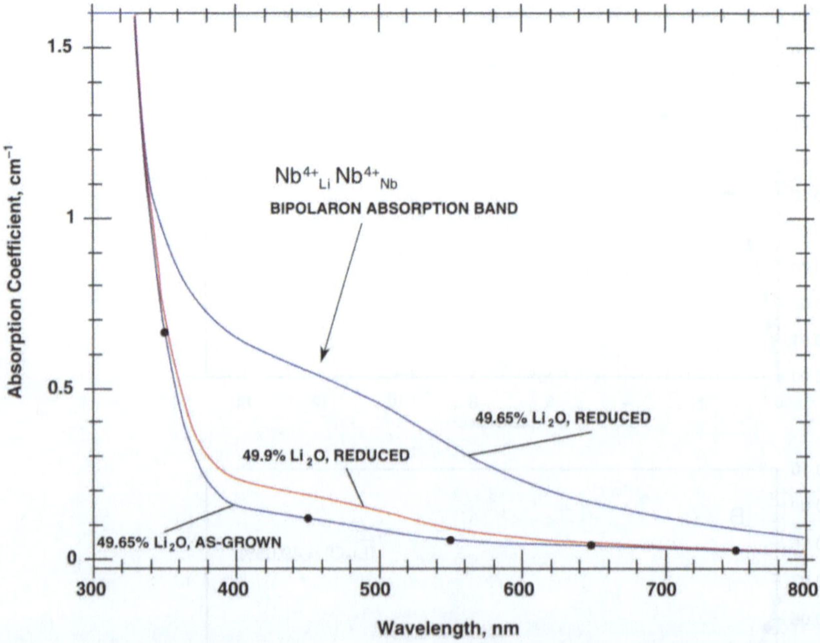

Fig. 9. Bipolaron absorption induced by reduction of lithium niobate

30 min (in dry argon). Crystals nearer the stoichiometric compositions (e.g. 49.9 mol% Li_2O) require lighter reducing conditions.

A substantial increase in the blue-green absorption is seen in the reduced materials (Fig. 9). After the reduction the two-photon photosensitivity of near-stoichiometric material is about 20 times higher than in as-grown materials [39,40]. For undoped materials the sensitivity saturates (or, even decreases) for near-stoichiometric compositions with $[Li_2O] > 49.65$ mol%. Since the density of the defects ($Nb_{Li}Nb_{Nb}$) decreases as the composition becomes closer to exactly stoichiometric, the absorption of the gating light (and hence two-photon sensitivity) becomes somewhat smaller for more stoichiometric reduced crystals (see Figs. 7 and 9). The saturation index change (i.e. dynamic range), however, continues to grow with decreasing defects density.

Sensitivity measurements were performed in transmission recording geometry (30° between the writing IR beams in air) with extraordinarily polarized infrared 800 nm Ti:sapphire laser light for recording and an expanded argon ion laser beam for gating. For a constant near-infrared writing intensity the dependence of the two-photon photosensitivity S on the gating intensity I_g (Fig. 10) can be approximated as follows:

$$S(I_g) = S_{max} \frac{I_g}{I_g + I_{S,1/2}}, \tag{6}$$

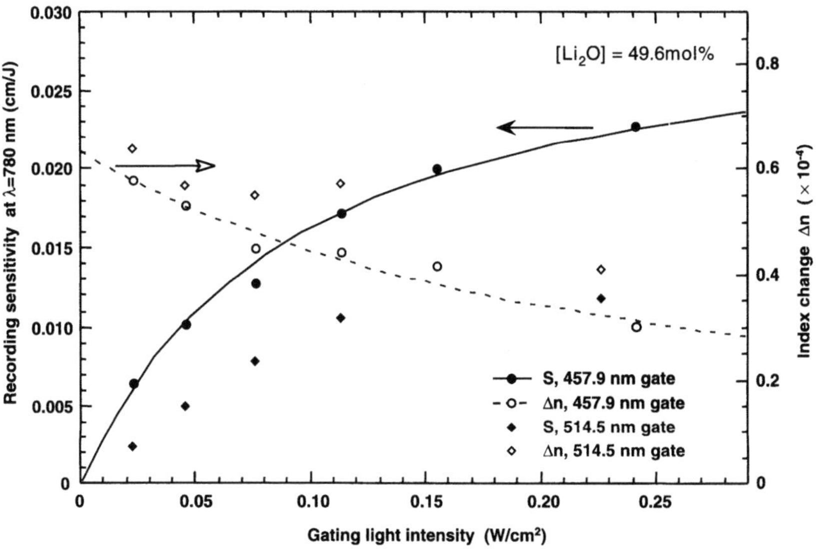

Fig. 10. Two-photon photosensitivity and saturation index change versus 457.9 nm and 514.5 nm gating light intensity in reduced lithium niobate crystal. Writing intensity is 10 W/cm^2

where $I_{S,1/2}$ is the gating intensity at which the sensitivity attains a value of 1/2 of that at saturation and $S_{\max}$ is the saturation sensitivity for high gating powers. The saturation in infrared sensitivity is attributed to the exhaustion of the bipolarons at high gating intensities. Increased gating intensity also leads to erasure and reduction of the fringe visibility of the grating recorded with near-infrared light. As a result the saturation index change Δn (and corresponding $M\#$, as they are linearly related to each other) falls off for high gating powers:

$$\Delta n(I_{\rm g}) = \Delta n_{\max} \frac{I_{\Delta n,1/2}}{I_{\rm g} + I_{\Delta n,1/2}} \tag{7}$$

where $I_{\Delta n,1/2}$ is the value of the gating intensity at which holographic index change Δn decreases by a factor 2 of its maximum value, which occurs at a zero gating intensity. It is important to mention that $I_{S,1/2}$ and $I_{\Delta n,1/2}$ are not necessarily equal, and both depend strongly on gating light wavelength.

Both sensitivity and the index change are enhanced in crystals with increased bipolaron concentration (Fig. 7). In the presence of the gating light, the induced absorption at the recording near-IR wavelength (800 nm) is of the order of 0.1 cm^{-1} (Fig. 8). Without gating, the erasure upon readout and IR one-photon sensitivity is determined essentially by the absorption tail of deep impurities in the near-IR. The sensitivity of the material is at least 50 to 1000 times less, which allows readout of the hologram thousands of times without substantial erasure and optically induced index damage. Dark decay of the hologram is determined by thermal excitation of electrons from deep bipolaron sites into the conduction band. Depending on the reduction state of the material, the room temperature lifetime can vary from several weeks to several months.

3.2 Doped Stoichiometric Lithium Niobate

Extrinsic dopants may play essentially the same role as deep bipolarons provided that their energy levels are sufficiently deep in the forbidden gap of the crystal (Fig. 11). Normally, the ionization absorption for dopants in $LiNbO_3$ is quite broadband, and therefore, in order to achieve high gating efficiency, it is important to separate the absorption peaks for deep donors and metastable polarons (~780 nm) as far as possible. Mn^{2+}, Fe^{2+}, and Ce^{3+} are appropriate deep donors because their energy levels lie below that of the bipolarons in the forbidden band gap [18], and therefore one should expect a smaller photoionization cross-section for near-IR light for these donors compared bipolarons. Weakly doped crystals exhibit higher gating efficiency of the two-photon recording process and, therefore, longer hologram lifetime upon continuous readout [40].

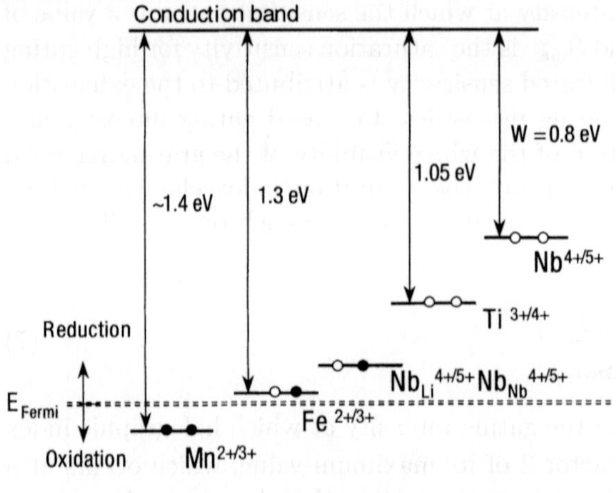

Fig. 11. Energy position of different impurities and intrinsic defect centers in $LiNbO_3$

Reduction converts the majority of the extrinsic acceptors (such as Fe^{3+} or Mn^{3+} ions) into donors (Fe^{2+} and Mn^{2+}). For 0.02 mol% Mn-doped samples, at least 90–95% of the Mn ions need to be converted (via reduction) into the Mn^{2+} valence state before any appreciable two-photon response can be measured. The optimum reduction for doped crystals requires stronger reducing conditions compared with undoped crystals. Typical processing conditions involve temperatures of about 950–1000°C and duration of about 30 min in a flow of dry pure argon. Similarly to undoped material, stronger reduction

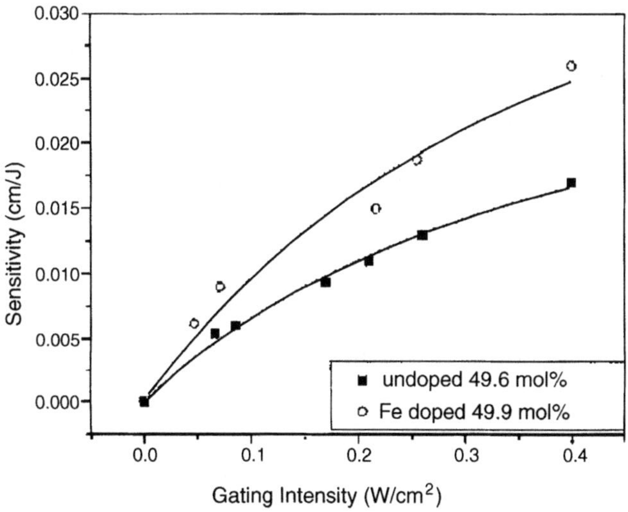

Fig. 12. Two-photon sensitivity vs. 457.9 nm gating intensity of reduced stoichiometric (49.9 mol% Li_2O) Fe-doped and reduced undoped (49.6 mol% Li_2O) $LiNbO_3$. Writing intensity is 10 W/cm^2

of Mn- and Fe-doped crystals gives rise to high dark conductivity and fast grating erasure immediately after recording.

Figure 12 shows the sensitivity versus gating intensity (at 457 nm) dependence in reduced Fe-doped stoichiometric crystal [44] ([Li_2O] = 49.9 mol% and [Fe] = 0.01 wt%). The two-photon sensitivity is about 30%, and the dynamic range is about 60% higher than that of the undoped crystals (with [Li_2O] = 49.6 mol%). An improvement in the sensitivity at 488 and 457 nm (compared with that at 514 nm) is more pronounced for the Fe-doped crystals than for the undoped crystals. Similar behavior with respect to gating wavelength was also found in Mn-doped crystals [40]. This enhanced sensitivity in doped lithium niobate is due to the Mn^{2+} and Fe^{2+} absorption bands being blue shifted as compared with the bipolaron band.

The largest dynamic range is found in a lightly reduced undoped material with the smallest defect density ($\Delta n \sim 0.6 \times 10^{-3}$ for undoped $LiNbO_3$ with [Li_2O] = 49.9 mol%). The undoped crystals of a composition near exact stoichiometry (with 49.9 mol% Li_2O content) have a dynamic range about twice that of Fe-doped crystal of the same composition but a sensitivity about three times lower owing to lower absorption of gating light (see Fig. 9).

Comparison of the Fe-doped and the undoped crystals also indicates that the erasure upon continuous IR readout is about three times slower in an Fe-doped sample than in an undoped crystal with 49.6 mol% Li_2O content, and is comparable to that in an undoped 49.9 mol% sample. The latter, however, has lower ($\times 3$) sensitivity than the Fe-doped samples.

In both Mn- and Fe-doped samples the two-photon sensitivity decrease (by a factor of 2) was seen when the dopant content was increased from 0.01 mol% to 0.03 mol%, which indicates that an optimal doping concentration exists for a high sensitivity and decreased IR erasure rate. With increasing concentration of acceptors in the crystal the recombination [45] from the small polarons into the deep traps (i.e. Fe^{3+} or Mn^{3+}) traps becomes more pronounced, which decreases the lifetime of the small polaron concentration, and thereby decreases the two-photon sensitivity.

3.3 Summary on Two-Photon Recording in $LiNbO_3$

Substantial improvement in two-photon recording sensitivity can be achieved in stoichiometric partially reduced lithium niobate. Two-photon sensitivity exhibits threshold-like behavior with respect to host crystal composition (at $[Li_2O] \sim 49.6$ mol% in the crystal). In undoped material small polarons are metastable shallow species responsible for the near-IR sensitivity, while bound bipolarons play the role of deep donors. By introducing deep level dopants, such as Fe^{2+} and Mn^{2+}, in reduced crystals, the gating efficiency and readout erasure time can be increased further compared with that of undoped material. The major role of reduction is in the control of the density of the deep acceptor levels in doped material. In undoped material reduction induces bipolarons, which provide efficient absorption for gating light. Further increase in sensitivity and recording dynamic range can be envisaged in stoichiometric crystals codoped with both extrinsic deep donors and extrinsic shallow acceptor levels. Choices of potential dopant pairs may include, e.g., Mn^{2+}/Ce^{4+}, Mn^{2+}/Fe^{3+} [37], and Cu^{+}/Ce^{4+}.

References

1. J.J. Amodei and D.L. Staebler, Appl. Phys. Lett. **18**, 540 (1971).
2. D. Kirillov and J. Feinberg, Opt. Lett. **16**, 1520 (1991).
3. G. Montemezzani and P. Günter, J. Opt. Soc. Am. B **7**, 2323 (1990).
4. V. Leyva, D. Engin, X.-L. Tong, M. Zhang, A. Yariv, and A. Agranat, Opt. Lett. **20**, 1319 (1995).
5. H. Vormann, G. Weber, S. Kapphan, and M. Wöhlecke, Solid State Commun. **57**, 543 (1981).
6. A. Yariv, S. Orlov, G. Rakuljic, and V. Leyva, Opt. Lett. **20**, 1334 (1995).
7. A. Yariv, S. Orlov, and G. Rakuljic, J. Opt. Soc. Am. B **13**, 2513 (1996).
8. G. Montemezzani, M. Zgonik and P. Günter, J. Opt. Soc. Am. B **10**, 171 (1993).
9. S. Orlov, D. Psaltis, and R.R. Neurgaonkar, Appl. Phys. Lett. **63**, 2466 (1993).
10. M. Carrascosa and F. Agullo-Lopez, J. Opt. Soc. Am. B **7**, 2317 (1990).
11. R. Matull and R.A. Rupp, J. Phys. D **21**, 1556 (1988).
12. V.V. Kulikov and S.I. Stepanov, Sov. Phys. Solid State **21**, 1849 (1979).

13. P. Hertel, K.H. Ringhofer, and R. Sommerfeldt, Phys. Status Solidi A **104**, 855 (1987).
14. W. Bollman and H.J. Stöhr, Phys. Status Solidi A **39**, 477 (1977).
15. L. Arizmendi, P.D. Townsend, M. Carrascosa, J. Baquedano and J.M. Cabrera, J. Phys. Cond. Matter **3**, 5399 (1991).
16. M. Carrascosa and L. Arizmendi, J. Appl. Phys. **73**, 2709 (1993).
17. R. Müller, L. Arizmendi, M. Carrascosa, and J.M. Cabrera, J. Appl. Phys. **77**, 308 (1995).
18. O.F. Schirmer, O. Thiemann, and M. Wöhlecke, J. Phys. Chem. Solids **52**, 185 (1991).
19. T.R. Volk, N. Rubinina, and M. Wöhlecke, J. Opt. Soc. Am. B **11**, 1681 (1994).
20. S. Klauer, M. Wöhlecke, and S. Kapphan, Phys. Rev. B **45**, 2786 (1992).
21. J.R. Herrington, B. Dischler, A. Rauber, and J. Schneider, Solid State Commun. **12**, 351 (1973).
22. A. Yariv, V. Leyva, and G.A. Rakuljic, "Relaxation and lifetime of 'fixed' charge holograms," in *Technical Digest, 1994 IEEE Nonlinear Optics, Materials, Fundamentals, and Applications*, Waikoloa, Hawaii, post-deadline paper PD6.
23. D.L. Staebler, W.J. Burke, W. Phillips, and J.J. Amodei, Appl. Phys. Lett. **26**, 182 (1975).
24. A. Mehta, E.K. Chang, and D.M. Smyth, J. Mater. Res. **6**, 851 (1991).
25. P.F. Bordui, R.G. Norwood, D.H. Jundt, and M.M. Fejer, J. Appl. Phys. **71**, 875 (1992).
26. D.H. Jundt, M.M. Fejer, R.G. Norwood, and P.F. Bordui, J. Appl. Phys. **72**, 3468 (1992).
27. S.C. Abrahams and P. Marsh, Acta Crystallogr. Sec B **42**, 61 (1986).
28. U. Schlarb and K. Betzler, Phys. Rev. B **48**, 15613 (1993).
29. L. Kovàcs and K. Polgár, in *Properties of Lithium Niobate*, EMIS Datareview series No 5, p. 109, RN-16037 (1989).
30. F.H. Mok, G.W. Burr, and D. Psaltis, Opt. Lett. **21**, 896 (1996).
31. F. Micheron and G. Bismuth, Appl. Phys. Lett. **23**, 71 (1973).
32. F. Micheron and G. Bismuth, Appl. Phys. Lett. **20**, 79 (1972).
33. A.G. Chynoweth, Phys. Rev. **102**, 705 (1956).
34. R. Landauer, J. Appl. Phys. **28**, 227 (1957).
35. D. von der Linde, A.M. Glass, and K.F. Rodgers, Appl. Phys. Lett. **25**, 155 (1974).
36. Y.-S. Bai and R. Kachru, Phys. Rev. Lett. **78**, 2944 (1997).
37. D.L. Staebler, W. Phillips, Appl. Phys. Lett. **24**, 268 (1974).
38. K. Buse, A. Adibi, and D. Psaltis, Nature **393**, 665 (1998).
39. S.S. Orlov, A. Akella, L. Hesselink, and R.R. Neurgaonkar, "High sensitivity two-photon non-volatile recording in lithium niobate," *Conference on Lasers and Electro-Optics '1997 (Optical Society of America)*, Baltimore, MD, post-deadline paper CPD29.
40. L. Hesselink, S.S. Orlov, A. Liu, A. Akella, D. Lande, and R.R. Neurgaonkar, Science **282**, 1089 (1998).
41. D. Lande, S.S. Orlov, A. Akella, L. Hesselink, and R.R. Neurgaonkar, Opt. Lett. **22**, 1722 (1997).
42. P. Nagels, *The Hall Effect and its Applications*, edited by C.L. Chien and C.R. Westgate, Plenum, New York, 1980.
43. J. Koppitz, A.I. Kuznetsov, O.F. Schirmer, M. Wöhlecke, and B.C. Grabmaier, Ferroelectrics **92**, 233 (1989).

44. Stoichiometric lithium niobate crystals with varying Fe concentrations grown by continuous charge double crucible Czochralski (DCCZ) method were provided to us by the National Institute for Research in Inorganic Materials in Japan (Dr. K. Kitamura).
45. F. Jermann, M. Simon, and E. Kratzig, J. Opt. Soc. Am. B **12**, 2066 (1995).

Two-Color Holography in Lithium Niobate

R. Macfarlane, H. Guenther, Y. Furukawa, and L. Kitamura

The development of a really satisfactory recording material for holographic data storage applications remains perhaps the most important barrier to practical implementation of the technology [1]. As discussed in the chapter entitled "Media Requirement for Digital Holographic Data Storage," the ideal material must simultaneously possess many properties such as good sensitivity, large dynamic range, long data retention times and excellent optical quality. In addition the material must be stable under repeated read cycles. This is easier to achieve in write-once-read-many (WORM) storage systems, since the material can be permanently deactivated after the writing process. In this application, irreversible chemical modification such as photochromism, photopolymerization etc. can be used. For reversible media the situtation is more difficult because the "fixing" process must be reversible, allowing rewriting immediately after an earlier recording or reading step. The requirement of reversibility often makes it more difficult to achieve long dark data retention times. Three main schemes for providing nondestructive readout in reversible photorefractive media have been proposed. The first was thermal fixing in lithium niobate [2,3], where a copy of the stored index gratings is made by thermally activating proton diffusion, which creates an optically stable complementary proton grating. Because of the long times required for thermal fixing and the need to fix large blocks of data at a time, thermally fixed media somewhat resemble reusable WORM materials. Electrical fixing [4] involves ferroelectric domain reversal near the Curie temperature. It is not of general applicability, and there are limits on the spatial frequencies that can be recorded. The third class of fixing process uses two wavelengths of light. One uses a different wavelength of light for recording and reading [5], but for storage applications this suffers from increased cross-talk and restrictions on the spatial frequencies that can be recorded. The most promising two-color scheme is "photon-gated" recording in a photorefractive material, in which charge generation occurs via a two-step process [6]. Coherent object and reference beams at a wavelength λ_1 record information in the presence of gating light at a wavelength λ_2. The gating light can be incoherent or broadband such as a white light source or LED [7]. Reading is done at λ_1 in the absence of gating light. Here the gating light can be thought of as sensitizing the material, in which case it is desirable that the sensitivity decays after the writing cycle, or as producing ionization of centers, in which a temporary grating is written at λ_1. Figure 1 shows a schematic level scheme comparing

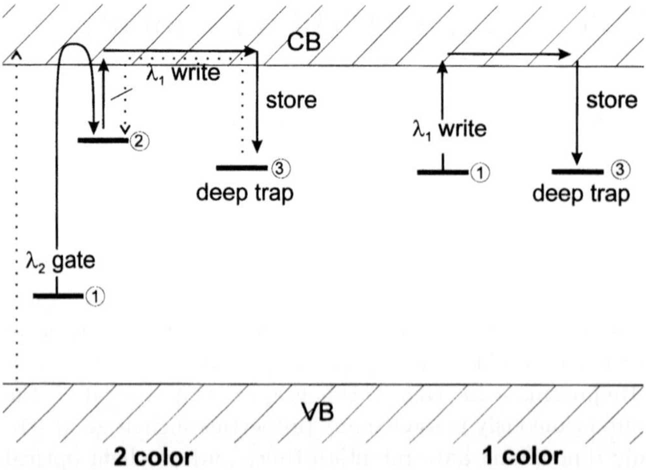

Fig. 1. Schematic level diagram of the two-color and one-color photorefractive effect. CB, conduction band; VB, valence band. In stoichiometric lithium niobate, level 1 is attributed to a Nb bipolaron state or Fe^{2+}/Fe^{3+} state, level 2 to a Nb_{Li} antisite polaron, and level 3 to an Fe^{3+} trap

the two-color and one-color cases for a photorefractive material with localized centers in the band gap. Photon-gated holography introduces another parameter, the gating ratio, or ratio between the sensitivity of the material in the presence and absence of gating light. Early work on two-color holography used short-pulse, high peak-power lasers since it was initially thought that the relevant intermediate states were short lived. Recent work [7–10] has shown that these states can be long lived, and that good sensitivity can be achieved with cw lasers and modest power densities of $\sim$1 W/cm^2 comparable to those used in the one-color case. We will concentrate here on the two-color photorefractive properties of reduced, near-stoichiometric lithium niobate, in which intrinsic defects appear to play the dominant role in the photorefractive sensitivity. Alternative schemes have been proposed for sensitization using double dopants such as Fe+Mn in $LiNbO_3$ [11,12], but in these cases the lifetime of the sensitized state is inconveniently long, and a desensitizing step is required. Reduced, stoichiometric lithium niobate shows one-color sensitivity in the blue-green spectral region and two-color sensitivity for writing in the near-IR and gating with blue-green light. From this it can be seen that the gating light also produces erasure. There is no fundamental reason for this, but it is a consequence of the broad spectral features of reduced or Fe-doped lithium niobate. Considerable progress is envisaged if a better separation of gating and erasing functions can be achieved by storing information in deeper traps and/or using wider band gap materials. Figure 2 compares one-color and two-color writing in a sample of reduced

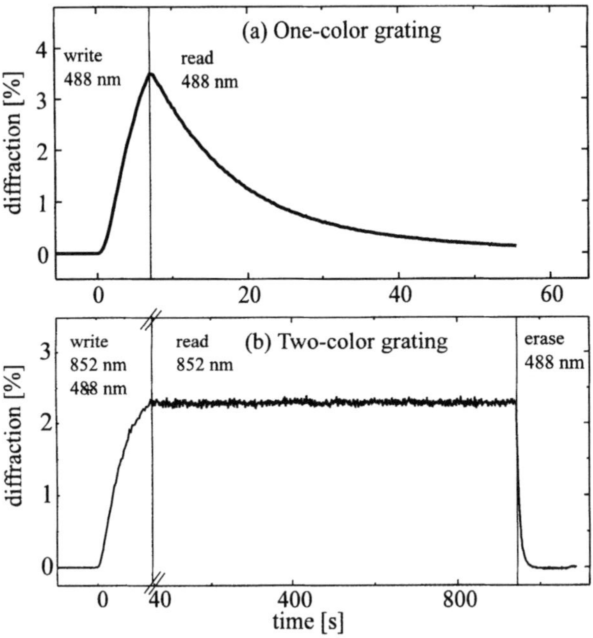

Fig. 2. Typical write–read–erase curve for holographic gratings in $LiNbO_3$ crystals: (**a**) one-color scheme, where an argon ion laser at 488 nm, 1 W/cm^2 is used for both writing (two beams) and reading (one beam); (**b**) two-color scheme, where a DBR laser diode at 852 nm (4 W/cm^2 total intensity) is used for writing and an argon ion laser at 488 nm, 1 W/cm^2 is used for the gating step. Nondestructive reading was done with one of the unattenuated writing beams (2 W/cm^2) and erasing with the gating light

near-stoichiometric lithium niobate to illustrate the nondestructive readout that can be achieved. The gating ratio in this case was in excess of 5000.

1 Materials

Crystals of stoichiometric lithium niobate (SLN) were grown using a double crucible technique [13]. The stoichiometry was measured from the phase-matching temperature for second harmonic generation of a Nd:YLF laser operating at 1047 nm. The crystal stoichiometry, expressed by the quantity C_{Li}=[Li]/([Li]+[Nb]), was equal to 49.7% in the nominally stoichiometric material. Photorefractive properties were compared with those of the congruent composition (CLN) for which $C_{Li} = 48.5\%$. Strong differences were observed, as shown in Table 1. Reduction of crystals was made in vacuum at 950°C for 1 hour at an oxygen partial pressure of about 10^{-3} mbar.

2 Experimental

Materials were evaluated in a plane-wave geometry in which two collimated 852 nm beams from a single-frequency DBR diode laser were incident on the sample at an external crossing angle of 20°. Gating light was provided either by an Ar^+ laser at 488 nm or by several GaN LEDs. The range of gating and writing light intensities used was generally in the range of 0.1–1 W/cm^2. Writing light was polarized parallel to the c-axis of the crystals, which were typically 8–10 mm in length. The time dependence of the diffraction efficiency was measured by a 650 nm diode laser Bragg-matched to the gratings in the sample and counterpropagating with the writing beams. The temperature of the sample could be varied between 12°C and 80°C using a thermoelectric cooler. Further details of the experimental setup have been published in [10].

3 Spectroscopy and Sensitization

Pure, unreduced lithium niobate is transparent, with an absorption onset below 340 nm. Reduction induces a broad visible absorption band (Fig. 3) attributed mostly to absorption by a bipolaron consisting of an electron trapped on a regular Nb site and another trapped at a Nb_{Li} antisite, together with a strong lattice distortion [14]. In addition there is some contribution from Fe^{2+} due to residual impurities. Irradiating with blue-green light is the gating or sensitizing step, which produces a transient absorption around 1.6 eV [15]. This absorption is assigned to a small polaron, or electron trapped at Nb_{Li}, produced by dissociation of the bipolaron [16], and it is responsible for the sensitivity at 852 nm. The lifetime of the sensitized state is an important quantity. When it is short, ~1 ms, very little population builds up in the intermediate state and sensitivity is low. For long lifetimes of tens of seconds, the material remains sensitive after writing, and also a significant fraction of the written gratings are stored in the transient level instead of being stored in a deep trap that is presumed to be Fe^{3+} impurities. The depth of the small

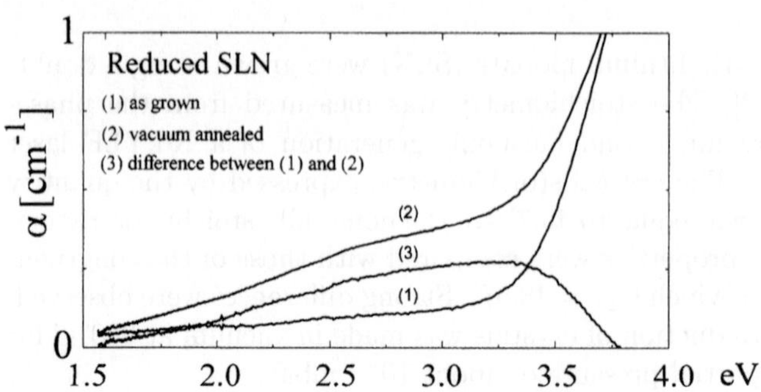

Fig. 3. Absorption spectrum for SLN crystal before and after reduction, showing the appearance of the broad bipolaron absorption band

polaron level below the conduction band was measured from the temperature dependence of the 1.6 eV induced absorption to be 0.7 eV.

4 Photorefractive Properties

The most important photorefractive properties for holographic data storage are sensitivity, $M\#$ or dynamic range, gating ratio, dark decay, and optical quality. Table 1 shows most of these properties for stoichiometric and congruent compositions compared with the behavior of conventional one-color Fe-doped lithium niobate. Optical quality is not quantified here, but in general congruent lithium niobate exhibits excellent quality, providing error-free imaging of 1Mpel 9 μm pitch data masks and scattered light levels equivalent to diffraction efficiencies of a few $\times 10^{-7}$ [17]. Stoichiometric material shows slightly inferior performance, but it can still exhibit high optical quality.

Table 1. Summary of data and comparison of two-color and one-color results

Material	Reduced SLN	Reduced SLN + Fe	CLN	Reduced CLN + Fe
Recording scheme[a]	2-color 852+488	2-color 852+488	2-color 852+488	1-color 488 nm
Fe conc. ppm	1.0	100	residual	200
α (cm^{-1}) gate at 488 nm	0.35	0.95	0.08	N/A
α (cm^{-1}) write at 852 nm	0.05	0.06	< 0.001	1.8[b]
10^3 $S_{\eta 2}$ (incident) (cm/J)	8	9	0.02	100
10^3 $S_{\eta 1}$ (absorbed) (cm^2/J)	160	150	> 20	170
M# /cm	0.8	0.5	0.05	24
Gating ratio at 852 nm	1600	10,000	[c]	N/A
Metastable lifetime (s)	16	0.15	0.1	N/A
Dark decay t_d (yr.)	0.07	0.03	0.32	~0.3

[a] $I_w = 4$ W/cm^2, 852 nm; $I_g = 1$ W/cm^2, 488 nm; $\Lambda = 6$ μm; E parallel to c-axis.
[b] Writing and reading wavelength 488 nm. [c] Not measured

4.1 Sensitivity

Photorefractive sensitivity for two-color recording in lithium niobate is linear in I_g only at low values of I_g [7,10] apparently because of competition between gating and erasing. Hence the sensitivity is defined in a similar way to that for one-color processes but for a fixed and reasonably low value of I_g=1 W/cm^2 at 488 nm. The growth of the square root of the diffraction efficiency during writing can be written as $A_0\ (1 - e^{-t/\tau_r})$ or $(A_0/\tau_r)t$ at small times, where τ_r is the recording time constant. In terms of *incident* intensities:

$$S_{\eta 2} = \frac{A_0}{\tau_r}\ \frac{1}{l\ I_w}, \qquad I_g = 1\ \mathrm{W/cm^2} \tag{1}$$

where I_g is the gating intensity, I_w is the total writing intensity in the two beams, l is the crystal length, and A_0/τ_r is the writing slope or initial time derivative of the square root of the diffraction efficiency, $\partial\sqrt{\eta}/\partial t|_{t=0}$.

The sensitivity for two-color recording, $S_{\eta 2}$ (expressed in units of cm/J), depends on a number of factors. The approximate gains available from each of these, with respect to undoped congruent material, were found to be: crystal stoichiometry (15×), degree of reduction [9] (20×) as illustrated in Fig. 4, temperature (5× upon cooling from 20°C to 0°C), gating wavelength (10× for a reduction in λ_{gate} from 520 nm to 400 nm) [7], and gating intensity, which is linear at low intensity and starts to saturate above about 1 W/cm^2 . The sensitivity increase with reduction comes partly from increased absorption at the gating wavelength and partly from an increase in the lifetime of the metastable state from which writing occurs. When this lifetime is too long there is an adverse effect, since the intermediate level then competes with the ultimate deep trap, and part of the grating is written to this level. This leads to a short-term component of dark decay, as shown in Fig. 4. As we will see below iron doping reduces this lifetime while maintaining the sensitivity.

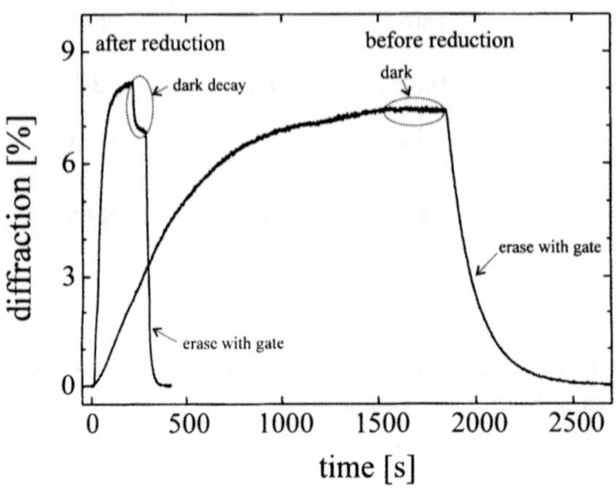

Fig. 4. Write–dark–erase curves for unreduced and reduced SLN crystals, showing that the saturation efficiency is essentially unchanged by reduction, while the sensitivity is greatly increased

Finally we note that the sensitivity in terms of *absorbed* power is $S_{\eta 1} = S_{\eta 2}/\alpha$, where α is the absorption coefficient at the writing wavelength. In terms of this sensitivity, all samples studied including the single photon Fe-doped material written at 488 nm, are almost equally sensitive (Table 1). This suggests that the sensitivity is determined by the amount of light that can be absorbed at the writing wavelength that is effective in charge generation. So far the maximum absorption of writing light that we have found in reduced SLN is 6% for $I_g = 1\ \mathrm{W/cm^2}$.

4.2 Gating Ratio

This parameter expresses the essence of two-color holography, i.e. the resistance to erasure during readout. Rather than measuring this by a small change in the diffraction efficiency with reading time, we have preferred the zero-background measurement of the ratio of the sensitivities in the presence of gating light to that in the absence of gating light. The writing wavelength we chose of 852 nm is close to the peak of the 1.6 eV (induced) absorption band. Typical values of the gating ratio in undoped, reduced crystals of stoichiometry 49.7% are in the range 2000–5000, while Fe-doped samples show values as high as 10 000. This reflects the ability of iron to reduce the occupation of levels contributing to absorption at 852 nm in the unsensitized state and with it the single-photon red sensitivity. At shorter writing wavelengths the gating ratio decreases as one-color and self-gated contributions become appreciable. For example at 670 nm the gating ratio is only 5.

4.3 Dynamic Range

The dynamic range, which determines how many holograms can be multiplexed in a single volume, can be expressed in terms of an $M\#$, defined [18] in terms of the diffraction efficiency for the Mth hologram stored in a common volume as $\eta_M = (M\#/M)^2$. This can be written as a product of the erasure time τ_e and the writing slope as $M\# = (A_0/\tau_e)\ \tau_r$. In a one-color (one-photon material), the $M\#$ is proportional to the modulation index $m = 2\sqrt{I_1 I_2}/(I_1 + I_2)$. In a two-color material, on the other hand, erasure is caused by the gating light but not by the writing light. This results in an $M\#$ proportional to $\sqrt{I_1 I_2}$, and hence for equal intensity object and reference beams the $M\#$ is proportional to the writing intensity (see Fig. 5). While this provides a way of increasing the dynamic range in a two-color material, the writing power requirements in the present material system become rather high. The $M\#$ shows a strong temperature dependence, which is expected since the sensitivity is much more strongly temperature dependent than is the erasure.

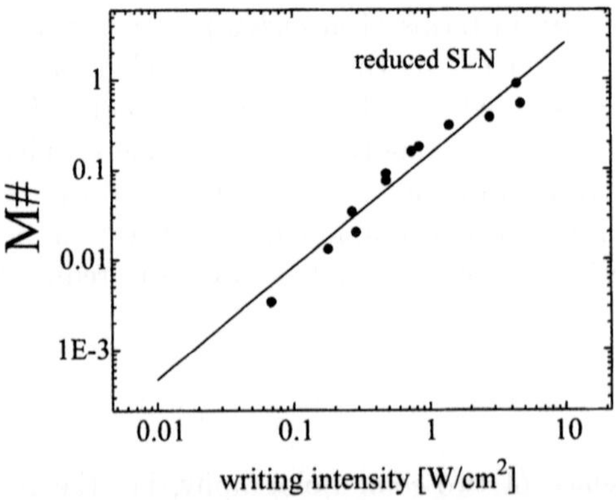

Fig. 5. Dependence of the $M\#$ or dynamic range on writing intensity. This is in contrast to one-color holography, where the $M\#$ depends on the modulation index but not on the writing intensity. The crystal length was 9.8 mm

4.4 Dark Decay

The dark decay time of holograms (t_d) determines the lifetime of the stored information, and it usually depends primarily on the electrical conductivity σ, through the relationship $t_d = \epsilon\epsilon_0/\sigma$. Congruent lithium niobate, in addition to showing low sensitivity for two-color writing, also shows rapid conductivity increases upon reduction, so that the dark decay of strongly reduced congruent samples is often found to be only a few minutes or even seconds. Stoichiometric material, on the other hand, shows dark decay times that are orders of magnitude longer. The temperature dependence of the dark decay times, measured in the range 35–80°C, was extrapolated to room temperature and gave values of the order of several months (see Table 1). The temperature dependence also gives the activation energy, which lies in the range 0.9–1.2 eV. This is associated with the formation of ionic gratings due to proton diffusion since these energies correspond well with those obtained from thermal fixing of single-color holograms in Fe-doped material [3]. The deep trap in which the electronic grating is initially stored is likely to be the Fe^{2+}/Fe^{3+} level, which has been shown from thermal fixing studies to have a slightly higher activation energy than do the protons.

4.5 The Role of Iron

Iron impurities are almost always present in lithium niobate at several ppm levels, and as noted above the presence of some Fe doping is believed necessary to provide a deep trap for the storage of the holograms. In general, Fe doping decreases conductivity due to trapping out shallow levels. The most interesting effect from the two-color holography standpoint is that in reduced, stoichiometric crystals trapping by iron shortens the lifetime of the intermediate level from which writing occurs. For example the addition of 60 ppm of Fe reduces the lifetime from 3.5 s to 0.18 s, with a similar result for 30 ppm doping. This quenching of the intermediate state lifetime eliminates the short-term dark decay seen in Fig. 4 because an insignificant part of the grating is then written in the intermediate level. The transfer of electrons from the Nb_{Li}^{4+} small polaron to Fe^{3+} is thus seen to be rather efficient. The concentration of iron should not be so high, however, that the Fe absorption becomes strong at the gating wavelength, since this would limit the usable thickness of material and also cause undesirable heating and photovoltaic damage effects. An interesting and important result is that while the short-term dark decay is quenched by Fe doping, the absorption coefficient at the writing wavelength, and hence the sensitivity, is maintained as shown in Table 1. In other words the shortening of the lifetime of the intermediate state by more than an order of magnitude does not significantly reduce its population. A possible reason is that the filling rate increases owing to population of the small polaron by direct photoexcitation of Fe, which absorbs efficiently at the gating wavelength.

5 Conclusion

Two-color, photon-gated holography provides a promising solution to the long-standing problem of destructive readout in read/write digital holographic storage. The basic process in near-stoichiometric lithium niobate is one based on two-step photoionization using a metastable (small polaron) level, which is the charge-generating step in the photorefractive process. This scheme has been studied for a number of years, mostly using high peak power lasers. We and others [7–10] investigated two-color holographic grating formation in the more practical low-power regime using cw lasers. Here we reported the use of diode lasers as writing sources and gating light of up to ~1 W/cm^2 to study the issues affecting the main parameters of interest for holographic storage: sensitivity, gating ratio, dynamic range and dark decay for doped and undoped lithium niobate crystals of different stoichiometry and degree of reduction. Photorefractive properties are attributed to both intrinsic defects and iron impurities. Two-color sensitivity per incident photon is still somewhat less than that of the one-photon process in Fe-doped lithium niobate at 488 nm, although the sensitivity per absorbed photon is approximately the same. Optimization of the sensitivity requires control over stoichiometry, degree of reduction, temperature, gating wavelength, and gating intensity. Two-color materials differ fundamentally from one-color materials in that the dynamic range or $M\#$ can be increased by using higher writing intensity, and the sensitivity can be increased with higher gating intensity. Another route to increasing the $M\#$ would be to find a material that exhibits a two-color erase process. Substantial progress has been made in recent years in the field of two-color holography, and further progress can be expected on this complex and challenging problem.

References

1. G.T. Sincerbox, Opt. Mater. **4**, 370–375 (1995).
2. J.J. Amodei and D.L. Staebler, Appl. Phys. Lett. **18**, 540–542 (1971).
3. A. Yariv, S.S. Orlov, and G.A. Rakuljic, J. Opt. Soc. Am. B **13**, 2513–2523 (1996).
4. F. Micheron, G. Bismuth, Appl. Phys. Lett. **20**, 79–81 (1972).
5. H.C. Kuelich, Opt. Commun. **64**, 407–411 (1987).
6. D. von der Linde, A.M. Glass, and K.F. Rodgers, Appl. Phys. Lett. **25**, 155–157 (1974).
7. H. Guenther, G. Wittmann, R.M. Macfarlane, and R.R. Neurgaonkar, Opt. Lett. **22**, 1305 (1997).
8. Y.S. Bai and R. Kachru, Phys. Rev. Lett. **78**, 2944–2947 (1997).
9. D. Lande, S.S. Orlov, A. Akella, L. Hesselink, and R.R. Neurgaonkar, Opt. Lett. **22**, 1722–1724 (1997).
10. H. Guenther, R.M. Macfarlane, Y. Furukawa, K. Kitamura, and R.R. Neurgaonkar, Appl. Opt. **37**, 7611 (1998).
11. D.L. Staebler and W. Phillips, App. Phys. Lett. **24,** 268 (1974).
12. K. Buse, A. Adibi, and D. Psaltis, Nature **393**, 665 (1998).
13. Y. Furukawa, M. Sato, K. Kitamura, and F. Nitanda, J. Cryst. Growth **128**, 909–914 (1993).
14. O.F. Schirmer, S. Juppe, and J. Koppitz, Cryst. Latt. Def. and Amorph. Mater. **16**, 353–357 (1987).
15. J.L. Ketchum, K.L. Sweeney, L.E. Halliburton, and A.F. Armington, Phys. Lett. **94A**, 450–453 (1983).
16. O.F. Schirmer, H.-J. Reyher, and M. Woehlecke, in *Insulating Materials for Optoelectronics: New Developments*, ed. F. Agullo-Lopez, World Scientific (1995) Chap. 4 and references therein.
17. M.-P. Bernal, G.W. Burr, H. Coufal, R.K. Grygier, J.A. Hoffnagle, C.M. Jefferson, R.M. Macfarlane, R.M. Shelby, G.T. Sincerbox, and G. Wittmann, Mater. Res. Soc. Bull. **21**, 51–60 (1996).
18. F.H. Mok, G.W. Burr, and D. Psaltis, Opt. Lett. **21**, 896–898 (1996).

Overview of Photorefractive Polymers for Holographic Data Storage

B. Kippelen

While photorefractivity during the 1970s and 1980s was studied in many different classes of inorganic materials [1], organic photorefractive (PR) materials emerged only in the 1990s. The quest for organic photorefractive materials was driven by their electronic optical nonlinearity, which leads to materials that combine high electro-optic coefficient and low dielectric constant, a property that improves the figure of merit for photorefractivity $Q = n^3 r/\varepsilon$, where n is the refractive index, r is the electro-optic (or Pockels) coefficient, and ε is the low-frequency dielectric constant.

During recent years, this new field of research experienced several milestones: a new and unexpected form of orientational photorefractivity was discovered; consequently high dynamic range was obtained with low-power semiconductor laser diodes; and the polymers were successfully used in different applications, including holographic storage. With their plasticity and other unique properties, they constitute a strategic class of materials because they enable the mass production at low cost of new devices with low weight, and high performance. Thus such materials are expected to have a significant impact on dynamic holographic technologies.

1 Brief History of Photorefractive Polymers

Photorefractivity in an organic material was first reported [2] in 1990 by the ETH Zurich group in an organic single crystal. The same year a group at Eastman Kodak developed a multifunctional polymer that showed photoconductivity and electro-optic activity simultaneously [3]. Holographic recording and retrieval was not carried out in this material, but was demonstrated shortly after in another polymer composite at IBM Almaden [4]. While these two composites were obtained by doping an electro-optic polymer with charge transport molecules, another approach was developed independently at the University of Arizona [5] and at the University of Chicago [6], which consisted of the synthesis of a fully functionalized side-chain polymer with multifunctional groups. Several new polymers were synthesized at that time [7], and it became rather obvious that the guest/host approach (see Fig. 1) was much faster and easier to implement and would provide a better way to test different combinations of polymers and molecules with photosensitivity, transport, and electro-optic activity.

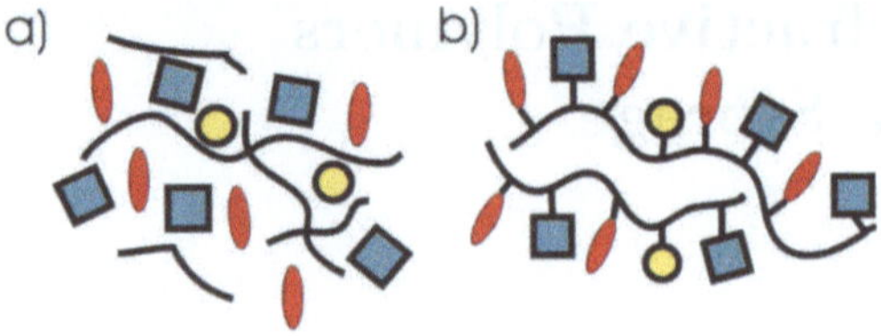

Fig. 1. Illustration of different chemical designs: (**a**) guest/host, (**b**) fully functionalized approach

A first milestone in this field occurred with the report by the IBM group of net gain and a diffraction efficiency of 1% in 125-µm-thick samples of the composite FDEANST:PVK:TNF [8]. Shortly after, the Arizona group reported on a composite based on the photoconductor PVK:TNF doped with the chromophore DMNPAA, and that exhibited 6% diffraction efficiency in 100-µm-thick films [9]. That significant improvement was due to an additional plasticizer ECZ, which was added to the composite and which reduced the glass transition temperature, enabling the orientation of the electro-optic chromophores with an applied field at room temperature. By improving the sample fabrication conditions, enabling higher fields to be applied, these composites exhibited nearly 100% diffraction efficiency, net gain coefficients of 200 cm^{-1}, and fully reversible index modulation amplitudes of 0.007 with a response time of 100–500 ms [10].

The report of these high diffraction efficiencies was very puzzling because the electro-optic properties of the dopant chromophores could not account for the high dynamic range that was observed. A possible explanation was proposed by the IBM group. They attributed the high dynamic range to an orientational birefringence contribution caused by the spatially modulated orientation of the chromophores by the total field in the polymer (Fig. 2). The observation of a change in coupling direction when changing the polarization of the beams from 'p' to 's' in DMNPAA:PVK:ECZ:TNF samples was a clear evidence that the dynamic range was indeed dominated by orientational birefringence.

The year 1994 was a turning point for the development of photorefractive polymers since it changed the design of chromophores completely: the Pockels effect was no longer the main driver but rather the orientational birefringence. A new form of photorefractivity was born: orientational photorefractivity, in which the index of a material is changed through the control

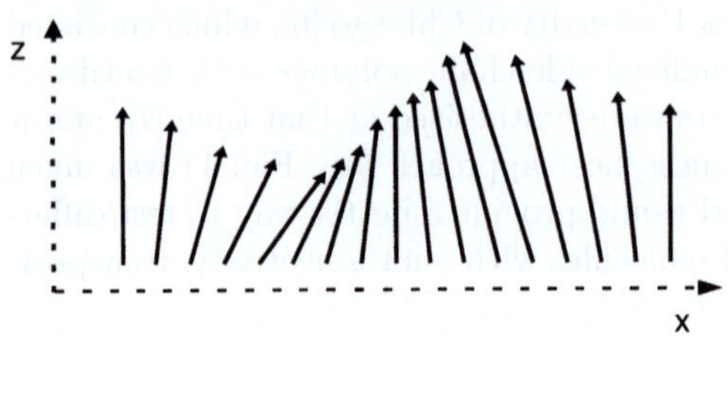

Fig. 2. Periodic orientation of the chromospheres in low-T_g polymers

of the orientation of optically anisotropic dopant molecules with a permanent dipole moment, by the internal space charge field that is produced as in a traditional photorefractive material by absorption followed by charge separation and trapping.

2 Physics and Chemistry of Photorefractive Polymers

In the subsections below, we briefly summarize the main physics and chemistry of photorefractive polymers. Their properties differ from those of inorganic photorefractive materials in many different ways: (i) the basic properties such as photogeneration, transport and index modulation are different; (ii) polymers are amorphous materials as opposed to crystalline for most of the inorganic photorefractive materials; (iii) the mechanism for index change is different, as discussed in the previous section.

2.1 Photogeneration

The space-charge build-up process can be divided into two steps: the electron-hole generation process followed by the transport of one carrier species. The creation of a correlated electron-hole pair (or exciton) after absorption of a photon can be followed by recombination. This process limits the formation of free carriers that can participate in the transport process, and is therefore a loss for the formation of the space-charge. These properties result in low photogeneration efficiency unless an electric field is applied to counteract recombination. The quantum efficiency for carrier generation is therefore strongly field-dependent, and increases with the applied field. A theory developed by Onsager [11] for the dissociation of ion pairs in weak electrolytes under an applied field has been found to describe reasonably well the temperature and field dependence of the photogeneration efficiency in some of the organic photoconductors [12].

2.2 Transport

After photogeneration of free carriers, the next step in the build-up of a space-charge is their transport from brighter regions of the interference pattern, where they are generated, to the darker regions, where they get trapped. In contrast to inorganic photorefractive crystals with a periodic structure, photorefractive polymers have a nearly amorphous structure. The local energy level of each molecule/moiety is affected by its nonuniform environment. The disorder in amorphous photoconductors splits the conduction bands of molecular crystals into a distribution of localized electronic states (Fig. 3). As a result, transport can no longer be described by band models but is attributed to intermolecular hopping of carriers between neighboring molecules or moieties [12].

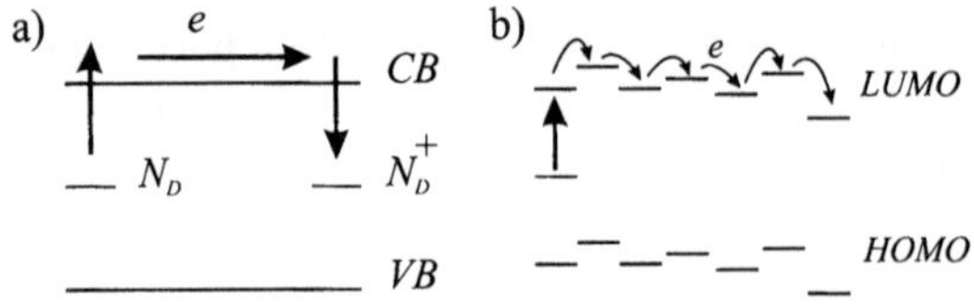

Fig. 3. Schematic representation of transport (**a**) in crystals, and (**b**) in amorphous polymers

In recent years, a model based on disorder due to Bässler and co-workers describes the transport phenomena of a wide range of different materials, and has emerged as a solid formalism to describe the transport in amorphous organic materials. This so-called disorder formalism is based on the assumption that charge transport occurs by hopping through a distribution of localized states with energetical and positional disorder. So far, most of the predictions of this theory agree reasonably well with the experiments performed in a wide range of doped polymers, main-chain and side-chain polymers and in molecular glasses [12].

In the Bässler formalism, disorder is separated into diagonal and off-diagonal components. Diagonal disorder is characterized by the standard deviation σ of the Gaussian energy distribution of the hopping site manifold (energetical disorder), and the off-diagonal component is described by the parameter Σ, which describes the amount of positional disorder. Results of Monte Carlo simulations led to the following universal law for the mobility when $\Sigma \geq 1.5$ [13]:

$$\mu_\mathrm{t}(E,T) = \mu_0 \exp\left[-\left(\frac{2\sigma}{3kT}\right)^2\right] \exp\left\{C\left[\left(\frac{\sigma}{kT}\right)^2 - \Sigma^2\right] E^{1/2}\right\}, \quad (1)$$

where μ_0 is a mobility prefactor and C is an empirical constant. Equation (1) is valid for high electric fields (a few tens of V/μm) and for temperatures $T_\mathrm{g} > T > T_\mathrm{c}$ where T_g is the glass transition temperature and T_c is the dispersive to nondispersive transition temperature. For a more detailed description the reader is referred to the original work by Bässler *et al.* and the recent review by P. Borsenberger [12].

Numerous simulations and experimental work have shown that the width of the density of states σ in (1) is strongly dependent on the polarity of the polymer host and the dipole moment of the dopant molecule. It was found that the total width σ comprised a dipolar component σ_D and an independent Van der Waals component σ_VdW [14]:

$$\sigma^2 = \sigma^2_\mathrm{VdW} + \sigma^2_\mathrm{D}. \quad (2)$$

Another important parameter that affects hole transport in guest/host systems is the relative position of the HOMO (highest occupied molecular orbital) energy level of the different dopant moieties. Dopant molecules with a lower HOMO energy will act as shallow traps.

2.3 Index Change: Electro-Optic and Orientational Effects

Like macroscopic systems, molecules exhibit second-order nonlinearities only if the centrosymmetry is broken. This can be achieved in conjugated molecules by connecting each end of the conjugated path with groups that have a different electronic affinity. This results in asymmetric polarization, and permits second-order nonlinear optical properties on a molecular level.

When a poling field is applied, orientation of these chromophores leads to macroscopic electro-optic properties but also to birefringence. In the oriented gas model and for a poling field applied along the Z axis these two effects can be described by [15]

$$\Delta n_Z^{(1)} = \frac{2\pi}{n} N F^{(1)} \Delta\alpha (\langle \cos^2\theta \rangle - 1/3), \tag{3}$$

$$r_{ZZ} = \frac{-8\pi}{n^4} N F^{(2)} \beta \langle \cos^3\theta \rangle, \tag{4}$$

where $F^{(1)}$ and $F^{(2)}$ are local field correction factors, $\Delta\alpha$ is the polarizability anisotropy of the chromophore, i.e. the difference between the polarizability along the molecular axis and the polarizability in a perpendicular direction, β is the first hyperpolarizability, N is the density of molecules, and θ is the polar angle between the poling field direction and the dipole moment of the molecule. The averaged quantities $\langle \cos^n\theta \rangle$ are given by the Langevin functions $L_n(x)$ with $x = \mu E/kT$ and where μ is the dipole moment and kT the thermal energy. In the low poling field limit, the Langevin functions can be approximated to $L_2(x) \approx 1/3 + 2x^2/45$, and $L_3(x) \approx x/5$. With these approximations and from (3) and (4), a possible molecular figure of merit for the chromophore for orientational photorefractivity becomes [16]

$$Q_{\mathrm{OP}} = N_{\mathrm{max}}(A(T)\Delta\alpha\mu^2 + \beta\mu), \tag{5}$$

where N_{max} is the maximum density of chromophores that can be doped or functionalized in the low glass transition polymer composite before chromophore aggregation or interaction occurs, and $A(T)$ is a scaling factor. The subscript OP in Q_{OP} stands for *orientational photorefractivity*. Note that the first term in the RHS of (5) scales with the dipole moment squared. This quadratic dependence on dipole moment is due to the Kerr-like character of the orientational birefringence. Consequently, the total refractive index change has a quadratic dependence on the total electric field, which is the superposition of the internal space-charge field and the applied field. Owing to this quadratic dependence in electric field of the orientational Kerr effect, the writing of a grating with vector $\boldsymbol{K}$ leads simultaneously to a secondary grating with grating vector $\boldsymbol{2K}$ [15]. This property can have important implications for multiplexing.

Note that all the orientational photorefractive properties are contingent on the ability of the chromophores to orient at the operation temperature that is, when the temperature is close to or above the glass transition temperature T_g. As discussed in the next section, the most efficient polymers to date are based on such orientational photorefractivity. However, polymers with high T_g can be designed in which poling (orientation of the chromophores) is achieved at high temperature and frozen in the material, leading to a spatially uniform electro-optic coefficient. In this case, photorefractivity is described by the Pockels effect, as in inorganic photorefractive crystals.

3 Performance of Current Photorefractive Polymers

In this section, we will briefly discuss some selected materials and applications to illustrate the properties discussed previously. Owing to the restricted length of this chapter, the following sections do not provide an exhaustive description of the numerous contributions to the field. These can be found in recent reviews [17–19]. Examples of compounds used in photorefractive polymers are shown in Table 1.

3.1 Spectral Sensitivity

In contrast to inorganic crystals, the spectral sensitivity of photorefractive polymers can be changed by using different dopant sensitizers. During recent decades, numerous organic sensitizers and photoconductors have been developed for xerography [20]. Due to their unique mechanical properties, organic photoconductors are widely used in photocopiers and printers. Over the years, several classes of sensitizers have been developed, including thiapyrylium salts, perylene-, azo-, squaraine-, and phthalocyanine-based compounds, some of which have been applied recently to photorefractive polymers.

So far, the spectral sensitivity of photorefractive polymers has been tuned through the entire visible spectrum (488–690 nm), and the near infrared (up to 830 nm).

3.2 Dynamic Range

Of great interest for holographic storage is the dynamic range or the maximum refractive index modulation amplitude. Dynamic range is generally measured by degenerate four-wave mixing experiments conducted in a tilted geometry that is imposed by the symmetry of the sample and by the physics of photorefractivity in organics. Samples consist of a thick polymer film sandwiched between two glass slides coated with transparent electrodes of indium tin oxide (ITO). This geometry results in an applied field with direction normal to the sample surface. Since the transport and charge separation occurs through field-assisted drift, the sample has to be tilted with respect to the bisector of the writing beams in order to have a nonvanishing component of the electric field along the grating vector of the grating (see Fig. 4).

Table 1. Chemical structure of some compounds used in photorefractive polymers. Sensitizers: C_{60} [7], TNF [8], TNFDM [22]. Plasticizers: ECZ [9], BBP [26], DIP [28]. Transport molecules and polymers: PVK [8], PSXCZ [30], DMNPAA [9], PDCST [26], DHADCMPN [22]

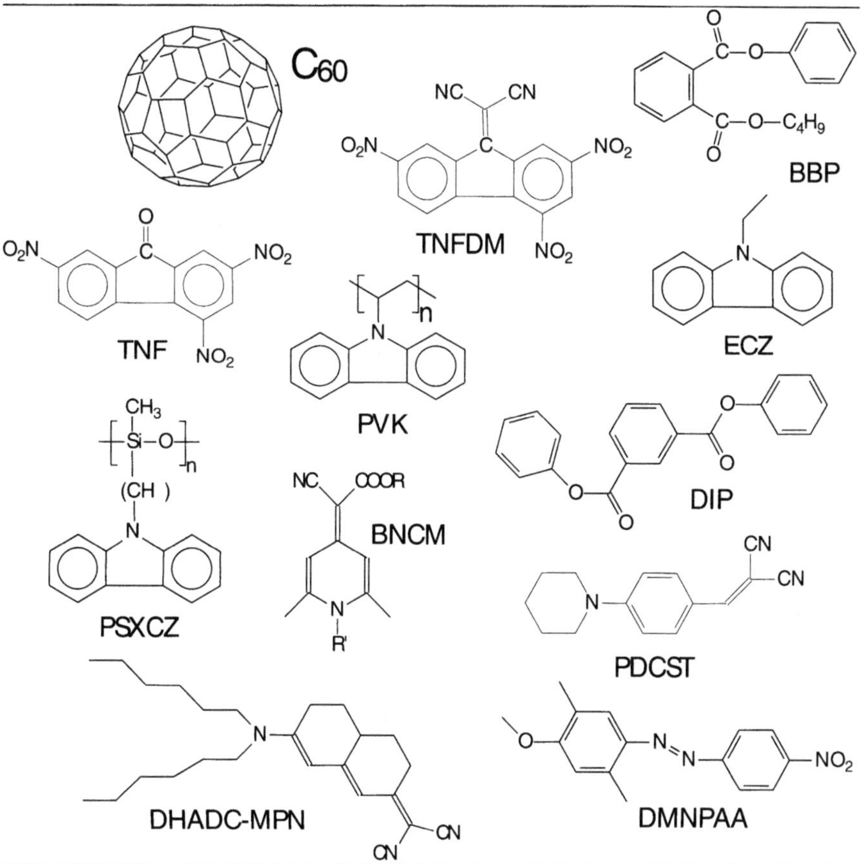

Dynamic range Δn is calculated using Kogelnik's coupled-wave theory. This theory neglects the anisotropy of the material and fringe bending effects due to the coupling of the writing beams but provides a good framework to compare the performance of different materials. Since orientational photorefractivity is dominant in current polymers, the index change is given by

$$\Delta n = pE^2 = pV^2/d^2, \tag{6}$$

where E is the applied field, and V and d are the applied voltage and sample thickness, respectively. Since Δn depends on the applied field, a better way to characterize dynamic range is to define a material parameter p as in (6).

For a long time, the polymer composite DMNPAA:PVK:ECZ:TNF was considered a benchmark [10] with $\Delta n = 0.007$ at $E = 90$ V/µm, correspond-

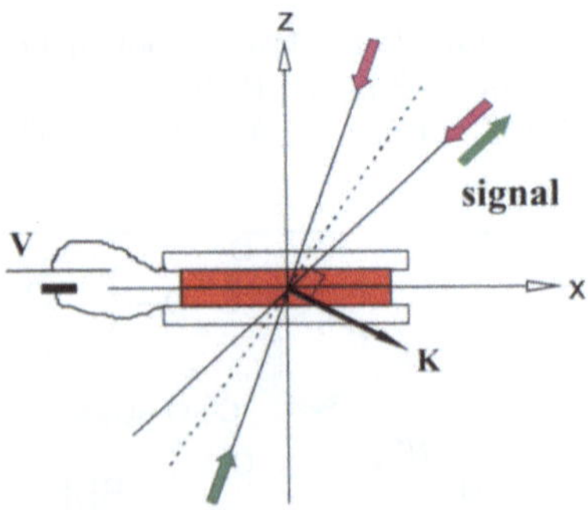

Fig. 4. Tilted geometry used to write holograms in photorefractive polymer samples

ing to $p = 86\ \mathrm{cm}^2\, V^{-2}$. The first material that provided a higher dynamic range was an organic glass formed with a molecule [21] that provided simultaneously transport and index change leading to a dynamic range that was 1.5 times higher. In recent polymer composites [22] based on PVK and the chromophore DHADCMPN the parameter p is $333\ \mathrm{cm}^2\, V^{-2}$.

3.3 Material Stability

While early guest/host systems based on PVK and dopant chromophores such as FDEANST or DMNPAA had limited shelf lifetimes of a few days to a few months depending on the fabrication conditions, some recent composites show excellent phase stability. Improved stability could be achieved in several ways: (i) by modifying good performing chromophores by adding groups such as ethylhexyl to impart better solubility [23]; (ii) by designing chromophores that do not crystallize but form amorphous glasses [21]; (iii) by using eutectic [24] or isomeric mixtures of chromophores [25]; (iv) by using liquid plasticizers [26]; (v) by designing monolithic structures such as multifunctional carbazole trimers [27]. In some cases, the compatibility of the polymer matrix and the dopant chromophore is so good that the composite does not show any signs of degradation even after accelerated aging tests at 85°C for a week [28]. Such samples have shelf lifetimes of several years at room temperature.

3.4 Speed

The speed of hologram formation in photorefractive polymers is rather complicated because it depends on several factors, including photogeneration efficiency, drift mobility, and field-induced orientational dynamics of the chromophores. The speed of current materials is spread over several orders of magnitude, ranging from minutes for photorefractive glasses to a few ms for the fastest materials [29]. Speed is strongly field dependent, as expected from the strong field dependence of the photogeneration efficiency and the mobility. Recent studies have shown that the response time in the fastest polymers is not limited by the orientation of the chromophores but rather by the photoconducting properties of the polymer composite. These studies indicate that sub-ms response times should be possible.

3.5 Applications

Holographic storage has been a strong driver during the early development of photorefractive polymers. Early demonstration of hologram recording and retrieval was performed in DMNPAA:PVK:ECZ:TNF samples, and was made possible by the high diffraction efficiency measured in 105-μm-thick samples. As a reconfigurable recording medium, photorefractive polymers have also been used for time-average interferometry, optical correlation [30], and imaging [22].

More detailed assessments of their potential for holographic storage were conducted with the holographic storage testbed developed at IBM. High-contrast 64-kbit digital data pages could be stored in 130-μm-thick DTNBI: PMMA:C60 samples placed near the Fourier plane in a *4-f* recording geometry. Owing to the limited thickness of the sample, only single data pages were stored. A global absolute threshold resulted in a readout BER of 1.5×10^{-4} [31]. By using polysiloxane polymers substituted with carbazole side-groups and doped with FDEANST and TNF, the IBM group demonstrated digital data recording at a density of 0.52 Mbit/cm^2 [32]. In samples of the composite 2BNCM:PMMA: TNF, single digital data pages could be stored and retrieved using only 1 part in 10^6 of the total dynamic range [21].

4 Trends and Outlook

To be practical for holographic storage, materials should fulfill the following requirements: (i) thickness >1 mm; (ii) high Δn; (iii) 670–850 nm spectral response; (iv) high optical quality; (v) response time <1 ms; and (vi) non-destructive read-out. Within a few years of research, photorefractive polymers have met most of these requirements. However, two major challenges remain: to (i) increase the current thickness of ≈ 100 μm to 1 mm, and (ii) develop non-destructive read-out schemes. To increase the thickness, stratified samples have been proposed [33]. It is not clear, however, how these multiple interfaces and ITO layers will affect the optical quality of such stratified samples. With the structural flexibility of organic materials, future breakthroughs are anticipated and current limitations will be overcome. The major uncertainty remains the good timing of these future materials advances with those of holographic storage technologies.

Acknowledgments

The author gratefully acknowledges his past and current colleagues for their fruitful collaboration. This work was made possible through the support by

AFOSR, NSF, ONR through the MURI Center CAMP, by travel grants through NATO and an NSF/ CNRS international collaboration.

References

1. P. Günter and J.-P. Huignard (Eds.): *Photorefractive materials and their applications*, Vols. 61 and 62 (Springer-Verlag, Berlin, 1988).
2. K. Sutter and P. Gunter: J. Opt. Soc. Am. B **7**, 2274 (1990).
3. J.S. Schildkraut: Appl. Phys. Lett. **58**, 340 (1990).
4. S. Ducharme, J.C. Scott, R.J. Twieg, and W.E. Moerner: Phys. Rev. Lett. **66**, 1846 (1991).
5. B. Kippelen, K. Tamura, N. Peyghambarian, A.B. Padias, and H.K. Hall Jr.: Phys. Rev. B **48**, 10710 (1993).
6. L. Yu, W. Chan, Z. Bao, and S.X.F. Cao: Macromolecules **26**, 2216 (1993).
7. Y. Zhang, Y. Cui, and P.N. Prasad: Phys. Rev. B **46**, 9900 (1992).
8. M.C.J.M. Donckers, S.M. Silence, C.A. Walsh, F. Hache, D.M. Burland, and W.E. Moerner: Opt. Lett. **18**, 1044 (1993).
9. B. Kippelen, Sandalphon, N. Peyghambarian, S.R. Lyon, A.B. Padias, and H.K. Hall Jr.: Electron. Lett. **29**, 1873 (1993).
10. K. Meerholz, B.L. Volodin, Sandalphon, B. Kippelen, and N. Peyghambarian: Nature **371**, 497 (1994).
11. L. Onsager: Phys. Rev. **54**, 554 (1938).
12. P.M. Borsenberger and D.S. Weiss: *Organic photoreceptors for imaging systems* (Marcel Dekker, Inc., New York, 1993).
13. H. Bässler: Adv. Mater. **5**, 662 (1993).
14. P.M. Borsenberger, E.H. Magin, M.B. O'Regan, and J.A. Sinicropi: J. Polym. Sci. B: Polym. Phys. **34**, 317 (1996).
15. W.E. Moerner, S.M. Silence, F. Hache, and G.C. Bjorklund: J. Opt. Soc. Am. B **11**, 320 (1994).
16. B. Kippelen, F. Meyers, N. Peyghambarian, and S.R. Marder: J. Am. Chem. Soc. **119**, 4559 (1997).
17. W.E. Moerner and S.M. Silence: Chem. Rev. **94**, 127 (1994).
18. B. Kippelen, K. Meerholz, and N. Peyghambarian: in *Nonlinear optics of organic molecules and polymers*, H.S. Nalwa and S. Miyata (Eds.) p. 465 (CRS, Boca Raton, 1997).
19. W.E. Moerner, A. Grunnet-Jepsen, and C.L. Thomson: Ann. Rev. Mater. Sci. **27**, 585 (1997).
20. K.Y. Law: Chem. Rev. **93**, 449 (1993).
21. P.M. Lundquist, R. Wortmann, C. Geletneky, R.J. Twieg, M. Jurich, V.Y. Lee, C.R. Moylan, and D.M. Burland: Science **274**, 1182 (1996).
22. B. Kippelen, S.R. Marder, E. Hendrickx, J.L. Maldonado, G. Guillemet, B.L. Volodin, D.D. Steele, Y. Enami, Sandalphon, Y.J. Yao, J.F. Wang, H. Röckel, L. Erskine, and N. Peyghambarian: Science **279**, 54 (1998).
23. M. Cox, R.D. Blackburn, D.P. West, T.A. King, F.A. Wade, and D.A. Leigh: Appl. Phys. Lett. **68**, 2801 (1996).
24. K. Meerholz, R. Bittner, Y. De Nardin, C. Bräuchle, E. Hendrickx, B.L. Volodin, N. Kippelen, and N. Peyghambarian: Adv. Mater. **9**, 1043 (1997).

25. E. Hendrickx, B.L. Volodin, D.D. Steele, J.L. Maldonado, J.F. Wang, B. Kippelen, and N. Peyghambarian: Appl. Phys. Lett. **71**, 1159 (1997).
26. A. Grunnet-Jepsen, C.L. Thompson, R.J. Twieg, and W.E. Moerner: Appl. Phys. Lett. **70**, 1515 (1997)
27. L. Wang, Y. Zhang, T. Wada, and H. Sasabe: Appl. Phys. Lett. **69**, 728 (1996).
28. E. Hendrickx, J. Herlocker, J.L. Maldonado, S.R. Marder, B. Kippelen, A. Persoons, and N. Peyghambarian: Appl. Phys. Lett. **72**, 1679 (1998).
29. D. Wright, M.A. Díaz-García, J.D. Casperson, M. DeClue, W.E. Moerner, and R.J. Twieg: Appl. Phys. Lett. **73**, 1490 (1998).
30. L. Volodin, B. Kippelen, K. Meerholz, B. Javidi, and N. Peyghambarian: Nature **383**, 58 (1996).
31. C. Poga, P.M. Lundquist, V. Lee, R.M. Shelby, R.J. Twieg, and D.M. Burland: Appl. Phys. Lett. **69**, 1047 (1996).
32. P.M. Lundquist, C. Poga, R.G. DeVoe, Y. Jia, W.E. Moerner, M.-P. Bernal, H. Coufal, R.K. Grygier, J.A. Hoffnagle, C.M. Jefferson, R.M. Macfarlane, R.M. Shelby, and G.T. Sincerbox: Opt. Lett. **21**, 890 (1996).
33. J.J. Stankus, S.M. Silence, W.E. Moerner, and G.C. Bjorklund, Opt. Lett. **19**, 1480 (1994).

25. E. Hendrickx, B.L. Volodin, D.D. Steele, J.L. Maldonado, J.F. Wang, B. Kippelen, and N. Peyghambarian: Appl. Phys. Lett. 71, 1159 (1997).
26. A. Grunnet-Jepsen, C.L. Thompson, R.J. Twieg, and W.E. Moerner: Appl. Phys. Lett. 70, 1515 (1997).
27. L. Wang, Y. Zhang, T. Wada, and H. Sasabe: Appl. Phys. Lett. 69, 728 (1996).
28. E. Hendrickx, J. Herlocker, J.L. Maldonado, S.R. Marder, B. Kippelen, A. Persoons, and N. Peyghambarian: Appl. Phys. Lett. 72, 1679 (1998).
29. D. Wright, M.A. Díaz-García, J.D. Casperson, M. DeClue, W.E. Moerner, and R.J. Twieg: Appl. Phys. Lett. 73, 1490 (1998).
30. B. Volodin, B. Kippelen, K. Meerholz, B. Javidi, and N. Peyghambarian: Nature 383, 58 (1996).
31. C. Poga, P.M. Lundquist, V. Lee, R.M. Shelby, R.J. Twieg, and D.M. Burland: Appl. Phys. Lett. 69, 1047 (1996).
32. P.M. Lundquist, C. Poga, R.G. DeVoe, Y. Jia, W.E. Moerner, M.-P. Bernal, H. Coufal, R.K. Grygier, J.A. Hoffnagle, C.M. Jefferson, R.M. Macfarlane, R.M. Shelby, and G.T. Sincerbox: Opt. Lett. 21, 890 (1996).
33. J.J. Stankus, S.M. Silence, W.E. Moerner, and G.C. Bjorklund: Opt. Lett. 19, 1480 (1994).

Photopolymer Systems

R.T. Ingwall and D. Waldman

1 Introduction

Photopolymers, in the context of this chapter, are systems of organic molecules that rely on photoinitiated polymerization to record volume phase holograms. Characteristics such as good light sensitivity, real-time image development, large dynamic range, good optical properties, format flexibility, good image stability, and relatively low cost make photopolymers one of the most promising materials for write-once, read-many (WORM) holographic data storage applications. We first present a brief description of the chemistry of photopolymers. This is followed by a discussion of photopolymer properties that are particularly relevant to the requirements of recording holograms for data storage. The intent of this chapter is to provide the nonspecialist with an appreciation of the possibilities and promise offered by photopolymers as candidate materials for holographic data storage, and to highlight areas where additional material development is desired. Readers interested in other aspects of photopolymers in holography can consult reviews by Lessard and Manivannan [1], Lougnot [2], and Colburn [3].

2 Chemistry of Photopolymer Systems

Photopolymer systems for recording holograms typically comprise one or more monomers, a photoinitiation system, and an inactive component often referred to as a binder. Other components are sometimes added to control a variety of properties such as pre-exposure shelf life, and viscosity of the recording medium. The resulting formulation is typically a viscous fluid, or a solid with a low glass transition temperature, that is prepared for exposure either by coating on a solid or flexible substrate, or by containing it between two transparent solid substrates.

2.1 Monomers

Vinyl monomers, such as acrylate and methacrylate esters, are used in most photopolymer systems. These monomers polymerize through a free radical mechanism as indicated in Fig. 1.

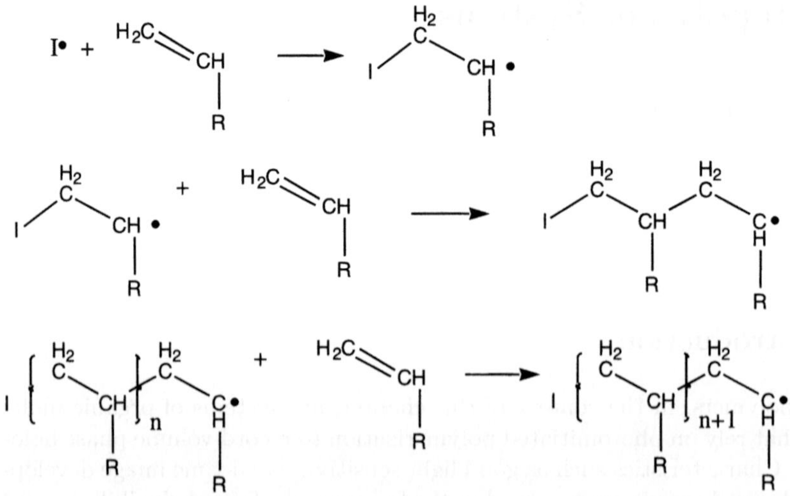

Fig. 1. Schematic of free radical polymerization of vinyl monomer

The polymerization is initiated when a monomer combines with a reactive radical species to produce the first unit in an actively growing chain. Chain growth then proceeds through successive addition of monomers to the active chain end. Growth stops when one of several possible termination reactions deactivates the radical at the growing chain end.

A wide variety of vinyl monomers are available, and can be included in recording systems to control important holographic properties [4]. Monomers that contain more than one vinyl group, for example, are often added to produce a molecular architecture comprising a cross-linked network of polymer chains that improves dimensional stability and image fidelity. Other properties that can be influenced by monomer choice are recording sensitivity, dynamic range, shrinkage, and environental stability of the exposed hologram.

Many comprehensive treatments of free radical polymerization are available, including a textbook by Odian [5], and a three-volume series devoted to radiation curing of polymers that is edited by Fouassier and Rabek [6]. In this chapter the discussion of free radical polymerization is confined to aspects that are particularly relevant to holographic data storage, such as initiation and termination reactions, polymerization rates, and copolymerization.

Monomers capable of polymerizing by a cationic ring-opening mechanism (CROP) have recently been applied to volume holographic recording [7]. An example of a useful monomer in this category is shown in Fig. 2. A general polymerization mechanism is also presented in Fig. 2; here only the active part of the monomer is shown.

Chain growth is initiated in this case when a strong acid protonates the epoxy oxygen. A second monomer reacts with this activated monomer

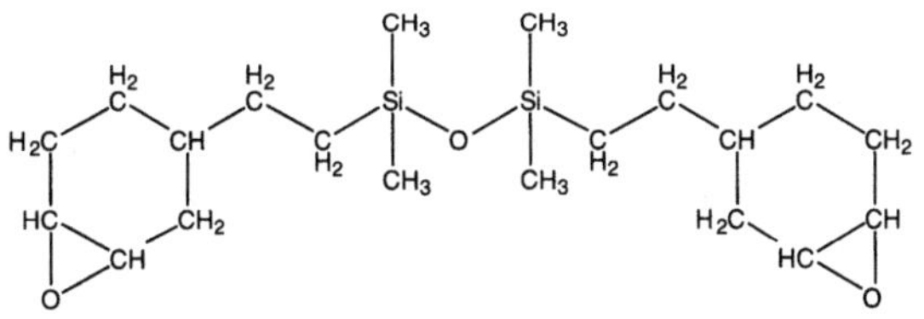

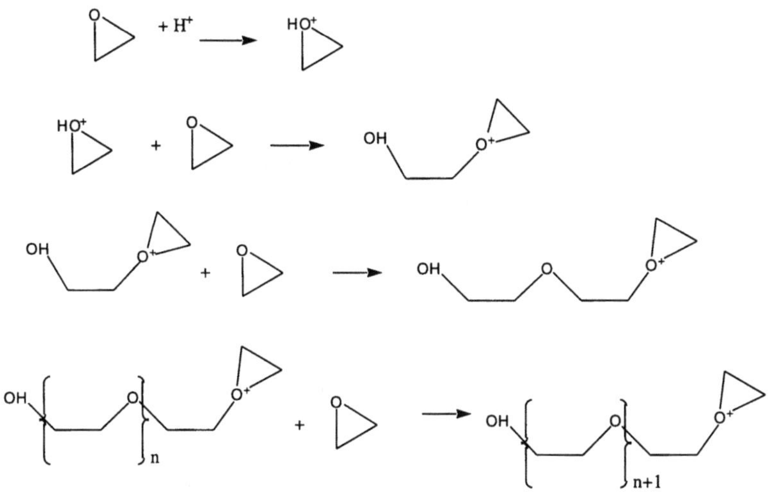

Fig. 2. Schematic of cationic ring-opening polymerization

through a ring-opening mechanism, typically forming a tertiary oxonium ion and the beginning of a growing chain. Polymerization continues as additional monomers add through the indicated ring-opening reaction sequence to form a chain with $(n + 1)$ repeat units. The predominant termination process is likely to be reaction of the growing chain end with nucleophilic impurities in the system. Many CROP monomers are available for incorporation into holographic recording systems. Specific choices are made according to the demands of the recording material [7]. Cationic polymerization is discussed in considerable detail by Odian [5] and in a book edited by Matyjaszewski [8].

2.2 Photoinitiation Systems

Holographic photopolymer systems typically use at least two different molecules to form a photoinitiation system that is sensitive to the visible wavelengths commonly used in holography. A photosensitizer molecule absorbs the imaging light, and in its excited state interacts with a radical generator or acid generator molecule, either through energy transfer or through a redox reaction, to produce the initiating species. Many photosensitizers that act through redox reactions are bleached during hologram recording and produce holograms that do not absorb light at the recording/reading wavelength. Useful discussions of sensitized vinyl photopolymerization are presented in Vol. II of [6], and in the review by Monroe and Weed [9]. Photosensitization of cationic polymerization in particular is reviewed by Pappas et al. [10].

With the proper choice of photosensitizer, holograms can be recorded throughout the spectral range from ~400 nm to ~650 nm. Du Pont, for example, offers photopolymer systems with blue, green, red [11] or panchromatic photosensitivity [12]. The availability of low cost, high performance diode lasers in the far-red and near-infrared makes this spectral region particularly attractive for data storage applications. Unfortunately, it has proven difficult to develop near-infrared photosensitization systems because of the relatively low energy of the imaging photons. At least two approaches have been described to overcome this limitation. Lougnot [13] and colleagues have described several cyanine dyes that act as photosensitizers upon direct excitation with near-infrared light. Braeuchle et al. [14] and Lougnot et al. [15] have explored systems that use two photons, a so-called gating photon and a recording/reading photon. Braeuchle et al. use a two-photon, four-level system in which the gating photon, produced by an incoherent UV source, excites an α-diketone, such as biacetyl, and populates its first excited triplet state. The imaging photon produces a transition from the first to the second excited triplet state, which results in molecular fragmentation and the production of initiating radicals. Diffraction efficiencies as large as 10% were formed using imaging sources in the 750–1100 nm range and with poly-α-cyanoacrylate as the host matrix. Lougnot et al. use the laser photon to produce an image-wise distribution of photoinitiator molecules through dye-sensitized generation of singlet oxygen. Subsequent irradiation with a uniform, incoherent UV source develops the hologram by producing polymer only in the regions containing photoinitiator.

Although these two-photon systems were developed to extend the spectral range of photopolymers, they offer an additional advantage for holographic data storage in which many holograms are multiplexed in the same volume of recording material. Since hologram recording requires both gating and recording/reading photons, data can be read from incompletely multiplexed volumes, without interfering with subsequent data recording in these volumes, by using only the recording/reading photon.

2.3 Binders

Monomers and the photoinitiating system are usually mixed with a third component, traditionally called a binder, to form a photopolymer system. The binder is sometimes a polymer that is included to modify the viscosity of the formulation, to aid sample preparation, and to enhance holographic exposure. Binders can also be small molecules or oligomers that are required for the development of the holographic image, as described in Section 3.

Systems with polymeric binders include Du Pont's HRS recording materials [16], while Polaroid's ULSH photopolymer employs either a small-molecule or polymeric binder [7].

3 Recording Characteristics of Photopolymers

For some holographic applications photopolymerizable materials are coated on a flexible plastic substrate in preparation for recording. The demands of data storage applications, however, require the use of high-quality optical substrates with excellent flatness, small wedge, low scatter, and often with an antireflection coating. The usual way to meet these requirements is to coat a fluid-like photopolymerizable material between two solid glass substrates. The glass is chosen to meet the specific optical requirements of the application and to closely match the refractive index of the recording material so that reflections at internal surfaces are minimized. Solid spacers are generally used to separate the substrates and to determine the thickness of the recording layer; the size and shape of the sample are limited only by substrate availability.

Photopolymer systems can spontaneously develop their holographic image during recording without requiring post-exposure processing steps. This real-time recording characteristic eliminates the need for complicated development procedures, and is particularly useful for applications, such as data storage, that cannot conveniently be carried out in a specialized laboratory. Systems that exhibit real-time image development, however, are susceptible to the recording of noise gratings, particularly during long exposures or during multiplexing. Apparently, unwanted gratings, formed by interference patterns derived from stray light or from partial reconstruction of previously recorded gratings, develop during recording and produce additional stray light. Noise gratings will scatter light and increase bit error rate in data storage holograms. Careful choice of multiplexing methods can substantially reduce these effects [17].

A blanket exposure with incoherent UV or visible light polymerizes unreacted monomer that remains after recording and consequently fixes the holographic image. Readout with the original coherent reference beam will also fix data holograms, but may cause recording of additional noise gratings. Such data readout will limit additional hologram recording, and thus it should be attempted only after completing a multiplexed exposure that consumes the entire dynamic range of the recording medium. Photopolymerization is, in general, not a reversible process. Photopolymer holograms, therefore, can not be erased and reused; they are suitable for WORM applications only.

We illustrate photopolymer recording characterisitcs by using the Polaroid ULSH photopolymer system as a representative example [18]. Plane-wave transmission gratings were recorded in 100 μm thick samples; the results are shown in Fig. 3. The CROP formulations comprised a functionalized reactive

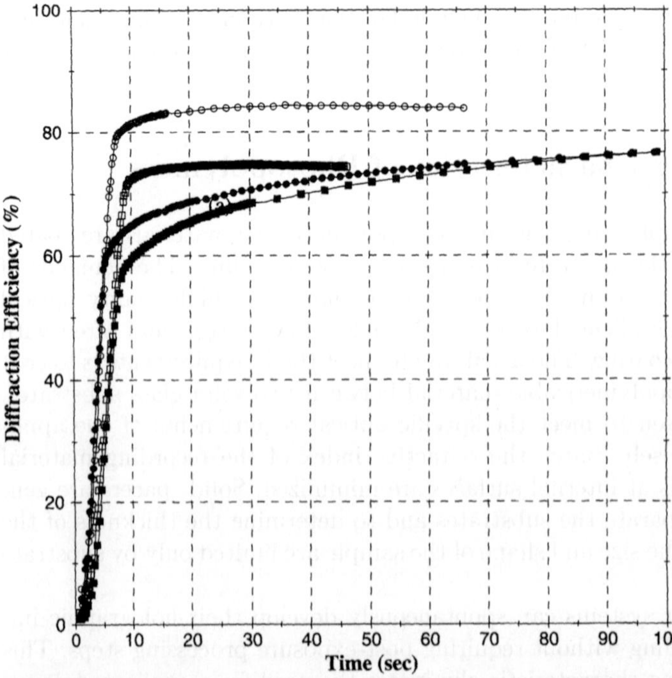

Fig. 3. Recording response of 100 μm thick coatings prepared from ULSH-500 formulations comprising a reactive copolymer. The real time growth in diffraction efficiency for plane wave gratings is plotted as a function of time for a write irradiance of 4.85 mW/cm^2 and an exposure fluence of 25 mJ/cm^2 (•), 34 mJ/cm^2 (○), 45 mJ/cm^2 (■), and 49 mJ/cm^2 (□)

copolymer in concentration ratios of about 2.0 (w/w) (•,○) and 1.0 (w/w) (■,□), respectively, relative to that of a tetrafunctional epoxide monomer, while other constituents of the medium were maintained at constant concentration. The recording exposure was carried out with an Ar^+ laser at $\lambda = 514.5$ nm using a write irradiance of 4.85 mW/cm^2. The growth in diffraction efficiency, η, was monitored during and subsequent to exposure with a He:Ne probe beam at the appropriate read angle. Diffraction efficiency is plotted as a function of time for an exposure fluence of about 25 mJ/cm^2 (•) and 34 mJ/cm^2 (○), and for 45 mJ/cm^2 (■) and 49 mJ/cm^2 (□), respectively. The wavelength of the probe beam does not overlap the absorption band of the sensitizing dye and therefore does not interfere with the grating exposure. There is an initial, or threshold, exposure of about 4.0 mJ/cm^2 (•,○) and 8.0 mJ/cm^2 (■,□), respectively, that does not produce detectable holographic activity ($\eta < 0.1\%$) which remains stable. After the threshold is exceeded the diffraction efficiency rises rapidly with additional exposure, and then saturates at a high value which is near 100% depending upon the magnitude of the exposure fluence. Recording sensitivity, about 3.2 cm/mJ (•,○)

and 2.9 cm/mJ (■,□), respectively, is influenced by the compositional ratios of the recording medium's constituents. Sensitivity is calculated by dividing the maximum slope of the square root of the diffraction efficiency plotted against exposure fluence by the sample thickness. Some additional growth in diffraction efficiency can occur subseqent to exposure (dark growth) over a time period of seconds to minutes. The magnitude of dark growth, as shown in Fig. 3, depends upon the composition of the recording medium, as well as the exposure conditions (i.e. intensity and fluence) and the physical state of the medium.

The high value obtained for the diffraction efficiency strongly suggests a volume-phase mechanism in which the holographic interference pattern is recorded as an image-wise variation of refractive index throughout the entire volume of the photopolymer. Accordingly, we apply Kogelnik's coupled wave theory [19] and obtain a value for the refractive index modulation, Δn, of 0.00275 for the the case of less than fully exposed holograms that attain diffraction efficiencies of about 95% and do not exhibit evidence of overmodulation.

4 Recording Mechanism

4.1 Refractive Index Changes

Tomlinson and Chandross have discussed possible photochemical mechanisms for refractive index changes in organic systems [20]. The most relevant for hologram recording with photopolymers include alternation in molecular electronic structure, density changes, and spatial segregation of system components. Examination of the Lorentz–Lorenz relationship of (1) for an isotropic, ideal mixture can provide estimates of these effects [21]:

$$\frac{n^2 - 1}{n^2 + 2} = \sum_i \frac{\rho_i R_i}{M_i} . \tag{1}$$

Here n is the average refractive index, ρ_i is the density, M_i is the molecular weight, and R_i is the molar refraction of the ith component of the mixture. Differentiation of (1) for a single component reveals the effects that small changes in ρ and R have on the refractive index [20]:

$$\Delta n = \left\{ \frac{(n^2 + 2)(n^2 - 1)}{6n} \right\} \left\{ \frac{\Delta R}{R} + \frac{\Delta \rho}{\rho} \right\} . \tag{2}$$

Change of electronic structure upon polymerization is accounted for by changes in molar refraction when a monomer is incorporated into the polymer chain. Generally, there is limited experimental data, and thus values for molar refraction are usually estimated from group contribution theory [21]. During vinyl polymerization the double bond of the monomer is converted to a single bond in the backbone structure of the polymer. This change in bond order can be responsible for a significant change in molar refraction, as indicated by Tomlinson and Chandros [20]. They estimate contributions to refractive index change from changes in molar refraction upon polymerization to be ~ -0.05 for the vinyl monomers methyl methacrylate and styrene. CROP monomers, however, polymerize with little change in their electronic structure. Molar refraction changes are, therefore, unimportant for holographic recording with these monomers.

Before polymerization monomers cannot approach closer to one another than the sum of their individual van der Waals radii. After polymerization monomers are linked by covalent bonds and thus the distance between connected monomers is shortened. Polymerization, therefore, almost always produces a decrease in volume and a concomitant increase in density [22]. Shrinkage is partially compensated for during CROP by the volume increase that accompanies ring opening, as two of the previously bonded atoms in the ring adopt positions with separation distances of greater than their individual van der Waals radii. Polymerization-induced volume shrinkage for vinyl and cationic ring-opening monomers is compared in Fig. 4 as a function of increasing monomer molecular weight [22].

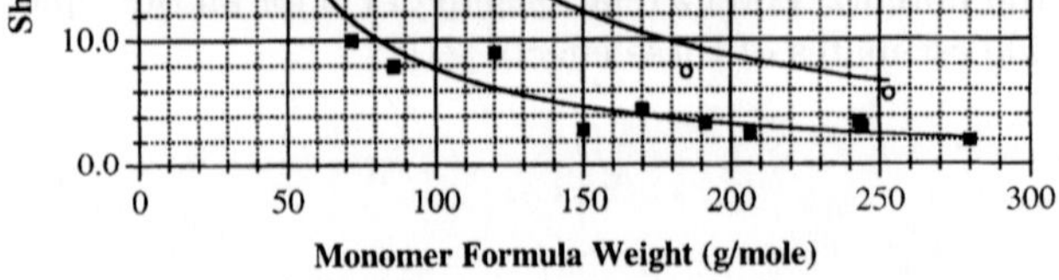

Fig. 4. Comparison of the volume shrinkage upon polymerization of vinyl (○) and cationic ring-opening monomers (■) as a function of molecular weight or epoxy equivalent weight for multifunctional chemical structures

The volume decrease accompanying polymerization becomes a smaller fraction of the total volume as the size of the monomer increases. This accounts for the results in Fig. 4 that show volume shrinkage decreases as monomer molecular weight increases. The difference in shrinkage between vinyl and CROP monomers is most important for the low and intermediate sizes that are often used in photopolymer recording systems. For a molecular weight of 100, for example, the two curves of Fig. 4 give shrinkage of 18% and 11% for vinyl and CROP monomers, respectively. We estimate the effects of density changes during polymerization on refractive index modulation by substituting these values into (2) and assuming a value of 1.5 for the refractive index of both monomers. The resulting values for Δn are 0.11 for the vinyl monomer and 0.07 for the CROP monomer.

A density increase in these essentially incompressible materials must occur by mass transport. During exposure contiguous volume elements, undergoing relatively slow and relatively fast polymerization, are established in the recording material by the spatial variation of light intensity in the holographic interference pattern. Density increases in the rapidly polymerizing volume elements require material influx from the surrounding volumes. Spatial variations in polymerization rates produce monomer concentration gradients that also drive monomer diffusion. These two effects combine to produce spatial segregation of the various components of the photopolymer system. If the components have different refractive indices their spatial segregation will make a significant contribution to the required refractive index modulation.

Equation (1) can be evaluated with simplifying assumptions to illustrate contributions to refractive index changes that are possible from component segregation. For example, consider a two-component, incompressible fluid comprising a monomer and a binder with equal bulk densities of 1.0 gm/cm^3. This system has only one independent compositional variable, which we take as the monomer weight fraction, w_1. Refractive indices of 1.470 for the monomer and 1.579 for the binder were chosen to match the corresponding components of Polaroid's ULSH photopolymer. From (1) it can be shown that a change in w_1 from 0.5 to 0.6 produces a Δn of about 0.01 in this model system.

It is apparent that each of the three mechanisms discussed above can produce significant changes in refractive index. Component segregation is required, however, to form stable holograms. The medium is fixed after holographic exposure by illumination with a flood post-exposure of UV or visible light to complete uniform monomer polymerization without producing additional component segregation. Post-exposure polymerization immobilizes the remaining reactive monomers by bonding them into a rigid, three-dimensional, polymer network and stabilizes the image-wise component segregation that was produced during hologram recording. Those regions of the hologram that correspond to the bright parts of the original fringe pattern will be relatively rich in polymer and relatively poor in other components such as binder. The opposite composition will exist in the regions that correspond to the dark portions of the interference pattern. Polymer-rich and polymer-poor regions will, in general, differ in density, average molar refraction and composition. Although each of the three mechanisms will, therefore, contribute to the refractive index modulation, they all derive from component segregation. Without component segregation post-exposure illumination would produce a featureless, uniformly polymerized sample with no holographic activity.

The ideal photopolymer system would rely solely on component segregation for developing refractive index changes. Molar refraction changes will produce changes in the average refractive index between the exposed and unexposed photopolymer that may distort reconstructed images. Density changes will cause volume changes that may also distort holographic images. The ideal material will record only component segregation that is not accompanied by electronic or density changes. Although current materials rely both upon component segregation and on electron and density changes, they can produce data page holograms with low bit error rate, as discussed in the last section of this chapter.

4.2 Component Segregation

The mechanisms producing component segregation during photopolymer recording are much more complicated than those that produce refractive index changes. Zhao and Mouroulis [23], Lougnot et al. [24], and Colvin et al. [25] have all published theoretical descriptions of the time and spatial variation of monomer and polymer concentrations during hologram recording. Diffusion constants for mobile components decrease significantly with increasing recording time as the photopolymer changes from a fluid to a solid. Since the extent of polymerization varies in correspondence with the interference pattern, the diffusion constants will also be a function of position in the recording material. The rate of conversion of monomer to polymer depends upon light intensity, the concentration of actively growing polymer chains, the concentrations of monomer, sensitizer dye and initiator molecules, and the physical state of the recording medium. All of these parameters will vary with time during recording and with position in the hologram. Although the three theoretical approaches referenced above account for these effects in somewhat different ways, they all emphasize the importance of the relative rates of polymerization and diffusion. Zhao and Mouroulis [23], for example, introduce a parameter R that is equal to the ratio of the initial diffusion rate to the initial maximum polymerization rate. They compare predicted refractive index profiles for recording with large and small values of R. For recording with large R they expect the refractive index profile in the finished hologram to match closely the light intensity variation in the original interference pattern. This situation results in linear recording, and produces holograms with the largest refractive index modulation and best image fidelity. For recording with small R they predict a relatively poor correspondence between the hologram's refractive index profile and the original interference pattern. Excess polymer collects at the boundary between the bright and dark regions of the interference pattern when monomer diffusion rate is much smaller that the polymerization rate. Polymer concentration is consequently depleted in the interior of the bright regions. This distorted refractive index profile may produce holograms with diminished dynamic range and degraded image fidelity. The use of formulations with low initial viscosity that contain relatively small monomers in relatively low concentrations, the incorporation of inert plasticizers in the formulation, and the recording of images with small fringe spacing, will all act to increase the monomer diffusion rate and increase R. Low light intensity during exposure will reduce monomer polymerization rate and thus will also increase R.

5 Recording Thick Photopolymer Holograms

A particularly useful format for holographic data storage with photopolymers comprises a layer of the active recording material held between two disk-shaped substrates [26,27]. Since the storage capacity of these disks increases nearly linearly with recording layer thickness [26] the capability of recording high-performance holograms in relatively thick photopolymer media is essential. Layer thickness should be 500 μm or greater to produce a competitive storage capacity [27]. Recording thick holograms in photopolymers is made difficult both by the absorption of light by the photosensitizer, and by the low viscosity of the photopolymer before exposure.

5.1 Light Absorption

Light attenuation as the exposing beams pass through the thickness of the sample produces a greater exposure at the front of the recording layer than at the back [28]. For light absorption that follows Beer's law, polymerization rates will decrease exponentially with depth of penetration through the hologram. Nonuniform exposure often results in distorted reconstructed images [29] and increased bit-error rates in multiplexed holograms. It is necessary, therefore, to reduce either the photosensitizer concentration or the absorption cross-section of the photosensitizer in order to reduce exposure nonuniformities in thick holograms to acceptable levels. Our experience with CROP-based recording materials, with thicknesses greater than 100 μm, is that significant image distortions result if a maximum optical density at the recording wavelength of ~0.2 (or minimum light transmission of ~60%) is exceeded. Adjusting the sensitizer concentration to produce the desired optical density is usually sufficient for holograms that are less than about 200 μm thick. Extremely low sensitizer concentrations, however, are often required to maintain acceptable optical densities for holograms that are thicker than about 500 μm. In many cases sensitizer molecules are converted to an inactive form during the photochemical process that leads to polymer chain initiation. One sensitizer molecule is, therefore, capable of initiating at most only a few polymer chains. In this case the minimum degree of conversion of monomer to polymer that is needed to produce high-quality, stable holograms cannot be achieved in formulations with very low sensitizer concentrations. The recording of thick holograms therefore requires special care in the choice of sensitizer. A high concentration of a sensitizer with a small extinction coefficient will allow the necessary degree of polymerization to occur during holographic exposure without exceeding the desired maximum light absorption in thick samples.

Waldman et al. [30] demonstrated that uplift of the nulls from zero intensity in Bragg angular detuning profiles of plane-wave transmission gratings is caused by a nonuniform grating profile. Gratings were recorded in samples of various thicknesses using the same CROP photopolymer formulation. These samples had the same sensitizer dye concentration and absorption coefficient, but the thinner samples exhibited reduced absorbance, and thus a more uniform grating profile was expected. Accordingly, for a given recording intensity the number of initiated reactions per unit volume is maintained in each coating thickness through a depth corresponding to the thinnest coating investigated. In thinner samples, however, a diminished amount of light would be absorbed before it reached the back plane of the sample, and thus the amount of light absorbed throughout the sample thickness would be more uniform during holographic recording. Similarly, the sensitivity of the recording medium, as a function of depth from the front surface, is expected to be more uniform for thinner holograms. Examples of results for measured angular selectivity profiles of plane-wave, non-slanted holograms recorded in different coating thicknesses are shown in Figs. 5a, b, and Fig. 6 for hologram thicknesses of about 30 μm, 45 μm, and 140 μm, respectively. The conventional prediction from Kogelnik's [19] coupled wave analysis is also shown in Fig. 5. It is apparent that the measured profiles increasingly depart from sinc^2 function behavior as the coating thickness of the hologram is increased. Note that the background uplift in the region of the first nulls is barely discernible for the thinnest hologram recorded, is more readily apparent in the profile of the slightly thicker hologram, and is significant for the hologram of about 140 μm thickness.

Waldman et al. [30] assumed an exponentially decaying $n_1(z)$ as a simple model for a non-uniform grating:

$$\frac{n_1(z)}{n_0} = \frac{n_{10}}{n_0}\mathrm{e}^{-\alpha_z}, \tag{3}$$

where n_{10} is the value of n_1 at the front plane ($z = 0$) of the hologram, and α is a constant that represents the extent of the exponentially decreasing index modulation through the thickness z. Although the solution to the coupled wave equation for $n_1(z)$ can be expressed in terms of Bessel functions [31], the expression is complicated. The nulls of calculated Bragg selectivity curves, obtained using the coupled wave analysis, are no longer at zero diffraction efficiency when the contribution from $n_1(z)$ is included. Additionally, as the half value of the decline in grating strength, defined for thickness L as $a = \alpha L/2$, increases, the magnitude of the nulls increases and their position moves further from the Bragg peak position at $u = 0$. If it is assumed that the decay of n_1 with depth is exponential (Fig. 3), then the theoretical results can be fitted to the experimental data when the amount of decay, αL, is small. An example case is shown in Fig. 6, where the reconstruction angles

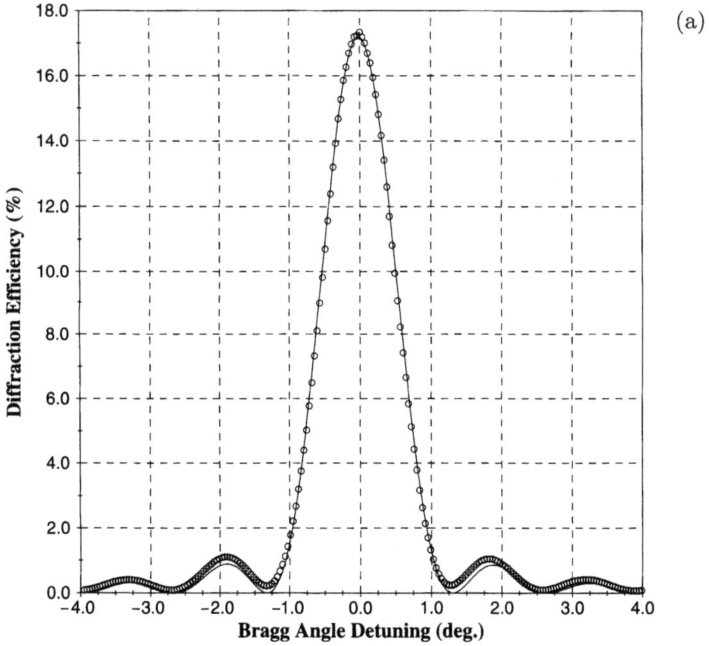

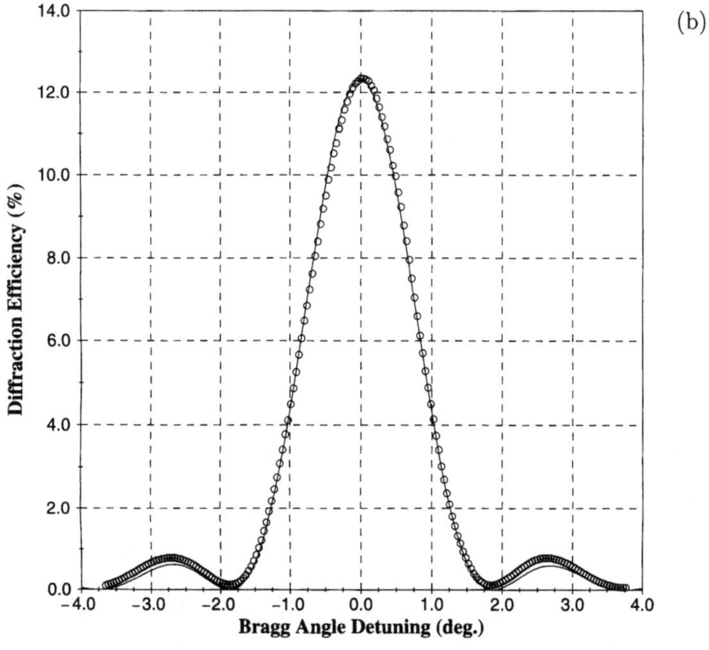

Fig. 5. Angle selectivity profiles of plane-wave, non-slanted holograms (○) recorded from the same CROP recording medium but in different coating thicknesses. Theoretical profiles according to the classical Kogelnik analysisis (—) for holograms with a thickness of 45 μm and $n_1 = 1.80 \times 10^{-3}$ for (**a**), and a thickness of 31 μm and $n_1 = 2.22 \times 10^{-3}$ for (**b**)

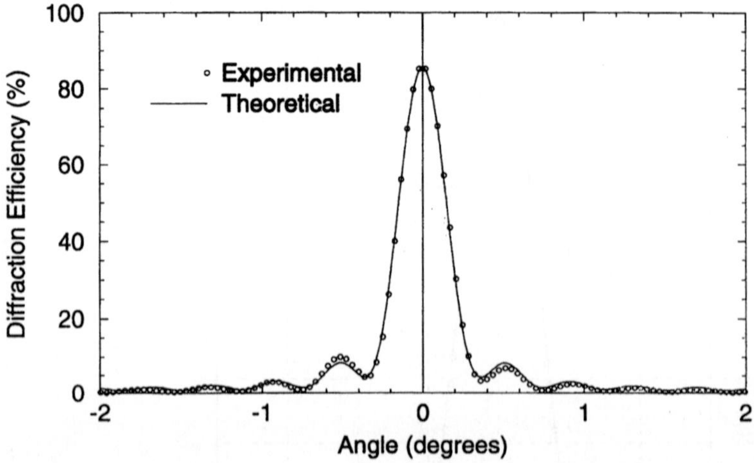

Fig. 6. Comparison between theoretical result (—) from coupled wave analysis and measured angle selectivity curve (○), where $\Omega_{1_{ext}} = \Omega_{2_{ext}} = 31°$ (external angle), $n_0 = 1.5$, and $\lambda = 633$ nm. Data fit to theory by using $a = \alpha L/2 = 0.62$, $L = 154$ µm, and $n_{10} = 0.0013$

were $\Omega_{1_{ext}} = \Omega_{2_{ext}} = 31°$ (external angle), $n_0 = 1.5$, and $\lambda = 633$ nm. The data can be fitted to theory by taking $a = \alpha L/2 = 0.62$, $L = 154$ µm, and $n_{10} = 0.0013$, whereas the value of material thickness obtained from direct measurement (using a profilometer) was $L = 136$ µm. The discrepancy in thickness could be because direct measurement of L was performed at the edge of the hologram sample, whereas the angle selectivity curve was measured at the center of the sample, suggesting that variations might exist in the coating thickness across the sample. It is more likely, however, that the assumption of n_1 exhibiting a exponential decline with depth is only approximate.

5.2 Low Viscosity

Holographic fringe structures are not stable in fluids. Problems associated with fringe instability in fluid media can be particularly severe when recording thick holograms. The initial fluid state of the photopolymers must, therefore, be converted to a form that supports stable fringe patterns before useful holograms can be recorded. This transformation is often effected by a pre-exposure processing step that partially polymerizes the monomer and produces a gel or rubbery material that supports fringes without completely inhibiting monomer diffusion. Heating at elevated temperature for a controlled time, or exposure with incoherent light, are common pre-exposure processing procedures. Pre-exposure processing must be carefully controlled, particularly for thick holograms, so that undistorted images are recorded and the dynamic range of the recording material is not seriously reduced.

The examples shown below using Polaroid's ULSH-500 photopolymer illustrate the effects of both sensitizer concentration and pre-exposure conditions on recording thick holograms.

Angular selectivity profiles of plane-wave, non-slanted holograms recorded in coating thicknesses of 100 μm and 200 μm are shown in Fig. 7a and b, respectively. A pre-imaging exposure was used in both cases, followed by an exposure fluence of about 35 mJ/cm^2 for (a) and 40 mJ/cm^2 for (b). Reconstruction of the Bragg selectivity of each hologram, at increments of $\Delta\theta = 0.01°$, was carried out with a HeNe laser at $\lambda = 632.8$ nm and with an intensity of 50 μW to prevent additional recording during readout. In neither case was the dynamic range of the recording medium consumed, as additional co-locational holograms could be multiplexed. For example, in the latter case nine holograms were co-locationally recorded using peristrophic multiplexing. The average diffraction of these was 25%, and the cumulative exposure fluence of 325 mJ/cm^2 was less than the total necessary to consume the entire dynamic range. The concentration of sensitizer had been adjusted to provide equivalent optical path length for the two coating thicknesses. It is apparent that in both cases the measured profile closely follows sinc^2 function behavior with mimimal background uplift in the region of the first "nulls" and with minimal asymmetry in the nearest satellite peak intensities.

6 Image Quality in Photopolymer Holograms

High capacity data storage demands near-perfect fidelity of the reconstructed image. The original image, therefore, must be recorded faithfully and must be free of distortions produced either by changes in the bulk refractive index of the recording material or by hologram shrinkage. Light scattering and other sources of noise, such as hologram-to-hologram or pixel-to-pixel crosstalk, and spurious gratings, must also be minimized. The best image fidelity is achieved by recording in the so-called linear regime, where the developed refractive index profile closely matches the light intensity variation of the original interference pattern [23–25]. Recording within the middle part of the dynamic range, beyond the exposure threshold and before complete saturation, should produce the best linear recording in most photopolymers. In the CROP medium, recording sensitivity is likely to be in the range of 0.5–25 cm/mJ in the linear regime, whereas sensitivity can decrease substantially in the regime nearing saturation. Changes in the bulk refractive index are reduced by minimizing shrinkage and by using monomers, such as CROP monomers, that do not change bond order upon polymerization.

6.1 Shrinkage

Shrinkage upon hologram recording is observed in essentially all photopolymer systems. Each time a monomer adds to a growing polymer chain the volume

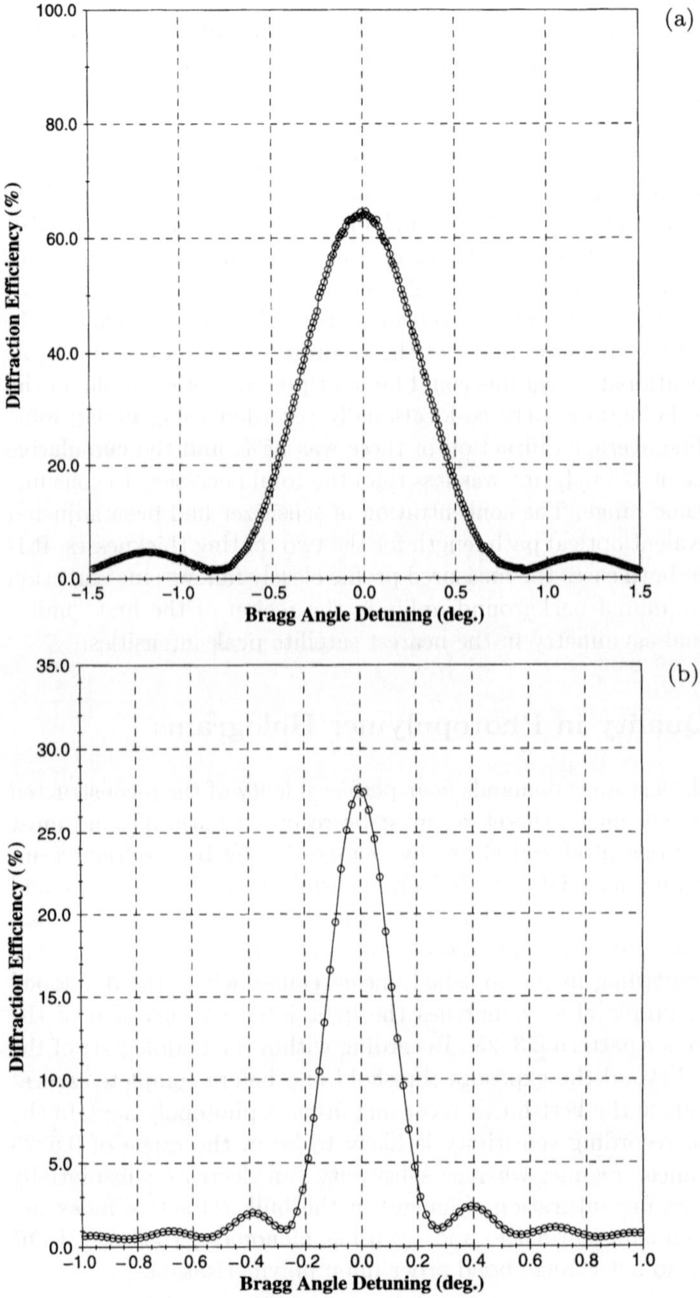

Fig. 7. Angle selectivity profiles obtained with HeNe probe laser at $\lambda = 632.8$ nm for non slant fringe plane-wave holograms recorded in (**a**) 100 μm; and (**b**) 200 μm thick ULSH-500 media, where recording was carried out subsequent to a pre-imaging exposure. Accuracy of angle positioning was 0.001° and data was collected at an angle increment of $\Delta\theta = 0.01°$. The interbeam angle was (**a**) 34°; and (**b**) 49°; for recording wavelengths of $\lambda = 532$ nm and 514.5 nm, respectively, and the exposure fluence was (**a**) 35 mJ/cm^2; and (**b**) 40 mJ/cm^2

of the system decreases as a covalent bond replaces a near van der Waals contact. In the case of CROP this volume reduction is partially compensated by a volume increase that occurs upon ring opening. See Fig. 4 for a comparison of vinyl and CROP monomer shrinkage. The polymerization of some classes of monomers, such as spiro orthocarbonates and spiro orthoesters, is reported to occur with no shrinkage or even a slight expansion [22]. It has been our experience, however, that these monomers polymerize too slowly to be practical for holographic data storage. For photopolymer systems that include multifunctional monomers, such as most holographic photopolymers, the degree of volume shrinkage is usually not directly proportional to the degree of conversion of monomer to polymer. Rapid polymerization produces a kinetic lag between polymer conversion and volume shrinkage [32]. This effect is particularly pronounced during the latter stages of polymerization after the system has solidified. Apparently, for highly cross-linked, very viscous systems, relatively rapid polymerization produces excess free volume at a greater rate than is dissipated by diffusion and thermal relaxation. This free volume excess allows continued monomer diffusion, and promotes a greater degree of polymer conversion for rapid reaction compared with slow reaction [33]. These considerations suggest that the free volume excess present in freshly recorded holograms will relax to a stable volume over a time period that varies from minutes to days. In substantially solidified systems such relaxiation can effect degradation of image fidelity over time.

Volume shrinkage, and consequent image distortion, will be reduced for holograms exposed after the recording fluid has been solidified by partial polymerization. Pre-imaging processing, such as exposure to incoherent light or by thermal treatment, can produce the desired partial polymerization. We show in the preceding section that pre-imaging partial polymerization is also required to form undistorted fringe structures in thick holograms. The initial shrinkage, manifested immediately after exposure, may slowly increase over periods of several days, as the fully exposed photopolymer relaxes to its equilibrium volume. We observe with CROP monomers that the resultant total shrinkage decreases, for the same degree of polymer formation, as the extent of pre-imaging polymerization increases. Unfortunately, increasing the pre-imaging polymerization also reduces the dynamic range of the recording material, and additionally may effect a decrease in sensitivity. A practical compromise between dynamic range and shrinkage must be established for each application without a significant attendant reduction in recording sensitivity.

The slant angles of volume phase gratings recorded in photopolymers media are altered by anisotropic volume shrinkage accompanying photopolymerization reactions. Analysis of slant fringe holograms, therefore, is a particularly convenient means of assessing the shrinkage that accompanies hologram formation in photopolymers. Shrinkage can be complicated by anisotropic volume changes that are produced by interactions at the photopolymer/substrate

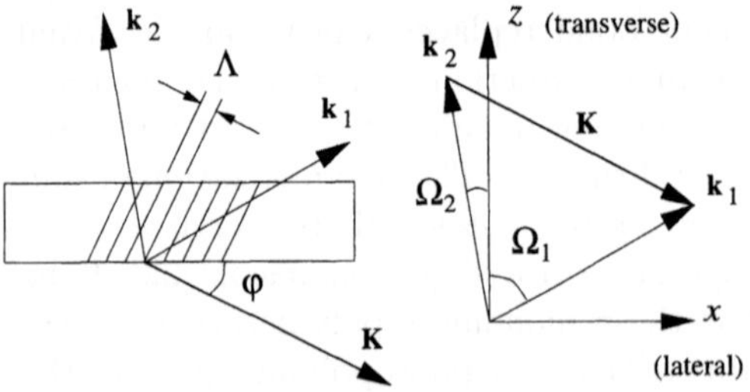

Fig. 8. Schematic of grating vector components, k_1 and k_2, in the xz-plane, for internal signal and reference beam angles Ω_1 and Ω_2, respectively, where ϕ is the internal slant angle and the isophase fringe planes are separated by the periodic distance Λ

interface. A significant degree of anchoring can occur in the lateral or film plane direction depending upon the medium thickness, and thus shrinkage in the transverse or thickness direction predominates. Waldman et al. have presented an analysis of slant fringe holograms that treats both transverse shrinkage and shrinkage in the sample plane, refered to as lateral shrinkage [30]. Their analysis considers the grating vectors and writing beam geometries shown in Fig. 8.

The grating vector components along the transverse (z-axis or thickness direction) and lateral directions (x-axis in film plane), $K_{z,i}$ and $K_{x,i}$, respectively, at the initial state of recording, can be expressed in terms of the external write beam angles, $\Omega_{1_{\text{ext}}}$ and $\Omega_{2_{\text{ext}}}$, the refractive index, n, and the recording wavelength, λ, as

$$K_{z,i} = \frac{2\pi}{\lambda}\left(\sqrt{n^2 - \sin^2 \Omega_{1_{\text{ext}}}} - \sqrt{n^2 - \sin^2 \Omega_{2_{\text{ext}}}}\right), \tag{4}$$

$$K_{x,i} = \frac{2\pi}{\lambda}\left(\sin \Omega_{1_{\text{ext}}} + \sin \Omega_{2_{\text{ext}}}\right). \tag{5}$$

Similarly, the internal grating slant angle after shrinkage can be expressed, without the assumption of surface anchoring, in terms of measurables (see Fig. 8) comprising $\Omega_{1_{\text{ext}}}$ and $\Omega_{2_{\text{ext}}}$, the differential angles (respective angular deviations from the Bragg matching condition of the two write beams), $\Delta\Omega_{1_{\text{ext}}}$ and $\Delta\Omega_{2_{\text{ext}}}$, and the refractive index change that takes place during and subsequent to recording, Δn, as

$$\phi' = 0.5\left\{\sin^{-1}\left[\frac{\sin(\Omega_{2_{\text{ext}}} + \Delta\Omega_{2_{\text{ext}}})}{n + \Delta n}\right] - \sin^{-1}\left[\frac{\sin(\Omega_{1_{\text{ext}}} + \Delta\Omega_{1_{\text{ext}}})}{n + \Delta n}\right]\right\} \tag{6}$$

Values of $\Delta K_x/K_{x,i}$, $\Delta K_z/K_{z,i}$ and $\Delta V/V$ can be determined by using (4) and (5) for an exact solution by adding $\Delta\Omega_{1_{\text{ext}}}$ to $\Omega_{1_{\text{ext}}}$, $\Delta\Omega_{2_{\text{ext}}}$ to $\Omega_{2_{\text{ext}}}$, and Δn to n, to obtain K_x and K_z after recording. The relative changes in the transverse and lateral components of the grating vector, $-\Delta K_z/K_{z,i}$ and $-\Delta K_x/K_{x,i}$, respectively, are equal to the physical material shrinkage along the respective directions independent of slant angle, ϕ, where $\Delta K_z/K_z = -\Delta\Lambda_z/\Lambda_z$ and

$\Delta K_x/K_x = -\Delta\Lambda_x/\Lambda_x$ for the grating period components in the transverse, Λ_z, and lateral, Λ_x, directions, respectively. To assess the relative volume change, $\Delta V/V$, during recording requires an assumption of homogeneity in the lateral directions (i.e. $\Delta x/x = \Delta y/y$), which is reasonable for volume hologram formation in photopolymers considering that the final recorded state is a glassy material. The relative volume shrinkage can thus be expressed as

$$\frac{\Delta V}{V} = -\left(\frac{\Delta K_z}{K_{z,i}} + \frac{2\Delta K_x}{K_{x,i}}\right). \tag{7}$$

The extent of transverse and lateral shrinkage was determined explicitly for a range of slant angles and imaging conditions. Angle multiplexing schemes for holographic data storage require holograms with low diffraction efficiency, typically less than about 0.1% [34], and which exhibit excellent angular selectivity and image quality. It was shown that both properties can be achieved for slant fringe holograms recorded with low diffraction efficiency in CROP photopolymer, with a concomitant reduction in material shrinkage, by utilizing a modicum of pre-imaging exposure that preserved significant dynamic range [30]. Holograms were recorded to low diffraction efficiency in the photopolymer holographic recording material, ULSH-500, and then flood-exposed subsequent to imaging and prior to measurement, in order to simulate the increase in density that occurs when holograms are multiplexed until the entire dynamic range of the medium is consumed. The angular deviations measured according to this experimental protocol, and which increased nonlinearly with slant angle, conservatively estimate the maximum values exhibited when recording is carried out with angle multiplexing schemes (i.e. the first few holograms recorded will display larger shifts than later recorded holograms owing to densification effected by later recording events).

Angular deviations for grating slant angles of $5° \leq \phi \leq 20°$ are shown in Fig. 9 for a range of pre-imaging exposure conditions. Two data points are plotted for selected slant angle and pre-exposure conditions to indicate the range of values obtained for angular deviations across multiple experiments. Figure 10 shows angular selectivity profiles of $\Omega_{1_{\text{ext}}}$ and $\Omega_{2_{\text{ext}}}$ for an internal slant angle $\phi \cong 20°$, where the pre-imaging exposure fluence was 20 mJ/cm^2 at $\lambda = 514.5$ nm and holographic recording provided a diffraction efficiency of about 0.2%. Deviations from the Bragg matching condition were $-0.28°$ and $0.39°$ for $\Delta\Omega_{1_{\text{ext}}}$ and $\Delta\Omega_{2_{\text{ext}}}$, respectively, and sinc^2 function like behavior was exhibited. The hologram in Fig. 10 was recorded in a medium comprising an increased ratio of binder to monomer. The corresponding angular deviations measured for holograms recorded with $\phi \cong 10°$ and $20°$, for a pre-imaging exposure fluence of 20 mJ/cm^2, are shown in Fig. 9 by the inner and outer pairs of thick vertical bars, respectively. Decreasing the monomer content thus reduced shrinkage further for the same level of pre-exposure and without sacrificing dynamic range.

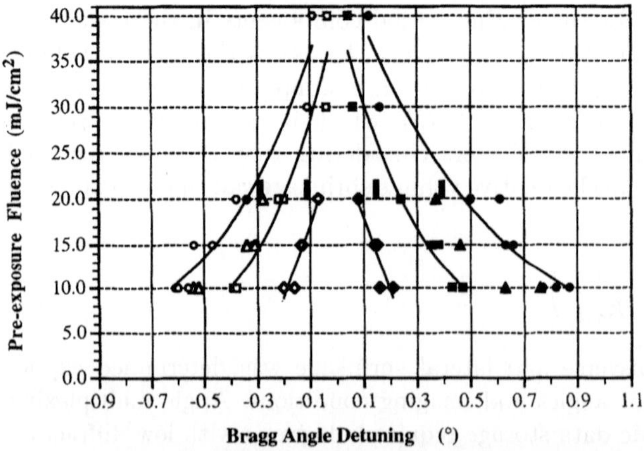

Fig. 9. Plot showing angular deviations from Bragg matching condition, $\Delta\Omega_{1_{\text{ext}}}$ (open symbols) and $\Delta\Omega_{2_{\text{ext}}}$, (filled symbols), as a function of internal slant angle, ϕ, and pre-imaging exposure fluence, for slant fringe plane-wave volume holograms recorded to low diffraction efficiency, and subsequently flood-exposed at $\lambda = 532$ nm to consume the full dynamic range of the CROP photopolymer, ULSH-500. The recording internal slant angle ϕ was about 5° (◊), (♦); 10° (□), (■); 15° (△), (▲); and 20° (○), (•), respectively. Solid lines represent curve fits

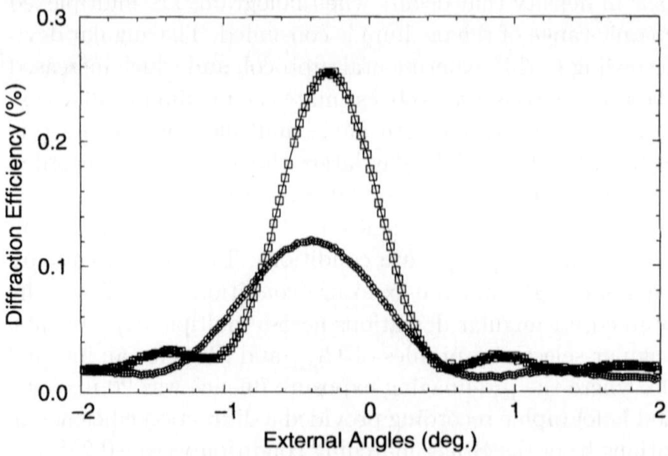

Fig. 10. Angle selectivity profiles of $\Omega_{1_{\text{ext}}}$ (□) and $\Omega_{2_{\text{ext}}}$ (○) showing angular deviations, $\Delta\Omega_{1_{\text{ext}}}$ and $\Delta\Omega_{2_{\text{ext}}}$, from Bragg matching condition for slant fringe hologram with internal slant angle, ϕ, of $\cong 20°$ recorded in a formulation with decreased monomer content. Pre-imaging exposure fluence was 20 mJ/cm^2 at $\lambda = 514.5$ nm, and angle shifts are steady-state values after a post-imaging exposure. Resolution of data collection was 0.001° and increment was 0.02°

A pre-imaging exposure fluence between 30 and 40 mJ/cm^2 was sufficient to significantly reduce the magnitudes of the angular deviations to $\leq 0.1°$, even for an internal grating slant angle of $\phi \cong 20°$. The calculated values for the relative change in the transverse (perpendicular to film plane) component of the grating vector, $-\Delta K_z/K_{z,i}$, for $\phi \cong 20°$ decreased to $\leq 0.2\%$ for a pre-imaging exposure fluence ≥ 30 mJ/cm^2. For $\phi \cong 10°$ the values of $-\Delta K_z/K_{z,i}$ decreased by factors of 7.5 and 11.2 for pre-exposure fluence of 30 and 40 mJ/cm^2, respectively, relative to that displayed at a pre-imaging exposure of 20 mJ/cm^2. The magnitude of the shrinkage exhibited by CROP photopolymer recording materials in the transverse direction can thus be rapidly diminished to acceptably low levels by using a pre-imaging fluence prior to angle multiplexing holograms of low diffraction efficiency. The calculated material volume shrinkage for holograms recorded with $\phi \cong 10°$ and $\phi \cong 20°$ was reduced to 0.11% and 0.09%, respectively, for a pre-imaging exposure fluence of 40 mJ/cm^2. Moreover, angular selectivity profiles of these low diffraction efficiency holograms exhibited the expected sinc function behavior with minimal background uplift for pre-exposure fluence of ≥ 10 mJ/cm^2, and good image uniformity was observed over the diffraction area.

Shrinkage in photopolymers is managed both by controlled pre-imaging polymerization and by optimizing chemical components and component ratios. The preceding section emphasized effects of pre-imaging processing. Photopolymer formulations can be optimized for reduced shrinkage by choosing large, low-shrinkage monomers, such as CROP monomers, and by including a high content of inert components.

7 Data Storage in Photopolymer Holograms

7.1 Multiplexing

Multiplexing many holograms in the same volume of recording material is necessary for high-capacity data storage [26,27]. Since hologram recording in photopolymers relies primarily on component segregation through diffusion to produce needed refractive index changes, the choice of a multiplexing scheme for data storage in photopolymers should consider requirements for optimized component segregation. Image-wise polymerization can establish preferred directions for monomer diffusion by producing polymerized fringe planes with high local viscosity and low diffusion coefficients. Some angle multiplexing procedures, for example, produce a polymerized fringe structure that blocks monomer diffusion in a direction parallel to the incident plane of the writing beams while allowing perpendicular diffusion. Since angle multiplexing requires monomer diffusion in a fixed incident plane, the polymerized fringes significantly limit the number and efficiency of holograms that can be recorded in this manner. The peristrophic multiplexing procedure of Curtis et al. [35] avoids this complication, and when combined with angle multiplexing [34–37] by the procedure of Pu et al. [38], the full dynamic range of the recording material can be utilized.

The number of usable holograms that can be multiplexed in the same sample volume is directly related to the recording material's maximum refractive index modulation, n_1. A convenient way to determine n_1 for film thicknesses in which holograms can be over-modulated (i.e. typically thickness ≥50 μm) is to carry out a determination of the cumulative or saturation grating strength, ν_M, as a function of exposure energy, described by Pu et al. [38] and defined here as

$$\nu_M = \sum_{i=1}^{M} \sqrt{(\eta_i)} \text{ for } M \text{ multiplexed holograms,} \tag{8}$$

where

$$\sqrt{\eta_i} = \sin \nu_i \simeq \nu_i \tag{9}$$

for the case of low diffraction efficiency in a photopolymer recording medium, and

$$\nu_i(\lambda) = \frac{\pi n_1(\lambda) T}{\lambda \cos \theta_{\text{int}}} \tag{10}$$

for diffraction efficiency at the Bragg condition, with the additional consideration of wavelength dispersion [18] in n_1. Moreover, the recording behavior of photopolymers (and thus n_1) is additionally dependent upon both the grating slant angle and the grating period.

Figure 11 shows the growth in cumulative grating strength, $\Sigma\eta_i^{0.5}$, for 1050 plane-wave holograms recorded sequentially and co-locationally in a 200 μm thick coating of the ULSH photopolymer comprising a functionalized copolymer. The pre-imaging fluence was 20 mJ/cm^2 at an irradiance level of 0.4 mW/cm^2 and the recording irradiance was 4.85 mW/cm^2, both at $\lambda = 514.5$ nm. A calculated exposure schedule was used for multiplexing the 1050 holograms, where the exposure times were increased monotonically from about 7 ms for hologram #1 to 10 ms by hologram #300, to 26 ms by #600, to 164 ms by #900, and 1.1 seconds for #1000, after which recording times were increased by coarser and nonscheduled increments. The initial value of the sample plane rotation, θ, from the bisector condition was −16.0°, and subsequent values, rotated after each peristrophic cycle by increments of $\Delta\theta = 5°$, were −11.0°, −6.0°, −1.0°, +4.0°, +9.0°, and +14.0°. Peristrophic rotation was carried out with constant increments of $\Delta\varphi = 1.2°$ for each grating angle condition, corresponding to 150 co-locational recordings per cycle. The manifold of the cumulative grating strength ν_M, shown in Fig. 11b, exhibited a desirable linear growth behavior as a function of sequential recordings, attaining a substantial value of 15.7 for measurement with an argon ion laser at $\lambda = 514.5$ nm. The linear growth in $\Sigma\eta_i^{0.5}$ exhibited in Fig. 11b establishes

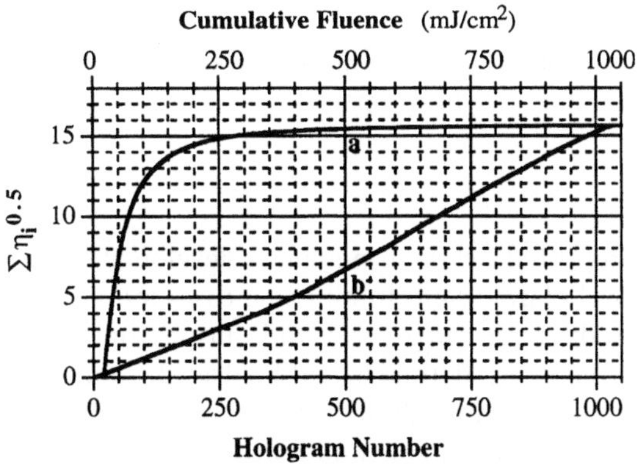

Fig. 11. Growth in cumulative grating strength, $\Sigma\eta_i^{0.5}$, as a function of the number of plane-wave holograms recorded co-locationally *a*; and cumulative fluence *b*; using peristrophic and angle multiplexing in a coating of 200 μm thickness for ULSH-500 formulation comprising polyfunctional copolymer. The pre-imaging exposure was 20 mJ/cm^2, the recording irradiance was 4.85 mW/cm^2, and a calculated exposure schedule was applied for holograms #1–#1000. Measurement of η_i was at $\lambda = 514.5$ nm

that the recording characteristics (i.e. sensitivity, etc.) of the CROP medium are predictable over its reaction history, which extends from a gelatinous state at low extents of reaction to a polymeric glass physical state at high extents of reaction. The growth in ν_M as a function of cumulative fluence, shown in Fig. 11a, exhibits rapid nonlinear growth characteristics, attaining 95% of its final value after a fluence of 250 mJ/cm^2, at which point 973 holograms had been co-locationally multiplexed. The 50% value for I_{wr}, defined as the recording fluence required to attain 50% of the cumulative grating strength, was 32.7 mJ/cm^2, and the total exposure fluence was about 1060 mJ/cm^2 inclusive of the pre-imaging exposure. The manifold of recording sensitivity, determined by differentiating data in Fig. 11a, peaked at about 18.5 cm/mJ at the beginning of the recording schedule and declined in monotonic fashion to a still reasonably sensitive level of about 1.0 cm/mJ by hologram #900, where $\Sigma\eta_i^{0.5} = 13.7$. By hologram #975, where 95% of the cumulative grating strength had been attained, the sensitivity had decreased to about 0.3 cm/mJ, which is still considered adequate for holographic data storage. The high levels of recording sensitivity, exhibited when utilizing the procedure of peristrophic and angle multiplexing for the case of recording holograms with low diffraction efficiency (i.e. $\eta \sim 0.001$), are not observed when carrying out either method for recording holograms with moderate levels (i.e. $0.05 \leq \eta \leq 1.0$) of diffraction efficiency. The extent of localized microstructural hardening, which can impede diffusion, is apparently more significant for the case of one or several grating vector recording conditions, when holograms of moderate or strong diffraction efficiency are recorded and the dynamic range is consumed, than for the case when many holograms are recorded co-locationally and each has both low diffraction efficiency and a grating vector with different directionality in K-space.

7.2 Data Page Recording

Recording data pages has been reported for photopolymers developed by Du Pont [27], by Lucent [39], and by Polaroid [40]. In all these cases storage capacity is limited by thickness and dynamic range. The Du Pont photopolymer was available in a maximum thickness of 100 μm. Pu and Psaltis [27] obtained a storage density of 10 bits/mm^2 by recording 32 multiplexed holograms with 590 kbits per hologram. The storage capacity was limited by the small thickness of the recording material. They reported a maximum recording rate of 0.7 Mbits/s using a total writing intensity of 2 mW/cm^2. Readout signal-to-noise ratio, SNR, dropped as the storage density increased, and obtained a value of 3.5 for their maximum density of 10 bits/mm^2. Shrinkage, estimated to be ~3%, proved to be an important noise source. Raw bit error rates (BER) were calculated to be $\sim 10^{-4}$.

Dhar et al. [39] recorded 480 kbit digital data pages in a Lucent photopolymer; samples with thickness between ~250 mm and ~500 mm were used. Hologram writing employed a frequency-doubled YAG laser operating with a total incident intensity of 16.5 mW/cm^2. Raw BERs of $\sim 10^{-4}$ to 10^{-3} were observed for 10 to 20 multiplexed holograms. A rise in BER with increasing number of multiplexed holograms was attributed to shrinkage, ~0.3%, and changes in average refractive index, ~0.0018. The dynamic range and sensitivity of the material were not reported.

Ingwall et al. [40] examined a Polaroid photopolymer using a specially designed and constructed holographic test bed [41]. Co-locational multiplexing of 70 holograms of 64 kbit data pages was carried out in 200-μm-thick samples, and the entire dynamic range was not consumed. Recording sensitivity, identified as the slope of the diffraction efficiency plotted against exposure fluence times sample thickness, varied from ~3000 cm/J at the begining of the multiplex exposure series to ~20 (cm/J) at the end. A total fluence of 700 mJ/cm^2 was required for the 70 hologram exposure set. An exposure schedule was chosen to produce approximately equal diffraction efficiencies of 10^{-4} for all 70 holograms. The recorded data pages were reconstructed with raw BERs less than 10^{-4} throughout the multiplex run. More recently, Waldman et al. [42] co-locationally multiplexed 75 to 100 holograms of 262 kbit data pages in 200-μm-thick samples of Polaroid's ULSH-500 recording medium. The diffraction efficiencies, as a result of exposure scheduling, were typically about 10^{-4} without any compensation for possible Bragg mismatch. The raw BER, with application of a global threshold, was better than 10^{-6} for 75% of the 75 hologram exposure set, and was $\leq 10^{-3}$ for all but two holograms which in particular exhibited low diffraction efficiencies of 10^{-6}.

8 Summary

Photopolymer systems have many characteristics that are required or useful for WORM holographic data storage. They are sensitive to light throughout the visible spectral region and into the near infrared. Their high light sensitivity allows hologram recording with nanosecond range exposures and modest laser powers. Image quality is limited by the shrinkage that accompanies recording for many photopolymers. Uniform partial polymerization before exposure reduces shrinkage and permits the recording of high quality data-page images with good SNR and low BER. This pre-exposure processing, however, can significantly reduce the dynamic range and ultimate storage capacity of those systems that are particularly susceptible to polymerization-induced shrinkage.

Recording in relatively thick photopolymer samples, ~500 μm, has been achieved, and should allow recording data densities near 100 bits/μm^2. Photopolymers are inexpensive, and can be readily incorporated into practical recording formats such as CD-style disks. Photopolymers are light and heat sensitive before recording, and must be stored carefully. Almost all photopolymers undergo slow, dark reactions that limit their pre-exposure shelf life. Data lifetime is expected to be very good based on the chemical and physical durability of most fully exposed photopolymers. For many systems, however, this expectation has not yet been tested adequately.

References

1. R.A. Lessard and G. Manivannan, Eds., *Selected Papers on Photopolymers*, **MS 114**, SPIE Opt. Eng. Press, Bellingham, Washington (1996).
2. D.J. Lougnot, "Self-processing photopolymer materials for holographic recording", Crit. Rev. Opt. Sci. Technol., **CR63**, 190–213 (1996).
3. W.S. Colburn, J. Imaging Sci. Technol, **44**, 443-456 (1997).
4. M.L. Schilling, V.L. Colvin, L. Dhar, A.L. Harris, F.C. Schilling, H.E. Katz, T. Wysocki, A. Hale, L.L. Blyer, and C. Boyd, "Acrylate oligomer-based photopolymers for optical storage applications," Chem. Mater., **11**, 247–254 (1999).
5. G. Odian, *Principles of Polymerization*, John Wiley and Sons, New York, Second Ed. (1981).
6. J.P. Fouassier and J.F. Rabek, *Radiation Curing in Polymer Science and Technology*, Elsevier Applied Science, London, (1993).
7. D.A. Waldman, R.T. Ingwall, P.K. Dal, M.G. Horner, E.S. Kolb, H.-Y.S. Li, R.A. Minns, and H.G. Schild, "Cationic ring-opening photopolymerization methods for holography," Proc. SPIE, **2689**, 127–141 (1996).
8. K. Matyjaszewski, Ed., *Cationic Polymerizations*, Marcel Dekker, New York (1996).
9. B.M. Monroe and G.C. Weed, *Photoinitiators for Free-Radical-Initiated Photoimaging Systems*, Chem. Rev., **93**, 435–448 (1993).
10. P.S. Pappas, B.C. Pappas, L.R. Gatechair, and J.H. Jilek, "Photoinitiated cationic polymerization IV. Direct and sensitized photolysis of aryl iodonium and sulfonium salts," Polym. Photochem., **5**, 1–22 (1984).

11. A.W. Weber, W.K. Smothers, T.J. Trout, and D.J. Mickish, "Hologram recording in Du Pont's new photopolymer materials", Proc. SPIE, **1212**, 30–39 (1990).
12. S.H. Stevenson, "Du Pont multicolor holographic recording film," Proc. SPIE, **3011**, 231–241 (1997).
13. N. Noirett, C. Meyer, and D.J. Lougnot, "Photopolymers for holographic recording: V. Self-processing systems with near infrared sensitivity," Pure Appl. Opt., **3**, 55–71 (1994).
14. C. Braeuchle, U.P. Wild, D.M. Burland, G.C. Bjorklund, and D.C. Alvarez, "Two-photon holographic recording with continuous-wave lasers in the 750–1100 nm range," Opt. Lett., **7**, 177–179 (1982).
15. D.J. Lougnot, D. Ritzenthaler, C. Carre, and J.P. Fouassier, "A new gated system for two-photon holographic recording in the near infrared," J. Appl. Phys., **63**, 4841–4848 (1988).
16. W.K. Smothers, B.M. Monroe, A.M. Weber, and D.E. Keys, "Photopolymers for holography," Proc. SPIE, **1212**, 20–29 (1990).
17. D.A. Waldman; unpublished results.
18. D.A. Waldman, H.-Y.S. Li, and E.A. Cetin, "Holographic recording properties in thick films of ULSH-500 photopolymer," Proc. of SPIE, Diffractive and Holographic Device Technologies and Applications V, **3291**, 89–103 (1998).
19. H. Kogelnik, "Coupled wave theory for thick hologram gratings," Bell Syst. Tech. J., **48**, 2909 (1969).
20. W.J. Tomlinson and E.A. Chandross. "Organic photochemical refractive-index image recording systems," Advances in Photochemistry, **12**, 201–276 (1980).
21. D.W. van Krevelen, *Properties of Polymers*, Elsevier, Amsterdam, 1990, Chap. 10, pp. 287–296.
22. R.K. Sadhir and R.M. Luck, *Expanding Monomers*, CRC Press, Boca Raton, 1992.
23. G. Zhao and P. Mouroulis, "Diffusion model of hologram formation in dry photopolymer materials," J. Mod. Opt., **41**, 1929–1939 (1994).
24. D.J. Lougnot, P. Jost, and L. Lavielle, "Polymers for holographic recording VI. Some basic ideas for modeling the kinetics of the recording process," Pure Appl. Opt., **6**, 225–245 (1997).
25. V.L. Colvin, R.G. Larson, A. Haris, and M.L. Shilling, "Quantitative model of volume hologram formation in photopolymers," J. Appl. Phys., **81**, 5914–5923 (1997).
26. H.-Y.S. Li and D. Psaltis, "Three-dimensional holographic disks," Appl. Opt., **33**, 3764–3774 (1994).
27. A. Pu and D. Psaltis, "High-density recording in photopolymer-based holographic three-dimensional disks," Appl. Opt., **35**, 2389–2398 (1996).
28. A.R. Schultz and M.G. Joshi, "Kinetics of photoinitiated free-radical polymerization," J. Polym. Sci., Polym. Phys. Ed., **22**, 1753–1771 (1984).
29. L. Solymar and D.J. Cooke, *Volume Holography and Volume Gratings*, Academic Press, New York, (1981).
30. D.A. Waldman, H.-Y.S. Li and M.G. Horner, "Volume shrinkage in slant fringe gratings of a cationic ring-opening volume hologram recording material," J. Imaging Sci. Technol., **41**, *5*, 497–514 (1997).
31. N. Uchida, "Calculation of diffraction efficiency in hologram gratings attenuated along the direction perpendicular to the grating vector," J. Opt. Soc. Am., **63**, 280 (1973).

32. J.E. Dietz and N.A. Peppas, "Reaction kinetics and chemical changes during polymerization of multifunctional (meth)acrylated for the production of highly cross-linked polymers used in information storage systems," Polymer, **38**, 3767–3781 (1997).
33. J.G. Kloosterboer, "Network formation by chain crosslinking photopolymerization and its application in electronics," Adv. Polym. Sci., **84**, 1–62 (1988).
34. F.H. Mok, "Angle-multiplexed storage of 5000 holograms in lithium niobate," Opt. Lett., **18**, *11*, 915 (1993).
35. K. Curtis, A. Pu, and D. Psaltis, "Method for holographic storage using peristrophic multiplexing," Opt. Lett., **19**, 993–994 (1994).
36. D.L. Staebler, W.J. Burke. W. Phillips, and J.J. Amodei, Appl. Phys. Lett., **26**, 182–184 (1975).
37. F.H. Mok, M.C. Tackitt, and H.M. Stoll, Opt. Lett., **16**, *11*, 605 (1991).
38. A. Pu, K. Curtis, and D. Psaltis, "Exposure schedule for multiplexing holograms in photopolymer films," Opt. Eng., **35**, *10*, 2824–2829 (1996).
39. L. Dhar, K. Curtis, M. Tackitt, M. Schilling, S. Campbell, W. Wilson, A. Hill, C. Boyd, N. Levinos, and A. Harris, "Holographic storage of multiple high-capacity digital data pages in thick photopolymer systems," Opt. Lett., **23**, 1710–1712 (1998).
40. R.T. Ingwall, D.A. Waldman, and R.M. Shelby, Optical Society of America, Fall Meeting, Baltimore (1998).
41. M.-P. Bernal, H. Coufal, R.K. Grygier, J.A. Hoffnagle, C.M. Jefferson, R.M. Macfarlane, R.M. Shelby, G.T. Sincerbox, P. Wimmer, and G. Wittmann, "A precision tester for studies of holographic storage materials and recording physics," Appl. Opt., **35**, 2360–2374 (1996).
42. D.A. Waldman, R.T. Ingwall, H.-Y.S. Li, and R.M. Shelby, International Society for Optical Engineering, Annual Meeting, Denver (1999).

Photopolymers for Digital Holographic Data Storage

L. Dhar, M.G. Schnoes, H.E. Katz, A. Hale, M.L. Schilling, and A.L. Harris

Holographic data storage has long promised storage densities and data transfer rates far beyond what is achievable by traditional magnetic and optical technologies. One of the major obstacles to the development of this promising technology has been the lack of an adequate recording material [1–3]. Media must satisfy a set of demanding requirements, some of which are listed in Table 1. (Many of the requirements depend on the holographic storage system; the quantities we list are general guidelines.)

The requirements for the media arise from the pagewise recording and recovery of digital data and the three-dimensional nature of holography. Media must: (i) be optically flat and of low scatter so that pages of data (arrays of bits) can be imaged through the material, recorded and recovered with low probabilities of error; (ii) have adequate dynamic range to support the overlap of large numbers of holograms, each with sufficiently high diffraction efficiencies to insure high data read rates; (iii) be thick enough to support the independent storage of large numbers of holograms to yield high densi-

Table 1. Requirements for media for digital holographic data storage

Material parameters	Requirements
Refractive index contrast	$\Delta n > 5 \times 10^{-3}$
Thickness of media	>500 μm
Dynamic range ($M\#^*$)	≥10
Optical scatter	<~10^{-5} of reference beam power
Optical flatness	<~$\lambda/2$
Photosensitivity	(in the visible to near-infrared)
	100–1000 mJ/cm^2 to achieve full dynamic range
Dimensional stability	within <0.5%
Processing for readout	Solvent/heat-free
Readout	Nonvolatile
Archival life of stored data	>5–10 years
	Environmental stability
	Thermal stability

$^*M\# \propto (\Delta n)$ (thickness of medium); $M\# = \sum_{i=1}^{N} \sqrt{\eta_i}$, where N is the number of holograms in the same volume of material and η_i is the diffraction efficiency of each hologram.

ties; (iv) exhibit high photosensitivity to yield high data write rates; and (iv) undergo limited changes in their dimensions and bulk refractive index upon recording to ensure the fidelity of data recovery.

Photopolymer materials have been considered to be attractive candidates for media for holographic storage because they can be designed to exhibit high dynamic range and photosensitivity, and can be easily processed [4–12]. Photopolymers record permanent holograms and therefore are suited for write-once-read-many (WORM) applications. Most of the currently available holographic photopolymers, however, have been optimized for display rather than storage applications. Typically, these materials can be used only as thin (≤100–200 μm) layers, and often exhibit non-negligible dimensional and bulk refractive index changes upon recording.

In this chapter, we focus upon strategies [13,14] that seek to address the needs of high-density digital data storage in WORM applications. We discuss methods that have enabled the fabrication of thick, optically flat, high dynamic range photopolymer media with low light scatter that support digital data storage and recovery.

1 Hologram Formation in Photopolymer Systems

Photopolymers typically consist of a photosensitive species dispersed in a matrix. In holographic recording, two coherent beams spatially overlap within the recording medium. The optical interference pattern initiates a pattern of polymerization: polymerization is induced in the light intensity maxima of the interference pattern while no polymerization occurs in the nulls. The concentration gradient that results from this patterned polymerization leads to diffusion of the unpolymerized species, creating a refractive index modulation that is determined by the difference between the refractive indices of the photosensitive component and the matrix [15–22] (see Fig. 1).

The polymerization reactions that occur during recording, however, can induce unwanted changes in the volume and bulk refractive index of the material. Changes such as these can distort the imaging process, degrade the fidelity with which holographic data can be recovered, and ultimately limit the density of data the material can support [23–25]. The design of photopolymer media for holographic storage applications must therefore balance the advantages of a polymerization mechanism (photosensitivity and large refractive index modulations) against the material changes that accompany the chemical reactions.

2 Photopolymer Materials

We have designed a new type of polymer system [14] that is composed of two independently polymerizable and compatible chemical systems: low refractive index matrix precursors and high refractive index photopolymerizable

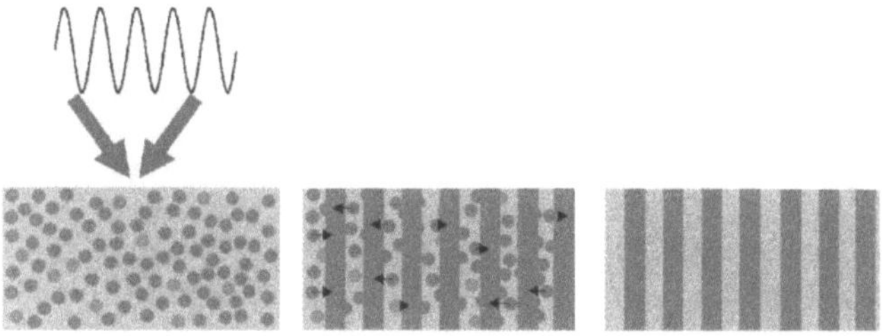

Fig. 1. Mechanism of hologram formation in photopolymer systems. The optical interference pattern generated during recording induces a spatial pattern of polymerization of the writing monomers. The unreacted monomers diffuse to reestablish a uniform concentration of the monomers in the media creating a compositional or refractive index gradient that mimics the optical interference pattern

monomers. The matrix of our media is formed by an in-situ polymerization to yield a cross-linked network in the presence of the photopolymerizable monomers, which remain dissolved and unreacted. The most important aspects of this strategy, which yield high performance holographic storage media, are: (i) preforming a polymerized matrix in-situ, which allows media to be shaped into the required thick and flat formats; (ii) the creation of a cross-linked matrix as a support structure for stable holographic gratings; (iii) the choice of compatible matrix and monomer systems to yield media with good optical clarity (low levels of light scattering); and (iv) the design of independent matrix and monomer systems so as to avoid cross-reactions that dilute the refractive index contrast. The fourth point ensures that the low refractive index of the "background" matrix is not detrimentally raised by the premature polymerization of the high refractive index monomer. Media with sufficiently high refractive index contrast can be fabricated using small amounts of the high index monomer, thereby minimizing the photopolymerization-induced dimensional and bulk refractive index changes that occur during holographic recording.

3 Formation of Thick, Optically Flat Media

Figure 2 shows an interferogram of a typical 1-mm-thick (polymer thickness) medium. The sample is flat to within 500 Å over its diameter. To form the disk, a resin consisting of the matrix precursors, photopolymerizable monomers, and a visible-light-sensitive photoinitiator was dispensed between two optically flat glass slides. The in-situ formation of the matrix allows routine fabrication of high-optical quality media with polymer thicknesses between 200 μm and 1.5 mm.

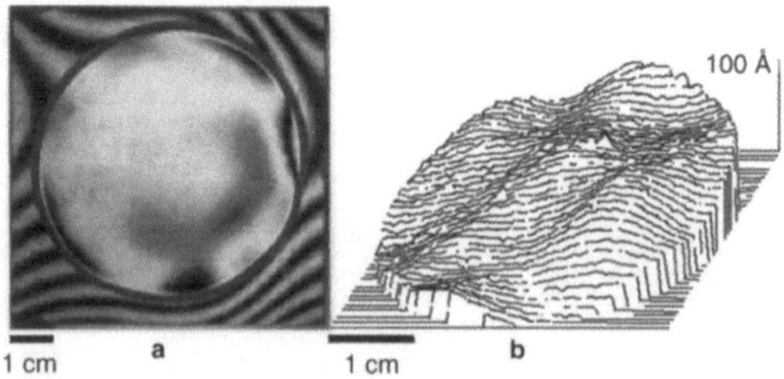

Fig. 2. (**a**) Transmission interferogram of a typical 1-mm-thick photopolymer medium. (**b**) Surface plot of the variation in the optical thickness of the inner 4 cm of the sample. The optical flatness of the sample varies by less than $\lambda/10/\mathrm{cm}$ (~500 Å) over the entire area of the medium

4 Holographic Characterization of Photopolymer Media

The holographic performance of a material depends on an interrelated set of parameters. In photopolymers, the dynamic range depends on the relative volume fraction of the writing monomer in addition to the difference between the refractive indices of the writing monomer and matrix. The amount of monomer, however, also determines the extent of the unwanted dimensional and bulk refractive index changes that occur during recording.

In this section, we focus on the effects of the material changes associated with polymerization, and outline how the dynamic range of photopolymer media can be optimized within the constraints of limiting these changes.

4.1 Recording-Induced Bragg Detuning

The changes in dimension and bulk refractive index accompanying the polymerization reactions that occur during holographic recording can degrade the fidelity of data recovery because they rotate or detune [23] the Bragg angle (the angle of incidence at which light is efficiently diffracted from a volume grating). In general, high-fidelity data readout requires that the levels of detuning be less than the widths of Bragg peaks characteristic of the storage medium. For example, since it is believed that media at least 500 μm thick will be needed to enable compelling storage densities and transfer rates, the detuning levels should be less than the angular separation between the first nulls of a typical Bragg peak in this thickness (~0.25°).

The detuning level of a volume grating depends on the grating tilt, increasing as the tilt angle increases. Volume holograms recorded in digital storage systems typically consist of a distribution of gratings with varying tilt angles due to the bandwidth of the images. The Bragg detuning levels should, therefore, be measured at the tilt angles typically encountered in digital recording systems.

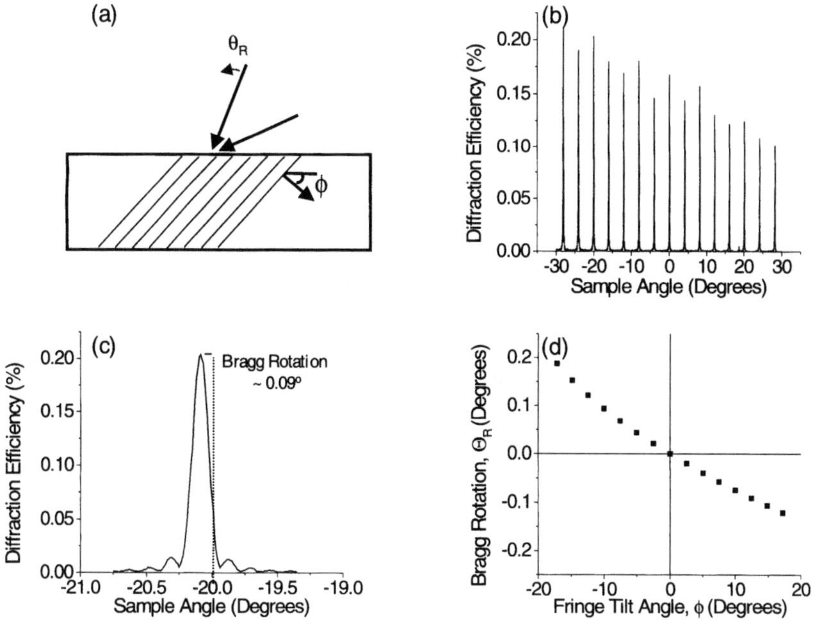

Fig. 3. (**a**) Geometry used to determine Bragg detuning levels of a recording medium. Holograms with different grating tilt angles, ϕ, are angle-multiplexed by rotating the medium with respect to the recording beams. The experiment is described in detail in [23]. (**b**) Angular position of a set of multiplexed holograms after the material has been flood-exposed to consume the writing monomer remaining after recording. (**c**) Readout position of one of the peaks of the multiplexed set shown in (**b**). The dotted line corresponds to the recording position. (**d**) Bragg detuning levels for a set of multiplexed holograms. In this example, the medium underwent decreases of ~0.35% in thickness and ~2.1×10^{-3} in the bulk refractive index upon full reaction of the photosensitive monomers

In Fig. 3a, we describe the results of an experiment that characterizes the Bragg detuning of a medium. Fifteen weak (diffraction efficiencies ~0.01%) plane-wave holograms were angle-multiplexed within a photopolymer medium to generate a set of gratings with varying tilt angles. The medium was then flood-exposed to consume the remaining photoactive species. The rotation of the angle of incidence of one of the readout beams from the original recording position required to achieve the maximum diffraction efficiency for each of the recorded holograms yielded the Bragg detuning levels [23].

4.2 Dynamic Range

The materials design strategy described in Sect. 2 enables us to optimize the dynamic range of the media while keeping the Bragg detuning levels constant by incorporating controlled amounts of writing monomers of increasing re-

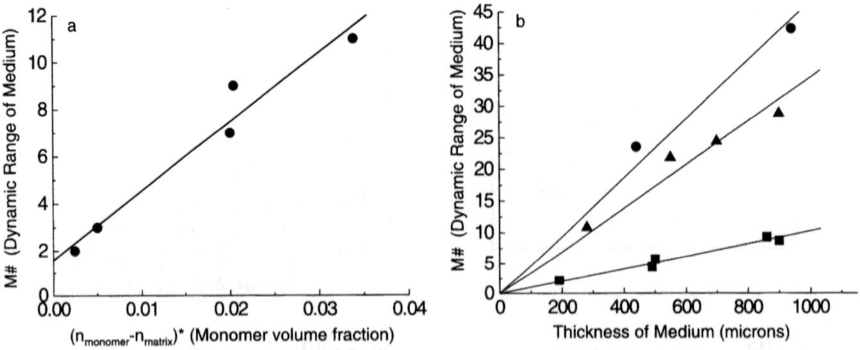

Fig. 4. (**a**) Dynamic range of a series of 200-µm-thick photopolymer media with different high refractive index monomers designed to exhibit the same levels of Bragg detuning upon recording. The dynamic range was measured by angle-multiplexing 35 plane-wave holograms within the samples using a frequency-doubled Nd:YAG laser (532 nm) as the recording source. The recording exposure schedules [6] were adjusted to consume all of the photoactive species and yield holograms of nearly equal diffraction efficiency. After recording, the samples were flood-exposed to bleach any excess photoinitiator. The $M\#$s of the media were calculated by summing the square roots of the diffraction efficiencies of the recorded holograms. All of the media exhibited Bragg detuning levels equivalent to those shown in Fig 3. (**b**) $M\#$ versus thickness for photopolymer media fabricated with a typical writing monomer: ■, media exhibit Bragg angle rotations of $-0.03°$ and $0.06°$ at internal grating tilt angles of $17.3°$; ▲, media exhibit Bragg angle rotations of $-0.13°$ and $0.19°$; •, media exhibit Bragg angle rotations of $-0.16°$ and $0.25°$. The $M\#$s in these experiments were measured by recording >250 holograms using angle and peristrophic multiplexing methods

fractive index. The independence of the reactions that form the matrix from those involved with holographic recording allows us to take full advantage of the refractive index difference between the two systems.

In Fig. 4a, we show the dynamic range ($M\#$ [26,27]) of a series of 200-µm-thick media that were fabricated using the same matrix but incorporating writing monomers of varying refractive index and varying size where the concentrations of the monomers were adjusted to yield the Bragg detuning levels described in Sect. 4.1. Increases in $M\#$ from 2 to 11 were realized while maintaining the same level of effective dimensional stability of the media.

Media with high $M\#$ were obtained by fabricating thick samples of our photopolymer materials. In Fig. 4b, we show how the $M\#$ scales with thickness in media fabricated with a typical writing monomer. Data from three sets of samples are shown, with each set formulated with a different concentration of the monomer and therefore exhibiting a different level of Bragg detuning. The gains in $M\#$ with increasing thickness are possible because of the low levels of light absorption and light scatter and the high level of effective dimensional stability of the media.

5 Holographic Digital Data Storage in Photopolymer Media

In this section, we illustrate the suitability of our media for high-density applications by describing the imaging characteristics of high-capacity digital data pages through our photopolymer materials. Descriptions of the holographic storage and recovery of data, specifically experiments that have demonstrated densities as high as 48 channel bits/μm^2 in moderate response versions of these media, are given in the chapter entitled "High-Density, High-Performance Data Storage via Volume Holograph."

The baseline performance of an optical system is measured by the "straight-through" image or the data page imaged through the storage medium and the optics of the object arm. A "straight-through" image in the optical system of Fig. 5 utilizing a 250 μm thick photopolymer sample is shown in Fig. 6a. An enlarged view of a section of the transmitted data page is shown in Fig. 6b.

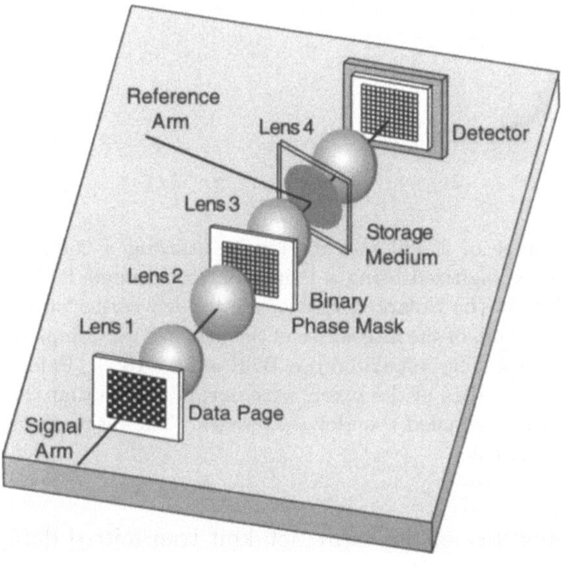

Fig. 5. Schematic of holographic storage system. The laser source for these experiments is a frequency-doubled diode-pumped YAG. The object arm is a 4f system. An 800 × 600 bit chrome on glass amplitude mask with random opaque and transmitting pixels of 24 μm pitch is imaged (pixel matched) onto a random binary phase mask with a pixel pitch of 24 μm. The phase mask serves to distribute the intensity of the Fourier transform of the data page at the recording plane. The beam spotsize at the recording plane is 5 × 4 mm. The data page is imaged at a magnification ratio of 1:1 onto a Princeton Instruments 1024 × 1024 CCD camera with 24 μm pixels

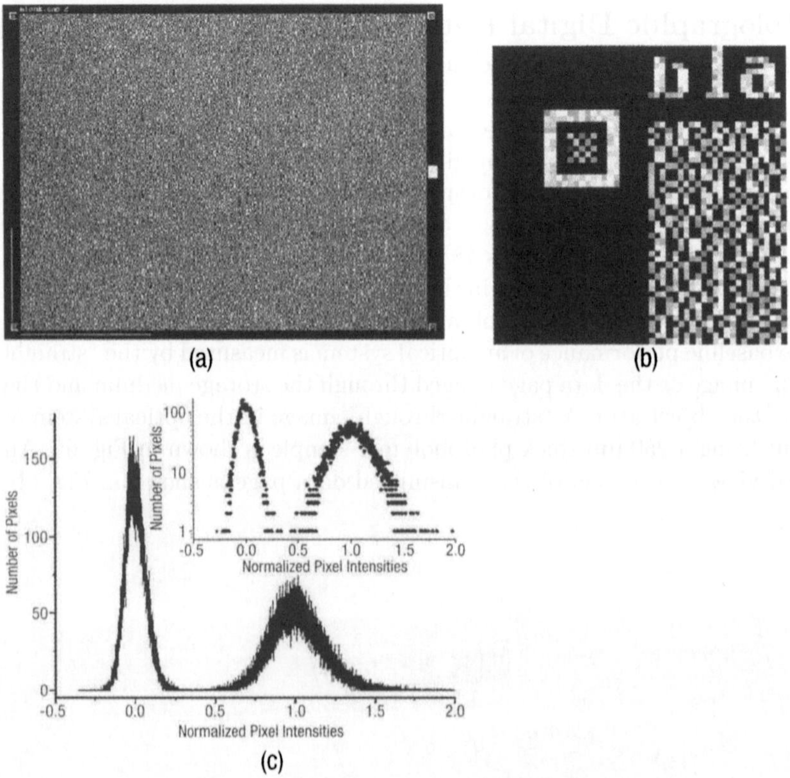

Fig. 6. (**a**) Straight-through image of a 480 kbit data page utilizing a 250-µm-thick sample. The intensities were digitized using a Princeton Instruments ST138 CCD camera with 16-bit resolution. (**b**) Enlarged view of the top left corner of the straight-through image. (**c**) Histogram of the intensities of the pixels of the complete 480-kbit straight-through data page. The calculated raw BER is 2.6×10^{-6}. (Before calculating the histogram, the intensities of the pixels were normalized so that the local averages of the on and off bits equaled the global averages.) The inset shows the data plotted on a logarithmic scale

A histogram of the pixel intensities of the entire 480 kbit transmitted data page is shown in Fig. 6c.

A raw bit error rate (BER) of 2.6×10^{-6} was calculated for the straight-through image by fitting the distributions of the off and on intensities of the histogram in Fig. 6c to Gaussians and computing the probability of overlap. Current error correcting strategies [6] allow error-free recovery of data pages with raw BER up to 5×10^{-3}, well above that of the straight-through image. Straight-through images through thicker media yield equivalently low raw bit error rates.

The ability to obtain high-fidelity straight-through images reflects the high optical quality and low level of scatter in our materials. Experiments that describe the storage and recovery of digital data in the media are described in [14] and [28] and in a later chapter of this book. These results are important steps in establishing the practicality of high-density data storage in photopolymer media.

6 Summary

We report substantial advances in the design and fabrication of photopolymer media for digital holographic data storage. Using a strategy based on two independently polymerizable and compatible chemical systems, we have fabricated thick, optically flat, high dynamic range recording media that exhibit low levels of light scatter with demonstrated capabilities for high-density storage. The ultimate success of photopolymer systems for data storage applications, however, rests upon a few important issues. Ongoing work is focused on the long-term archival life of stored data and on the effects of changes in temperature of the polymer media [23] on the fidelity of data readout. With these considerations, we believe the results described here greatly enhance the promise of polymer materials for holographic data storage.

Acknowledgments

We would like to thank H. Bair, L. Blyler, C. Boyd, S. Campbell, M.J. Cardillo, E.A. Chandross, V. Colvin, K. Curtis, M. Galvin-Donoghue, A. Hill, N. Levinos, S. Patel, S. Popielarski, X. Quan, E. Reichmanis, F. Schilling, M. Tackitt, W. Wilson, T. Wysocki, and P. Wiltzius for their invaluable contributions to this work.

References

1. G.T. Sincerbox: Proc. SPIE. **2866**, 130 (1996).
2. M.-P. Bernal: MRS Bull. **21**, 51 (1996).
3. G.T. Sincerbox: Opt. Mater. **4**, 370 (1995).
4. W.J. Tomlinson, E.A. Chandross, H.P. Weber, G.D. Aumiller: Appl. Opt. **15**, 95 (1976).
5. B.M. Monroe, W.K. Smothers, D.E. Keys, R.R. Krebs, D.J. Mickish, A.F. Harrington, S.R. Schicker, M.K. Armstrong, D.M.T. Chan, C.I. Weathers: J. Imaging Sci., **35**, 19 (1991).
6. B.M. Monroe: J. Imaging Sci., **35**, 25 (1991).
7. R.T. Ingwall, H.L. Fielding: Proc. SPIE, **523**, 306 (1985).
8. D.A. Waldman, H.-Y.S. Li: Proc. SPIE, **3010**, 354 (1997).
9. D.J. Lougnot, C. Turck: Pure Appl. Opt. **1**, 251 (1992).
10. R.T. Ingwall, H.L. Fielding: Proc. SPIE, **253**, 306 (1985).
11. A. Pu, D. Psaltis: Appl. Opt. **35**, 2389 (1996).
12. K. Curtis, D. Psaltis: Appl. Opt. **31**, 7425 (1992).

13. M. Schilling, V. Colvin, L. Dhar, A. Harris, F. Schilling, H. Katz, T. Wysocki, A. Hale, L. Blyler, and C. Boyd: Chem. Mater., **11**, 247 (1999).
14. L. Dhar, A. Hale, H. Katz, M. Schilling, M. Schnoes, F. Schilling: Opt. Lett. **24**, 487 (1999).
15. W.S. Colburn, K.A. Haines: Appl. Opt. **10**, 1636 (1971).
16. B.L. Booth: Appl. Opt. **14**, 593 (1975).
17. V.L. Colvin, R.G. Larson, A.L. Harris, M.L. Schilling: J. Appl. Phys. **84**, 5913 (1996).
18. C.R. Kagan, T.D. Harris, A.L. Harris, M.L. Schilling: J. Chem. Phys. **108**, 6892 (1998).
19. R.R. Adhami, D.J. Lanteigne, D.A. Gregory: Microw. Opt. Techol. Lett. **4**, 106 (1991).
20. G. Zhao, P. Mouroulis: J. Mod. Opt. **41**, 1929 (1994).
21. S. Piazzoll, B.K. Jenkins: Opt. Lett. **21**, 1075 (1996).
22. D.J. Lougnot, P. Jost, L. Lavielle: Pure Appl. Opt. **6**, 225 (1997).
23. L. Dhar, M. Schnoes, T. Wysocki, H. Bair, M. Schilling, C. Boyd: Appl. Phys. Lett., **73**, 1337 (1998).
24. S. Campbell, S.-H. Lin , X. Yi, P. Yeh: J. Opt. Soc. Am. B., **13**, 2218 (1996).
25. J.T. Gallo, C.M. Verber: Appl. Opt., **33**, 6797 (1994).
26. F.H. Mok, G.W. Burr, D. Psaltis: Opt. Lett. **21**, 896 (1996).
27. A. Pu, K. Curtis, D. Psaltis: Opt. Eng. **35**, 2824 (1996).
28. L. Dhar, K. Curtis, M. Tackitt, M. Schilling, S. Campbell, W. Wilson, A. Hill, C. Boyd, N. Levinos, A. Harris: Opt. Lett. **23**, 1710 (1998).

Photoaddressable Polymers

T. Bieringer

Polymers are the perfect materials for a variety of applications in almost every field of technical as well as human life. Because of their macromolecular architecture there are a lot of degrees of freedom in the synthesis of polymers. Owing to the change of their functional composition, they can be tailored even for quite difficult demands. Since a whole industry deals with the processing of polymers, cheap production lines have been developed for almost every polymer. This is the reason why not only the molecular composition but even the price of polymers has been optimized. Therefore these materials can be considered as encouraging components even in highly sophisticated areas of applications.

As far as the application of holographic data storage is concerned, several polymers have been synthesized that can be used as recording material. Most of them are photopolymers, in which a polymerization and/or a ring opening reaction is initiated by the illumination (*see "Photopolymer Systems"* and "*Photopolymers for Digital Holographic Data Storate*"). Since the requirements for holographic data storage are stringent, there are some drawbacks that hinder the introduction of these polymers in commercial mass production. So far, shrinkage during illumination is the most severe problem of photopolymers.

It is still an open question whether there are recording materials or even polymeric recording materials that fulfill all the requirements of holography. However, there are materials that are very likely to fulfill most of those requirements. One class of these materials is the class of photoaddressable polymers: polymers that react to light with a change in their molecular configuration. A considerable advantage of these polymers is the fact that the light-induced reaction is a local effect. Therefore no diffusion processes take place, and the change of the optical parameters is related only to the molecular orientation.

1 Photoaddressable Polymers

In principle all materials that react to light with a change of specific properties can be described as photoaddressable polymers. In order to induce photo-optical changes, there is the possibility of using different dichroic dyes, such as spiropyranes, spirooxazines, stilbenes, fulgides, and other compounds [1–8]. In the following section the term *photoaddressable polymer* (PAP) is used as a synonym for a class of polymers with azo dyes as antennae for the incident light. These PAPs are basically azobenzene containing liquid crystalline copolymers. The aim of this chapter is to give an overview of the optical and photophysical properties of these polymers, and to report on the most important experimental results performed with PAPs.

1.1 Photochemistry of Azobenzene

The photophysical reactions of azobenzene are well studied. It can be demonstrated that owing to controlled light-induced reactions of azo chromophores, the properties of the whole system incorporating the dye can be modified. These properties include for example viscosity [9,10], solubility [9,11,12], mechanical parameters [13], bioactivity [14], and optical constants [15]. Additionally, azobenzene molecules can be used as a probe for their molecular surroundings. In this way a detailed study of polymeric parameters can be performed by monitoring the photochemical behavior of the azo dyes. The success of azo dyes in all these applications can be explained with detailed knowledge of their light-induced photoreactions. Azobenzene chromophores exist in two isomeric states: the long rodlike *trans* form and the bent *cis* configuration. The isomerization can be induced by light in both directions, from *trans* to *cis* and from *cis* to *trans*, whereas the *cis* isomer can also undergo a thermal backrelaxation to the thermodynamically more stable *trans* isomer (Fig. 1).

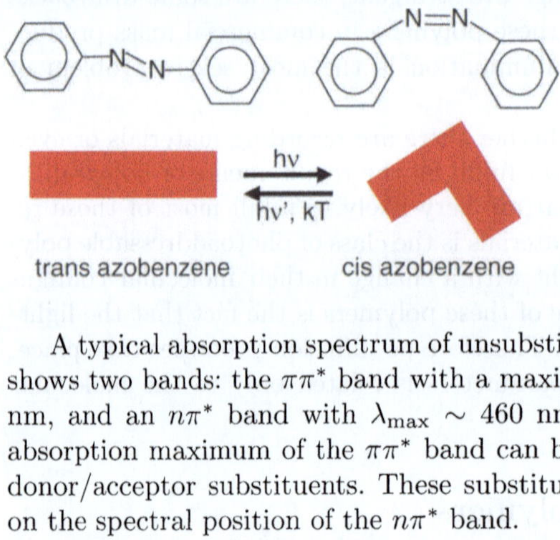

Fig. 1. *Trans–cis* isomerization of azobenzene

A typical absorption spectrum of unsubstituted azo chromophores (Fig. 2) shows two bands: the $\pi\pi^*$ band with a maximum absorbance at $\lambda_{\max} \sim 360$ nm, and an $n\pi^*$ band with $\lambda_{\max} \sim 460$ nm. The spectral position of the absorption maximum of the $\pi\pi^*$ band can be shifted by the substitution of donor/acceptor substituents. These substituents do not have any influence on the spectral position of the $n\pi^*$ band.

It must be emphasized that the spectrum of Fig. 2 is a superposition of the spectrum of the *trans* and the *cis* isomers.[1] Illumination leads therefore to a series of *trans–cis–trans* isomerization cycles, resulting in a photosta-

[1] Without illumination the photostationary equilibrium will be located on the side of the thermodynamically more stable *trans* isomers.

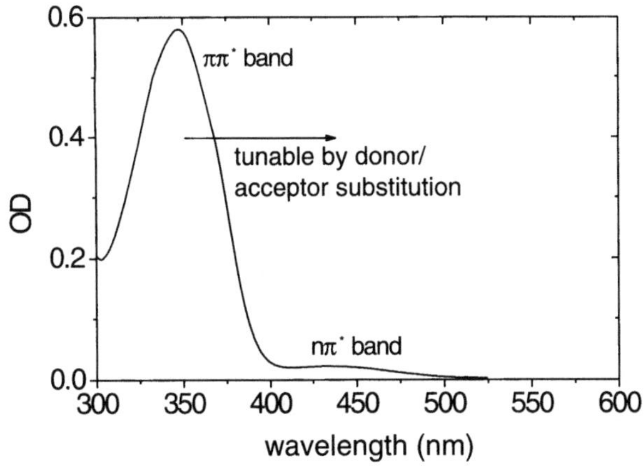

Fig. 2. Absorption spectrum of unsubstituted azobenzene in solution

tionary equilibrium that depends only on the wavelength of the actinic light and the temperature of the sample [16,17]. In the case of unsubstituted azo chromophores a majority of *trans* isomers are created by illumination within the $n\pi^*$ band, whereas illumination within the $\pi\pi^*$ band results in a photostationary equilibrium, in which more *cis* isomers are present.

There are different models describing the molecular mechanisms of photoisomerization. Most research groups argue with two possible reaction schemes: a three-dimensional rotation (rotation mechanism) and a planar reorientation of a phenyl ring of the chromophore (inversion mechanism). It is generally accepted that within the $\pi\pi^*$ regime both rotation and inversion mechanisms can occur, whereas in the $n\pi^*$ regime and during thermal reorientations the inversion mechanism dominates [18].

1.2 Azobenzene Containing Polymers

In principle there are three ways to incorporate chromophores into a polymer:

- In guest host systems chromophore guests are doped into a polymeric host. In these systems the chromophore concentration cannot exceed a specific value, because highly concentrated chromophores tend to phase separation and crystallization.
- This disadvantage can be avoided by attaching the chromophores to the backbone as side chains,
- or by fixing them directly as part of the main chain.

The first investigations of the isomerization kinetics of azo chromophores in polymers were performed in 1972 [19], and were focused mainly on spectroscopic measurements [20–29]. All these experiments can be summarized with the experimental observation that the isomerization of azo dyes in a polymeric environment is possible, even at temperatures below the glass transition temperature. This result applies to side chain as well as to main chain polymers.

Since the isomerization results in a change of the molecular shape of the azo chromophores, as described above, there is a specific demand of free volume to enable this reaction.[2] In solution this free volume condition is fulfilled. Spectroscopic measurements show that photochemical reactions in solution are monoexponential. In polymers, however, there may be some steric constraints due to the inhomogeneous free volume distribution: For some chromophores there are local environments with large free volume, where the isomerization can be performed just as in solution. But there are also configurations where the free volume is not sufficient for the isomerization of the dyes. Therefore strong deviations of the photochemical reactions in polymers from the monoexponential behavior occur. Most groups use a biexponential approach to describe the photochemical reactions [9,30,31]. In the case of thermal *cis–trans* relaxation in polymers Paik [19] and Eisenbach [20,21] use a biexponential decay, whereas in some recent publications three exponents or stretched exponential functions [24–26] are used. It is generally accepted for a polymeric system that there are not only slow relaxation processes as in solution (with time constants of, typically, a few thousand seconds), but also further processes with faster time constants. Paik explained those fast time constants in terms of the occurrence of stressed *cis* molecules [19], which immediately relax back to the *trans* form after the actinic light beam has been switched off. In the meantime this explanation has gained acceptance. However, all groups agree that the observed time constants result from a variety of different molecular reorientation factors, only one of them being the stressed *cis* form.

Beside these experiments there has been a variety of research a with azobenzene as molecular probe for different polymeric environments. Reviews of the most important results can be found in [24,32,34].

1.3 Liquid Crystalline Side Chain Polymers

Liquid crystalline side chain polymers are polymers with different types of functional side groups. Figure 3 shows a sketch of a typical polymer under investigation: via flexible spacer units a chromophoric and a mesogenic side chain are attached to the polymer backbone. Azobenzene or derivatives of azobenzene are used as chromophores, acting as antennae for the actinic light. Being illuminated with visible light, a typical mesogenic side chain doesn't show any absorbance. Mesogens of the rigid-rod class are characterized by their long-shaped molecular geometry and their tendency to spontaneous self-organization in a specific temperature range. This temperature range is located between the solid and the liquid phase the cor-

[2] In the range of 10^{-1} nm^3 in the case of unsubstituted azos [19,30,33]

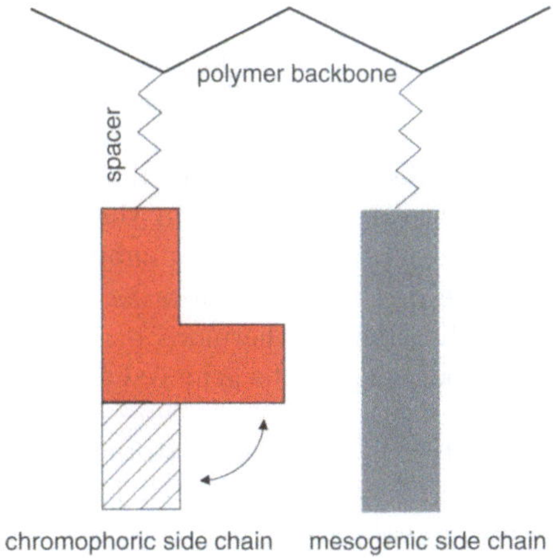

Fig. 3. Scheme of an azobenzene containing liquid crystalline copolymer

responding phase is therefore called the *mesogenic*[3] phase. The transition temperature between the solid and the liquid crystalline phase is called the *glass transition temperature* T_g; the temperature between the liquid crystalline and the liquid phases is called the clearing temperature T_{cl}. This nomenclature results from the fact that in the liquid crystalline phase the molecules tend to form ordered polydomains, acting as scattering centers for the incident light. At the clearing temperature the polymers become liquid, and the polydomain structure breaks down. The result is a transparent, clear film. In contrast to a crystal with a long-range order even in the position of the single components, the molecules in the liquid crystalline phase follow an order principle only in the orientation. In the case of nematic liquid crystals, for example, the majority of the molecules are aligned along a special direction. Details of the liquid crystalline phase can be found in [35–40]. Within this chapter only the tendency of the liquid crystalline molecules to align in concert in the same direction is of importance.

The motivation for the development of polymers following the scheme of Fig. 3 can be summarized as follows:

- The dye acts as an antenna for the incident light. In order to optimize the response to light, adequate dye systems must be used. Azo dyes are the perfect chromophores for this purpose because of their well-known photochemistry.
- It is the task of the mesogenic groups to stabilize and amplify the reorientation of the chromophores. The physical reason for this can be explained the intrinsic tendency of the mesogens to spontaneous organization in an oriented alignment.

[3] Greek: in between

The Spacer Concept. The movement of the polymer main chain is mostly caused by the tendency to form a configuration where the entropy reaches a maximum, e.g. where the system is mostly disordered. This collides with the task of the mesogens to order along the same direction, as described above. The movement of the backbone can be decoupled from the movement of the side chains by attaching chromophores and mesogens to the backbone via flexible methylene (CH_2)- spacer groups. The spacers supply the chromophores and the mesogens with the needed flexibility in order to undergo the light-induced orientation and stabilization. Furthermore, even large structural changes in the glassy state of the polymer can be achieved [41–44].

Orientational Hole Burning. For application in optical data storage photoaddressable polymers have to be illuminated with polarized light. This leads to the formation of optical anisotropy and therefore to birefringence of the illuminated sample. Figure 4 explains the occurrence of the optical axes during illumination. In the following, the quantum-mechanical description of the excitation will be performed in the so-called dipole approximation. The isomerization probability of an azo dye is a function of the angle Θ between the polarization direction of the actinic light and the dipole momentum of the chromophore[4] [45]:

$$W \propto |\langle|\boldsymbol{\mu}\varepsilon|\rangle|^2 \propto \cos^2\Theta .$$

In the $n\pi^*$ band the dipole momentum of the *trans* azo dye lies parallel to the rod-like axis of the dye for *cis* molecules $\boldsymbol{\mu}$ lies almost parallel to the

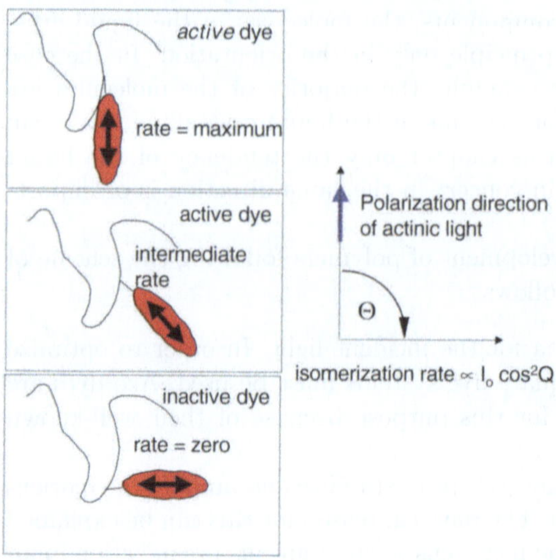

Fig. 4. Orientational hole burning by irradiation of azodyes with polarized light

[4] The expression $\langle|\boldsymbol{\mu}\varepsilon|\rangle$ is the transition dipole momentum of the excitation from the initial electronic state to the excited state.

N=N double bond. Since all isomerization cycles result in a *trans* form, the following explanation can be restricted to the trans form of the dye: Dyes with their long axis parallel to the polarization direction of the incident light will preferably undergo an isomerization ($\cos^2 \Theta = 1$). Dyes with an intrinsic orientation parallel to ε (upper picture) will lose this orientation after a *trans–cis–trans* isomerization cycle.[5] As long as the angle between the polarization direction and the long axis of the dye in the resulting orientation reach 90°, this azo dye will undergo further isomerization cycles (middle picture, $\cos^2 \Theta \neq 0$). If an isomerization results in a dye orientation perpendicular to ε (lower picture), this dye can no longer be addressed, since the excitation probability is equal to zero ($\cos^2 \Theta = 0$). As every dye with an orientation perpendicular to the polarization direction is trapped and will not be able to take part in further excitation steps, illumination with polarized light results in an increased number of chromophores oriented in this direction. After a series of cycles the macroscopic orientational distribution function in the direction parallel to the polarization is decreased (orientational hole burning).

A measure of the efficiency of this orientational hole burning is the light-induced birefringence (Fig. 5). Starting from an isotropic alignment of the chromophores and dyes, the dyes are oriented in a direction perpendicular to the polarization[6] after illumination with polarized light. The different molecular configurations parallel and perpendicular to the polarization direction

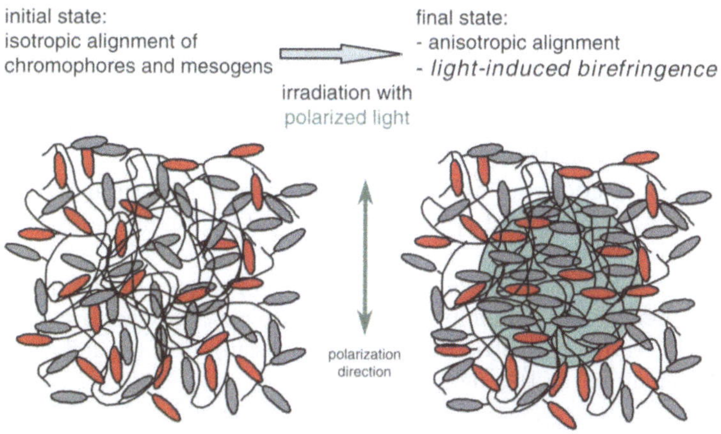

Fig. 5. Formation of light-induced birefringence with photoaddressable polymers

[5] This behavior is on the one hand a result of the flexible attachment of the dyes via the spacers. On the other hand, the chromophore loses the memory to the starting configuration after a whole isomerization cycle.

[6] In Fig. 5 a system is sketched, where the mesogenic (grey) side chain follow the orientation of the (red) chromophores (cooperative motion).

result in two different values of the refractive index along these directions: the medium is birefringent.

2 Materials Under Investigation

In the 1990s, the synthesis of polymers with different optical and thermodynamical properties made it possible to learn details about the relationship between polymeric structure and the behavior of the polymers under laser illumination. Polymers with a strong tendency to form a liquid crystalline phase are difficult to prepare for optical applications, since the liquid crystalline polydomain texture has to be suppressed in order to avoid light scattering. For a good thermal stability of the light-induced molecular reorientation, the glass transition temperature must be well above room temperature. In this way the different relaxation modes of the systems are frozen in. These requirements have been combined for the first time in amorphous high-T_g polymers [32]. Even though the purely amorphous high T_g polymers show excellent properties, it was realized in our group that these polymers show problems in the long-term stability of the light-induced reorientation. Therefore systems with a liquid crystalline phase are used. The challenge is on the one hand to produce polymers with an inherent mesogenic potential. The presence of the liquid crystalline phase should on the other hand have no negative influence on the optical properties of the polymers: There should be no formation of light scattering polydomain structures, and the sample preparation should be easy. It can be shown that it is possible to produce liquid crystalline copolymers where the liquid crystalline phase can be quenched by simple methods, e.g. rapid cooling of the polymer melt or other preparation techniques such as spin coating [46,47].

In the following sections an overview of the relevant structural factors affecting the thermodynamic and optical behavior and their influence on the orientation process is given.

Figure 6 shows a typical chemical structure of the polymers under investigation. The content of the chromophoric and mesogenic components can be varied over a wide range. The structural elements discussed in the following sections can be identified in this figure.

2.1 The Choice of the Main Chain

The choice of the main chain has a major influence on the value of the glass transition temperature T_g. Polyacrylates and polymethacrylates (PMMA) [47–49], polysiloxanes [50], polycarbonates [51], polyurethanes [68], polyimides [83], and aliphatic polyesters [52–58] have been investigated. In this way, the mechanical and thermodynamical properties of the systems can be changed over a wide range. The rigidity of a PMMA backbone favors the storage stability and archivability of the polymer. T_g values well above 100°C can

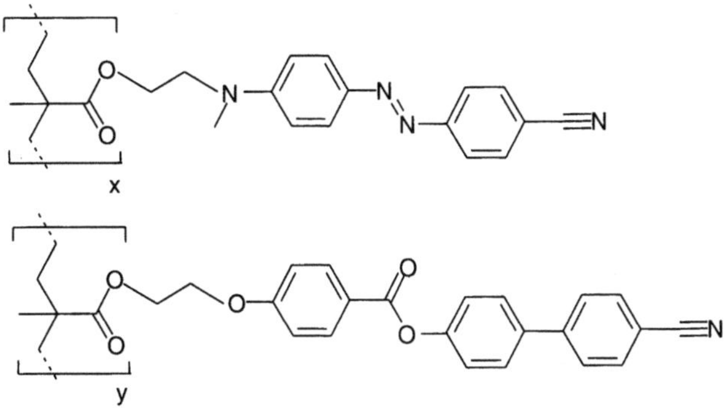

Fig. 6. Typical chemical structure of a photoaddressable polymer

be realized, whereas the light-induced reorientation steps are induced at room temperature. At room temperature large long-range translations of the main chain do not take place, and the backbone constitutes a matrix, stabilizing the molecular reorientation processes.

2.2 The Spacer

As discussed above, spacer groups between the functional groups and the main chain are used to decouple the side chain motion from the rigidity of the main chain. The spacers directly affect the motion of the functional groups, since a rigid link may suppress the local reorientation processes necessary for achieving a macroscopic net orientation. A shorter spacer on the chromophore side has a double effect: Since the coupling to the backbone is more rigid, higher energy is required to achieve orientation. On the other hand the rigid coupling produces a better stability of the light-induced orientation, after the actinic light has been switched off. If mesogenic side chains are tethered to the main chain by shorter spacer units the loss of flexibility can hinder the formation of a liquid crystalline alignment. In this way, the ability to enable cooperative movement might even be affected, and amorphous polymers are produced with the described advantages in the preparation of optically transparent, nonscattering samples [59,60]. From a thermodynamic point of view, a shortening of the spacer length induces an increase in the glass transition temperature due to the loss of flexibility in the whole polymer system.

2.3 The Choice of the Azo Group

A variety of azo substitution can be found in the literature. In most cases, the chromophore is connected to the spacer via an ether or amino bridge (donor type). By changing the para-substitution of the other side of the chromophore, donor-donor or donor-acceptor dyes are realized. In particular,

Disperse Red 1 and p-cyano substituted dyes are very popular [61,62]. A change of the substitution has an influence on the absorption characteristics of the dye and therefore on the location of the photostationary equilibrium. The rate of the light-induced birefringence and the time constants of the thermal *cis–trans* backrelaxation are also affected by the substitution. The majority of research is concentrated on donor-acceptor substituted azobenzenes, since their representatives show the fastest isomerization rates, the largest values of birefringence, and the fastest thermal *cis* to *trans* backrelaxation times [18].

Not only the polarity of the substitution, but also the anisotropy of the resulting azo group have an important impact on the level of the light-induced birefringence. Long-shaped azo groups are able to achieve much higher birefringence values than unsubstituted shorter chromophores. Since bulkier groups need, however, more free volume to isomerize, in some cases it may be harder for them to achieve the orientation: i.e. the isomerization rate can decrease if long-shaped chromophores are used.

2.4 The Choice of the Mesogenic Group

In the literature many different mesogenic groups are mentioned. The most popular are aromatic esters or amides [63,64], biphenyls, and other liquid crystalline molecules. As already mentioned above, the task of the mesogenic side groups is to stabilize and amplify the orientation of the chromophores without being detrimental to the optical quality of the polymer (non scattering samples). The choice of an adequate liquid crystalline neighbor strongly depends on the chromophore under investigation. Since the mesogenic partner has to follow the reorientation of the dye, there has to be an interaction between both groups either by steric or, more efficient, by dipolar interaction. Therefore the substitution of the mesogenic unit has to be adapted to the chromophoric group. Additionally, it has to be pointed out that the most effective cooperative reorientation between chromophores and mesogens is induced in systems where both groups have a similar molecular shape. Taking this molecular architecture plan into account, our group was able to synthesize copolymers with birefringence values up to $\Delta n = 0.5$, whereas typical homopolymers, containing solely chromophoric groups, only reach Δn values of about 0.1.

2.5 The Azo Group Concentration

The composition of the copolymer determines the intrinsic tendency of the liquid crystalline moieties to organize into oriented domains. A pure liquid crystalline polymer with no azo group attached to the backbone would form a turbid sample of macroscopically highly ordered polydomains. The thermodynamic tendency towards mesogenic order can be gradually suppressed by incorporating chromophoric comonomers. This is due to the fact that most chromophores are not able to build up a liquid crystalline phase and therefore act as disturbance for oriented domains of the liquid crystals. At a polymer-specific azo concentration this disturbance reaches a point where the formation of a mesogenic phase is not possible any longer. Experiments in our group showed that polymers with a concentration inducing a phase behavior at the border between the liquid crystalline and amorphous phases show the best orientation properties. They are already amorphous or have a rather weak liquid crystalline phase: thus there are no problems with light-scattering and sample preparation. However, these polymers still remember the tendency of the liquid crystalline phase to align in concert. The largest birefringence values can be achieved with these polymers owing to cooperative reorientation mechanism.

The true challenge in the synthesis of photoaddressable polymers is to choose all the listed polymeric properties in order to tailor a system for a specific application. In the next section a few experimental results are summarized that underline the very promising and versatile features of photoaddressable polymers in the field of optical as well as holographic data storage.

3 State of the Art in the Literature

In order to test the optical properties of photoaddressable polymers a lot of different experimental methods are used: e.g. IR spectroscopy [65,66], UV/VIS spectroscopy [16,49,67,69–75], birefringence measurements [71–73,78–80], forced Rayleigh scattering experiments [81], holographic experiments [16,46–48,67,69,76,77,82,?,84–86], confocal Raman [87–89] and AFM measurements [90–92], etc.

Following the pioneering work of Shibaev et al., the first holographic measurements with liquid crystalline side chain polymers were performed by Wendorff et al. in 1987 [82]. Those first experiments focused on liquid crystalline polymers, prepared as monolayers. Since the liquid crystalline polymers under investigation tend to form polydomain structures, acting as scattering centers for the actinic light, a special sample preparation technique was used. By a combination of electromagnetic field and surface treatment, the molecular dipoles were perfectly aligned and produced a transparent and non-scattering polymer sample. In the following period mainly German research groups focused on photoaddressable polymers: whereas Spieß et al. [65,66] and Stumpe et al. [71–73,78–80] followed the idea of a cooperative movement of both liquid crystalline and mesogenic molecules in the glassy state, Wendorff claimed a molecular orientation of only the chromophoric groups without the movement of the liquid crystalline moieties [70]. In recent works [93,112–129] Natansohn et al. have investigated the problem again and proved that both explanations are right: The possibility of a cooperative motion of both functional groups depends strongly on the liquid crystalline nature and polarity of the functional groups. In liquid crystalline polymers the cooperative motion is caused by the steric effect of forming organized domains. This explanation applies to systems where the liquid crystalline phase is well established. In amorphous systems, however, where the liquid crystalline nature of the mesogenic side chains is not dominant, other reorientation mechanisms may occur. A detailed study of different representatives of these amorphous polymers shows that the main driving force for the molecular reorientation is the interaction between mesogenic and chromophoric side chains due to dipolar forces: Mesogenic and chromophoric units move in concert when there is a strong dipolar coupling between them. Therefore the polarity of the functional groups is the most important parameter in systems with an instrinsic tendency of the liquid crystalline groups, which is not well established or even totally suppressed. If the substitution of the functional groups does not allow any dipolar interaction, there is no way for the liquid crystalline environment to take part in the light-driven interaction. This example shows that a lot of different factors are responsible for the response of the individual molecules to light.

Furthermore, Natansohn and Rochon have demonstrated that the light-induced birefringence can be locally erased by illuminating a spot with circularly or unpolarized light. In this way, the light-induced anisotropy can be randomized again, producing an isotropic sample [32,94,95].

Recently a lot of different research groups all over the world have started to investigate the system of photoaddressable polymers. Most of these groups deal with the interpretation of the molecular effects responsible for the light-induced birefringence and the application of these polymers in the field of optical data storage, as command surfaces in liquid crystal cells [97–104], as well as the use of azo-polymers as functional devices in electronics,[7] and in nonlinear optics. The molecular working principle of all these devices is based on the reorientation of the azo groups, as described above. Since the structural reorientation of the dye can modify the local molecular environment, the dyes can be used to control other molecules[8] or even the thermodynamic phase behavior of the whole macroscopic system.[9]

It has further been shown that under certain polarization conditions of the incident light a holographic grating experiment can lead to the formation of a surface grating [105–108]. The mechanisms responsible for the formation of those gratings are not yet fully understood [131–134]. Since these experiments are performed at room temperature, well below the glass transition temperature (T_g up to 150°C), a mass transport is not expected. It was however possible to demonstrate that the presence of a free surface is a necessary condition for the formation of these gratings [109]. Furthermore, there are po-

[7] For example as waveguides that can be generated by exposure to light [130].

[8] As in command surfaces the surrounding low molar liquid crystalline molecules are aligned.

[9] Controlled *trans–cis–trans* cycles can be used to switch a liquid crystalline copolymer from the isotropic to the liquid crystalline phase [110,111] in a reversible way.

larization configurations of the two actinic laser beams where the grating is not established very effectively (for example two s-polarized writing beams). Therefore there are ways to suppress these surface gratings and to maintain the good optical quality of films used for holographic data storage.

4 Photoaddressable Polymers from Bayer

Five years ago, the Central Research Department of Bayer started to work on the field of photoaddressable polymers. In a first three-year period around 300 polymers were synthesized, and their optical parameters were characterized in a BMBF-funded project.[10]. As far as the application in optical data storage is concerned, those polymers were optimized in the following way:

- The light-induced birefringence can be induced with a high spatial resolution. Holographic grating experiments with a grating distance of several thousand lines per mm could be performed. Furthermore scanned lines with a diameter smaller than 400 nm could be exposed by using a scanning laser confocal microscope.
- The value of the light-induced birefringence is as high as $\Delta n \sim 0.5$. These high birefringence values are induced with the blue or green light of an argon-ion laser ($\lambda = 488$, 514 nm) and detected with the red light ($\lambda =$ 633 nm) of an HeNe laser or the infrared light ($\lambda = 820$ nm) of a laser diode.
- With typical power densities in the 100 mW/cm^2 range of the argon laser, these high birefringence values can be induced within only a few seconds.
- One big advantage of these PAPs is the long-term stability of the light-induced birefringence. Even at temperatures as high as 160°C there is no decay in the maximum birefringence.
- The polymers are amorphous,[11] therefore they can easily be processed. The optical quality of these amorphous polymers is excellent.
- Since the corresponding physical properties are polarization-dependent molecular orientation mechanisms, no further chemical development steps are necessary. PAPs are dry recording materials.
- Furthermore, the light-induced reorientation is reversible, and the material can, in principle, be used as a rewritable medium.

[10] Together with the Humboldt University of Berlin (Stumpe et al.), the Philipps University of Marburg (Wendorff et al.), Moscow State University (Shibaev et al.), Agfa AG in Munich (Vedder et al.), the Technical University of Berlin (Rübner et al.), the Institute of Applied Chemistry in Berlin, Adlershof (Ruhman et al.), and the University of Bayreuth (Haarer et al.)

[11] More exactly, as discussed above, the best polymers have a phase behavior at the border of the amorphous and the liquid crystalline phase.

5 Photoaddressable Polymers Used in Holographic Data Storage

All the experiments discussed above were performed with samples of a thickness in the range of only a few microns. So far, thicker samples could not be used for optical experiments, since the optical densities of the films were too high. Therefore the incident light could not penetrate the samples of higher thickness.

In order to test the holographic properties of these thin films, first holographic measurements in IBM's PRISM tester [96] had been performed. Photoaddressable polymers with Δn values in the order of 0.3 were investigated.

Because of the large Δn value, the dynamic range of even 1 μm-thick-samples reaches values of $M\# \sim 1$. With the experimental writing conditions used at IBM ($\lambda = 514$ nm, $I \sim 140$ mW/cm^2, holographic contrast ratio of $m \sim 0.33$) holograms could be written within 100 ms, resulting in a sensitivity of these films of $S \sim 400$ cm/J. One big advantage of these amorphous films is their outstanding optical quality: the scattering efficiency is as low as $\eta_{\text{scat}} \sim 10^{-7}$. Furthermore these first experiments reveal excellent properties of photoaddressable polymers in terms of bit error rate (BER). The occurrences of the dark and bright spots in a hologram of a 256 k-mask (in the IBM PRISM tester) in Fig. 7 results in a BER of about 10^{-9}. This measurement was performed with a typical photoaddressable polymer film with a thickness of 40 μm, sandwiched between two glass plates.

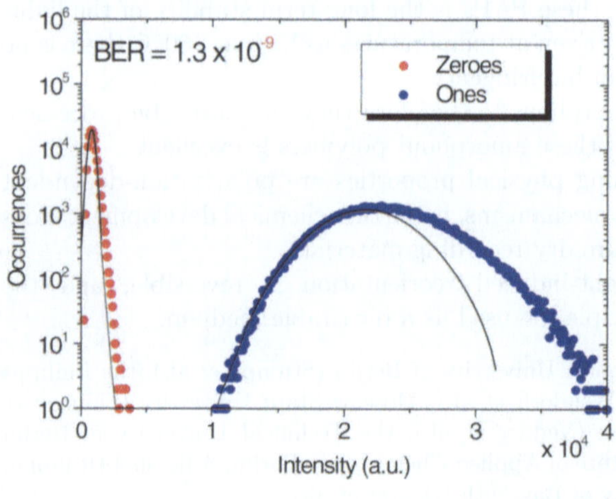

Fig. 7. Bit error rate of the hologram of a 256 k-mask

6 Open Questions and Outlook

The true advantage of holographic data storage is the use of all three dimensions of the recording medium. With films of a thickness of only a few microns one cannot really take advantage of the third dimension. The resulting holograms must be categorized into the class of thin holograms. The Bragg angle of these thin holograms is in the range of a few tens of degrees, so no true angular multiplexing is possible.

By choosing a modified chromophore system with an adapted absorption behavior and sophisticated dilution methods, it was possible in recent work to synthesize photoaddressable polymers with a thickness of several millimeters and moderate optical densities. First holographic experiments in IBM's PRISM tester have been performed, underlining the excellent optical quality of those films. With these films, Bragg angles in the range of $10^{-1\circ}$ can be realized. So far, an increase in the thickness by diluting the absorbing chromophore systems has caused the disadvantage of losing dynamic range and sensitivity. The current polymer development in our group has the goal of choosing the right polymer composition of chromophoric and mesogenic groups to avoid this loss. In the last few months, the best chromophoric groups seem to be identified, which allow storage in cw applications even in the low millisecond range. In current works, adapted mesogenic groups for these chromophores are tailored that will allow the stabilization and even amplification of the dyes, light-induced orientation by cooperative motion.

Experience in the development of photoaddressable polymers in the 1990s showed that there are many degrees of freedom in the synthesis of PAPs and, therefore, a variety of demands can be performed by choosing an adequate molecular architecture. It is still an open question whether all the requirements of holographic data storage can be fulfilled with photoaddressable polymers. The answer will be given by the experimental results in coming years.

References

1. V.A. Barachevskii, Ed., *Properties of Light Sensitive Materials and Their Applications in Holography*, Leningrad: Nauka (1987).
2. C.B. McArdle, Ed., *Applied Photochromic Polymer Systems*, London: Blackie (1992).
3. V.P. Shibaev, Ed., *Polymers as Electrooptical and Photooptical Active Media*, Berlin: Springer (1996).
4. I. Cabrera, V. Krongauz, and H. Ringsdorf, Angew. Chem. Int. Ed. Engl., **26**, 1178 (1987).
5. S. Yitzchaik, I. Cabrera, F. Buchholtz, and V. Krongauz, Macromolecules, **23**, 707 (1990).
6. L. Natarajan, V. Tondiglia, T. Bunning, R. Crane, and W. Adams, Adv. Mater. Opt. Electron. **1**, 293 (1992).

7. I. Cabrera, A. Dittrich, and H. Ringsdorf, Angew. Chem., **103**, 106 (1991).
8. P. Darcy, H. Heller, P. Strydom, and J. Whittall, J. Chem. Soc., Perkin Trans., **I**, 202 (1981).
9. G. Sudesh Kumar and D.C. Neckers, Chem. Rev., **89**, 1915 (1989).
10. M. Patel, R. Patel, V. Patel, and S. Maiti, J. Polym. Mater., **8**, 67 (1991).
11. H. Yamamoto, A. Nishida, T. Takimoto, and A. Nagai, J. Polym. Sci., Part A: Polym. Chem., **26**, 67 (1990)
12. H. Yamamoto and A. Nishida, Bull. Chem. Soc. Jpn., **61**, 2201 (1988).
13. L. Mateika, N.E. Ilavsky, K. Dusek, and O. Wichterle, Polymer, **22**, 1511 (1981).
14. I. Willner, S. Rubin, and A. Riklin, J. Am. Chem. Soc., **113**, 3321 (1991).
15. S. Barley, A. Gilbert, and G. Mitchell, J. Mater. Chem., **1**, 481 (1991).
16. M. Eich. *PhD thesis.* Technische Hochschule Darmstadt (1987)
17. G. Zimmermann, L. Chow, and U. Paik, J. Am. Chem. Soc., **80**, 359 (1979).
18. H. Rau, *Photochemistry and Photophysics*, J.F. Rabek. Ed., CRC, Boca Raton, FL, Vol. 2, p. 119 (1990).
19. C.S. Paik and H. Morawetz, Macromolecules, **5**, 171 (1972).
20. C.D. Eisenbach, Makromol. Chem., **179**, 2489 (1978).
21. C.D. Eisenbach, Ber. Bunsenges. Phys. Chem., **84**, 680 (1980).
22. C.S. Paik Sung, I.R. Gould, and N.J. Turro, Macromolecules, **17**, 1447 (1984).
23. E. Dubini-Paglia, P.L. Beltrame, B. Marcandalli, P. Carniti, L. Seves, and A. Vicini, J. Appl. Polym. Sci., **31**, 1251 (1986).
24. K. Horie and L Mita, Adv. Polym. Sci., **88**, 77 (1989).
25. I. Mita, K. Horie, and K. Hirao, Macromolecules, **22**, 558 (1989).
26. T. Naito, K. Horie, and I. Mita, Eur. Polym. J., **26**, 1295 (1990).
27. T. Naito, K. Horie, and I. Mita, Polym. J., **23**, 809 (1991).
28. T. Naito, K. Horie, and I. Mita, Macromolecules, **24**, 2907 (1991).
29. P. Uznanski, M. Kryszewski, and E.W. Thulstrup, Eur. Polym. J., **27**, 41 (1991).
30. L. Lamarre and C. Sung, Macromolecules, **16**, 1729 (1983).
31. I. Mita, K. Horie, and K. Hirao, Macromolecules, **22**, 558 (1989).
32. A. Natansohn, S. Xie and P. Rochon, Macromolecules, **25**, 5531 (1992).
33. T. Naito, K. Horie, and I. Mita, Polym. J., **23**, 809 (1991).
34. L. Läsker, T. Fischer, J. Stumpe, S.G. Kostromin, and R. Ruhmann. Freiburger Arbeitstagung Flüssigkristalle (1994).
35. A.L. Tsykalo. *Thermophysical Properties of Liquid Crystals*, Gordon and Breach Science Publishers (1991).
36. P.J. Collings, *Liquid Crystals*, Princeton University Press, Princeton, New Jersey (1990).
37. S. Chandrasekhar, *Liquid Crystals*, Cambridge University Press (1992).
38. G.W. Gray, *Thermotropic Liquid Crystals*, John Wiley and Sons (1987).
39. G.R. Luckhurst and G.W. Gray, *The Molecular Physics of Liquid Crystals*, Academic Press (1977).
40. W.H. de Jeu, *Physical Properties of Liquid Crystalline Materials*, Gordon and Breach Science Publishers (1980).
41. H. Ringsdorf and R. Zentel, Makromol. Chem., **183**, 1245 (1982).
42. E.T. Samulski, Physics Today, 40 (1982).
43. H. Finkelmann, Angew. Chem., **99**, 840 (1987).
44. H. Finkelmann, H. Ringsdorf, and J.H. Wendorff, Makromol. Chem., **179**, 273 (1978).

45. B.H. Bransden and C.J. Joachain, *Physics of Atoms and Molecules*, Wiley and Sons (1991).
46. Th. Bieringer, R. Wüttke, D. Haarer, U. Geßner, and J. Rübner, Macromol. Chem. Phys., **196**, 1375 (1995).
47. S.J. Zilker, Th. Bieringer, D. Haarer, R.S. Stein, and J.W. van Egmond, Adv. Mater., **10**(11), 855 (1998).
48. M. Eich and J. Wendorff, Makromol. Chem., Rapid Commun., **8**, 467 (1987).
49. K. Anderle, R. Birenheide, M. Eich and J. Wendorff, Makromol. Chem., Rapid Commun., **10**, 477 (1989).
50. R. Ortler, C. Braeuchle, A. Millera, and G. Riepl, Makromol. Chem., Rapid Commun., **10**, 189 (1989).
51. M. Eich, J. Wendorff, B. Reck, and H. Ringsdorf, Makromol. Chem., Rapid Commun., **8**, 59 (1987).
52. S. Hvilsted, F. Andruzzi, and P. Ramanujam, Opt. Lett., **17**, 1234 (1992).
53. S. Hvilsted, F. Andruzzi, C. Kulinna, H. Siesler, and P.S. Ramanujam, Macromolecules, **28**, 2172 (1995).
54. N. Holme, P.S. Ramanujam, and S. Hvilsted, Appl. Opt., **35**, 4622 (1996).
55. N.Holme, P.S. Ramanujam, and S. Hvilsted, Opt. Lett., **21**(12), 902 (1996).
56. S. Hvilsted and P.S. Ramanujam, Curr. Trends Polym. Sci., **1**, 53 (1996).
57. L. Nikolova, T. Todorov, M. Ivanov, S. Hvilsted, and P.S. Ramanujam, Appl. Opt., **35**, 3835 (1996).
58. R.H. Berg, S. Hvilsted, and P.S. Ramanujam, Nature, **383**, 505 (1996).
59. T. Fischer, L. Laesker, J. Stumpe, and S.J. Kostromin, Photochem. Photobiol. A: Chem., **80**, 453 (1994).
60. X. Meng, A. Natansohn, and P. Rochon, J. Polym. Sci., Part B. Polym. Phys., **34**, 1461 (1996).
61. I. Zebger, C. Kulinna, H. Siesler, F. Andruzzi, M. Pedersen, P.S. Ramanujam, and S. Hvilsted, Macromol. Symp., **94**, 159 (1995).
62. M. Ho, A. Natansohn, C. Barrett, and P. Rochon, Can. J. Chem., **73**, 1773 (1995).
63. J. Stumpe, L. Mueller, D. Kreysig, G. Hauck, H. Koswik, R. Ruhmann, and J. Ruebner, Makromol. Chem., Rapid Commun., **12**, 81 (1991).
64. S. Ivanov, I. Yakovlev, S. Kostromin, V. Shibaev, L. Laesker, J. Stumpe, and D. Kreysig, Makromol. Chem., Rapid Commun., **12**, 709 (1991).
65. U. Wiesner, N. Reynolds, C. Boeffel, and H.W. Spieß, Makromol. Chem., Rapid Commun., **12**, 457 (1991).
66. U. Wiesner, N. Reynolds, C. Boeffel, and H.W. Spieß, Liq. Cryst., **11**, 251 (1992).
67. M. Eich and J.H. Wendorff, J. Opt. Soc. Am. B, **7**(8), 1428 (1990).
68. O. Watanabe, M. Tsuchimori, A. Okada, and H. Ito, Appl. Phys. Lett., **71**(6), 750 (1997).
69. R. Birenheide and J.H. Wendorff, Angew. Makromolek. Chem., **183**, 167 (1990).
70. K. Anderle, R. Birenheide, M.J.A. Werner, and J.H. Wendorff, Liq. Cryst., **9**, 691 (1991).
71. L. Läsker, T. Fischer, J. Stumpe, S. Kostromin, S. Ivanov, V. Shibaev, and R. Ruhmann, Mol. Cryst. Liq. Cryst., **253**, 1 (1994).
72. L. Läsker, T. Fischer, J. Stumpe, S. Kostromin, S. Ivanov, V. Shibaev, and R. Ruhmann, Mol. Cryst. Liq. Cryst., **246**, 347 (1994).

73. T. Fischer, L. Läsker, J. Stumpe, and S. Kostromin, J. Photochem. Photobiol. A, **80**, 453 (1994).
74. M. Dumont, G. Froc, and S. Hosotte, Nonlinear Opt., **9**, 327 (1995).
75. R. Loucif-Saibi, K. Nakatani, J. Delaire, M. Dumont, and Z. Sekkat, Chem. Mater., **5**, 229 (1993).
76. T. Todorov, L. Nikolova, and N. Tomova, Appl. Opt., **24**, No. 6, 785 (1985).
77. T. Todorov, N. Tomova, and L. Nikolova, Opt. Commun., **47**, 123 (1983).
78. J. Stumpe, L. Müller, L. Läsker, D. Kreysig, G. Hauck, S. Kostromin, and V. Shibaev, Freiburger Arbeitstagung Flüssigkristalle (1991).
79. S. Ivanov, I. Yakovlev, S. Kostromin, V. Shibaev, L. Läsker, J. Stumpe, and D. Kreysig, Makromol. Chem., Rapid Commun., **12**, 709 (1991).
80. J. Stumpe, L. Müller, D. Kreysig, G. Hauk, H.D. Koswig, R. Ruhmann, and J. Rübner, Makromol. Chem., Rapid Commun., **12**, 81 (1991).
81. U. Wiesner, M. Antonietti, C. Boeffel, and H.W. Spiess, Makromol. Chem., **191**, 2133 (1990).
82. M. Eich, J.H. Wendorff, B. Reck, and H. Ringsdorf, Makromol. Chem., Rapid Commun., **8**, 59 (1987).
83. Z. Sekkat, J. Wood, W. Knoll, W. Volksen, R.D. Miller, and A. Knoesen, J. Opt. Soc. Am. B, **14**(4), 829 (1997).
84. R. Wuttke, PhD Thesis. Universität Bayreuth (1994).
85. R. Birenheide, PhD Thesis. Technische Hochschule Darmstadt (1991).
86. T. Todorov, L. Nikolova, and N. Tornova, Appl. Opt., **23**, No. 24, 4588 (1984).
87. F. Lagugné Labarthet and C. Sourisseau, J. Raman Spectrosc., **27**, 491 (1996).
88. F. Lagugné Labarthet, Buffeteau and C. Sourisseau, J. Phys. Chem. B, **102**, 5754 (1998).
89. F. Lagugné Labarthet, PhD Thesis, University of Bordeaux I (1998).
90. P. Rochon, E. Batalla, and A. Natansohn, Appl. Phys. Lett., **66**(2), 136 (1995).
91. D.Y. Kim, L. Li, X.L. Jiang, V.Shivshankar, J. Kumar, and S.K. Tripathy, Macromolecules, **28**, 8835 (1995).
92. P.S. Ramanujam, N. Holme, and S. Hvilsted, Appl. Phys. Lett., **68**(10), 1329 (1996).
93. A. Natansohn, P. Rochon, X. Meng, Ch. Barrett, T. Buffeteau, S. Bonenfant, and M. Pézolet, Macromolecules, **31**, 1155 (1998).
94. P. Rochon, J. Gosselin, A. Natansohn, S. Xie, Appl. Phys. Lett., **60**(1), 4 (1992).
95. A. Natansohn, P. Rochon, J. Gosselin, and S. Xie, Macromolecules, **25**, 2268 (1992).
96. M.-P. Bernal, H. Coufal, R.K. Grygier, J.A. Hoffnagel, C.M. Jefferson, R.M. Macfarlane, R.M. Shelby, G.T. Sincerbox, P. Wimmer, and G. Wittmann, Appl. Opt., **35**(14), 2360, (1996).
97. W.M. Gibbons, P.J. Shannon, S.T. Sun, and B.J. Swetlin, Nature, **351**, 49 (1991).
98. S.T. Sun, W.M. Gibbons, and P.J. Shannon, Liq. Cryst., **12**(5), 869 (1992)
99. C. Jones and S. Day, Nature, **351**, 15 (1991).
100. Y. Kawanishi, T. Tamaki, T. Seki, M. Sakuragi, and K. Ichimura, Mol. Cryst. Liq. Cryst., **218**, 153 (1992).
101. K. Aoki, T. Seki, M. Sakuragi, and K. Ichimura, Makromol. Chem., **193**, 2164 (1992).
102. K. Aoki, T. Seki, Y. Suzuki, T. Tamaki, A. Hosoki, and K. Ichimura, Langmuir, **8**, 1007 (1992).

103. K. Aoki, T. Tamaki, T. Seki, Y. Kawanishi, and K. Ichimura, Langmuir, **8**, 1014 (1992).
104. Y. Kawanishi, T. Tamaki, M. Sakuragi, T. Seki, Y. Suzuki, and K. Ichimura, Langmuir, **8**, 2601 (1992).
105. C. Barrett, P. Rochon, and A. Natansohn, J. Chem. Phys., **109**(4), 1505 (1998).
106. D.Y. Kim, S.K. Tripathy, L. Li, and J. Kumar, Appl. Phys. Lett., **66**(10), 1166 (1995).
107. N. Holme, L. Nikolova, P.S. Ramanujam, and S. Hvilsted, Appl. Phys. Lett., **70**, 1518 (1997).
108. A. Liebmann-Vinson, L.M. Lander, M.D. Foster, W.J. Brittain, E.A. Vogler, C.F. Majkrzak, S. Satija, Langmuir, **12**, 2256 (1996).
109. N.K. Viswanathan, S. Balasubramanian, L. Li, J. Kumar, and S.K. Tripathy, J. Phys. Chem., B **102**, 6064 (1998).
110. T. Ikeda, T. Miyamoto, T. Sasaki, S. Kurihara, and S. Tazuke, Mol. Cryst. Liq. Cryst., **188**, 235 (1990).
111. T. Ikeda, S. Horiuchi, D.B. Kranjit, S. Kurihara, and S. Tazuke, Macromolecules, **23**, 36 (1990).
112. A. Natansohn and P. Rochon, Progress in Pacific Polymer Science, Volume 3, Berlin: Springer (1994).
113. P. Rochon, J. Mao, A. Natansohn, and E. Batalla, Polym. Prep. (Am. Chem. Soc.), Div. Polym. Chem., **3**(2), 154 (1994).
114. S. Xie, A. Natansohn, and P. Rochon, Chem. Mater., **5**, 403 (1993).
115. P. Rochon, D. Bissonnette, A. Natansohn, and S. Xie, Appl. Opt., **32**(35), 7277 (1993).
116. P. Rochon, J. Mao, A. Natansohn, and S. Xie, SPIE, **2042**, 347 (1994).
117. A. Natansohn, P. Rochon, M. Pezolet, P. Audet, D. Brown, and S. To, Macromolecules, **27**, 2580 (1994).
118. C. Barrett, A. Natansohn, and P. Rochon, Macromolecules, **27**, 4781 (1994).
119. P. Rochon, E. Batalla, and A. Natansohn, Appl. Phys. Lett., **66**(2), 136 (1995).
120. M.S. Ho, A. Natansohn, C. Barrett, and P. Rochon, Can. J. Chem., **73**, 1773 (1995).
121. A. Natansohn, P. Rochon, C. Barrett, and A. Hay, Chem. Mater., **7**, 1612 (1995).
122. A. Natansohn, P. Rochon, M.S. Ho, and C. Barrett, Macromolecules, **28**, 4179 (1995).
123. D. Brown, A. Natansohn, and P. Rochon, Macromolecules, **28**, 6116 (1995).
124. M.S. Ho, A. Natansohn, and P. Rochon, Macromolecules, **28**, 6124 (1995).
125. M.S. Ho, A. Natansohn, and P. Rochon, Macromolecules, **29**, 44 (1996).
126. X. Meng, A. Natansohn, C. Barrett, and P. Rochon, Macromolecules, **29**, 946 (1996).
127. X. Meng, A. Natansohn, and P. Rochon, J. Polym. Sci., B **34**, 1461 (1996).
128. C. Barrett, A. Natansohn, and P. Rochon, Chem. Mater., **7**, 899 (1995).
129. T. Buffeteau, A. Natansohn, P. Rochon, and M. Pézolet, Macromolecules, **29**, 8783 (1996).
130. P. Rochon, A. Natansohn, C.L. Callender, and L. Robitaille, Appl. Phys. Lett., **71**(8), 1008 (1997).
131. T.G. Pedersen, P.M. Johansen, N.C.R. Holme, P.S. Ramanujam, and S. Hvilsted, Phys. Rev. Lett., **80**(1), 89 (1998).

132. J. Kumar, L. Li, X.L. Jiang, D-Y. Kim, T.S. Lee, and S. Tripathy, Appl. Phys. Lett., **72**(17), 2096 (1998).
133. P. Lefin, C. Fiorini, and J.-M. Nunzi, Pure Appl. Opt., **7**, 71 (1997).
134. C. Barrett, A. Natansohn, and P. Rochon, J. Phys. Chem., **100**, 8836 (1996).

Part III

Components

Laser Sources

B. Pezeshki and S.S. Orlov

Though a variety of lasers have been used for holography over the years, a data storage system that is commercially viable requires a compact, efficient and ultimately low-cost source. The main requirements are coherence and wavelength compatibility with the material. Conventional semiconductor lasers, though compact and efficient, generally lack the coherence length required for holography. Similarly, gas lasers such as argon ion can be used in the laboratory to characterize materials, but the need for external cooling, the large size, and the immense power requirements preclude their use in real systems. In this section we first examine the general requirements for a laser suitable for practical holographic data storage, and then concentrate on two particular laser systems that meet our criteria: diode-pumped solid-state lasers and wavelength-stabilized semiconductor lasers. The two lasers that are particularly relevant are the diode-pumped frequency-doubled Nd:YAG, and the InGaP DFB lasers.

1 Laser Requirements

Optical holography requires coherent laser sources with high spectral and spatial brightness. Volume holography and holographic storage records the interference pattern between two mutually coherent beams within the volume of recording media and subsequently reconstructs the wavefront of the signal beam using the illumination of the reference beam on the recorded hologram.

The basic requirements of optical sources for holography stem from the coherent nature of the interaction between the two beams inside the recording volume. First, the temporal coherence of the optical source, which is a characteristic of the short-term wavelength stability, has to be sufficient to ensure maximum contrast of the interference pattern throughout the full volume of the medium. The second requirement is related to the long-term stability of the laser output wavelength. Volume holograms, particularly in reflection and widely used 90° recording geometry, exhibit substantial wavelength selectivity. This critical dependence on wavelength is in fact exploited in other applications including holographic wavelength multiplexing [1] and narrowband reflection filters [2]. Long-term changes in laser emission wavelength after the recording may reduce the diffraction efficiency and cause cross-talk in wavelength-multiplexed holograms.

Additional requirements for laser sources used in holographic data storage can be understood if we consider a realistic system. The laser source used for recording needs to comply with the corresponding spectral response of the media. The majority of the currently available holographic recording materials exhibit photosensitivity in the visible (blue-green or red) region of the spectrum. Most of the digital holographic data storage architectures require near-diffraction limited single spatial mode operation of the source laser. This is required to ensure the spatial uniformity of illumination light used in high-resolution holography. Single spatial mode lasers also provide superior wavelength and power stability.

In holographic data storage the diffracted light from the hologram is detected by a two-dimensional array, and decoded into bit streams. The parallelism of detecting all bits in the page simultaneously gives rise to an extremely high data transfer rate. The tradeoffs between the capacity, transfer rate and the reading laser power can be captured by considering the following approximate equation [3]:

$$\#\text{pages} = \left(\frac{2(M\#)^2 \Theta QEP \eta_{\text{opt}} \eta_{\text{fix}} t_{\text{read}}}{MNp_{\text{min}}} \right)^{1/2}, \tag{1}$$

where #pages is the number of data pages stored collocationally, $M\#$ is a variable based on system and materials parameters, Θ is the number of photons per watt, QE is the quantum efficiency of the detector array, η_{opt} is the optical efficiency of the imaging system, η_{fix} is the fixing efficiency of the material, t_{read} is the read time of the recalled data, M and N are the number of rows and columns on the data page, p_{min} denotes the number of photoelectrons needed to get the required SNR, and P is the power of the readout beam. For fixed parameters of the holographic storage system and fixed capacity the transfer rate (which is inversely proportional to the integration time t_{read}) is proportional to the read laser power P. Alternatively, for fixed transfer rate, the available storage capacity (which is proportional to the number of superimposed data pages #pages) increases with the available laser power. Most holographic data storage systems built so far use lasers with greater than 500 mW of cw power.

Additional requirements of laser sources used in holographic data systems include size (compactness), power efficiency, reliability, and the cost of the device. The two laser systems that meet most of the above criteria are the diode-pumped frequency-doubled solid state lasers (e.g., Nd:YAG), and visible wavelength semiconductor DFB lasers.

2 Diode-Pumped Solid-State Lasers

Solid-state lasers are among the oldest types of lasers, and have been a workhorse of holography for many years. Efficient and low-cost semiconductor pump sources have begun to replace flashlamps as the excitation source in these lasers, and have led to compact and high-power components useful for holography.

The term *solid-state laser* generally refers to a family of lasers where rare earth or transition metal ions are inserted into a transparent host crystal, usually either an oxide or a fluoride. For example, a small Cr^{3+} ion easily replaces Al^{3+} in Al_2O_3 to provide gain in a ruby laser, while Nd^{3+} replaces Y^{3+} in a $Y_3Al_5O_{12}$ (yttrium aluminum garnet or YAG) laser. Rare earth ions can also substitute for Y^{3+} in a fluoride host such as YLF, or a transition element for Al^{3+} in $LiSrAlF_6$ (abbreviated LiSAF) or $LiCaAlF_6$ (abbreviated LiCAF). Generally the oxides are harder, with better mechanical and thermomechanical properties, while fluorides have better thermo-optic behavior. Though oxide glasses such as SiO_2 and P_2O_5 are easy to melt, and can thus lend themselves to simpler fabrication, they exhibit worse thermo-optic properties when used as laser media.

When a rare earth element is doped into a host material, the two $6s$ electron states and one of the $4f$ electrons are used for ionic bonding: thus the element becomes triply ionized. All the optical properties of the laser are determined by the remaining $4f$ transitions, that are in fact forbidden in an isolated ion owing to parity considerations. In the host material, the interaction with the crystal field, though weak, creates a nonzero transition probability. Thus lifetimes of the excited states are very long, allowing considerable power to be stored for pulsed operation and Q-switching. The $5s^2$ and $5p^2$ electrons also screen the $4f$–$4f$ transitions, further isolating the state, leading to sharp transition lines and a weak nonradiative recombination that enhances efficiency. In transition metals such as Cr, Ti, Co, and Ni, electrons from the $4s$ and $3d$ are ionized, while the remaining $3d$ electrons rearrange themselves in a large number of states. Optical transitions occur between these $3d$–$3d$ orbitals. Since there is no screening, the orbitals interact strongly with the crystal, leading to wide absorption and emission bands. Though the transition is once again allowed only due to crystal fields, these are stronger and lifetimes are consequently shorter than those of rare earth ions (a few microseconds rather than hundreds of microseconds).

The number of solid-state lasers is large, and devices are used for a variety of applications. For example Er-doped silica fibers are now widespread as lasers and amplifiers for the 1.55 μm telecommunication band, while Ti: sapphire lasers are used for very short pulse and tunable applications. For holography, either direct or frequency-doubled emission is required at relatively short wavelengths. The ruby laser was the first laser demonstrated [4], and on account of its visible wavelength and short pulse characteristics was used extensively in the early days of holography. The emission wavelengths of 692.9 nm and 694.3 nm are a good match for many holographic materials however, the short absorption wavelength of 420 nm and 550 nm is not readily achievable with semiconductor pumps, and thus devices are still excited with flashlamps. Perhaps the most interesting lasers for holography are frequency-doubled Nd:YAG and Nd:YVO4 (vanadate) lasers, emitting at 532 nm and 523 nm respectively. These lasers are readily pumped by semiconductor lasers at 800 nm, and the high emission powers at 1.064 μm allow for doubling into the green. Frequency-doubled solid-state laser systems with cw power of several watts and more in the green are now commercially available.

3 Semiconductor Lasers

The most efficient and lowest-cost method is to directly use a semiconductor laser for reading and writing holograms. The three main challenges have traditionally been the inability to produce high power pulses, the generally long wavelength of the lasers, and the poor coherence. Unfortunately, very little can be done about the first factor. Since the radiative recombination times of semiconductor materials are on the order of nanoseconds, there is very little energy storage, and very high-power pulsed operation is not possible. For most applications, the lasers are either operated cw, or electrically pulsed at close to the same power levels. The two other challenges, however, are gradually being met. Though conventional red Fabry–Pérot lasers, such as those used in laser pointers, lack the coherence length and wavelength stability necessary for holography, more sophisticated structures emitting in the 630–680 nm band are now possible with internal grating stabilization, and provide high-power coherent beams.

In a direct-gap semiconductor, the optical transition generally occurs between the conduction band and the valence band of the material. Thus the emission wavelength is determined largely by the bandgap of the active material. At the shortest wavelength, the GaN family of materials (AlGaInN) grown on a sapphire or SiC substrates have very recently been developed for lasers at about 420 nm [5]. Owing to the immaturity of the material system, only low-power devices are currently available. The main application appears to be standard optical storage, where the shorter wavelength translates to a smaller spot size for higher data densities. At the blue-green parts of the spectrum ZnSSe materials grown on GaAs substrates originally showed some promise in the late 1980s and early 1990s, but could not deliver on lifetime. There is currently no semiconductor laser available at the center of the visible band, though GaN may be extended into this range in the near future with the addition of sufficient amount of indium. At the red wavelengths, suitable for holography in photorefractive polymers, InGaAlP material has been very successful. With an emission wavelength of 630–690 nm, the devices can be internally grating stabilized to deliver up to 400 mW. The infrared wavelengths are less interesting for holography with photosensitive materials, but the 780 nm–1.06 μm band can be addressed with the InGaAs system and the 1.3–2 μm band with the InGaAsP on InP substrates.

In the red materials, low-power Fabry–Pérot lasers emitting in the 630–680 nm range in GaInP/AlInP are now commonplace in laser pointers, supermarket scanners, and for DVD optical storage. These lasers typically have a waveguide with a cross-section of 4 μm × 1 μm are single spatial mode, and can be focused to a diffraction, limited spot. The waveguide length, however, is more than a thousand wavelengths long, on the order of 500 μm, and thus cannot provide stable lasing in the longitudinal direction. The output spectrum of such a laser generally consists of multiple longitudinal modes, and the beam can be unstable in power and wavelength. Broad-area lasers in the same material system have a waveguide width of about 100 μm instead of 4 μm, and are therefore multimode both longitudinally and spatially.

Narrow-waveguide semiconductor lasers can be easily stabilized using a longitudinal grating, as shown in Fig. 1. In this case, the epitaxial growth of the waveguide lasers is interrupted for the fabrication of a diffraction grating in the guiding layer. Once the grating is etched, the wafer is placed back in the reactor for the growth of the upper cladding and contact layers. Figure 1 shows a distributed Bragg reflector (DBR) laser, where the grating is placed at the rear of the cavity and acts as the back mirror. Since the reflectivity of the grating is narrow-band, it selects only one longitudinal mode, and the device becomes spectrally stable. The device can be tuned about 3 nm by injecting current into the rear grating and heating the device to vary the refractive index of the mode.

Recently a number of groups have demonstrated red lasers whose longitudinal mode is stabilized internally by a diffraction grating [6–14]. These different devices all have roughly the same epitaxial structure as the common red Fabry–Pérot lasers, except for the addition of the grating. The gain is provided by strained InGaP quantum wells, where the strain and well width are adjusted for the desired polarization and wavelength. The quantum wells

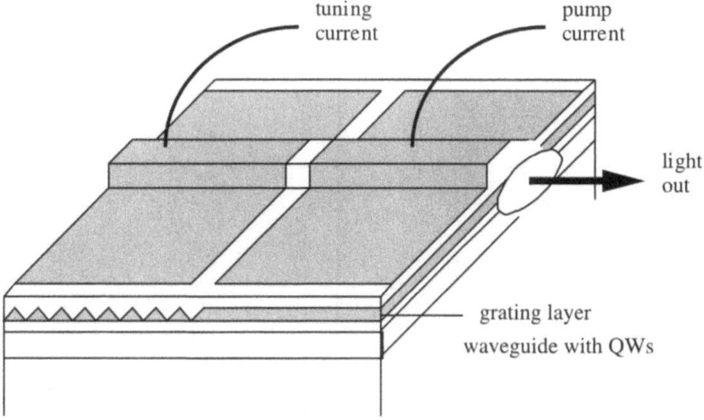

Fig. 1. Schematic diagram of a tunable DBR structure

are contained within an InGaAlP waveguide and between AlInP cladding regions. The grating layer is also composed of AlGaInP, but with the aluminum content minimized to ease regrowth. In fact the greatest problem with the fabrication of such lasers is good quality regrowth over the aluminum-containing grating layer, which can be eased by *in-situ* cleaning steps.

In addition to placing the grating in the rear of the device, as shown in Fig. 1, the grating can also be continuous and placed throughout the cavity to form a distributed feedback (DFB) laser [6–9]. Unlike the DBR [8,10], which needs a strong rear reflector, in the DFB the grating is made relatively weak, with a total ("cold") reflectivity of only about 10–30% along the length of the device. The back mirror is cleaved and coated for high reflectivity, while the front facet is antireflection coated to allow the built-in grating to act as the front mirror.

Figures 2 and 3 show the light–current curve and spectrum of an optimized 662 nm DFB laser. Since the DFB has only a single contact, no independent electrical tuning is possible. The wavelength changes only with temperature – which can be adjusted directly by controlling the heatsink TE cooler, or indirectly by varying the operating current. All tuning occurs continuously with no mode jumps. The tuning rate is 0.05 nm/°C, corresponding to the change in refractive index with temperature. Note that in Fabry–Pérot structures, the wavelength shifts much faster with temperature as it is the change of the bandgap with temperature and not the change in index that is important.

These DFB devices have demonstrated high power and good lifetimes. Accelerated lifetest at 40 mW, 50°C for 1500 hours showed no failures out of 20 devices. The higher temperature accelerates degradation by about 10 times compared with room temperature operation. The devices have also been

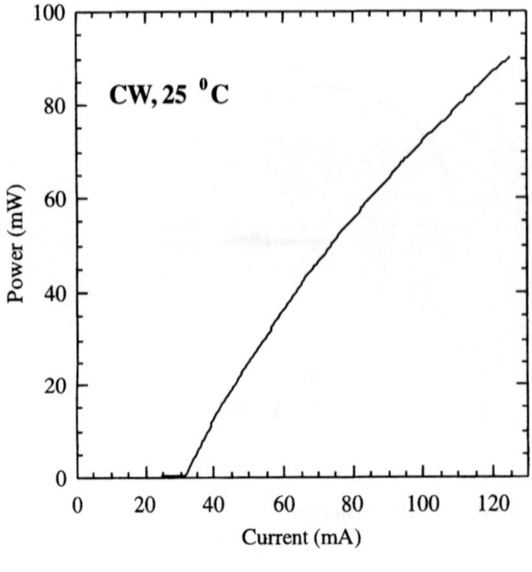

Fig. 2. DFB light–current characteristics

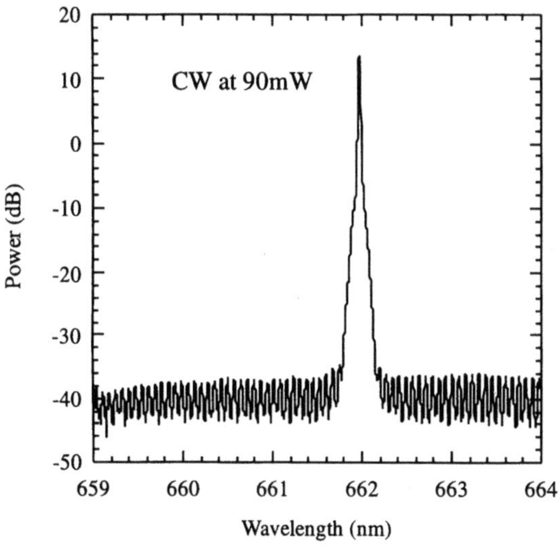

Fig. 3. Spectrum of DFB

demonstrated at the shorter wavelength of 635 nm [14], suitable for direct replacement of He-Ne lasers. The wavelength is generally stable to better than 1 part in 10^5, usually limited by the control over the temperature. Since the wavelength tunes slightly with temperature, one can even compensate for shrinkage of holographic materials. For example one may write a hologram at one wavelength, fix the hologram, and compensate for the shrinkage during fixing by reading the information at a slightly shorter wavelength.

To obtain higher powers from a semiconductor laser, one must increase the stripe width without allowing higher-order lateral modes. This can be done in a monolithic fashion by using a flared amplifier in an MOPA configuration [15] (master oscillator, power amplifier) or by using a lateral grating. In the same way that a longitudinal grating can stabilize the longitudinal mode, an angled grating can stabilize both the longitudinal and the lateral modes of a broad area laser. The angled-grating DFB, also known as α-DFB, delivers high single-frequency power. It is a broad-area structure, and thus has a much lower optical energy density for a given output power. Such devices have been demonstrated at a wavelength of 1 µm in the InGaAs/AlGaAs material system, delivering more than 1 W [16]. The same configuration has recently been demonstrated at visible wavelengths [17] at 400 mW.

Figures 4 and 5 show the light–current curve and the spectrum of such a laser with a stripe width of 60 µm and cavity length of 750 µm. The threshold is 600 mA with a linear slope efficiency of 0.75 W/A. On account of its much wider stripe width, the α-DFB can achieve much higher power levels than the conventional DFB lasers, emitting 400 mW at a current of 1.14 A. The spectrum at this power level is shown in Fig. 5 with an instrument resolution limit of 0.05 nm. Since the emitting aperture is wide and thin, the beam diffracts into a line, with a FWHM laterally of about 0.3° and vertically with

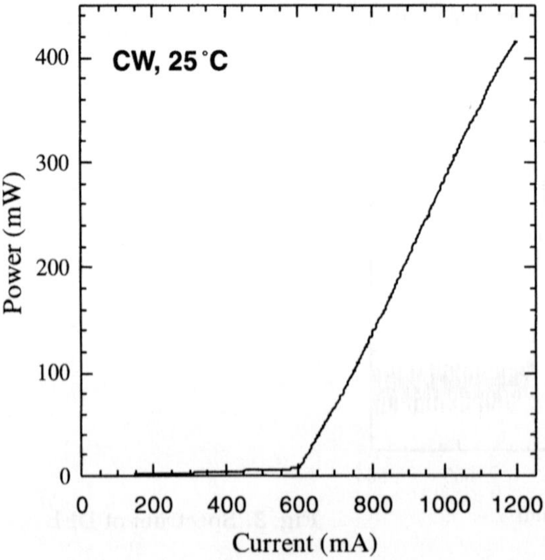

Fig. 4. *L–I* characteristics of α-DFB

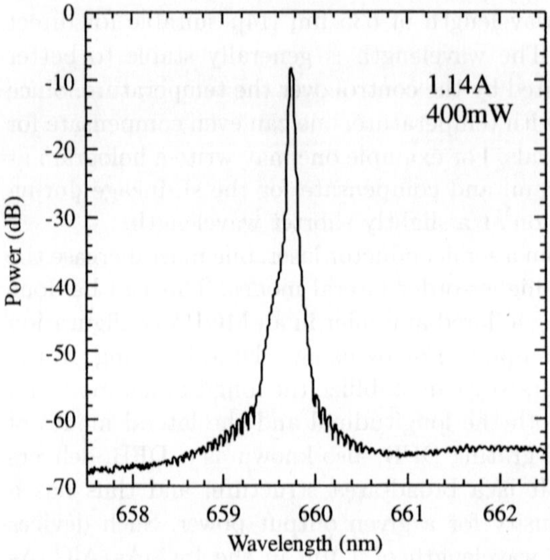

Fig. 5. Spectrum of α-DFB

a FWHM of about 35°. The beam is easily circularized with a cylindrical lens close to the emitting aperture.

Acknowledgment

The work was partially supported by the Defense Advanced Project Agency (DARPA) in the Holographic Data Storage System (HDSS) consortium, under contract MDA972-95-3-0004.

References

1. G.A. Rakuljic, V. Leyva, and A. Yariv, "Optical data storage by using orthogonal wavelength-multiplexed volume holograms," Opt. Lett., **17**, 1471 (1992).
2. G.A. Rakuljic and V. Leyva, "Volume holographic narrowband optical filter," Opt. Lett., **18**, 459 (1993).
3. L. Hesselink, paper presented at OSA Annual Meeting, Rochester, NY, 20 to 24 October 1996.
4. T.H. Maiman, "Stimulated optical radiation in ruby masers," Nature, **187**, 493 (1960).
5. S. Nakamura and G. Fasol, *The blue laser diode, GaN based emitters and lasers*, Springer, 1997.
6. T. Chong and K. Kishino, "660 nm wavelength GaInAsP/AlGaAs distributed feedback lasers," Electron. Lett., **24**, 416 (1988).
7. M. Korn, T. Korfer, A. Forchel, and P. Roentgen, "Fabrication and optical characterization of first-order DFB GaInP/AlGaInP laser structures at 639 nm," Electron. Lett., **26**, 614 (1990).
8. D.-H. Jang, Y. Kaneko, and K. Kishino, "Shortest wavelength (607 nm) operations of GaInP/AlInP distributed Bragg reflector lasers," Electron. Lett., **28**, 428 (1992).
9. H.-P. Gauggel, C. Geng, H. Schweizer, F. Barth, J. Hommel, R. Winterhoff, and F. Scholz, "Fabrication and operation of first-order GaInP/AlGaInP DFB lasers at room temperature", Electron. Lett., **31**, 367 (1995).
10. B. Pezeshki, J.S. Osinski, H. Zhao, A. Mathur A., and T.L. Koch, "GaInP/AlGaInP 670 nm singlemode DBR laser," Electron. Lett., **32**, 2241 (1996).
11. H.-P. Gauggel, H. Artmann, C. Geng, F. Scholz, and H. Schweizer, "Wide-range tunability of GaInP-AlGaInP DFB lasers with superstructure gratings," IEEE Photon. Technol. Lett., **9**, 14 (1997).
12. H.-P. Gauggel, J. Kuhn, C. Jerichow, C. Geng, F. Scholz and H. Schweizer, "Low threshold CW operation of GaInP/AlGaInP DFB lasers at 680 nm," Electron. Lett., **33**, 1385 (1997).
13. B. Pezeshki, M. Zelinski, H. Zhao, and V. Agrawal, "40 mW 650 nm distributed feedback lasers," IEEE Photon. Tech. Lett., **10**, 36 (1998).
14. B. Pezeshki, M. Zelinski, and V. Agrawal, "635 nm 20 mW DFB laser," Electron. Lett., **34**, 987 (1998).
15. B. Pezeshki, J.S. Osinski, M. Zelinski, S. O'Brien, and A. Mathur, "660 nm 250 mW GaInP/AlInP monolithically integrated master oscillator power amplifier," Electron. Lett., **33**, 1314 (1997).
16. V.V. Wong, S.D. DeMars, A. Schoenfelder, and R. Lang, "Angled-grating distributed-feedback laser with 1.2 W CW single-mode diffraction limited output at 1.06 μm," in CLEO '98, Technical Digest, 34–35 (1998).
17. B. Pezeshki, M. Hagberg, M. Zelinski, S.D. DeMars, E. Kolev, and R.J. Lang, "400 mW single frequency 660 nm semiconductor laser," IEEE Photon. Technol. Lett., **11**, 791 (1999).

Beam Deflectors and Spatial Light Modulators for Holographic Storage Application

G. Zhou, F. Mok, and D. Psaltis

Beam deflectors and spatial light modulators (SLM) are key components in a holographic data storage system. The deflector is typically used to scan the reference beam to a particular angle or position for the purpose of multiplexing. The spatial light modulator is used to display the data to be stored in a two-dimensional page format. In this chapter we will discuss the use of appropriate SLMs and beam deflectors to achieve the target system performance for holographic data storage application.

A large array of applications for holographic memories have been proposed and studied [1–9]. To maintain the coherence of this chapter, we will focus our discussions on one particular system: the holographic optical disk [8,11,12]. The holographic optical disk is developed for ROM or WORM applications with the promise of terabit capacity and gigabit per second data rate. We will give a description of the holographic disk system at the beginning of this chapter. For other related information on holographic ROM, the reader is referred to the chapter entitled "Holographie Rend-Only Memorys." We will discuss the recording rate, recording density, SNR, readout rate, and image alignment issues for this system, and the emphasis will be on the impact of beam deflectors and SLMs to system performance. In most cases the results presented here can be easily extrapolated to other types of holographic memory systems.

1 Description of the Holographic Disk System

The holographic disk is a high capacity, disk-based data storage device that can provide the performance to meet the needs of next-generation mass data storage applications. With a projected capacity approaching 1 terabit on a single 12 cm platter, the holographic disk has the potential to replace CD-ROM jukeboxes, for data warehousing, multimedia, and interactive TV. The high readout rate of the holographic disk makes it especially suitable for network storage applications and to generate multiple, high-bandwidth datastreams to client computers. Because of the holographic nature of the stored data, the holographic disk can also offer unique advantages in various data mining applications.

Figure 1 is the conceptual diagram of data readout from a holographic disk. The system stores digital holographic images on a flat, disk-shaped medium. Stored images are organized in different tracks. The disk can spin

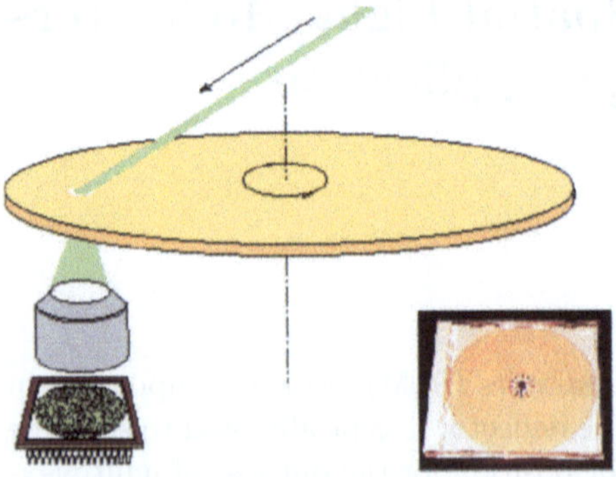

Fig. 1. Data readout from holographic ROM

continuously, and it can also move across tracks to allow the laser pickup to access the entire surface area of the medium. For each storage location, many angle-multiplexed holograms are recorded and overlapped using a suitable multiplexing method.

The complete holographic disk-based ROM system consists of three separate modules. The first is the recording module. This is where the holograms are recorded onto a fresh disk. This module includes both the reference and signal beam paths, and it contains components such as the SLM and the scanners. The recording module is similar to the CD-ROM mastering device, and as such it could be large and bulky if necessary. The second is the readout module, which was shown in Fig. 1. This is where the holographic images are retrieved from the disk and converted to electrical signals by an appropriate detector array. This module contains light modulators, scanners, and other electronics. The readout module should be compact and portable. The third module is a device to replicate the recorded master hologram and efficiently produce copy disks in volume.

Figure 2 is a schematic drawing of the recording module. The setup consists of a diode-pumped green laser source at 532 nm. In the signal arm, an SLM is illuminated by a plane wave, and the disk to be recorded is located at the Fourier plane of the SLM. A detector array is placed at the image plane of the SLM. In the reference arm, a beam deflector and a 4-f imaging system are used to scan the angle of the reference beam to implement angular multiplexing. During recording, the deflector scans the reference beam through all reference angles. At each reference angle a light exposure is given to the medium. Besides angular multiplexing, we can also translate and rotate the disk to implement peristrophic multiplexing [18]. After all multiplexed holograms are recorded, the disk moves to the next spot and the process starts again until the entire disk is recorded. We usually use photopolymer [10,15,17] as the storage medium. The disk we record is typically 12 cm in diameter (standard DVD platter size). The disk is fabricated by laminating a layer of 100 μm thick medium onto a glass substrate. We use a pair of f/1.1 lenses

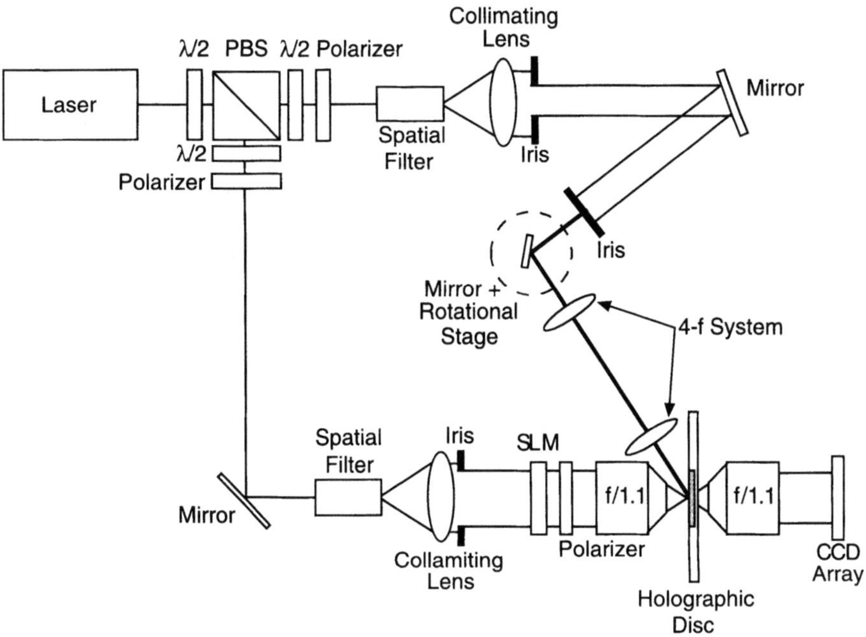

Fig. 2. Schematic diagram of the recording setup

in the system, and the recording spot size is about 0.6 mm. The number of pixels in each hologram is about 180 000 [7,8].

There are several options available when choosing beam deflectors for use in a system like the one shown in Fig. 2. Examples include acousto-optic deflector, liquid crystal scanner, and mechanical deflector. Acousto-optic cells deflect light by the acoustic wave inside a crystal such as tellurium dioxide. The main advantage of the acousto-optic deflectors is the high speed, typically at the microsecond timescale. They are driven by radio-frequency amplifiers, and the deflection angle is small compared with other deflectors. A user could include extra optical components in the system to amplify the deflection angle to get sufficient scanning range. Additionally, the AO deflector also introduces a frequency shift on the deflected beam, and this must be compensated in recording by using, for example, an electro-optic modulator in the signal path [6,16]. Equivalently, one could also use a pair of AO cells to offset and cancel the frequency shift, but this is at the expense of optical throughput. Liquid crystal (LC) scanners deflect a laser beam by modulating its wavefront to form an optical phased array [19]. The LC deflector is a nonmechanical device that is lightweight, precise, and can be used efficiently for random access. Although several companies have introduced LC scanner products, the number of selections available is still very limited at this time. The existing products are usually developed for applications in adaptive optics, which has a very different set of requirements. A mechanical scanner is another alternative to consider for holographic recording. The galvo mirror, for example, has sufficient accuracy and stability for holographic memory applications.

Spatial light modulator (SLM) technology has been driven by the flat-panel display market, and by the need for high-quality multimedia display for digital products such as head mount display, cellular phone, and hand-held computers. The most commonly used SLM in holographic recording is the liquid-crystal display [19–21]. Twisted-nematic liquid-crystal displays and ferroelectric liquid crystal (FLC) display are both usable, with the former having gray-scale and the latter providing higher speed. LC displays are inexpensive and easy to use, but they often need modifications to meet the stringent uniformity and other requirements for holographic memory. The digital (or deformable) mirror device (DMD) [22,23] from Texas Instruments is a high-quality alternative to consider as the SLM for holographic storage. This inherently digital micromechanical device has high resolution, good contrast, and extremely fast update rate.

In the following section we start by discussing the optimization of recording density and the considerations in deflector and SLM selection. The subsequent sections will continue the discussion recording rate, SNR, and the design of the readout module.

2 Recording Density

In this section we discuss the SLM and deflector selection criteria based on density consideration. The surface density of the holographic disk can be written as

$$\text{Density} = (N_{\text{P}}/A) \times M \quad (\text{bits}/\mu\text{m}^2), \tag{1}$$

where M is the number of multiplexed holograms per location, N_{P} is the number of pixels per hologram, and A is the area of the recording spot. The quantity N_{P}/A in the above equation is called the *density per hologram* here for obvious reason. It is determined by the SLM and by the imaging optics. The number of multiplexed pages M is determined by the medium thickness and the scanning range of the deflector [13]. To reach a given storage density, it is a good idea to maximize the density per hologram, therefore minimizing the required hologram number M. Since the diffraction efficiency of a hologram is proportional to $M-2$, a smaller M can significantly increase the number of diffracted photons into the detector, and consequently the data readout rate is much higher. For Fourier plane holograms, the recording area A is determined by the pixel pitch of the SLM. Specifically,

$$A = (2\lambda F/\delta)^2, \tag{2}$$

where λ is the recording wavelength, F is the focal length, and δ is the pixel pitch of the SLM. The quantity $2\lambda F/\delta$ is the size of the Airy disk of a square pixel having dimension δ. The above expression is valid for a thin disk-like medium; for a thick medium there is a correction term associated with thickness [11]. Given a lens with focal length F and diameter D, the number of pixels that can be imaged by the lens is

$$N_{\mathrm{P}} = (\pi/4) \times (D/\delta - 2\lambda F/\delta^2)^2. \tag{3}$$

From the above equations it is clear that

$$\text{Density per hologram} = \frac{\pi}{4}\left(\frac{D}{2\lambda F} - \frac{1}{\delta}\right)^2. \tag{4}$$

The density per hologram has only a weak dependence on the pixel pitch δ of the SLM. The dominating factor is the $f\#$ of the lens, which is the quantity F/D in (4). The density per hologram is almost inversely proportional to the square of the $f\#$ of the lens. Figure 3 is a plot of density dependence on SLM pixel pitch given the other parameters in our system. The effect of SLM pixel pitch is important only when $2\lambda \cdot (f\#)$ becomes comparable to δ. For example, if $\lambda = 532$ nm, $f\# = 1.1$, then the SLM pitch does not play a significant role as long as δ remains larger than 10 μm. Of course the SLM pixel pitch also affects other system parameters, such as readout rate and imaging noise. In practice it is best to use a small δ that does not sacrifice density, and which can be handled by the imaging system's resolving power and distortion specifications.

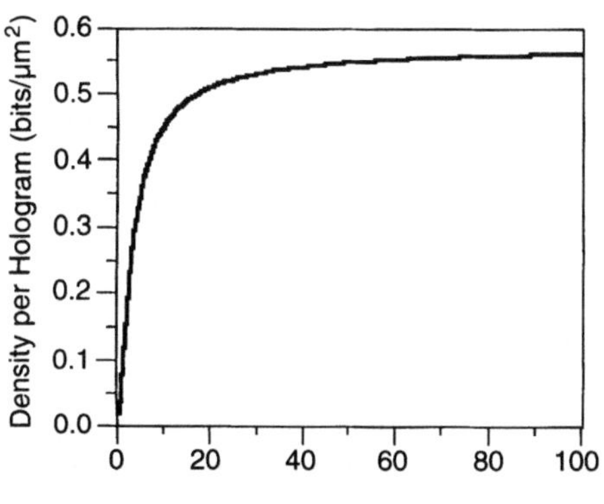

Fig. 3. Density per hologram as a function of SLM pitch

If the performance of imaging optics is not the limiting factor in the holographic storage system (as is the case for phase conjugate readout), or if system capacity, and not density, is the major concern, then it is desirable to use small-pitch SLMs. In these situations, a large-format, small-pitch SLM can increase the system capacity and therefore lower the cost per Mbyte by better amortizing the component costs.

In our HROM system the SLM pitch δ is 24 μm, and density per hologram is about 0.5 bits/μm^2. In other words, when only one hologram is stored per location, i.e. no multiplexing is done, the surface density is already about the same as that of a CD-ROM. By scanning the reference beam over a suitable range and making $M \gg 1$, the surface density of the holographic disk can be well above, for instance, DVD density.

To optimize the number of multiplexed pages M, the beam deflector must have a range and resolution determined by the Bragg selectivity of the medium. For a medium of 100-μm thickness, the angular separation between two successive holograms is of the order of 1°. To store enough holograms, the beam deflector needs to have typically 20–40° of scanning range and milliradiant resolution. The fine resolution is necessary to allow adjustment of beam angles to compensate for material deformation such as shrinkage in photopolymers [14,17]. In our present system we multiplexed 84 holograms to achieve a surface density of 40 bits/μm^2. More discussion on experimental results can be found in the chapter entitled "Holographic Read-Only Memory."

3 SLM Characteristics and System SNR

Multiple factors can affect the signal-to-noise ratio (SNR) of recorded holograms. Material nonuniformity, scattering, and deformation are some of the common causes of SNR degradation. On the systems side, the SLM determines the starting level of the SNR in the system, since it is the entry point of digital data into the optical module. Factors to consider when choosing a SLM include uniformity, contrast, efficiency, resolution, and frame rate.

The SLM uniformity is a measure of the amount of variation in pixel intensities across the entire display. A nonuniform display can cause some "on" pixels to appear darker than some "off" pixels and generate bit errors when the image is converted to digital data. There are ways to encode data that overcomes the nonuniformity problem [24], but this result usually comes at the expense of memory capacity. For example, the differential encoding method uses a pair of (bright, dark) pixels to represent digital 1 and a pair of (dark, bright) pixels to represent 0. The coding is robust against nonuniformity, but two SLM pixels are needed to represent one data bit. Independent of coding scheme, it is a good practice to try to reduce the nonuniformity in SLM before use in holographic recording. Anti-reflection coatings can reduce the optical fringe in the display caused by multiple reflections of the coherent illumination beam. For example, for uncoated transmissive SLM with the typical 4% reflectivity per surface, the fringe modulation depth in the image is about 0.08. If AR-coated to 0.5% reflectivity, the visibility drops to 0.01. The need for AR coating is especially high for reflective SLMs. For example, an uncoated surface with 4% reflectivity would produce a fringe modulation depth of 0.4, which is intolerable. The optical birefringence of the cover glass on SLMs can also cause some nonuniformity problems. For example, the optical window in some DMDs have shown excessive birefringence. For use in holographic recording the glass window must be annealed properly to release internal stress and eliminate optical anisotropy.

The resolution of the SLM is a measure of the total number of independent pixels that it can display. As the spatial frequency of the display image increases, some SLMs have difficulty maintaining the contrast of the display. As a result the resolution is less than the total number of fabricated pixels. Figure 4a shows the optical response of a Kopin liquid crystal display as a function of the spatial frequency of the displayed image. The curve indicates that the contrast of the display decreases from 10:1 to less than 2:1 as the spatial frequency approaches the Nyquist limit. In other words the modulation transfer function (MTF) of the display begins to drop significantly before the Nyquist frequency is reached. Note that the cross-section plot in Fig. 4a is for pixels on the same row. Pixels on the same column did not show this degradation in contrast with spatial frequency increase. The SLM's MTF problem

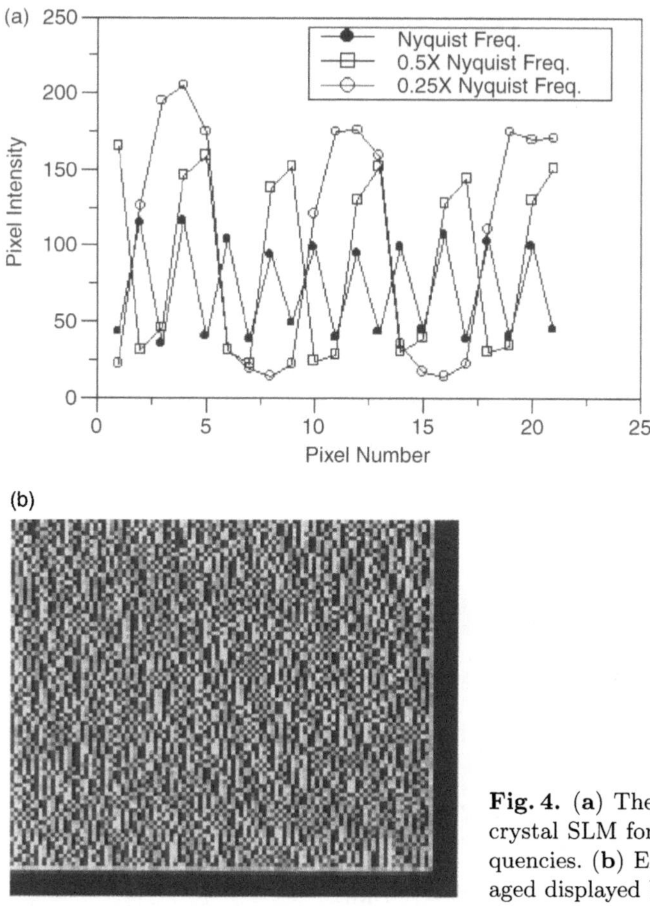

Fig. 4. (**a**) The response of a liquid crystal SLM for different spatial frequencies. (**b**) Example of binary imaged displayed by the SLM

increases image noise and bit error rate when random bits are displayed. Figure 4b is an example of part of a binary image displayed by the SLM above. The SNR for the image shown here is about 5. There is a significant variation of the pixel intensity across the image. The isolated "on" pixels are darker than they should otherwise be.

The DMD display has a much better MTF curve. Figure 5a shows the DMD optical response as a function of the spatial frequency. There is no observable degradation in the display contrast as the spatial frequency approaches the Nyquist limit. Figure 5b shows an example of a binary image displayed by the DMD. The SNR for the image is better than 9. The image quality is clearly better than that of Fig. 4b, owing to the better MTF characteristics.

The LCSLM usually has a limited fill factor [20]. Figure 6 shows the active area of the Kopin SLM. The fill factor is less than 20% in this case. A lenslet

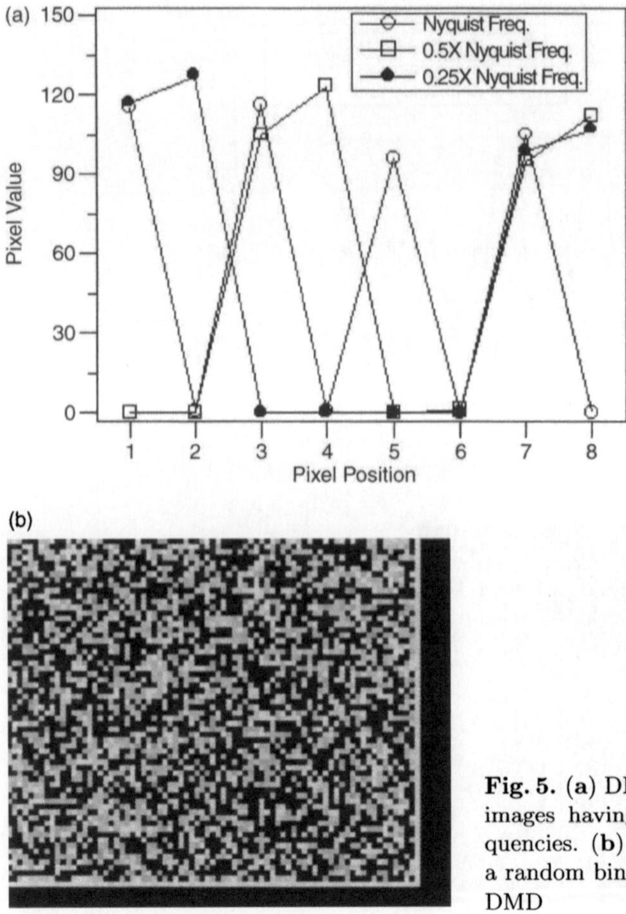

Fig. 5. (**a**) DMD response to display images having different spatial frequencies. (**b**) Example of a part of a random binary image displayed by DMD

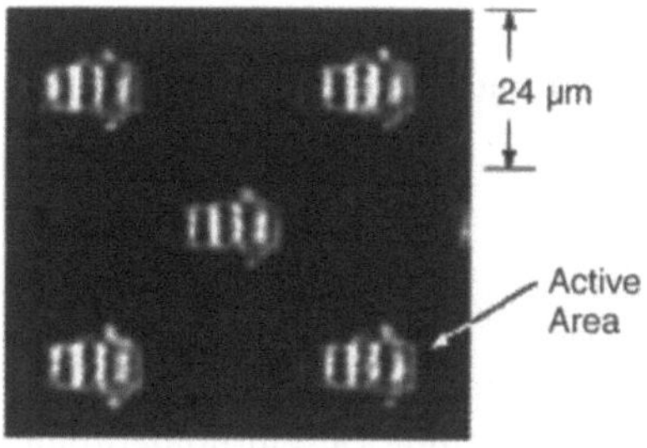

Fig. 6. Active area of the Kopin SLM

array can be used to improve the optical efficiency of the SLM. A scanning electron microscope profile of such a lenslet array is shown in Fig. 7a.

The total size of the fabricated array is 640 × 480 pixels (15.4 mm × 11.5 mm), and the pixel pitch (24 μm) matches that of the SLM we used. The lenslet array was fabricated by electron beam lithography, and then bonded to the SLM to form a monolithic piece. Each lenslet in the array focuses the corresponding beam of light onto the SLM's active area. Figure 7b shows the array of focused spots formed by the lenslet array. In this picture each focused spot is about 8 μm. By using the lenslet array we improved the SLM's optical throughput from less than 5% to about 15%. In addition to the improved efficiency, the lenslet array also functions as a random phase mask in front of the SLM. As seen in Fig. 7a, each lenslet in the array is recessed by a different amount (usually a fraction of the wavelength) to provide a random phase delay. Four different lenslet recess amounts are implemented, corresponding to the phase delays 0, $\pi/2$, π, and $3\pi/2$. By randomly distributing these phase delays over the entire array, the dc component in the Fourier plane of the

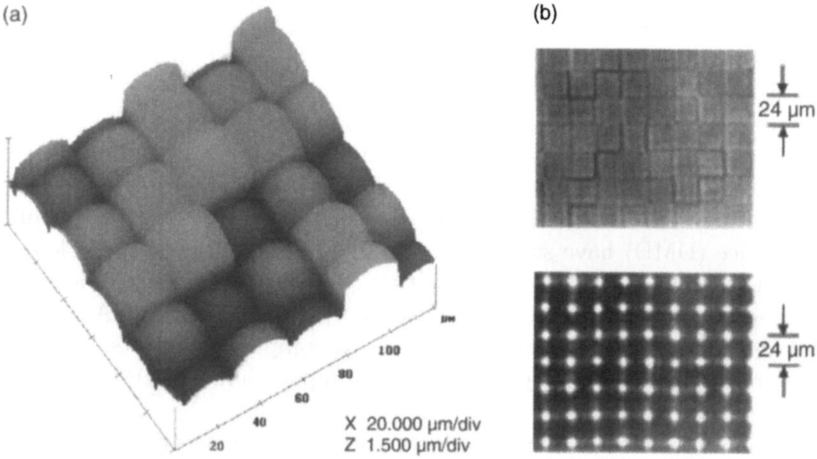

Fig. 7. (**a**) SEM profile of the fabricated 640 × 480 lenslet array; (**b**) Top: Optical image of the lenslet array; Bottom: Array of focused spots generated by the lenslet array

SLM is greatly reduced [5]. This gives rise to a more uniform signal intensity profile in the Fourier plane, and allows holographic recording to take place at exactly the Fourier plane with good SNR.

4 Recording Rate

When the holographic disk is used for WORM application, the recording rate becomes a critical system parameter. The data recording rate in holographic memory is given by

$$R_{\mathrm{REC}} = N_{\mathrm{P}}/T_{\mathrm{REC}}, \tag{5}$$

where N_{P} is the number of pixels per hologram and T_{REC} is the time it takes to record one hologram. T_{REC} usually consists of four parts:

$$T_{\mathrm{REC}} = T_{\mathrm{EXPO}} + T_{\mathrm{SLM}} + T_{\mathrm{DEFL}} + T_{\mathrm{TRANSIT}}, \tag{6}$$

where T_{EXPO} is the exposure time needed to achieve a sufficient diffraction efficiency given the medium, T_{SLM} is the SLM's update time, T_{DEFL} is the beam deflector settling time, and T_{TRANSIT} is the transit time for the disk to move to a new spot.

Given a target diffraction efficiency for each hologram, T_{EXPO} is obviously determined by the laser power and by the sensitivity of the medium. For DuPont polymer (HRF150-100), we have measured a recording sensitivity of about 2000 cm/J. This means that with a recording intensity of 2 W/cm^2, the time needed to achieve, for example, 10^{-3} diffraction efficiency is less than 1 ms for a 100-μm-thick medium. If the other time factors in the previous equation can be reduced to less than 1 ms, then we could reach a materials-limited recording rate of better than 180 Mbyte/s. This high recording rate is potentially useful for real-time movie recording, high-definition TV, and other high-bandwidth multimedia or defense applications. Filling up a 1 terabit disk at this recording rate takes about one and a half hours.

Table 1 is a list of spatial light modulators and their corresponding speed specifications. An example list of manufacturers is provided for reference. Obviously both the ferroelectric liquid crystal display (FLC) and the deformable mirror device (DMD) have sufficient speed. The FLC, however, has a lower resolution than the DMD. Higher-resolution FLC displays are available from other vendors (for example, Canon makes a FLC display at SVGA resolution), but those are usually targeted for use as computer monitors, and the pixel pitch is too large for holographic memory application. The DMD display is extremely fast with a 20-μs frame rate.

If the material response and the SLM speed are sufficiently fast, the remaining factors to optimize are the beam deflector speed and the transit time. Their relative significance depends on the way the holograms are recorded. There are two possible modes for recording the holograms on the holographic

Table 1. List of SLM types and their typical data rate

Device name	Manufacturer (example)	Resolution	Frame rate	Display data rate
Twisted nematic liquid crystal SLM (LCSLM)	Kopin	640 × 480 (VGA)	33 ms	10 Mpix/s
Ferroelectric liquid crystal SLM (FLC)	Displaytech	256 × 256	100 μs	650 Mpix/s
Deformable mirror device (DMD)	Texas Instruments	VGA or SVGA	20 μs	15–40 Gpix/s

disk. The first is the "on the fly" mode, in which the holograms are recorded while the disk is rotating continuously. Consecutive holograms are recorded at different locations along the same track on the disk. The second is the "stop and go" mode, in which the disk is rotated to a new location and stops completely before a series of multiplexed holograms are recorded at the same location.

In the first mode, the beam deflector speed is of minimal concern when optimizing recording rate, since the deflection angle changes only when one track has been completely recorded by a given reference beam. The recording rate is affected most by the transit time. With the typical parameters in the disk system (recording spotsize 0.6 mm, rotating at 600 rpm), the transit time is about 0.2 ms, and the recording rate is material-limited, as discussed above. For the "recording on the fly" to work, the laser pulse must be very short so that the disk movement is much less than a fringe spacing during recording. Q-switched lasers with nanosecond pulse width and sufficient coherence length can be used to implement this recording method.

In the second recording mode, the beam deflector speed has a direct impact on recording speed, whereas the effect of transit time is minimal. The advantage of this mode is that a medium power (for example, 100 mW) cw laser can be used. In order to achieve material-limited recording rate, the deflection time must be minimized. Acousto-optic deflectors have sufficient speed with microsecond-scale deflection time. Mechanical deflectors can also have submillisecond response time, and they are much easier to use. Figure 8a shows a picture of a galvo mirror deflector available from Cambridge Technology. The device is small, lightweight, and has a scanning range of 40° of mechanical angle (80° optical). Figure 8b is a frequency response curve measured for this galvo mirror, with an oscillation amplitude set at more than 10 mechanical degrees at dc. The large-amplitude response bandwidth of the device approaches 1 kHz. The small-angle settling time of this device can be as good as 0.3 ms. The small-angle speed is relevant because the Bragg angle separation in the holographic disk is typically less than 1° optical.

An experiment was set up to confirm the speed and stability of the galvo mirror for holographic recording. In this experiment the reference beam was

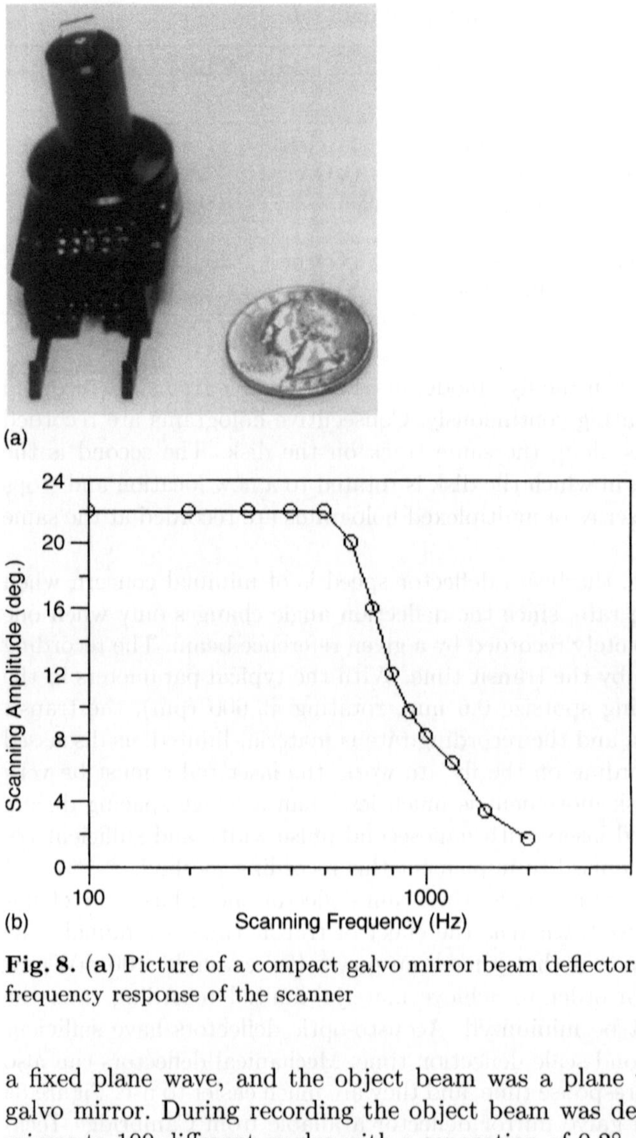

Fig. 8. (**a**) Picture of a compact galvo mirror beam deflector. (**b**) Large-amplitude frequency response of the scanner

a fixed plane wave, and the object beam was a plane wave deflected by a galvo mirror. During recording the object beam was deflected by the galvo mirror to 100 different angles with a separation of 0.2°. The total deflection range was 20°, and each hologram was recorded for 1 ms. The total recording time was 100 ms. The recording laser was a cw green laser, with a recording intensity of 500 mW/cm^2. Figure 9 shows the 100 reconstructed holograms from the experiment, where each dot represents one plane-wave hologram. The recording started from the right side, and the nonuniform intensity profile of the reconstruction is consistent with the exposure characteristics of the photopolymer used in the experiment.

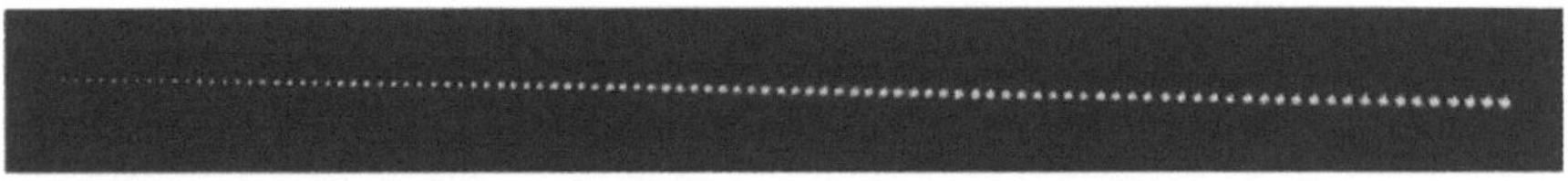

Fig. 9. Reconstruction of 100 plane wave holograms recorded by using the galvo mirror

5 Beam Deflector for Holographic Data Readout

The data readout module for the holographic disk is shown schematically in Fig. 10a. The laser beam (from the top) passes through the modulator and enters the scanner. The two beam deflectors in the scanner can change the laser beam propagation direction along two orthogonal axes in space. The scanner serves two purposes. First, it deflects the laser beam to the appropriate angles in order to address different multiplexed holograms. The entire disk is read out at a particular angle before the deflector changes to the next multiplexing angle. The second role of the scanner is as a compensator in a servo system to correct for image misalignments. This will be explained later in this section. The laser beam is directed by the scanner to the holographic disk which spins continuously at 600 rpm. At this rotation speed, the optical data rate from the holographic disk is 1 Gbit/s [7]. The disk and its spindle motor can also move linearly to allow the laser pickup to probe different tracks on the disk. A stationary image is read out from the moving disk by pulsing the laser at the appropriate time. A detector array integrates the readout image and converts the photons into electrical signals. We use a diode-pumped green laser as the light source. An external electro-optic modulator is used to pulse the laser. Figure 10b is a picture showing the two beam deflectors used in the holographic disk reader. The beam deflectors in the module are miniature galvo mirrors. The imaging lens is identical to the ones used in the recording setup. In the following we will focus on discussing the servo mechanism needed to maintain image alignment and good data registration.

When the holographic image is read out from the spinning disk and captured by the detector array, the image could be shifted or rotated with respect to the detector cells. The one-to-one match of thousands of pixels with the detector array is not a trivial task, especially when the medium is undergoing rapid mechanical motion. Misregistration of image pixels by the detector array can be caused by disk wobble and by the mechanical deformation in the disk substrate, for example. When these things happen, the reconstructed holographic image is misaligned with respect to the detector, and the bit error rate becomes high. In order to maintain image alignment and good pixel registration, a servo system can be incorporated that detects the readout error in the hologram and actively compensates for the alignment error in real time.

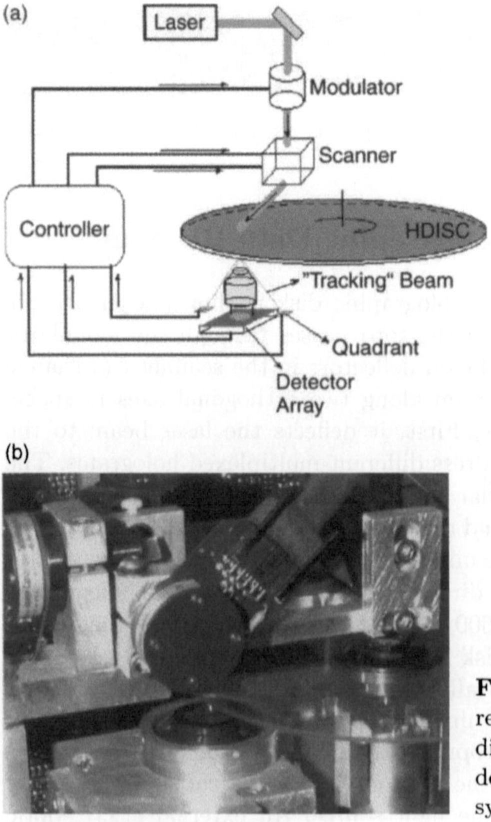

Fig. 10. (**a**) Block diagram of the readout module of the holographic disk system; (**b**) Picture of the beam deflectors and imaging lens in the disk system

The detection of alignment error is facilitated by the so-called control pixels and a pair of quadrant detectors in the system. A pair of control pixels were recorded alongside the main image for each hologram. These two extra pixels are directed toward two separate quadrant detectors as illustrated in Fig. 10a. When the probe laser is pulsed to read out the hologram, the signal from each quadrant detector is sampled by a controller. The hologram alignment errors are derived from the quadrant signals. The image shift errors (along two orthogonal directions) can be derived by the summation and difference operation of the signals from the four individual cells of a quadrant detector. The image rotational error is derived by comparing the signals from the two different quadrant detectors. Identical signals means no image rotation, a disparity indicates image rotation.

Once the alignment errors in the readout image are detected, they are compensated by the beam deflectors and by adjusting the laser pulse timing. Any rotational error in the detected image is corrected by either advancing or delaying the laser pulse timing. The shift errors are corrected by driving the two independent beam deflectors in a way that forces the image into alignment [25]. Figure 11 shows three pictures of the same hologram read out from the spinning disk under different control conditions. Figure 11a is obtained with the servo mechanism activated; it shows good pixel match and data registration. Figure 11b is obtained with the servo mechanism disengaged, and the disk tilted slightly. The readout image shows a uniform smear and poor contrast. Figure 11c shows the effect of timing error on the readout image. The captured image is slightly rotated relative to the detector array, and the incorrect registration by the detector produces several stripes of low contrast fringes in the image. These aliasing effects are indicative of bad alignment.

Figure 12 shows the amount of jitter in the readout image from the holographic disk. Here the intensity profile of a small part of a readout image is plotted. In this test, the same holographic image is readout once every 10 s for 20 times, and the intensity profile of each read out image is plotted on the same graph. It can be seen that the profile stays well defined, with a very small amount of jitter. The result demonstrates the good stability of the holographic disk system.

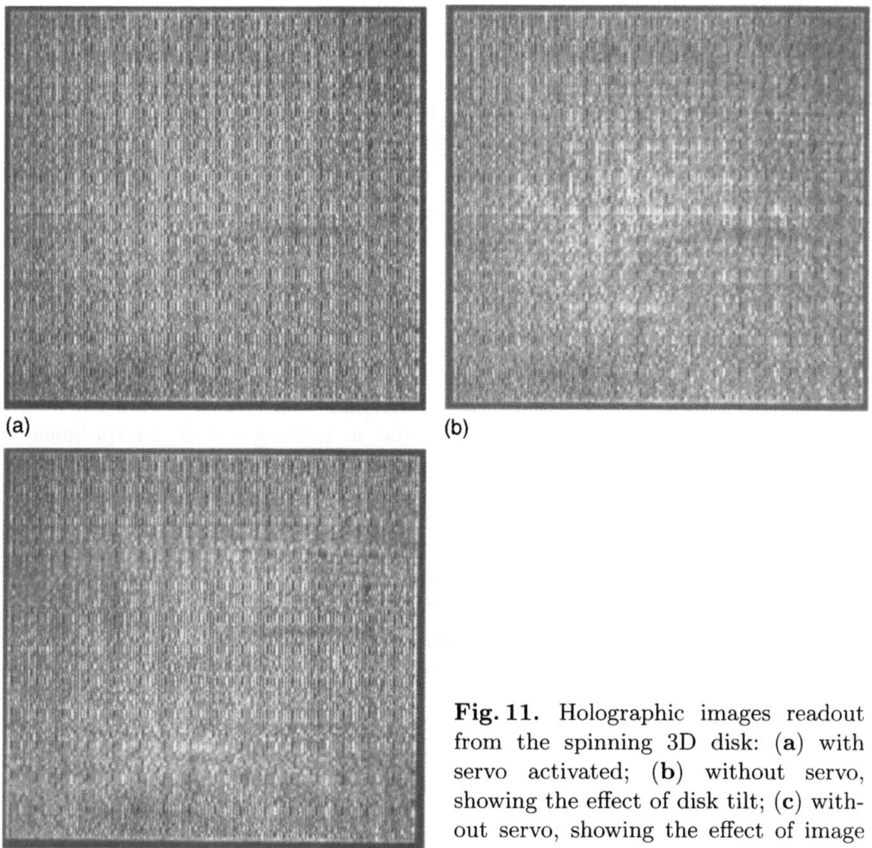

(a) (b) (c)

Fig. 11. Holographic images readout from the spinning 3D disk: (**a**) with servo activated; (**b**) without servo, showing the effect of disk tilt; (**c**) without servo, showing the effect of image rotation on data registration

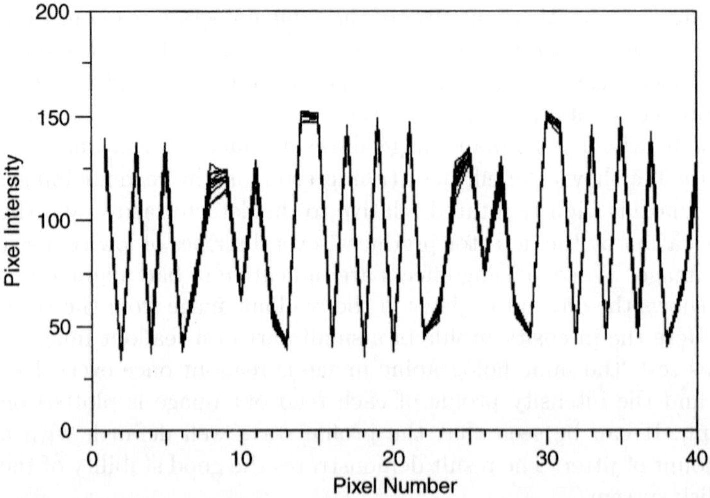

Fig. 12. Measurement of image jitter in readout hologram

Acknowledgments

We thank Allen Pu, Olga Ivanova, Ajay Chugh, Yong Qiao, Xin An, and Gaston Arraya for many discussions on the topics covered here, and for their assistance in the experimental work. We also thank Daniel W. Wilson of the Jet Propulsion Laboratory for the fabrication of the micro lens array.

References

1. D. Psaltis and F. Mok, "Holographic memories," Sci. Am., **273**(5) (1995) 70.
2. J. Heanue, M. Bashaw, L. Hesselink, "Volume holographic storage and retrieval of digital data," Science, **265**(5173) (1994) 749–752.
3. G. Burr, W. Chou, M. Neifeld, H. Coufal, J. Hoffnagle et al., "Experimental evaluation of the user capacity in holographic data storage systems," Appl. Opt., **37**(23) (1998) 5431–5443.
4. L. Dhar-L, K. Curtis, M. Tackitt, M. Schilling, S. Campbell, W. Wilson, et al., " Holographic storage of multiple high capacity digital data pages in thick polymer systems," Opt. Lett., **23**(21) (1998) 1710–1712.
5. J. Hong, I. Mcmichael, and J. Ma, "Influence of phase masks on cross talk in holographic memory," Opt. Lett., **21**(20) (1996) 1694–1696.
6. J. Hong and D. Psaltis, "Dense holographic storage promises fast access," Laser Focus World, **32**(4) (1996) 119.
7. G. Zhou, A. Pu, O. Ivanova, and F. Mok, "Output interface for holographic memories," in *Optoelectronic Interconnects*, J. Bristow and S. Tang, Editors, Proceedings of SPIE, Vol. 3632, (1999) 292–296.
8. G. Zhou, A. Pu, O. Ivanova, F. Mok, and D. Psaltis, "Implementation of holographic optical disk," Proc. Int. Symp. Opt. Mem. (1998), Tsukuba, Japan. 14–15.

9. G. Zhou, Y. Qiao, F. Mok, and D. Psaltis, "A holographic system for fingerprint identification," Opt. Photon. News, March (1996) 43.
10. W.J. Gambogi, A.M. Weber, and J. Trout, "Advances and applications of DuPont holographic polymers," SPIE Proc., **2043** (1993).
11. H.-Y.S. Li and D. Psaltis, "3D holographic disks," Appl. Opt., **33**(17) (1994) 3764.
12. D. Psaltis and A. Pu, "Holographic 3-D disks," Optoelectronics, **10**(3) (1995) 333.
13. H. Kogelnik, "Coupled-wave theory for thick hologram gratings," Bell Syst. Tech. J., **48** (1969) 2909.
14. U.-S. Rhee, H.J. Caulfield, C. Vikram, and J. Shamir, "Dynamics of hologram recording in DuPont photopolymer," Appl. Opt., **34**(5) (1995) 846.
15. G.J. Steckman, I. Solomatine, G. Zhou, and D. Psaltis, "Characterization of the PMMA-PQ photopolymer for holographic recording," Opt. Lett., **23**(16) (1998) 1310.
16. X. An and D. Psaltis, "Experimental characterization of an angle-multiplexed holographic memory," Opt. Lett., **20**(18) (1995) 1913–1915.
17. D.A. Waldman, H.-Y.S. Li, and M.G. Horner, "Volume shrinkage in slant fringe gratings of a cationic ring-opening holographic recording material," J. Imaging Sci. Tech., **41**(5) (1997) 497–513.
18. K. Curtis, A. Pu, and D. Psaltis, "Holographic recording using peristrophic multiplexing," Opt. Lett., **19**(13) (1994) 993.
19. D.P. Resler, D.S. Hobbs, R.C. Sharp, L.J. Friedman, and T.A. Dorschner, "High-efficiency liquid crystal optical phased-array beam steering," Opt. Lett., **21**(9) (1996) 689–691.
20. H. Chase, M.A. Handschy, M.J. O'Callaghan, and F.W. Supon, "Improvement of spatial light modulator optical input/output performance using microlens arrays," Opt. Lett., **20**(12) (1995) 1444–1446.
21. M.J. O'Callaghan and M.A. Hanschy, "Diffractive ferroelectric liquid crystal shutters for unpolarized light," Opt. Lett., **16**(10) (1991) 770–772.
22. L.J. Hornbeck, "Digital light processing and MEMS, reflecting the digital display needs of the networked society," SPIE Europto Proc., **2783**, Besancon, France, (1996) 2–13.
23. L.J. Hornbeck, "Deformable-mirror spatial light modulators," in *SLMs and Applications III*, SPIE Critical Reviews, **1150** (1989) 86–102.
24. X.P. Chen, K.M. Chugg, and M.A. Neifeld, "Near-optimal parallel distributed data detection for page-oriented optical memories," IEEE-S-T-QU, **4**(5) (1998) 866–879.
25. G. Zhou, D. Psaltis, F. Mok, and A. Pu, "Method and system to align holographic images," U.S. Patent: # 05982513, issued 11/9/1999.

Beam Conditioning Techniques for Holographic Recording Systems

R.K. Kostuk, M.P. Bernal Artajona, and Q. Gao

The Fourier transform hologram is a useful configuration for holographic storage. In this arrangement the most useful components of the frequency spectrum can be formed into a relatively compact object beam. This increases the information density and allows more effective use of the recording material. Another benefit is that Fourier transform holograms are less sensitive to misalignment and to imperfections in the recording material and optical system. However, when the object field has regular or periodic components the spectrum will contain a series of high-intensity peaks [1].

Typically a spatial light modulator (SLM) is used to encode data in the object field, and has a high degree of periodicity. The transmittance function of a two-dimensional SLM can be described as

$$t(x,y) = \sum_{m=1}^{N} \sum_{n=1}^{N} a_{mn} \operatorname{rect}(x/P - m) \operatorname{rect}(y/P - n),$$

where P is the SLM pixel dimension, a_{mn} is either a 1 or a 0 representing the digital value stored at a pixel, and N is the total number of pixels in each dimension. If the transmittance function is illuminated with a normally incident plane wave, the field in the back focal plane of the lens is the Fourier transform of this function:

$$u(f_x, f_y) = C \operatorname{sinc}(f_x P) \operatorname{sinc}(f_y P) \sum_{m,n=1}^{N,N} a_{mn} X \exp[-j2\pi(m f_x + n f_y)P].$$

This function has a large dc peak with amplitude $N^2/2$ centered on the recording plane.

Recording this spectrum in a material with a limited linear response characteristic will saturate the medium in the peak areas. This results in nonlinear grating formation and an increase in the noise level of the reconstructed image. The high-intensity regions also make it difficult to match the reference and object beam intensities, and reduce the diffraction efficiency for the higher spatial frequency components of the object spectrum.

Several techniques are available to reduce the large-intensity variations in the spectrum from a SLM. These include defocusing the object spectrum, using a phase mask, or using an axicon to change the relative form of the spectrum in the region of the recording material. Each of these approaches has certain advantages and disadvantages for accomplishing the intensity smoothing operation. Several criteria can be used to evaluate these different methods. These include: 1) the uniformity of the intensity spectrum in the region of the recording material; 2) the storage density; 3) the alignment tolerance; and 4) the complexity of designing and fabricating the corrective element and the increased complexity of the data storage system. The results of the first two criteria impact on the performance of the holographic storage system, while criteria 3 and 4 are more related to the practical considerations of implementing the techniques in an actual system.

1 Defocusing

This is the simplest technique for reducing the intensity variations in the transform plane [?]. The recording material is simply shifted away from the transform plane, as shown in Fig. 1. This works reasonably well for many situations; however, there are some major problems with this approach. To begin with, thick holographic recording materials are desirable for holographic storage systems. This allows high angular selectivity and angular multiplexing capability. However, an extended material thickness means that for defocusing to be effective the Fourier plane must be shifted a considerable distance. The use of thick holographic recording material also means that the Fourier spectrum exists at only one plane within the recording material. The intensity spectrum must now be considered as a Fresnel spectrum [1].

At the Fourier plane the intensity of a diffraction order is concentrated within the first zeros of the sinc^2 function. For a uniformly illuminated aperture the size of this region at the transform plane is

$$2\Delta x = 2.44\frac{\lambda}{D}f,$$

where λ is the wavelength, f is the focal length of the Fourier transform lens, and D is either the diameter of the Fourier transform lens or the aperture of the SLM, whichever of these two apertures is smaller. For small focal shifts relative to the focal length, the increase in size of this spot increases approximately linearly with the focal shift Δz as

$$\varepsilon = D\frac{\Delta z}{f} + 2\Delta x.$$

The intensity of the diffraction order decreases approximately with ε^2 as $I_{\text{F0}} \cdot (2\Delta x/\varepsilon)^2$, where I_{F0} is the intensity at the Fourier or focal plane. Therefore when $\Delta z = 2\Delta x \cdot (f/D)$ the area of the spot corresponding to the diffraction order increases by a factor of 2 and the intensity decreases by a factor of 4. Further reduction to the peak intensity can be achieved with larger amounts of defocus. However, this will result in lower storage density in the recording material.

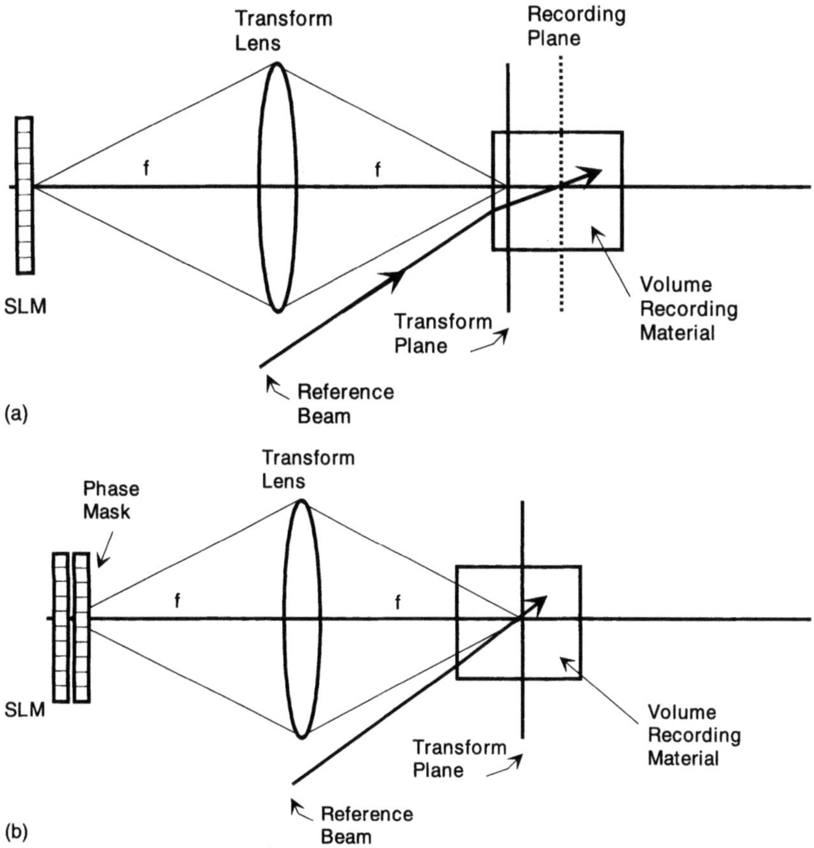

Fig. 1. (**a**) Focal shift method for reducing intensity peaks in the recording medium. The material is shifted relative to the plane of overlap between the reference and object beams. (**b**) An arrangement for incorporating a phase mask into a holographic recording

2 Random Phase Masks

An alternate approach for making the intensity in the transform plane more uniform is to use a phase mask [2–6]. In this method a mask with randomly varying phase elements is used to change the relative phase between pixels of the SLM.

The basic concept of the random phase mask is based on the following principle. The intensity peaks in the Fourier transform of the SLM have widths on the order of $2\Delta x$ as described in the previous section. This width is inversely related to D, the limiting diameter of the optical system. If a two-dimensional array of phase elements with pixel period and size matching those of the SLM is inserted next to the SLM, the transmittance function becomes

$$t(x,y) = \sum_{m=1}^{N}\sum_{n=1}^{N} a_{mn}\boldsymbol{E}^{j\phi(m,n)}\text{rect}(x/P-m)\text{rect}(y/P-n),$$

where $\phi(m,n)$ is the phase element at the (m,n) pixel in the array. The phase steps can have values of $2\pi/M$, where M is an integer. If it is assumed that the SLM is illuminated with a normally incident unit amplitude plane wave, the Fourier transform of the transmittance function gives the value of the field in the back focal plane of the lens,

$$u(f_x,f_y) = C\text{sinc}(f_xP)\text{sinc}(f_yP)\sum_{m,n=1}^{N,N} a_{mn} \times \exp[-\mathrm{j}2\pi(mf_x+nf_y)P+\mathrm{j}\phi(m,n)],$$

and the intensity is

$$I(f_x,f_y) = u \cdot u^*.$$

If the phase $\phi(m,n)$ is randomized the contribution of each pixel to the field is also randomized. The resultant aperture width for the transmittance function is now reduced to P, and the peak intensity at the center of the spectrum is reduced because the diffraction pattern is spread out. The intensity passing through the SLM is redistributed uniformly under the sinc^2 envelope function in the Fourier transform plane.

The design of a random phase mask is complicated by several factors. In particular the aberrations of the optical system and the nonlinear holographic recording process affect the design procedure. To examine the degradation resulting from the optical system, consider the noise caused by the interference between two adjacent pixels in the object field. For an imaging system with an arbitrary transfer function H, the amplitude of the image of two pixels is

$$u = [P_1 + P_2\mathrm{e}^{\mathrm{j}\phi}]\cdot H,$$

where P_1 and P_2 are two rect functions defining the location and size of the two pixels, and ϕ is the relative phase difference between the two pixels introduced by the phase mask. The resulting intensity pattern is

$$I_\phi = uu^* = (P_1\cdot H)^2 + (P_2\cdot H)^2 + 2\cos(\phi)(P_1\cdot H)\times(P_2\cdot H),$$

where the last term is the interference term between the two pixels. If this intensity is compared with the image intensity formed without the phase mask (i.e. $I_{\phi=0}$), the intensity variation due to the phase mask can be approximated as

$$\Delta I = I_0 - I_\phi = 2[1-\cos(\phi)](P_1\cdot H)(P_2\cdot H).$$

Since the phase difference between mask pixels is $\phi = 2\pi/M$, the noise introduced by the phase mask can be reduced by increasing the number of phase steps M. It can also be seen that $M = 2$ results in the largest value of noise produced by the phase mask. Increasing the number of phase steps also tends to reduce the alignment sensitivity between the phase mask and the SLM.

3 Pseudo-Random Phase Masks

An improvement to the storage density can be obtained when the phase difference between pixels in the phase mask is restricted. If the phase difference between adjacent pixels is constrained to vary by only one phase step, the spatial frequency bandwidth in the Fourier plane will be reduced. This results from lowering the maximum phase difference from 2π to $2\pi/M$, and allows hologram recording in a smaller region of the recording material. Figure 2 is an illustration of the difference in phase profiles between a completely random phase mask, and the constrained random phase mask, which is sometimes referred to as a pseudo-random phase mask [7,8]. Figure 3 shows experimental data for spectra formed with a six-level pseudo-random phase mask, a six-level random phase mask, and with the focal shift method. For comparison, the width of the defocused spectrum is kept equal to that of the spectrum resulting with the pseudo-random phase mask. The width of the pseudo-random phase mask spectrum is about 1/2 that of the random phase mask spectrum. The sharp dc component of the defocus spectrum is also reduced for both the random and pseudorandom phase masks. Figure 4a–c show a series of holographic images of a data mask reconstructed a) without focal shifting or use of a phase mask, b) with ~1 cm of focal shift, c) with a pseudo-random phase mask. For reference Fig. 4d is a direct image of the data mask used as the object beam for the experiments. Both the data mask and phase mask have 36-μm pixels. The data mask was imaged onto the phase mask with f/2.8, 1:1 imaging system, and the Fourier transformed into a $LiNbO_3$ photorefractive crystal. As indicated, the focal shift and phase mask provide a significant improvement to the image quality. However, the large focal shift

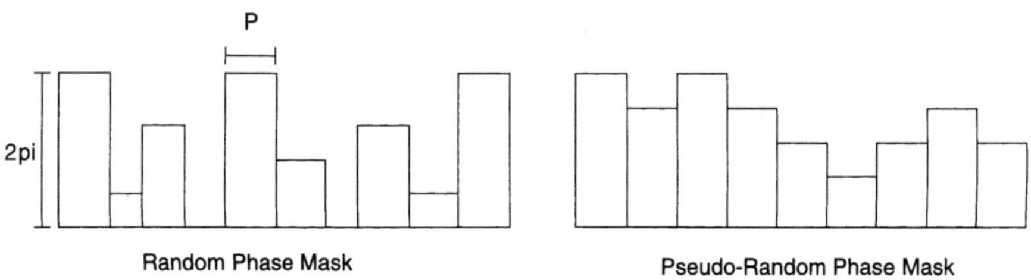

Fig. 2. Distinction between random and pseudo-random phase masks. The random mask allows random difference of phase change between adjacent pixels (P). The pseudo-random mask allows only one step in phase ($2\pi/M$) between adjacent pixels

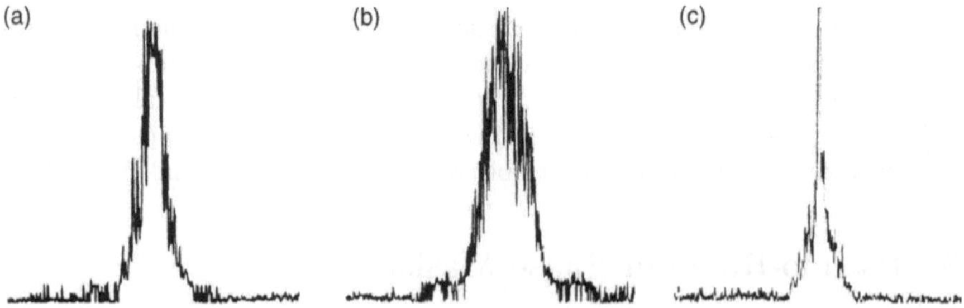

Fig. 3. Experimental spatial frequency spectra of a data mask in the Fourier plane formed with: (**a**) six-level pseudo-random phase mask; (**b**) six-level random phase mask; and (**c**) focal shifted detection plane

Fig. 4. (**a**) Image of a data mask without defocus or use of a phase mask. Image degradation is considerable. (**b**) Holographic image of a data mask with approximately 1 cm of defocus. (**c**) Holographic image of a data mask and pseudo-random phase mask. (**d**) Direct image of the data mask used for recording the holograms in Figs. 4a–c

required to smooth the intensity spectrum significantly reduces the storage density relative to the holographic image formed with a phase mask in the optical system.

4 Axicons

Although phase masks offer considerable improvement to the efficiency and image quality of Fourier transform holograms they require careful alignment between phase and data mask pixels. This can become a difficult task as the pixel dimensions are reduced and the number of pixels is increased. An alternative approach is to use an axicon [9]. This element avoids most of the difficulties of the phase mask and at the same time uniformly distributes the energy in the Fourier plane.

A basic section of an axicon is illustrated in Fig. 5. For this arrangement a point source located on the axis of revolution (radius R) does not focus to a single point but on a line of length determined by the axicon parameters in the focal plane of the lens. An axicon of length h, radius R, refractive index n, illuminated with wavelength λ, will have a phase distribution given by

$$\Psi(x,y) = k(n-1)h\left[1 - \frac{1}{R}\sqrt{x^2+y^2}\right],$$

where $k = 2\pi/\lambda$. With the axicon located prior to the Fourier lens in the holographic system, the transform in the back focus of this lens is a ring of radius

$$\rho = f \cdot \tan[(n-1)\alpha],$$

where f is the focal length of the transform lens and α is the angle of the axicon section. This distribution will occur provided that the data mask or SLM is located within a distance

$$d = \frac{R}{(n-1)\alpha} \text{ from the axicon.}$$

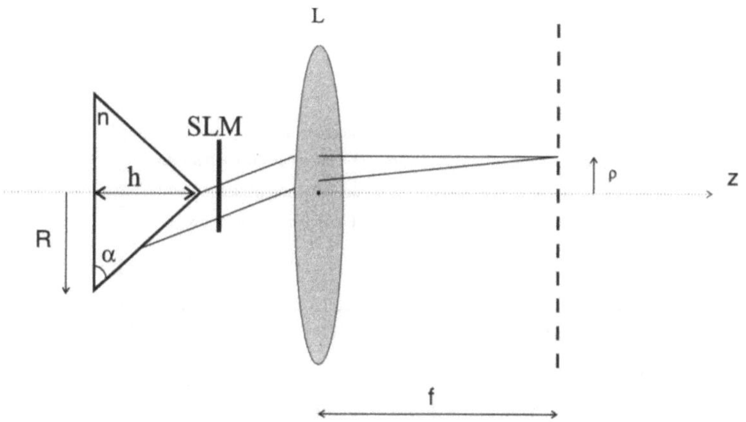

Fig. 5. An arrangement for using an axicon to vary the phase of a field illuminating an SLM

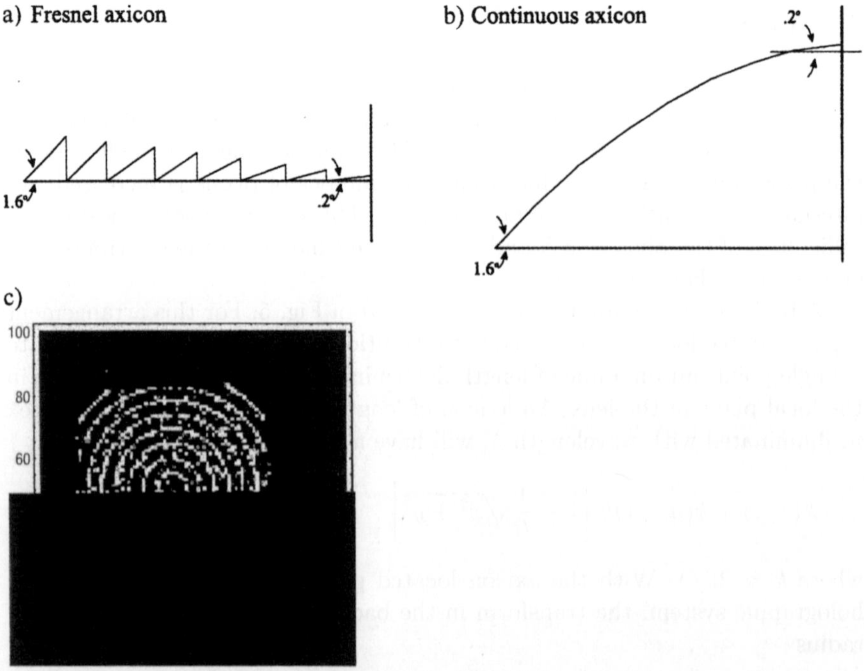

Fig. 6. Different types of axicons: (**a**) Fresnel axicon; (**b**) continuous axicon; (**c**) resulting far-field pattern from the axicon

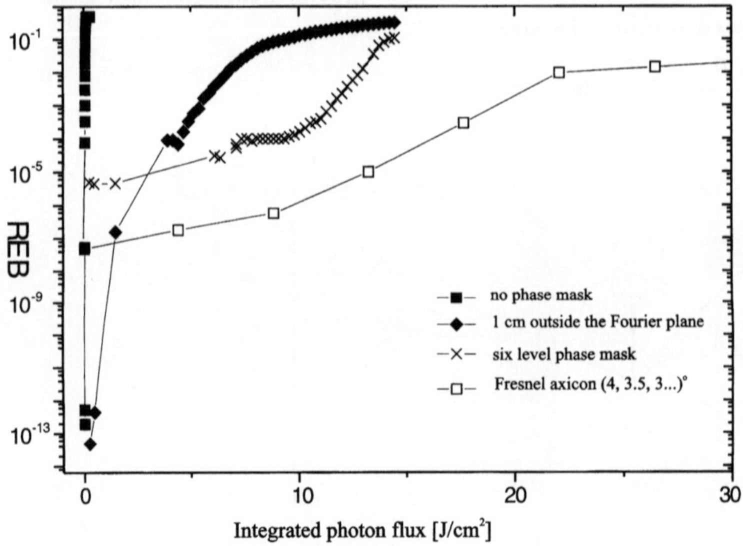

Fig. 7. BER as a function of integrated photon flux or exposure

Variations to the basic axicon form have been realized that distribute light more uniformly in the transform plane [10]. These are the continuous and Fresnel axicons shown in Fig. 6. Both designs consist of concentric axicons with decreasing angles as they approach the rotation axis, and improve the uniformity of intensity in the transform plane.

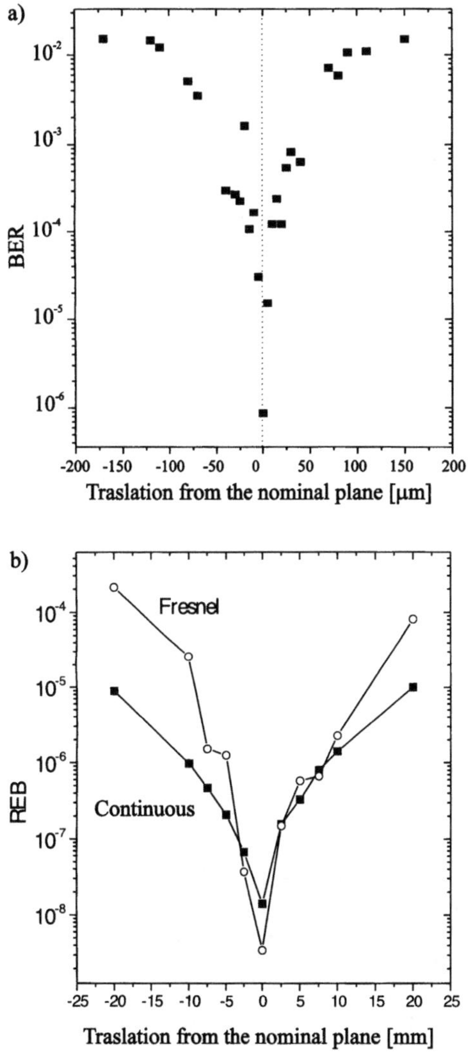

Fig. 8. Translation tolerances with (**a**) a six-level phase mask, and (**b**) an axicon

5 Discussion and Summary

In this chapter three different methods have been discussed for improving the diffraction efficiency and image quality of holographic images recorded using Fourier transform data storage systems. These include the use of defocusing, phase masks, and axicons to redistribute light more uniformly in the transform plane. In order to compare their performance a series of experiments were performed on a holographic storage system that allowed the use of each technique [11]. Measurements of the BER in the resulting images were obtained as a function of increasing the exposure level of the recording beams. Since the diffraction efficiency is proportional to the exposure level the measurements effectively relate BER to diffraction efficiency. A plot of this data is illustrated in Fig. 7. As indicated without using any corrective technique the image noise increases rapidly. Defocus provides some improvement, and the use of a six-level phase reduces the BER near 10^{-5} out to an exposure $\sim$10 J/cm^2. The axicon and in particular the Fresnel axicon provide a BER less than 10^{-5} to an exposure near 15 J/cm^2. Therefore the axicon provides a significant performance advantage over both defocusing and traditional phase masks.

Another indication of performance is the alignment sensitivity of the corrective element: the less tolerant, the more expensive the recording system. Comparative measurements between alignment of a six-level phase mask and an axicon are shown in Fig. 8 [11]. For these experiments the BER is measured as a function of translating the phase mask or axicon relative to the data mask. As indicated, the axicon has a much greater alignment tolerance than the phase mask.

References

1. J.W. Goodman, *Introduction to Fourier Optics*, 2nd ed., pp. 78–81, McGraw-Hill, New York, 1996.
2. C.B. Burckhardt, "Use of a random phase mask for the recording of Fourier transform holograms of data masks," Appl. Opt., Vol. 9, 695–700 (1970).
3. Y. Takeda, "Hologram memory with high quality and high information storage density - Hologram Memory - ," Jpn. J. Appl. Phys., Vol. 11, 656–665 (1972).
4. Y. Tsunoda and Y. Takeda, "High density image storage holograms by a random phase sampling method," Appl. Opt., Vol. 13, 2046–2051 (1974).
5. R. Brauer, F. Wyrowski, and O. Bryngdahl, "Diffusers in digital holography," J. Opt. Soc. Am. A., Vol. 8, 572–578 (1991).
6. S.N. Dixit, I.M. Thomas, B.W. Woods, A.J. Morgan, M.A. Henesian, P.J. Wegner, and H.T. Powell, "Random phase plates for beam smoothing on the NOVA laser," Appl. Opt., Vol. 32, 2543–2553 (1993).
7. Y. Nakayama and M. Kato, "Diffuser with pseudorandom phase sequence," J. Opt. Soc. Am., Vol. 69, 1367–1372 (1979).
8. Q. Gao and R.K. Kostuk, "Improvement to holographic data-storage systems with random and pseudorandom phase masks," Appl. Opt., Vol. 36, 4853–4861 (1997).

9. J.H. McLeod, "The axicon, A new type of optical element," J. Opt. Soc. Am. Vol. 44, 592–597 (1954).
10. M.-P. Bernal, et al., "Phase shifting element for optical information storing systems, AM9-97-084, U.S. Patent (1998).
11. M.-P. Bernal, et al., "Experimental study of the effects of a six-level phase mask on a digital holographic storage system," Appl. Opt., Vol. 37, 2094–2101 (1998).

Detector Arrays for Digital Holographic Storage Applications

S. Campbell and E.R. Fossum

This chapter discusses detectors for holographic data storage applications. While it may be possible to use a single detector or linear array of detectors, the focus here is on 2-D arrays of pixels. There are many considerations for the detector array when designing a complete system that arise from the intimate relationship between the holographic storage system, optical readout technique, and the sensor pixels. The chapter begins with a discussion of these considerations, based on the assumption that commercially viable digital holographic data storage (DHDS) systems must be "smaller than a breadbox," be affordable, have data capacities in the hundreds of gigabytes arena, and have readout rates in the regime of hundreds of megabytes per second. Today there are two primary technology choices for realization of the sensor array: charge-coupled devices (CCDs) and CMOS (complementary metal-oxide-semiconductor) active pixel sensors (APS). Both will be introduced and their relative merits presented [1–6]. In fact the CMOS APS, a much newer technology, will be shown to offer significant advantages over its predecessor, the CCD, when applied to digital holographic data storage systems. An example of a recently-fabricated CMOS APS for use in a DHDS is shown in Fig. 1.

Fig. 1. A 816 × 616 element, 17-μm pixel pitch CMOS APS detector array designed for DHDS applications

1 General Considerations for Detector Arrays

In digital holographic data storage (DHDS) systems, the sensor array plays the critical role of converting optical data signals into electronic data signals. These optical data signals generally come in the form of a "page," or two-dimensional array, of bright and dark spots of light. The sensor array must then be aligned and physically matched with the optical system such that each spot of incoming (readout) light is incident on one, or a small collection of, prespecified sensor array pixel(s). Such a criterion places stringent demands on the sensor array's linear and angular attitudes, as well as its pixel pitch. The sensor array must also be highly sensitive to the particular wavelength(s) of the readout light, as average light levels will typically be quite low ($< 10\,000$ photons per pixel per read). Furthermore, the sensor array must be capable of converting and delivering 10–100 *user* megabytes per second. These and other considerations will be addressed in the following three subsections [7].

1.1 Size, Power and Cost

Because it is preferred to have memory systems that are small, low power, and affordable, there are certain systems considerations that must be addressed in choosing or designing a sensor array for DHDS.

Perhaps the most inclusive system-based concern to consider is that of overall size. High-performance sensor array solutions must be found that are truly compact and/or self-contained (on the order of one to ten cubic centimeters).

In parallel with system size concerns are system power concerns. Most DHDS systems will want their complete sensor array subsystems to consume less than 1 or 2 W. Beyond the general system-level requirement for low power consumption is the vital requirement for low heat generation within the sensor array and its housing. Material dimensions are thermally dependent, and sensor array alignment must be maintained to within a few micrometers for good system performance. In most systems, just a few degrees of temperature change at or near the sensor array plane can shift the sensor array out of alignment with the incoming optical data page.

There is at least one point of relief for sensor array requirements. Since DHDS systems expect the "images" they are receiving to be rather noisy, containing perhaps hundreds to thousands of data bit errors in every page, pixel defects in the sensor array are not much of a problem. Embedded error correction codes (ECC) in the data pages will also correct for pixel defects. Thus the effective yield in the manufacturing of the sensors will be higher, reducing sensor manufacturing cost.

1.2 Number of Pixels, Readout Rate, and Pixel Size Considerations

Because it is preferred to have memory systems with high storage densities (for high capacity in small volumes) and fast data readout rates, there are certain data density and data rate considerations that must be addressed in choosing or designing the sensor array.

One element that must be considered in designing sensor arrays is their space bandwidth product (SBP). For a sensor array, the SBP is equal to the total number of pixels. (The "space" part has to do with the total photosensitive array area, and the "bandwidth" part has to do with the solid diffraction angle of the incoming light cone required to focus to a pixel-size spot). Typically, DHDS systems are designed for use with megapixel arrays (commonly, 1024×1024 arrays are used as a benchmark). There are two main reasons for this choice. First, accessing data consists of a seek step and a read step. If the seek step is relatively long, then one wants to read as much data as possible per step. The second reason has to do with storage density; the higher the sensor array's SBP, the more data bits one can store in a single hologram.

The readout rate of the sensor is also very important, and needs to be in the hundreds of megapixels/s range. It can be shown that with a megapixel sensor array, a good performance holographic storage material, a 100 mW laser source, and a holographic storage density on the order of 500 channel bits per μm^2, it is possible to achieve sustained user data recall rates in excess of 100 Mbytes/s. At these impressive storage densities and data rates (e.g., 300 Gbytes read out at 100 user Mbytes/s) it would still take nearly one hour to read a single disk.

Pixel size affects the optical design as well as the sensor cost. For a given SBP, sensor array cost scales approximately as the fourth power of the linear dimension of the pixel, since chip areal size affects not only how much silicon is used per chip but also chip yield. Thus smaller pixels generally result in a less expensive chip. On the other hand, once some critical dimension is reached (say, 5–10 μm for a megapixel image sensor) the cost of the optics starts to climb rapidly. Small 0.1% for megapixel sensors lens distortions could be disastrous in mass-produced systems. Furthermore, 1:1 pixel-size matching between the "write" SLM and the sensor pixels simplifies optical system design. Thus pixel sizes of the order of 5–20 μm are best. Of course, it may prove worthwhile in some systems to design sub-spot size sensor array pixels. In this manner, "oversampling" can be performed (detecting each data spot over several sensor array pixels).

1.3 Noise, Dynamic Range, and Analog-to-Digital Converter Resolution

One of the pacing items in DHDS design is the uniformity of the readout image. A variation of 10:1 in page intensity across the array is possible, as well as page-to-page intensity variations. In this case it is possible to have 5000 photons incident on one pixel, and 500 photons on another, resulting in signals in the 100–1500 electrons level. For detector arrays more photons is almost always better, but for DHDS systems this means high power dissipation and/or slower access times. Thus the read noise of the sensor needs to be in the sub-20 electrons rms level, with lower noise floors yielding lower power dissipation and/or faster access times.

In the real world, the optical data page emanating from the holographic storage medium will contain various forms of noise. These noise sources include interpage cross-talk, intrapage cross-talk, point defect scattering (due to pits, protuberances, dust, or smudges on the various optical surfaces, or bubbles, density gradients, dopant gradients, or debris in the bulk of the optical elements), and harmonics due to nonlinear effects in recording and/or readout. They will generate optical noise, which must be discriminated from the optical signal. Generally, these noise sources will contribute up to 50% of the total light reaching the sensor array. Because of these noise sources, and the page intensity non-uniformity, there is a useful limit on the resolution of the ADC used in the sensor array. Generally, that limit is roughly five to eight bits.

2 Detector Array Choices

There are several types of detector arrays that can be considered for DHDS systems. These are charge-coupled devices (CCDs), photodiode arrays, charge-injection devices (CIDs), and CMOS (complementary metal-oxide-semi-conductor) active pixel sensors (APS). Of these, the two leading contenders are CCDs and CMOS APS. The others suffer from a combination of high readout noise and slow speed [1–3].

In both CCDs and CMOS APS, incident photons are absorbed in the silicon and converted to electrons. The main difference between CCDs and CMOS APS is in the readout technique. In addition, CMOS APS, being fabricated from mainstream microelectronics technology, is readily integrable with other on-chip circuits to form a "camera-on-a-chip" that has many significant advantages over a CCD-based system.

For a CCD, after the signal is collected and integrated, it is shifted out in a series of thousands of steps, each shifting the collected electrons one pixel at a time until the signal reaches the corner of the array. There the electrons are converted to a voltage by an output amplifier that has a *conversion gain*, typically in the range of 10–20 μV per electron. Access to a particular pixel is serial: that is, to read a particular pixel, all pixels before it must be read out first. This concept is diagramed in Fig. 2. Furthermore, to avoid losing electrons in the shifting process, the readout speed is limited to typically 10 Mpixels/s, and a maximum value of approximately 30 Mpixels/s. The voltage

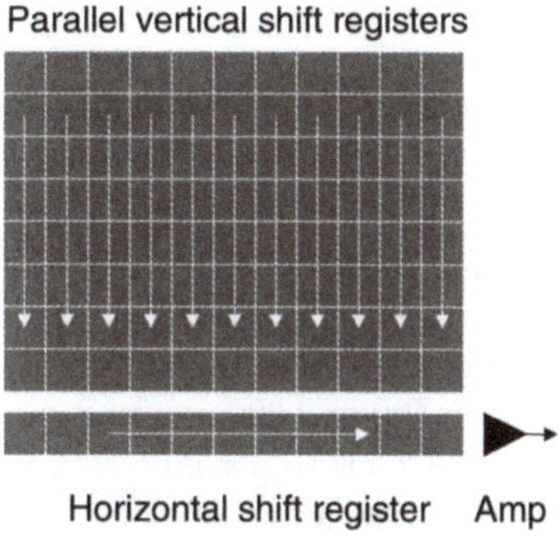

Fig. 2. CCDs work by shifting out photogenerated charge to a common amplifier

from the output amplifier is then driven off-chip to ancillary electronic chips that may perform noise reduction and analog-to-digital conversion (ADC).

In the CMOS APS, each pixel contains an amplifier that converts the electrons to a voltage within the pixel. The voltage signal is read out through a matrix of X-Y wires, allowing independent access to each pixel. Thus any pixel can be selected for readout. This concept is diagramed in Fig. 3. In practice, an entire row of pixels is addressed at a time and their output voltages transferred to an analog signal processor located at the bottom of the array, where noise reduction and ADC takes place. This on-chip circuitry in addition to on-chip timing and control circuitry permits the realization of an electronic camera-on-a-chip.

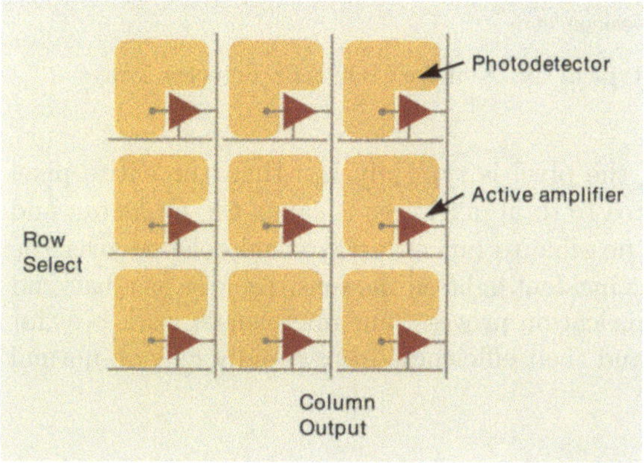

Fig. 3. Active pixel sensors contain an amplifier in each pixel, and the signal is read out over X-Y wires

2.1 Quantum Efficiency

Quantum efficiency (QE) measures the ratio of collected signal carriers to incident photons across an entire pixel. Thus, if the quantum efficiency is, say 30% at 550 nm for a pixel, and 1000 photons are incident on the pixel, then 300 electrons are collected. Since both CCDs and CMOS APS are made from the semiconductor silicon, they have similar quantum efficiencies and are generally responsive from ultraviolet to the near infrared (e.g. 300–1000 nm). The peak value of QE typically occurs between 500 and 600 nm, and can vary from as high as 70% in specialized (expensive) CCDs to perhaps 20% in commercial CCDs. CMOS APS devices typically peak at about 30–40%, but this can also vary significantly with design. A typical Q.E. curve is plotted in Fig. 4.

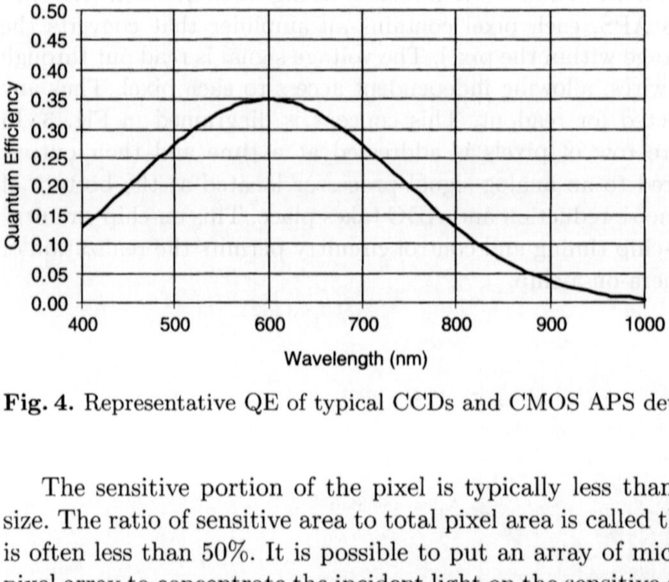

Fig. 4. Representative QE of typical CCDs and CMOS APS detector arrays

The sensitive portion of the pixel is typically less than the entire pixel size. The ratio of sensitive area to total pixel area is called the *fill factor*, and is often less than 50%. It is possible to put an array of microlenses over the pixel array to concentrate the incident light on the sensitive area as a backend step to the normal device fabrication process, but microlenses work best for normally incident photons, and their efficiency drops rapidly for non-normal incidence.

2.2 Noise

Noise is an important consideration, especially when power and throughput need to be optimized. Temporal noise in detector arrays comes from photon shot noise (inherent in the photonic signal incident on the array), which varies as the square root of the number of incident photons or collected electrons, and readout noise introduced in the readout and conversion of the signal to a voltage. The latter is often independent of the signal. In a CCD all the signal charge is converted by a single amplifier, so it must operate at a higher frequency and consequently its noise is higher, e.g. 20–40 electrons rms at 10–30 Mpixels/sec. (Very low noise, e.g. 1–5 electrons rms, can be obtained in scientific CCD systems, but only at very low readout rates of 0.05–0.5 Mpixels/s) In a CMOS APS, low noise of 5–15 electrons rms can be achieved even at high readout speeds because each column may have its own amplifier. This *column-parallel* architecture reduces the frequency of the amplifier, and the noise is lower. Additional noise can be introduced in the analog-to-digital converter (ADC). Again, a column-parallel architecture for on-chip ADC also reduces noise at the expense of complicated chip design. High-speed column-parallel ADCs with resolution up to 10 bits have been demonstrated with CMOS APS devices.

Another source of noise is dark current. This is the signal detected by the pixel in the absence of light, and is due to thermally generated electrons

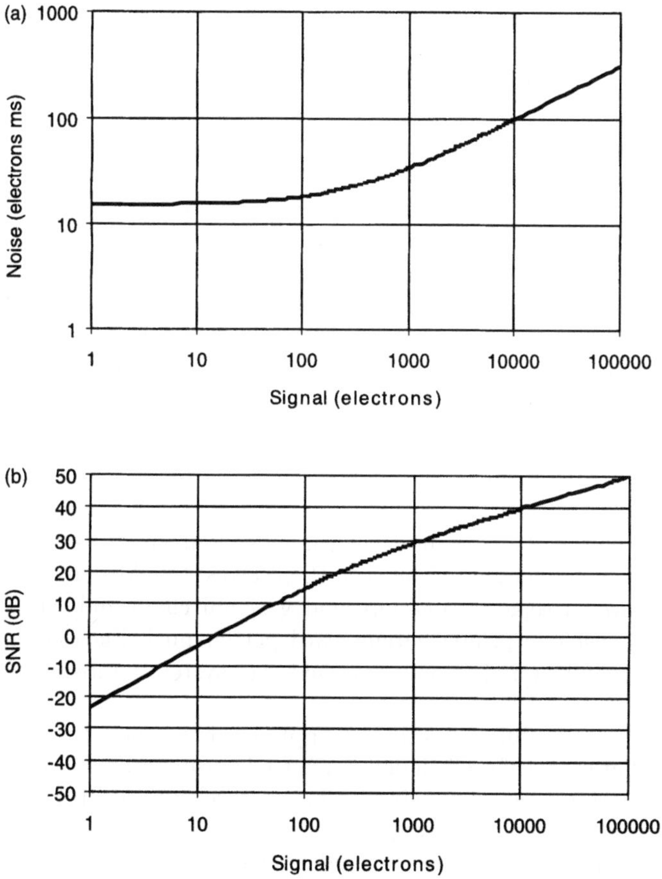

Fig. 5. (**a**) Noise as a function of signal. (**b**) SNR as a function of signal

rather than optically generated electrons. Like optical shot noise, the noise in this dark signal varies as the square root of the number of carriers. Dark signals in high-speed CCDs and CMOS APS are of the order of 500–5000 electrons/s/pixel, so for 1 ms integration times (1 kHz frame rates) dark noise is only of the order of 1–3 electrons rms, but can be substantially more for slower operation.

Figure 5 shows noise vs. signal and signal-to-noise ratio (SNR, expressed in dB as $20\log[\text{SNR}]$) vs. signal, where read noise is taken to be 15 electrons rms and dark noise 2 electrons rms. Each curve has two parts: for lower signals, noise or SNR is dominated by read noise, and for higher signal levels, noise and SNR are dominated by photon shot noise. Typically it is SNR that is important in detection.

3 Readout Rate

In a CCD, readout rate is typically limited to 10–30 Mpixels/s per readout channel. High-speed CCDs have been built with multiple readout channels per chip to achieve high frame rates. Kodak has built and demonstrated, in conjunction with IBM, a 1024×1024 element CCD operating at nearly 1000 frames/s using 64 parallel readout channels each operating at 15 Mpixels/s [5].

In CMOS APS, the analog signal chains (typically one per column) and ADCs (also one per column) are integrated on chip. To date, the highest throughput CMOS APS device has been demonstrated by Photobit Corporation [6]. That device, developed for the US Government, has 1024×1024 elements and has been operated at over 500 frames/s with 8 bits digital output (524 Mpixels/s) and dissipates approximately 0.350 W at 3.3 V operation. It is expected that 1000 Mpixels/s can be achieved in the next generation.

4 System Implementation

In a CCD, it is difficult to have so many parallel readout channels without the channels introducing crosstalk between themselves. In addition, each output channel requires a separate off-chip analog signal chain and ADC, corresponding to at least one printed circuit card per channel. Each channel needs to be well matched and to stay matched over system temperature variations.

CCDs also require many different voltage supply levels to allow them to operate with good charge transfer efficiency. Each CCD also requires a separate timing generator chip, and driver circuits for each "clock" signal – and often 10–16 different clock signals are required. Power dissipation of a high-speed CCD system is of the order of 200 W, and a separate cooling system is also required. Making a compact system (e.g. smaller than a breadbox) with this approach has not yet been demonstrated.

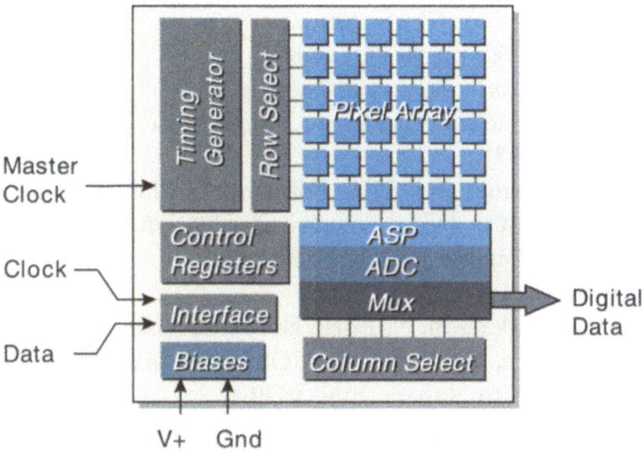

Fig. 6. The CMOS APS lends itself to integration with timing generators, ADCs and other functions to reduce system component count and power

For CMOS APS-based systems, the integration of the timing generator, drivers, analog signal chain, and ADC significantly advances the ability to build a compact DHDS system. The use of a single supply voltage (e.g. 5 V or 3.3 V) and the full digital interface to the chip (no analog signals) greatly enhances faster readout and system design ease, and increases system reliability. Furthermore, the integration of additional circuits to perform "smart" preprocessing of the output data (e.g. error correction) also greatly simplifies system design. A typical CMOS APS system is diagramed in Fig. 6.

On-chip signal processing is a feature that can be invaluable in DHDS sensor arrays. As mentioned, one form of on-chip processing is that of analog-to-digital conversion. In addition to ADC, numerous other processes can be implemented such as error correction codes (ECC), localized auto-exposure, data decoding, calculations of bit value statistics, auto-start triggering, and data formatting (in preparation for data export). For example, in the case of on-chip differential data decoding, it has been demonstrated that sensor array readout rates can increase by a factor of more than 100 [7].

5 Conclusion

While both CCDs and CMOS APS can be used for DHDS detector arrays, the actual and potential advantages of the CMOS APS technology are overwhelming. The major disadvantage of the high-speed CMOS APS technology is a lack of widespread availability, although high-speed, high-performance CCDs are also not widely available. The availability of CMOS APS technology through off-the-shelf supply as well as custom design is expected to improve greatly in the next few years.

References

1. A. Theuwissen, *Solid-State Imaging with Charge-Coupled Devices*, Kluwer Academic Publishers, Boston, 1996.

2. E.R. Fossum, "Active pixel sensors – are CCDs dinosaurs?" in *CCDs and Optical Sensors III*, Proc. SPIE vol. 1900, pp. 2–14, (1993).
3. E.R. Fossum, "CMOS image sensors - electronic camera on a chip," IEEE Trans. Electron Devices, Special Issue on Solid-State Image Sensors, October 1997 vol. **44**(10) pp. 1689–1698 (1997).
4. B. Mansoorian, H.-Y. Yee, S. Huang and E.R. Fossum, "A 250 mW, 60 frame/sec 1280 × 720 pixel 9 bits CMOS digital image sensor," *1999 Int'l Solid-State Circuits Conf. Dig. of Tech. Papers*, pp. 312–313 (1999).
5. Hans Coufal, IBM, private communication.
6. A. Krymski, D. Van Blerkom, A. Andersson, N. Bock, B. Mansoorian, and E.R. Fossum, "A high speed, 500 frames/s, 1024 × 1024 CMOS active pixel sensor," *1999 VLSI Circuits Symposium*, Kyoto, Japan, June 1999.
7. S. Campbell, K. Curtis, A. Hill, T. Richardson, M. Tackitt, and W. Wilson, "Digital holographic memory using a digital micromirror device SLM and a CMOS active pixel sensor camera," in *Optical Data Storage Technical Digest*, pp. 168–170, Aspen, Colorado, May 10–13, 1998.

Part IV

Channels

Modulation Codes for Holographic Recording

B. Marcus

In a data recording system, the goal of coding and signal processing is to reduce the bit error rate (BER) to a sufficiently low level while achieving such important figures of merit as high density and high data rate. This is accomplished by stressing the physical components of the system well beyond the point where the channel is error free, and then introducing coding and signal processing schemes to reduce the BER to levels acceptable to users.

In a holographic recording system, a page of digital (binary) data is input to a spatial light modulator (SLM) and recorded in a medium. The data is retrieved on a charge-coupled device (CCD). The retrieved data is an array of pixel intensity values. These values are essentially analog, and so a *detection scheme* is needed to convert the received values back to the digital data that was originally recorded.

The purpose of a modulation code is to constrain the digital data patterns that are recorded. Some patterns are more likely to be corrupted by the channel, while other patterns may be particularly helpful for a given detection scheme. So, a modulation code will forbid certain patterns from appearing and demand that other patterns do appear. The set of patterns allowed by a modulation code can be viewed as a *constraint.* Since modulation coding and detection schemes often go hand in hand, we will discuss them together.

We will focus on two kinds of constraints: balanced and lowpass. The former is intended to help the channel cope with variations in beam intensity across a data page. The latter is intended to help the channel cope with inter-pixel interference.

A modulation code is implemented via an *encoder* that encodes user data sequences into allowed patterns. We divide our treatment into two parts: block encoders and strip encoders. We will also refer to these as block codes and strip codes. Balanced codes are quite often implemented as block codes, the subject of Sect. 1, while lowpass codes are often implemented as strip codes, the subject of Sect. 2.

In the following, we use the term ON to mean a pixel representing a 1 and the term OFF to mean a pixel representing a 0.

1 Block Codes

A *block encoder* or *block code* encodes user data strings to (usually) rectangular arrays all of the same shape. The arrays are then patched together to tile the entire SLM. The encoding is required to be one-to-one, and the constraint must be satisfied across the borders of the arrays as well as within each array. If the user strings are of length p and the coded arrays are of size $m \times n$, then we say that the code is of *rate* $p : q$, where $q = mn$.

A block modulation code effectively adds redundancy to the data and so incurs an overhead. This overhead is usually described in terms of the code rate, $p/q < 1$. The capacity of a stack is $(p/q)MN$ where M is the number of coded holograms recorded in the stack and N is the number of pixels on the SLM. The "game" in coding (in particular, in modulation coding) is to exploit the tradeoff between code rate p/q and number of holograms recorded – the idea being that a more constrained modulation code, matched to a particular detection scheme, may allow more holograms to be recorded and retrieved at the same level of fidelity. For this reason, we use the term *capacity* to refer to the product $(p/q)M$ for a given block modulation code (and detection scheme).

1.1 Correlation Detection and Balanced Block Codes

The simplest detection scheme is *global threshold detection*, in which one chooses a threshold T and declares any CCD pixel with intensity above T to be a 1 and below T to be a 0. But it is not at all obvious how to choose a threshold, especially in the presence of spatial variations in intensity.

One method is to introduce *fiducial patterns* (i.e. known data patterns) within a data page. CCD pixel intensity patterns corresponding to these data patterns can then be used to give a reasonable estimate of a *local threshold* that can be applied to data recorded in the vicinity of the fiducial. Note that the introduction of fiducials essentially yields a (usually high rate) block modulation code. Part of the block is uncoded and part of the block is a fiducial.

Of course, the performance of this technique can be expected to improve with a higher density of fiducial patterns or more elaborate (in particular, bigger) fiducials. But it is inefficient to reserve lots of space for fiducials. So, each fiducial should be relatively small and should be used for a relatively large region of the page. Accuracy may be improved by an *adaptive threshold* scheme, where the threshold is updated using feedback as more and more data within a region is processed (such as in [1]). However, the feedback inherent in adaptive schemes can propagate errors.

Within a sufficiently small region, one might expect that there will not be much variation in pixel intensity. If the data page is divided into several such small regions, and within each region the data patterns are *balanced*, i.e. have an equal number of 0s and 1s, then detection can be accomplished without using a threshold. For instance, in *sorting detection* letting N denote the number of pixels in a region, one declares the $N/2$ pixels with highest intensity to be 1s and those remaining to be 0s. Thus, sorting detection combined with balanced modulation coding provides a means to obviate the inaccuracies inherent in threshold detection.

The balanced condition can be achieved by a block modulation code that encodes arbitrary data patterns into codewords represented as balanced arrays all of the same shape: such a code is called a *balanced block code*. The simplest example of this is *differential encoding*, where the page is divided into 1×2 arrays, each of which contains either 0 1 or 1 0. One encodes 0 to 0 1 and 1 to 1 0, and so the rate of this code is 1/2. A higher-rate code can be obtained by dividing the page into 2×4 arrays, each of which contains exactly four 1s and four 0s. Since (8 choose 4) $= 70$ and $2^6 = 64 < 70$, we can use 64 of the 70 balanced arrays to obtain a rate 6:8 code that encodes 6 user bits into a balanced 2×4 array. Two versions of such a code were developed at Optitek and IBM.

One problem with sorting detection is that the balanced array detected by sorting may not be a valid coded array. For instance, for the 6:8 code, the sorting detector may output one of the 6 ($= 70 - 64$) balanced arrays that are not part of the code. To handle such instances, one needs a procedure that transforms balanced arrays into valid coded arrays. This is not much of a problem when most balanced arrays of the given shape are valid coded arrays such as in the 6:8 balanced code. But for other codes, such as the 8:12 code discussed in Sect. 2.1 below, this process can introduce serious errors. For this reason, *correlation detection* was proposed in [2]. In this scheme, the detector chooses the coded array that maximizes correlation with the array of received pixel intensities. But as usual there is a tradeoff: correlation detection can perform better but tends to be more complex than sorting detection.

1.2 Sparse Block Codes

The arguments presented in the preceding section show that sorting or correlation detection does not really require balancing of coded arrays: all that matters is that each coded array has the same proportion of 1s and 0s. For instance, if in a 3×3 array exactly three pixels are ON, then sorting detection would declare the three pixels with highest intensity levels to be 1s and the remainder to be 0s.

Such codes are called *constant-weight codes*. Among constant-weight codes, balanced codes have maximal coding rate (at least on arrays of even area). So, why bother with constant-weight codes that are not balanced?

One reason is that 1s are more "expensive" than 0s to record: for a given budget of signal photons, one can record more pages if there are fewer 1s than 0s. Another is that 1s tend to provoke more noise than 0s (as revealed by typical histograms of received pixel intensity). Finally, it may be desirable to forbid a 1 surrounded by 0s or vice versa in order to mitigate the effects of inter-pixel interference (see Sect. 2.2); such a constraint would be rather incompatible with a balanced constraint but might be compatible with a constant-weight constraint. For these reasons, one might use a *sparse code*, which is defined to be a constant-weight code where the proportion of 1s is less than 1/2, instead of a balanced code. Sparse codes have been investigated by M. Neifeld [3] and A. Daiber. As an example, observe that since (9 choose 3) = 84, there is a rate 6:9 sparse code with sparseness = 3 (i.e. 3 pixels are ON and 6 pixels are OFF). Experimental and analytical evidence suggest that a sparseness on the order of 3 or 4 (i.e. 1/3 or 1/4 of the pixels are ON) may be optimal.

1.3 Parity Thresholding

Sorting detection and correlation detection depend critically on the use of a modulation code, balanced or more generally constant weight. The rate of the modulation code is a penalty worth bearing only if sufficiently many additional holograms can be recorded. *Parity thresholding*, introduced in [4], is a hybrid of correlation detection and threshold detection. This scheme effectively achieves substantially higher code rate but perhaps not quite as many holograms. In this scheme, a small coded array (using a balanced modulation code) is attached to a large uncoded array. The coded array carries only one piece of information: the number, k, of 1s in the uncoded array. The number k is recovered by correlation detection and then used to help set a threshold for detection of the uncoded array. If the size of the coded array is small compared with that of the uncoded array, then this essentially yields a very high-rate modulation code. Experimental evidence [4] shows that for a target raw BER on the order of 10^{-3} this scheme can be competitive in terms of capacity with correlation detection using balanced codes.

2 Strip Codes

Up to this point, we have focused on block codes: codes that can be implemented by patching together small arrays independently of one another. In this section, we introduce a more general kind of encoder.

In a *strip encoder* or *strip code*, the page is divided into nonoverlapping horizontal strips of a fixed height h, and within each strip, encoding is implemented via a finite-state machine. The coded array is a function of an internal state of the machine as well as the p-bit user data string, and the machine then proceeds to a next state; we assume that the coded arrays are rectangles of fixed size $A = h \times w$, and we encode a long data sequence into a long horizontal strip of height h by concatenating the coded arrays $A_1, A_2, \ldots$ side to side from left to right (see Fig. 1). The rate of the strip code is $p : hw$, and just as with block codes, we measure the capacity of a strip modulation code (together with detection scheme) as $(p/q)M$, where $q = hw$ and M is the number of coded holograms that can be recorded (in one stack) and retrieved with a given fidelity.

In order to limit error propagation, we require that the decoder be implemented as a *sliding block decoder*: a coded array A is decoded to a user

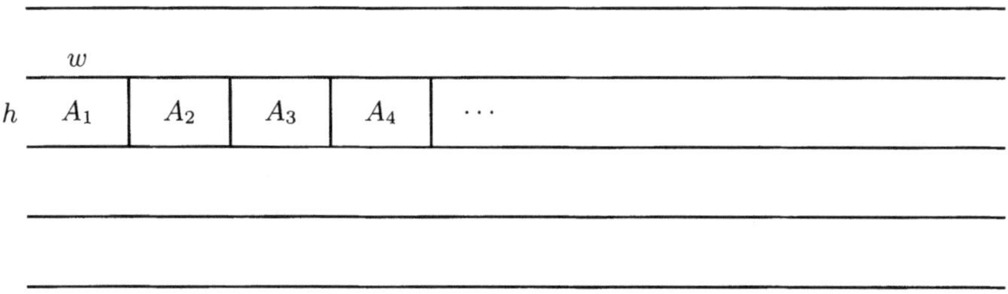

Fig. 1. Strip encoder

data string as function of A and a bounded number of coded arrays to the right (*anticipation*) and left (*memory*), within the same horizontal strip, of A. A *k-block decoder* is a sliding block decoder whose decoding window (i.e. memory + 1 + anticipation) consists of k coded arrays. In the special case where the decoder has no memory and no anticipation, we say that the code is *block-decodable.*

For both block codes and strip codes some means must be taken to ensure that the modulation constraint is satisfied globally, i.e. across the borders of the coded arrays in the case of a block code and across the borders of the strips in the case of a strip code. This can be done by imposing some additional (and often artificial) constraint on the blocks or strips. For a strip code, this additional constraint is called a *strip constraint.* If the original constraint is a balanced constraint (more generally constant-weight constraint), then the constraint holds within each strip and so there is no need to impose an additional constraint. But for other constraints, such as lowpass constraints disccussed below, the additional constraint is necessary.

Once such a constraint is imposed, the problem of designing a modulation encoder is reduced to a one-dimensional problem, which can be solved by the state-splitting algorithm [5] or other algorithms.

2.1 Balanced and Pseudo-Balanced Strip Codes

Recall that balanced codes are useful because they aid sorting/correlation detection. But they have another advantage: error-correction power, since no two balanced codewords can differ in only one bit. In other words, the minimum Hamming distance between codewords in a balanced code is at least 2. This distance is exploited by the correlation detection scheme, which makes use of the (analog) pixel intensities to determine the correct coded array.

It is possible to design balanced codes with even greater minimum distance. For example, there is a rate 8:12 balanced strip code that has minimum Hamming distance = 4 [2]. The 12-bit coded arrays are of size 2×6, and they are all balanced. The data page is divided into horizontal strips of height 2, each of which is divided into such coded arrays. A finite-state machine with only four states iteratively encodes each 8-bit user byte into a 2×6 array as a function of an internal state, and the encoding determines a new state, which is used to encode the next 8 bits. This is continued until an entire strip of height 2 is exhausted. Correlation detection is used at each state to determine the most likely candidate coded array, and then the results of this are fed into a Viterbi detector (see section 9.2 of [6]) to determine the overall most likely coded array. This code is block-decodable.

If the target BER is sufficiently low (in the range 10^{-5} to 10^{-7}), then the 8:12 code outperforms (i.e. has higher capacity than) the 6:8 code mentioned in Sect. 1.1; evidently, the lower code rate (2/3 versus 3/4) is more than compensated for by the enhanced minimum distance, which allows more holograms to be recorded in this range of BER [2,4]. On the other hand, correlation detection for the 8:12 code is more complex than that for the 6:8 code.

A generalization of the class of balanced codes is the class of *pseudo-balanced codes*, in which coded arrays are concatenated from left to right and the resulting rectangles are required to have bounded imbalance: that is, the accumulated imbalance has a fixed ceiling. Pseudo-balanced codes are naturally implemented via strip encoders rather than block encoders. Such codes offer a more general and sometimes simpler way of achieving performance that can be somewhat comparable to balanced codes (though typically not quite as good). Detection for pseudo-balanced codes is accomplished by a combination of correlation detection and a relatively weak reliance on an estimate of mean intensity. Codes C1 and C4 in Sect. 2.3 are pseudo-balanced.

2.2 Inter-Pixel Interference and Low-Pass Codes

A code that forbids one or more patterns of high spatial frequency is called a *low pass code*. Such codes constrain the allowed pages to have limited high spatial frequency content. For instance inter-pixel interference (IPI), which degrades the performance of the channel, tends to occur when an OFF pixel is "surrounded" by ON pixels or vice versa; specifically, IPI may be provoked by either of the following patterns:

$$\mathrm{IPI}_0 = \begin{matrix} & 1 & \\ 1 & 0 & 1 \\ & 1 & \end{matrix}$$

or

$$\mathrm{IPI}_1 = \begin{matrix} & 0 & \\ 0 & 1 & 0 \\ & 0 & \end{matrix} .$$

For IPI_1, a strip constraint can be readily defined. For instance, one can choose $h = 3$ and require that the pattern

$$\begin{array}{ccc} & 0 & \\ 0 & 1 & 0 \end{array}$$

not appear in the bottom two rows of the strip, and that the pattern

$$\begin{array}{ccc} 0 & 1 & 0 \\ & 0 & \end{array}$$

not appear in the top two rows of the strip. A stronger condition would be to require that within each coded array each 1 sits next to (either horizontally or vertically) a 1. We call such an array *1-protected*; similarly we have the notion of *0-protected*.

According to the model in [7], for certain realistic and relatively optimal choices of system parameters, the number of holograms that can be recorded and retrieved at a target BER $\approx 10^{-4}$ can be increased by about 15% if one forbids IPI_1 and IPI_0. Even more interesting is the fact that almost the same improvement can be achieved if one forbids only IPI_1. Evidently, in this setup IPI_0 does not cause much of a problem because ON pixels adjacent to an OFF pixel interfere destructively, while ON pixels adjacent to one another interfere constructively.

A code at rate $8 : 9 = 0.888\ldots$ that forbids IPI_1 is described in [8] (Code L2 below). Thus, this code allows an increase in M of 15%, which more than compensates for the 11% penalty incurred by the 8/9 code rate, yielding an overall improvement in capacity. Below are figures of merit for some lowpass codes. Codes L2, L3 and L4 are mentioned in [8].

Code L1: (see [9]). Rate 3:4 strip code (3 bits $\mapsto 2 \times 2$ array), forbids both IPI_0, and IPI_1, 4 encoder states, block decoder.
Code L2: Rate 8:9 strip code (8 bits $\mapsto 3 \times 3$ array), forbids only IPI_1, 3 encoder states, 2-block decoder.
Code L3: Rate 11:12 strip code (11 bits $\mapsto 3 \times 4$ array), forbids only IPI_1. Probably at most 10 encoder states and decoder is at most 3-block.
Code L4: Rate 15:16 strip code (15 bits $\mapsto 4 \times 4$ array), forbids only IPI_1. Probably complicated.

Of course, by switching 1s and 0s, codes L2, L3, and L4 above can be transformed to forbid IPI_0 instead of IPI_1.

2.3 Combined Constant-Weight Low-Pass Codes

If spatial variations in intensity and inter-pixel interference both pose a problem then it may be desirable to use a modulation code that is simultaneously constant-weight (e.g. balanced) and low-pass. But the cascade in series of two modulation codes satisfying two different constraints is not likely to satisfy both features simultaneously. Below are figures of merit from [8] for modulation codes that simultaneously satisfy both a balanced (or pseudo-balanced) constraint and a low-pass constraint.

Code C1: Rate 5:8 (5 bits $\mapsto 2 \times 4$ array) pseudo-balanced strip code, all coded arrays are 0-protected and 1-protected (and so forbids both IPI_0 and IPI_1), and have imbalance 0, +4, or −4 (by *imbalance* we mean the excess of 1s over 0s), 2 encoder states, 1-block decoder (memory = 0, anticipation = 1).
Code C2: Rate 5:8 (5 bits $\mapsto 2 \times 4$ array) balanced strip code, forbids both IPI_0 and IPI_1, 9 encoder states, 2-block decoder (memory = 0, anticipation = 1).
Code C3: Rate 8:12 (8 bits $\mapsto 2 \times 6$ array) balanced strip code, forbids both IPI_0 and IPI_1, 8 encoder states, 1-block decoder (memory = −1, anticipation = 1).
Code C4: Rate 9:12 (9 bits $\mapsto 3 \times 4$ array) pseudo-balanced strip code, forbids both IPI_0 and IPI_1, coded arrays have imbalance 0, +4, −4. Probably complicated (many encoder states, large decoder window).

Finally, we turn our attention to combined sparse/lowpass codes. Clearly the sparse condition is more compatible with the constraint that forbids IPI_0 than IPI_1. As mentioned in Sect. 2.2, there is some evidence that indicates it is more important to forbid IPI_1 than IPI_0. Nevertheless, there may be some situations where the reverse is true. Below are figures of merit taken from [8] for some codes that simultaneously satisfy a sparse constraint and forbid IPI_0.

Code C5: Rate 8:20 strip code (8 bits $\mapsto 5 \times 4$ array), forbids only IPI_0 and requires that 1s be isolated (both horizontally and vertically), sparseness = 5 (i.e. 4 pixels are 1 and 16 pixels are 0). Probably about 10–15 encoder states, and decoder is at most 3-block.
Code C6: Rate 6:9 strip code (6 bits $\mapsto 3 \times 3$ array), forbids only IPI_0, sparseness = 3 (i.e. 3 pixels are 1 and 6 pixels are 0), 3 encoder states, 2-block decoder.
Code C7: Rate 6:9 strip code (6 bits $\mapsto 3 \times 3$ array), forbids only IPI_0, sparseness = 2.25 (i.e. 4 pixels are 1 and 5 pixels are 0), 7 encoder states, 2-block decoder.
Code C8: Rate 9:12 strip code (9 bits $\mapsto 3 \times 4$ array), forbids only IPI_0, sparseness = 2.40 (i.e. 5 pixels are 1 and 7 pixels are 0), probably complicated (many encoder states, large decoder window).

In fact, Code C6 outperforms a wide variety of alternative modulation codes in experiments at low aperture [13] (we remark that Code C6 is indeed the rate 6:9 code used in these experiments, contrary to the erroneous statement in [13, top of p. 22] which describes the complement of Code C6 instead).

Final remarks: There are other modulation constraints that naturally arise in holographic recording. These include constraints for gray-level recording [10] and constraints to help the performance of a phase mask [11,12]. Also, one can imagine encoding schemes that are more genuinely two-dimensional than either block encoders or strip encoders. At this point, most such schemes have been avoided becuase of the need to rely on a large buffer.

References

1. X. A. Shen, A.-D. Nguyen, J.W. Perry, D.L. Huestis, and R. Kachru, "Time-domain holographic digital memory," Science, 278 (1997) 96–100.
2. G. Burr, J. Ashley, H. Coufal, R. Grygier, J. Hoffnagle, C.M. Jefferson, and B. Marcus, "Modulation coding for pixel-matched holographic data storage," Opt. Lett., 22(9) (1997) 639–641.
3. M. Neifeld, "Computer-generated holography for optical memory using sparse data words: capacity and error tolerance," Appl. Opt., 32(26) (1993) 5125–5134.
4. G. Burr, W. Chou, M. Neifeld, H. Coufal, J. Hoffnagle, and C.M. Jefferson, "Experimental evaluation of user capacity in holographic data storage system," Appl. Opt., 37 (1998) 5431–5443.
5. B. Marcus, P. Siegel, and J. Wolf, "Finite-State Modulation Codes for Data Storage," IEEE Journal on Selected Areas of Communication, 10 (1992) 5–37.
6. R.J. McEliece, *The Theory of Information and Coding*, Addison-Wesley, 1977.
7. M.-P. Bernal, G. Burr, H. Coufal, and M. Quintanilla, "Balancing inter-pixel crosstalk and detector noise to optimize areal density in holographic storage systems," Appl. Opt., 37 (1998) 5377–5385.
8. J. Ashley and B. Marcus, "Constant-weight/lowpass modulation codes for holographic recording," IBM Research Report RJ 10089 (91905), October 1997.
9. J. Ashley and B. Marcus, "Two-dimensional lowpass filtering codes for holographic storage," IEEE Trans. Commun., 46 (1998) 724–727.
10. G. Burr, M. Neifeld, G. Barking, H. Coufal, J. Hoffnagle, and C.M. Jefferson, "Gray-scale data pages for digital holographic data storage," Opt. Lett., 23 (1998) 1218–1220.
11. J. Hong, I. McMichael, and J. Ma, "Influence of phase masks on cross talk in holographic memory," Opt. Lett., 21 (1996) 1694–1696.
12. M.-P. Bernal, G. Burr, H. Coufal, R. Grygier, J. Hoffnagle, C.M. Jefferson, E. Oesterschulze, R. Shelby, G. Sincerbox, and M. Quintanilla, "Effects of multilevel phase masks on interpixel crosstalk in digital holgraphic storage," Appl. Opt., 36(14) (1997) 3107–3115.
13. G. W. Burr, B. Marcus, "Coding trade-offs for high-density holographic data storage," Proceedings of SPIE, Advanced Optical Data Storage: Materials, Systems, and Interfaces to Computers, 3802 (1999) 18–29.

Interleaving and Error Correction for Holographic Storage

M.A. Neifeld and W.-C. Chou

The potential advantages of volume optical storage were recognized almost 30 years ago, and since that time many research efforts have focused on improving the underlying materials and devices for use in such systems. Recent progress in these supporting technologies has made possible several system-level demonstrations of volume optical storage. These recent demonstrations have verified the feasibility of volume memory systems for offering large volumetric storage capacities, fast access times, and very high data transfer rates realized via the parallel two-dimensional (2-D) nature of the stored data. The success of these volume storage testbeds has served to ignite additional research into supporting two-dimensional or page-oriented interface technologies such as parallel 2-D data detection and error correction [1–14]. This application therefore provides a strong impetus to study traditional communication theoretic topics such as signaling, equalization, and coding in the context of highly parallel 2-D channels. In this chapter we will focus specifically on 2-D interleaving and error correction strategies and their impact on the capacity of volume holographic memory (VHM).

Conventional wisdom holds that the purpose of error correction coding (ECC) is to improve the fidelity of noisy, error-prone data. Before discussing the details of ECC, we will therefore review the sources of noise and errors in holographic storage as well as the utility of this conventional wisdom. Many noise sources have been detailed elsewhere in this volume, and from those discussions it is apparent that the storage and retrieval processes associated with holographic memory systems are quite complex [15–22]. We can classify errors into two categories. The first category comprises systematic errors from, for example, interpage and interpixel interference, photovoltaic damage, optical system misalignment, lens aberrations, and manufacturing defects associated with the SLM, CCD, lenses, and storage materials. The systematic error sources are assumed to be fixed; however, systematic errors can have a random component owing to the randomness of the stored data. Consider for example a small magnification error in the data imaging system. In this case, the (i, j) CCD pixel output depends not only on the (i, j) SLM pixel input data but also on data associated with neighboring SLM pixels. Consider a corner CCD pixel on an $N^2 = 512 \times 512$ pixel page corrupted by 0.10% magnification error. Assuming 100% fill factors for both CCD and SLM, 1 W total optical power, and 1 ms detector integration time, we find that the average '1' and '0' signal levels change from from 3.81 nJ to 2.96 nJ, and from 0 nJ to 0.85 nJ respectively, as a result of this magnification error. Storing multiple pages in the VHM further corrupts the retrieved signals. Using a simple scaling law for VHM systems, the diffracted intensity for each data page is inversely proportional to the square of the total number of pages (M) stored in the memory [23]. If we now consider storing 3000 data pages in the previously described system, the estimated '1' signal level becomes 0.33 fJ, corresponding to about 844 photons at 514 nm wavelength. This signal is comparable to the thermal noise associated with the CCD readout electronics, which must now be included in the noise analysis.

VHM performance is further corrupted by the second error category: non-systematic or "random" error sources. The thermal noise associated with photodetectors and readout electronics, photon shot noise, and coherent scattering or speckle noise are some examples of random error sources. In the presence of random error sources, repeated measurement of a particular CCD pixel yields fluctuating results. Once again considering a pagewise parallel interface in which the 512×512 pixels are read out simultaneously within one clock cycle (i.e. 1 ms detector integration time), a detector noise equivalent power of NEP $= 5.25 \times 10^{-16}$ W/$\sqrt{\text{Hz}}$ produces an estimated additive thermal noise standard deviation of about 0.0167 fJ.

In order to simplify this discussion of error correction and two-dimensional interleaving, we limit our study to a single random noise source combined with various systematic error sources. We select thermal noise because large SLM and CCD pixel arrays are desired to facilitate high data transfer rate, and thus the readout electronic circuits must operate over a wide bandwidth. The CCD signal probability density functions (pdfs) used in this study are non-Gaussian however, owing to the inclusion of systematic error sources. Once the magnitudes of the systematic error and thermal noise have been determined, these are assumed to remain unchanged with M, and the SNR per pixel of the VHM system can be modeled using the SNR $= \gamma/M^2$ scaling law. With the system parameters of the above example we find $\gamma = 2.304 \times 10^8$. Using this signal-scaling relation together with a configuration of systematic error it is possible to determine the maximum number of pages that can be stored while maintaining some minimum acceptable SNR. Given the non-Gaussian nature of the CCD signal pdfs, however, we prefer to utilize a maximum acceptable BER metric to determine the maximum number of stored pages for various coded and uncoded systems.

1 Capacity

Noise and error sources can be viewed as mechanisms through which information is lost during VHM storage and/or retrieval. Information theory offers us a mechanism through which we can quantify this loss and thus make an accurate estimate of the *information* capacity of VHM. Before discussing the use of ECC and two-dimensional interleaving in a VHM system, let us first define this information theoretic VHM storage capacity metric. Other researchers have studied the storage capacity bounds of VHM systems using various methods, such as counting the number of independently addressable voxels [24], representing the capacity as the maximum allowed number of pages given some minimum SNR requirement [25–27], or evaluating the limitations of conventional multiplexing technologies [28]. It is also possible to use the analogy between a VHM storage channel and a conventional communication channel to study the VHM storage capacity from an information theoretic point of view [29–31]. For the case of a serial communication channel, once the channel noise characteristics are known (i.e. the pdfs of the received signals), the mutual information $I(X;Y)$ between the channel input (X) and output (Y) is calculated using [32]

$$I(X;Y) = \int f(x,y) \log_2 \left(\frac{f(x,y)}{f(x)f(y)} \right) \mathrm{d}x\, \mathrm{d}y, \tag{1}$$

where $f(x,y)$ is the joint pdf of the input (X) and output (Y), and $f(x)$ and $f(y)$ are the marginal pdfs of X and Y respectively. Assuming a binary channel where X takes the values $x = 0$ and $x = 1$ with probabilities π_0 and π_1, the mutual information becomes:

$$I(X;Y) = \sum_{i=0,1} \pi_i \int f(y|x=i) \log_2 \left(\frac{f(y|x=i)}{f(y)} \right) \mathrm{d}y, \tag{2}$$

where $f(y|x=i)$ is the conditional pdf of Y given $X = i$. Under a specific signaling scheme, for example equal prior signaling with $\pi_i = 0.5$, $I(X;Y)$ represents the actual amount of information communicated through the channel each time a binary symbol is transmitted. The capacity ($R_{\max}$) of a binary channel is the maximum mutual information among all possible prior probabilities, π_0 and π_1, so that

$$R_{\max} = \max_{\pi_0,\pi_1} [I(X;Y)]. \tag{3}$$

We can perform an identical analysis to find the maximum storage capacity in a single-VHM pixel. The various noise sources are combined to yield single pixel conditional pdfs as described previously, and (2) is used to compute the corresponding capacity. In [31] we compute this single-pixel capacity and find the optimum prior probabilities to be very close to 0.5. Throughout the remainder of this discussion, therefore, we will assume equal priors.

For a two-dimensional array of communication channels, we wish to find the mutual information $[I(\underline{X};\underline{Y})]$ between the input page $\underline{X}$ and output page $\underline{Y}$, where $\underline{X}$ and $\underline{Y}$ are input and output random binary vectors with N^2 dimensionality. To calculate $I(\underline{X};\underline{Y})$ for an $N \times N$ pixel array, (2) becomes $2N^2$ multiple integrals, and the joint pdf between $\underline{X}$ and $\underline{Y}$ also has $2N^2$ dimensionality. In the absence of systematic error, it can be safely assumed that all pixels on a page are independent and identically distributed (iid), so that $I(\underline{X};\underline{Y}) = N^2 \times I(X;Y)$. In this case the capacity (C) of a VHM system storing M pages is simply

$$C = MN^2 R_{\max}(M), \tag{4}$$

where $R_{\max}(M)$ is the single-pixel capacity. Figure 1 provides a plot of VHM storage capacity as a function of the number of stored pages. We have used $\gamma = 2.304 \times 10^8$ for this plot. The solid curve in Fig. 1 depicts the storage capacity corresponding to (4). We notice that this capacity increases linearly for small M where $R_{\max}(M)$ is close to 1. As M gets larger, however, the signal levels drops as $1/M^2$ and the conditional pdfs of the received signals become broader so that $R_{\max}(M)$ becomes less than 1 and the storage capacity decreases accordingly. From Fig. 1 we see that there exists an optimum number of pages to maximize capacity. For these system parameters we find the peak capacity at 7632 pages corresponding to a 1.82-Gbits VHM storage capacity.

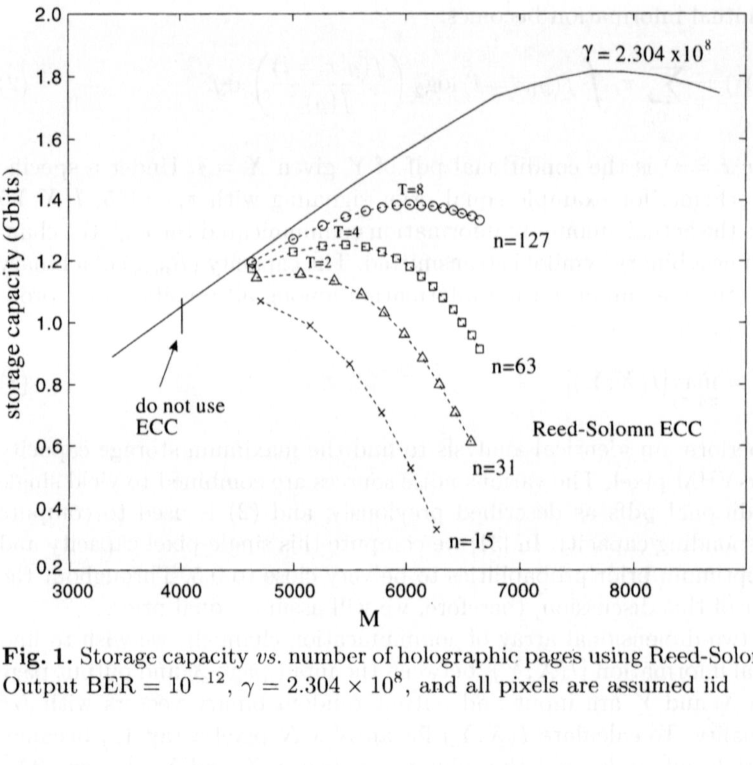

Fig. 1. Storage capacity *vs.* number of holographic pages using Reed-Solomon ECC. Output BER = 10^{-12}, $\gamma = 2.304 \times 10^8$, and all pixels are assumed iid

2 Error Correction

Error correction coding (ECC) has been used in both communication and memory systems to achieve acceptable BER in the presence of noise. ECC is a method of adding redundancy to the stored data in order to gain error tolerance and thus achieve some output BER goal. The overhead cost associated with ECC is characterized by the code rate ($R_{\rm ECC}$), where $R_{\rm ECC}$ is the ratio of user data (k symbols) to code word length (n symbols) $R_{\rm ECC} = k/n$. An alternate way to view ECC is as a method to relax the system raw BER requirement. In the context of VHM, allowing the raw BER to increase suggests that more data pages can be stored in the memory. As long as this increase in the number of pages results in a greater capacity gain than the associated ECC overhead cost, the use of ECC can actually *increase* the effective VHM storage capacity [11]. In this section we will investigate how the use of Reed–Solomon (RS) ECC can be used to access this increased capacity.

There are two basic types of error correction code: block codes and convolutional codes. These two methods are distinguished by the presence (convolutional) or absence (block) of memory within the encoding and decoding processes. Reed–Solomon codes are examples of linear block error correction codes, and for this reason we will restrict our attention to block codes throughout this discussion. There are numerous excellent textbooks on the subject of error correction coding, and the reader is referred to any one of these for a more general treatment of, or for more detail on, any of the topics discussed here [33–35]. In broad terms, however, the use of any forward error correction algorithm within a VHM system requires two complementary processes: encoding and decoding. The encoding process takes place in preparation for data recording, and is simply a mapping from an input space of dimension k to a code word space of dimension $n > k$. The code redundancy is measured by the dimensionality mismatch $n - k$. If we consider a binary code, the encoding process corresponds to collecting groups of k user bits and appending $n - k$ redundant bits to produce n-bit code words. Because there are only 2^k possible code words, this code word space is sparsely populated, and the Hamming distance between valid code words in this n-dimensional space can be greater than one. If the minimum distance between code words is large enough (i.e. $d_{\rm min} = 2t + 1$), then t bit errors can be tolerated before there will be any confusion as to which code word was originally stored. A powerful code, therefore, is one that can produce a large minimum distance between code words. It is this structure of the code word space that facilitates error correction during decoding.

Following error encoding, modulation encoding, and page composition, data is stored in the VHM. Noise associated with the storage and retrieval processes will result in some corruption of the original code word data. Upon retrieval from the VHM, equalization, signal detection, and/or modulation decoding take place and produce data with some raw BER. It is at this stage that the error decoding process takes place. The retrieved data is first collected into n-bit blocks. The decoding process can be conceptualized as one in which each retrieved binary n-vector (with errors) is mapped to the closest code word. Given that no more than t errors have occurred, this process will produce the correct stored code word, and the original data can be obtained by stripping away the redundant bits. Unfortunately, this conceptual algorithm is not feasible in practice owing to the large number of possible code words and the resulting complexity of the minimum distance search. The solution to this complexity problem is obtained by imposing some mathematical structure on the code word space, thus simplifying the tasks of encoding and decoding. This structure defines the essential differences among the numerous block codes that are available for error correction. RS codes represent one type of structure imposed on the larger class of linear, cyclic codes. The details of this algebraic structure are well beyond the scope of this discussion; however, we outline some of the salient features of RS codes below.

RS codes differ from the simplified ECC description provided above in that they are nonbinary codes. The encoding process, therefore, must first collect groups of m bits to create nonbinary data symbols. Groups of k symbols are then mapped to n-symbol code words as described earlier. A t-symbol error correcting RS code has the following parameters: $n = 2^m - 1$ and $2t = n - k$. Because RS codes are nonbinary, the decoding process generally takes place in three (as compared with two) steps. The first step involves computing a syndrome from the retrieved data (a quantity that depends only on the error pattern, not on the code word). The syndrome can be computed easily, using a linear feedback shift register. The next step utilizes the syndrome to compute an error locator. This step is typically accomplished using one of two popular algorithms: Berlekamp–Massey or Euclidean. Note that in the binary case, decoding would be complete following this step because a bit-error can only have one possible magnitude; however, in the case of a nonbinary code a third step is required. In this step the syndrome and the error locator are used to determine the magnitude of each symbol error. A Chien search is often used for realizing this step of the algorithm. Recent work has studied the computational and implementational complexity of various algorithmic choices within RS decoders, and several efficient parallel implementations have been developed [36–38].

The use of RS codes will provide some degree of error correction within a VHM system. The amount of correction is determined by the block length and the code rate of the RS code. Because the job of ECC is to achieve an acceptable decoded BER, we have elected to characterize the performance of several example RS codes using the data shown in Fig. 2. This figure represents the error correction capability of each code by examining the decoded BER as a function of AWGN noise variance. The dramatic difference between raw and decoded BER reflects the power of the RS codes, and it is clear that increased block length and lower code rate both result in improved ECC per-

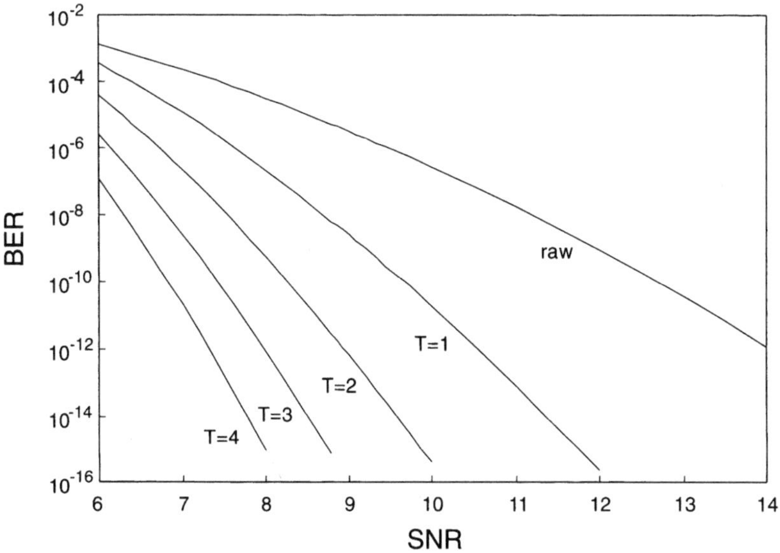

Fig. 2. Raw and coded bit error rate (BER) vs. signal to noise ratio (SNR) for additive Gaussian noise using Reed–Solomon $n = 31$ ECC

formance. In the spirit of the discussion of section 1, the BER gains can be translated into capacity gains. Referring once again to Fig. 1 we find that in order to achieve a 10^{-12} output BER in the absence of ECC, only 4043 pages can be stored in the memory. This corresponds to the value of M for which the raw BER equals 10^{-12} and yields a 1.06 Gbits capacity, which is only 58% of the maximum storage capacity. The four dashed curves in Fig. 1 represent the storage capacity associated with applying Reed–Solomon (RS) ECC to this VHM system. The ECC overhead cost is accounted for by calculating the storage capacity as $C_{\text{ECC}} = MN^2R_{\text{ECC}}(M)$, where $R_{\text{ECC}}(M)$ is the RS code rate required for each value of M (i.e. for each value of raw BER). Different curves correspond to different code word lengths (n) and for each code the number of bits per symbol (m) is related to code word length as $n = 2^m - 1$. The leftmost point on a curve corresponds to one-symbol error-correcting ability ($T = 1$). As the number of correctable symbol errors (T) increases, the error correction ability of the code increases while the code rate $R_{\text{ECC}} = (n - 2T)/n$ (for RS codes) decreases. It is apparent from Fig. 1 that the use of ECC facilitates storing more pages in the VHM (larger raw BER) while maintaining BER(out)=10^{-12}. From Fig. 1, we notice that longer code words produce capacities that are closer to the information theoretic capacity bound, and that the optimum number of pages depends upon the choice of ECC. For the $n = 127$ case, the optimum number of pages M_{opt} =6033. The resulting maximum storage capacity is 1.38 Gbits, which represents a 30% increase compared with not using ECC, and achieves 76% usage of the storage capacity bound.

3 Interleaving

The previous section was concerned primarily with the use of ECC as a method of combating random error sources within a VHM system. The use of ECC was shown to increase the user capacity of the VHM, allowing this user capacity to approach the information theoretic capacity bound. However, we have not yet included the effects of systematic errors. Systematic errors such as magnification error, CCD shift and rotation error, and lens aberrations result in correlated error patterns in the retrieved data page. Using *a priori* knowledge of these error correlations will allow us once again to design a coding scheme that approaches the information-theoretic limit. We begin this discussion with an analysis of the new capacity bounds computed in the presence of systematic errors.

Rather than assuming iid pixels over a page, we now include the effect of spatially variant BER. To simplify the calculation task, we establish a bound $\tilde{I}(\underline{X};\underline{Y})$ using

$$I(\underline{X};\underline{Y}) \geq \tilde{I}(\underline{X};\underline{Y}) = \sum_{i=1}^{N^2} I(\underline{X}_i';Y_i), \tag{5}$$

where

$$I(\underline{X}_i';Y_i) = \sum_{\underline{x}_i' \in \underline{X}_i'} \int p(\underline{x}_i') f(y_i) \log_2[\frac{f(y_i|\underline{x}_i')}{f(y_i)}] \mathrm{d}y_i, \tag{6}$$

and Y_i is a random variable representing the received signal level of the ith CCD pixel, $\underline{X}_i'$ is a set of all possible SLM input patterns that affect the ith CCD pixel output, and $p(\underline{x}_i')$ is the probability of input pattern $\underline{x}_i'$ occurring. Instead of including all 2^{N^2} possible SLM pixel patterns in the calculation of the mutual information for each CCD pixel [i.e. $I(\underline{X};Y_i)$], we consider all combinations of only those SLM pixels that contribute to the ISI of a certain CCD pixel. We include the effect of up to 16 SLM pixels, the local configuration of the interfering pixels being determined by the local point spread function associated with the 4F system. This effect of SLM neighborhood correlations is included in (5) while the correlation between CCD pixels is ignored. This procedure produces a lower bound on information, and Fig. 3 shows this storage capacity bound as a function of M for several configurations of systematic error. The square symbols in Fig. 3 represent the case of 0.1% magnification error, and the triangle symbols represent the case of 0.1% magnification combined with 10% CCD x-shift error. Both cases also include third-order lens aberrations. We can observe the loss of capacity due to systematic errors. Also we notice the optimum number of pages is reduced from M_{opt} = 7632 to M_{opt} =7439 for 0.1% magnification error, and to M_{opt} =7076 when 10% shift error is included. The corresponding page-average mutual information for the three curves in Fig. 3 are 0.9076, 0.8857,

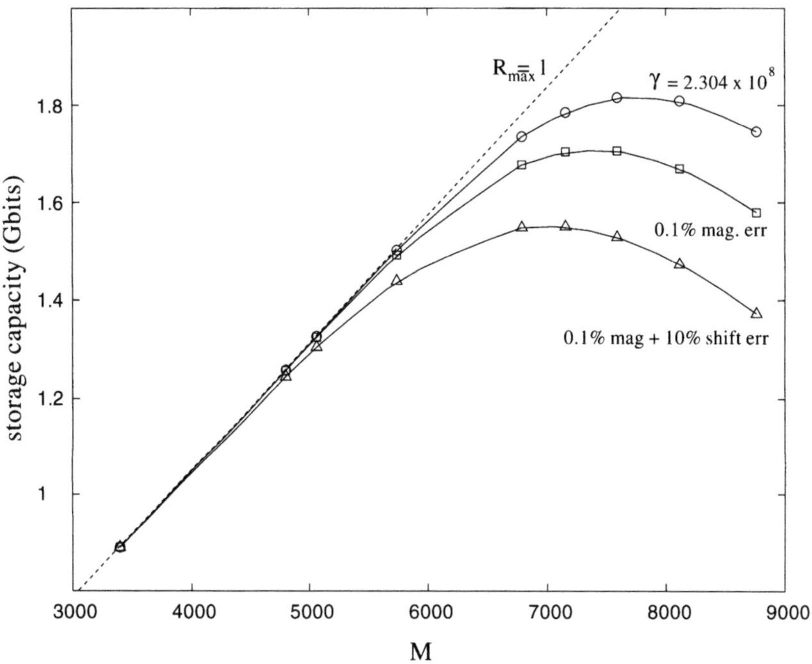

Fig. 3. Information-theoretic capacity vs. number of stored pages given no systematic error (○), 0.1% magnification error (□), and 0.1% magnification plus 10% shift error (△)

and 0.8373, verifying the degradation of memory capacity arising from larger misalignment errors.

In order to gain some intuition concerning the nature of the error correlations arising from systematic sources, Fig. 4 depicts the surfaces of constant BER for one of 4800 stored pages that have been corrupted by 0.1% magnification error and 10% CCD pixel shift error. In the presence of spatially correlated errors such as those arising in VHM, the system capacity associated with the use of ECC will depend not only on the power of the code, but also on how each code word is spatially composed (i.e. the two-dimensional interleaving scheme). Interleaving and the use of nonbinary symbols have long been utilized in communication and storage channels for mitigating the effects of burst errors [39,40]. Figure 5 schematically demonstrates these two methods for the case of $n = 7$ and $m = 3$. By using 3-bit symbols, a single 14-bit burst error results in a six symbol burst, requiring ECC with $T = 6$ if all erroneous symbols were to fall within a single code word. This situation is depicted in Fig. 5a. The use of three-code-word interleaving is shown in Fig. 5b, in which we see that after de-interleaving, the six symbol errors are distributed equally among three different code words. In this case, therefore, only $T = 2$ symbol ECC is required to tolerate the same 14-bit burst.

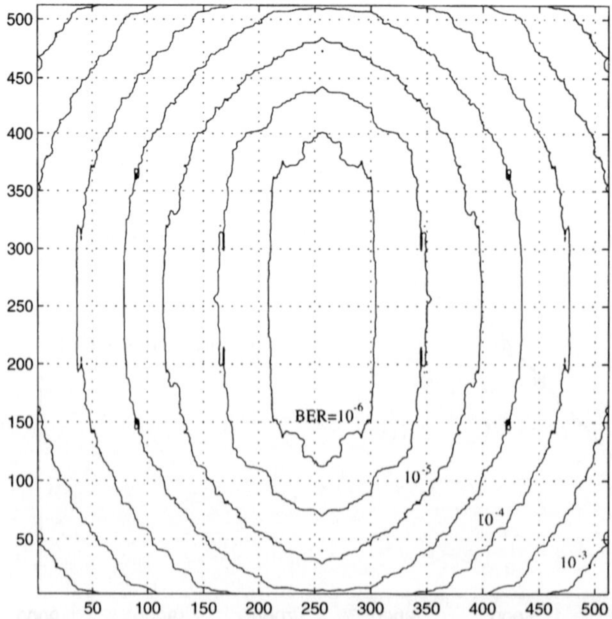

Fig. 4. Raw BER contours plot on a 512×512 pixel page given 0.10% magnification error, 10% shift error, and 4800 stored pages

The highly correlated BER pattern shown in Fig. 4 can be thought of as a mapping of spatial burst error probability in a two-dimensional communication channel. Within each contour region all pixels have similar raw BER, with pixels at the edge of a page more likely to be in error than those at the center. Following the same interleaving idea that was used for the serial channel, we can design a two-dimensional "matched interleaver" for spatial burst errors [18]. Using a RS(n, K, T) code we first divide the corrupted data page into n regions according to the raw BER contour plot, such that each region contains an equal number of pixels. Within each region, groups of m nearest-neighbor pixels are selected to form each symbol. Each code word is composed of one such symbol drawn from each of the n regions. This procedure can be thought of geometrically as arranging symbols "parallel" to the error correlations and arranging code words "orthogonal" to these correlations. The matched interleaver attempts to cause each bit in a symbol to have equal raw BER and each code word to have equal average symbol error probability. Using such a matched interleaver we expect the decoded BER to be lower on average and more uniform across the page.

Figure 6 shows the page-average output BER performance of matched and serial interleavers using RS $n = 31$ error correction codes as a function of magnification error. In this example we use the VHM system parameters from the previous section to store 4800 pages with $\gamma = 2.304 \times 10^8$. The

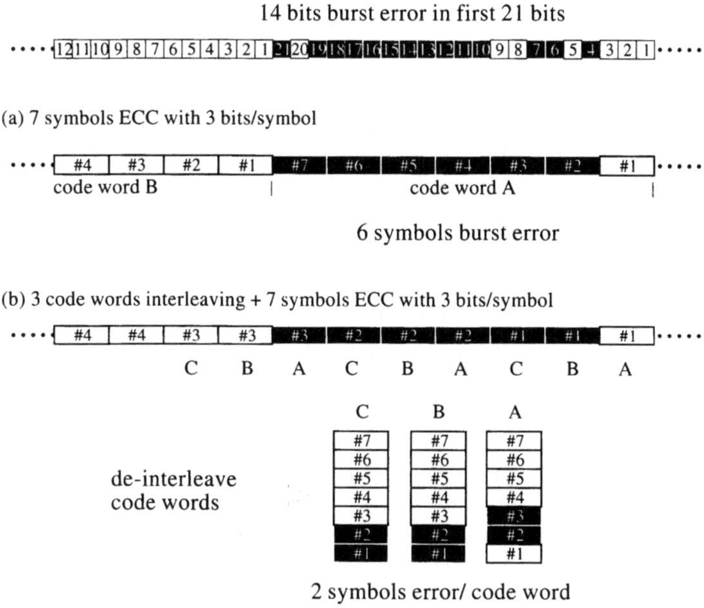

Fig. 5. Interleaving schematic for depicting a 14 bit burst error. ECC uses $n = 7$ symbols per code word with $m = 3$ bits/symbol: (**a**) serially transmitting code words; (**b**) interleaving with three code words

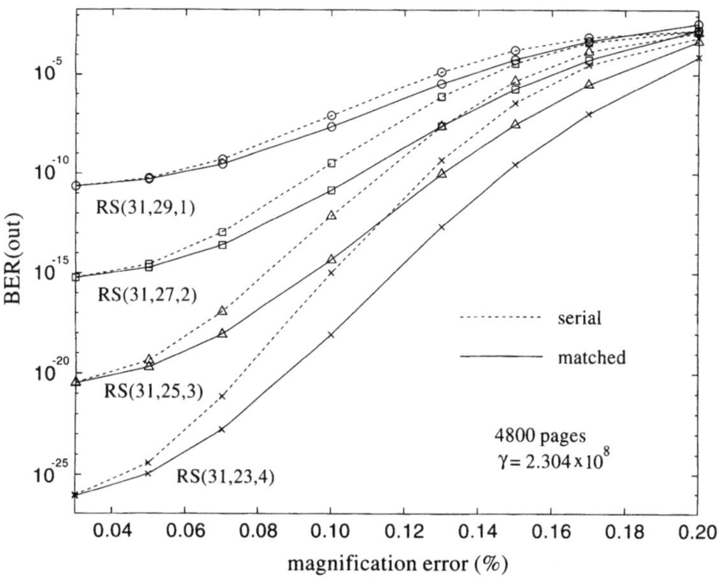

Fig. 6. Output bit error rate performance of serial (*dashed lines*) and matched (*solid lines*) interleavers for different magnification errors. A Reed–Solomon $n = 31$ code is used to store 4800 pages

solid curves depict matched interleaver performance, and the dashed curves represent the performance of serial interleavers (i.e. a simple rastering of the CCD data). Four curves are shown corresponding to $n = 31$ RS ECC with four different code rates. From this figure we can see that as magnification error increases, the average BER(out) of both matched and serial interleavers increases. For the same interleaver we observe that lower code rate (i.e. larger T) always yields lower average BER(out). We find that to store 4800 pages and simultaneously comply with the requirement of BER(out)=10^{-12}, we must use at least two-symbol error correction. When using an RS (31,23,4) code the system can tolerate up to 0.135% magnification error using matched interleaving as compared with 0.115% magnification error using a serial interleaver.

The BER gains depicted in Fig. 6 can once again be represented as gains in capacity. Figure 7 shows the storage capacity performance of using ECC and two-dimensional interleaving. This plot is based on 0.1% magnification error and a required BER(out)=10^{-12}. The uppermost solid curve is the information-theoretic capacity bound taken from Fig. 3. If no ECC is used then BER(out)=10^{-12} allows only 3471 pages (0.91 Gbits) to be stored in

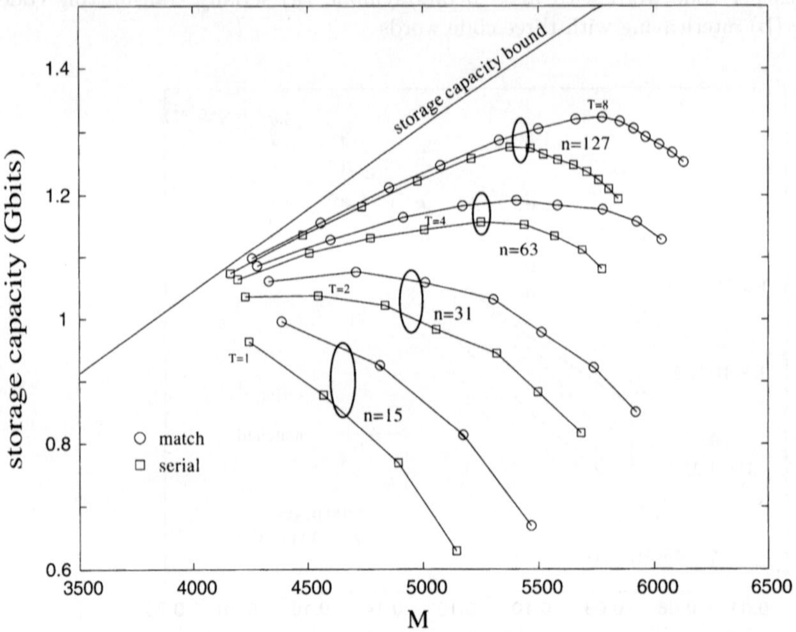

Fig. 7. Storage capacity of matched, serial, center, and mag-shift interleavers using various Reed–Solomon ECCs. This plot is based on 0.1% magnification error and a required output BER of 10^{-12}

the memory. Pairs of curves in Fig. 7 corresponding to serial and matched interleaver performance for four different code word length RS codes. Similar to Fig. 3, the leftmost point on any curve represents the $T = 1$ case. As T increases, the number of pages that can be stored in the memory increases while the ECC code rate (R_{ECC}) decreases. This tradeoff produces the capacity peaks shown in the figure. We can see that the use of matched interleaving always yields higher storage capacity than that of serial interleaving. The use of RS(127,111,8) ECC together with matched interleaving results in a 1.32 Gbit storage capacity, which is 68% of the information-theoretic bound.

4 Conclusions

The use of error correction within VHM systems can be viewed from two complementary perspectives. On the one hand, ECC involves the use of structured redundancy to improve the fidelity of retrieved data. The resulting BER reduction comes at the cost of some coding overhead in the form of non-user bits that must be stored in the VHM. On the other hand, the noise-limited information capacity of the VHM system can be achieved only through the use of advanced coding methods, and the use of RS ECC is but one technique that can be used to approach this capacity bound. In fact, both of these perspectives are accurate inasmuch as the use of ECC effectively increases the system BER requirement, thus facilitating the storage of additional holographic data pages. When capacity is defined in a fidelity-sensitive way (i.e. it is nonsense to specify a capacity without ensuring an acceptable BER), then these two perspectives become one and the same.

We have also discussed the incorporation of a certain type of prior knowledge within the error correction strategy of a VHM system. The various sources of systematic error within VHM produce correlated error patterns on and between retrieved data pages. Knowledge of these error patterns can be used to improve the performance of the ECC. In particular, we have outlined a method for designing a 2-D interleaving scheme based on measurements of the 2-D error correlations. Such a 2-D interleaver together with nonbinary ECC has been shown to achieve better capacity utilization than does the ECC alone.

References

1. J. Heanue, M. Bashaw and L. Hesselink, "Channel codes for digital holographic data storage," J. the Opt. Soc. Am. A, **12**, 2432–2439 (1995).
2. B. Olson and S. Esener, "Partial response precoding for parallel readout optical memories," Opt. Lett., **19**, 661–663 (1994).
3. B. Olson and S. Esener, "Multidimensional partial response for parallel readout optical memories," Proc. SPIE, **2297**, Optical Implementation of Information Processing, 331–344 (1994).

4. J. Heanue, K. Gurkan, and L. Hesselink, "Signal detection for page access optical memories with intersymbol interference," Appl. Opt., **35**, 2431–2438 (1996).
5. J.F. Hutton, G.A. Betzos, M. Schaffer, and P.A. Mitkas, "Error correcting codes for page-oriented optical memories," Proc. SPIE, **2848**, Materials, Devices, and Processing for Optoelectronic Processing, 146–156 (1996).
6. B.J. Goertzen and P.A. Mitkas, "Error-correcting code for volume holographic storage of a relational database," Opt. Lett., **20**, 1655–1657 (1995).
7. G.W. Burr, J. Ashley, H. Coufal, R.K. Grygier, J.A. Hoffnagle, C.M. Jefferson, and B. Marcus, "Modulation coding for pixel-matched holographic data storage," Opt. Lett., **22**, 639–641 (1997).
8. A. Vardy, M. Blaum, P.H. Siegel, and G. Sincerbox, "Conservative arrays: multidimensional modulation codes for holographic recording," *IEEE Transactions on Information Theory*, **42**, 227–229 (1996).
9. M.A. Neifeld and J.D. Hayes, "Parallel error correction for optical memories," J. Optical Memory and Neural Networks, **3**, 87–98 (1994).
10. M.A. Neifeld and M. McDonald, "Error correction for increasing the usable capacity of photorefractive memories," Opt. Lett., **19**, 1483–1485 (1994).
11. M.A. Neifeld and J.D. Hayes, "Error correction schemes for volume optical memories," Appl. Opt., **34**(35), 8183–8189 (1995).
12. K.M. Chugg, X. Chen, and M.A. Neifeld, "Two-dimensional linear MMSE equalization for page-oriented optical memories," *31st Annual Asliomar Conference on Signals, Systems, and Computers*, paper MP6-7, Pacific Grove, CA, November 2–5 (1997).
13. X. Chen, K.M. Chugg, and M.A. Neifeld, "Near-optimal parallel distributed data detection for page-oriented optical memories," J. Selected Topics in Quantum Electronics, **4**(5), 866–879 (1998).
14. K.M. Chugg, X. Chen, and M.A. Neifeld, "Two-dimensional equalization in coherent and incoherent page oriented optical memory," J. Opt. Soc. of Am. A, **16**(3), 549–562 (1999).
15. M.-P. Bernal, G.W. Burr, H. Coufal, R.K. Grygier, J.A. Hoffnagle, C.M. Jefferson, E. Oesterschulze, R.M. Shelby, G.T. Sincerbox, and M. Quintanilla, "Effects of multilevel phase masks on interpixel crosstalk in digital holographic storage," Appl. Opt., **36**, 3107–3115 (1997).
16. M.-P. Bernal, G.W. Burr, H. Coufal, R.K. Grygier, J.A. Hoffnagle, C.M. Jefferson, R.M. McFarlane, R.M. Shelby, G.T. Sincerbox, and G. Wittman, "Holographic data storage materials," MRS Bull., **21**, 51–60 (1996).
17. G.W. Burr, W.-C. Chou, M.A. Neifeld, H. Coufal, J.A. Hoffnagle, and C.M. Jefferson, "Experimental evaluation of optimal coded user capacity in holographic data storage systems," Appl. Opt., **37**(23), 5431–5443 (1998).
18. W.-C. Chou and M.A. Neifeld, "Interleaving and Error Correction in Volume Holographic Memory Systems," Appl. Opt., **37**, 6951–6968 (1998).
19. M.C. Bashaw, J.F. Heanue, and L. Hesselink, "Organization of data for monochromatic multiplexed volume holography," J. Opt. Soc. Am. A, **13**, 2174–2186 (1996).
20. C. Gu, J. Hong, I. McMichael, R. Saxena and F. Mok, "Cross-talk limited storage capacity of volume holographic memory," J. Opt. Soc. Am. A, **10**, 2547–2550 (1993).
21. X. Yi, P. Yeh and C. Gu, "Statistical analysis of cross-talk noise and storage capacity in volume holographic memory," Opt. Lett., **19**, 1580–1582 (1994).

22. E.S. Maniloff and K.M. Johnson, "Effects of scattering on the dynamics of holographic recording and erasure in photorefractive lithium niobate," J. Appl. Phys., **73**, 541–547 (1993).
23. D. Brady and D. Psaltis, "Control of volume holograms," J. Opt. Soc. Am. A, **9**, 1167–1182 (1992).
24. P.J. van Heerden, "A new optical method for storing and retrieving information," Appl. Opt., **2**, 387–392 (1963).
25. C. Gu, J. Hong, I. McMichael, R. Saxena, and F. Mok, "Cross-talk limited storage capacity of volume holographic memory," J. Opt. Soc. Am. A, **9**, 1978–1983 (1992).
26. K. Blotekjaer, "Limitations on holographic storage capacity of photochromic and photorefractive media," Appl. Opt., **18**, 57–67 (1979).
27. K. Rastani, "Storage capacity and cross-talk in angularly multiplexed holograms: two case studies," Appl. Opt., **32**, 3772–3778 (1993).
28. H.-Y.S. Li and D. Psaltis, "Alignment sensitivity of holographic three-dimensional disks," J. Opt. Soc. Am. A, **12**, 1902–1912 (1995).
29. C.E. Shannon, "Communication in the presence of noise," Proc. Inst. Radio Engineers, **37**, 10–21 (1949).
30. S.V. Miridonov, A.V. Khomenko, D. Tentori, and A.A. Kamshilin, "Information capacity of holograms in photorefractive crystals," Opt. Lett., **19**, 502–504 (1994).
31. M.A. Neifeld and W.-C. Chou, "Information theoretic limits to the capacity of volume holographic optical memory," Opt. Lett., **36**, 514–517 (1997).
32. T.M. Cover and J.A. Thomas, *Elements of Information Theory*, First Edition, Wiley and Sons, New York, NY (1991).
33. W.W. Peterson and E.J. Weldon, *Error Correcting Codes*, MIT Press, Cambridge, MA (1972).
34. E.R. Berlekamp, *Algebraic Coding Theory*, Mc-Graw Hill, New York, NY (1968).
35. S. Lin and D.J. Costello, *Error-Control Coding: Fundamentals and Applications*, Prentice-Hall, Englewood Cliffs, NJ (1983).
36. R.E. Blahut, *Theory and Practice of Error Control Codes*, Addison-Wesley, Reading, MA (1983).
37. R. Kotter, "A fast parallel implementation of a Berlekamp–Massey algorithm for algebraic-geometric codes," IEEE Trans. Inf. Theory, **44**, 1353–1368 (1998).
38. M.A. Neifeld and S. Sridharan, "Parallel error correction using spectral Reed–Solomon codes," J. Opt. Commun., **18**, 144–150 (1997).
39. J.G. Proakis, *Digital Communications*, third edition, McGraw-Hill, New York, NY (1995).
40. J.D. Roberts, A. Ryley, D.M. Jones, and D. Burke, "Analysis of error correction constraints in an optical disk," Appl. Opt., **35**, 3915–3924 (1996).

Equalization for Volume Holographic Data Storage Systems

B. V. K. Vijaya Kumar, V. Vadde, and M. Keskinoz

Equalization and detection are an integral part of any data storage system, and can potentially lead to significant improvements in storage density. Data detection refers to the method of converting retrieved analog values from the output detector array (referred to as a charge-coupled detector or CCD array in this chapter) to bits (i.e. 1s and 0s). The performance of any data detection method is affected by impairments such as noise and pixel spreading. A major goal of equalization is to process the CCD outputs in order to mitigate the effect of such impairments, and thereby produce improved bit error rates (BER).

Figure 1 illustrates the benefits of equalization. This figure shows BER before and after equalization as a function of the "best" percentage of a holographically imaged data page. The "best" refers to those portions of the data page that yield the smallest BER values. For a target BER of 5×10^{-3}, equalization allows an additional 35% of the data page to be used.

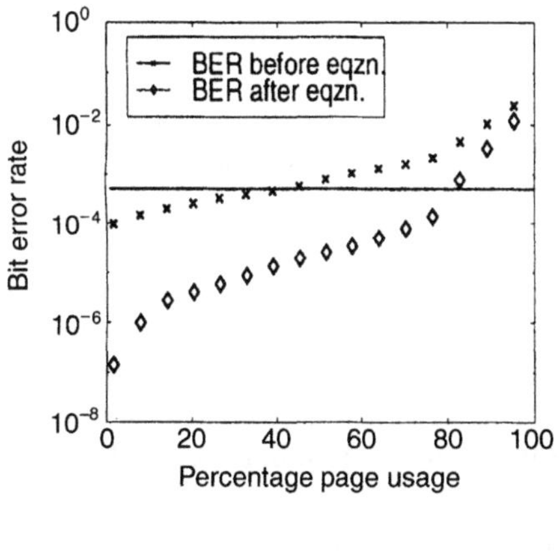

Fig. 1. Bit error rate (BER) as a function of the "best" percentage of the data page: ×, before equalization; ◊, after equalization

BER improvement obtained by applying a carefully designed equalization and detection method corresponds to a higher effective signal-to-noise ratio (SNR). An increase in effective SNR is equivalent to an increase in storage density. Equivalently, lower raw BERs can be obtained via equalization, when using the same fraction of a data page, and the smaller raw BERs ease the demands on the error-correcting codes (ECC) that lead to smaller ECC overhead. These storage density improvements through equalization come with virtually no data overhead, unlike improvements that come from the use of modulation and error-correction codes. In the following sections, we will discuss some promising equalization approaches.

1 Channel Modeling

To design and analyze effective equalization methods, we need a good storage channel model that captures the main factors affecting the BER. The BER in holographic data storage is significantly affected by the size of the aperture used in the frequency plane, noise, input spatial light modulator (SLM) contrast ratio, the fill-factors of the input SLM and the output CCD, and the mismatch between SLM and output CCD pixels. Good holographic storage system design requires pixel matching between the SLM and the CCD, and we will assume in this chapter that the pixels are indeed matched. As the contrast ratio decreases, the separation of the ON and OFF pixels decreases, making the task of data detection difficult. The SLM fill-factors appear to have a minor effect [1,2] on the BER in holographic storage. Higher CCD fill-factors are generally preferred as they lead to better SNR and potentially higher data rates (owing to the smaller integration times needed). Good noise models have been introduced by Gu et al. [3].

The aperture used in the frequency plane affects the intersymbol interference (ISI) as well as storage density. Smaller frequency plane apertures allow higher storage densities since the input data is stored in a smaller area (and hence smaller volume) of the recording medium. An aperture in the frequency plane acts as a spatial lowpass filter, thus blurring the pixel edges and causing interference between neighboring pixels. An aperture of size $W = \lambda f / \Delta x$, where Δx is the SLM pixel pitch, λ is the wavelength and f is the focal length, is known as the Nyquist aperture. At apertures smaller than the Nyquist aperture, the ISI dominates leading to a high BER. As we approach the Nyquist aperture, the BER improves owing to decreased ISI. Beyond the Nyquist aperture, greater ensquared energy in the detector plane helps decrease BER but the storage density decreases. A simple figure of merit ($\sqrt{\mathrm{SNR}}/d^2$) encapsulates the BER as well as density dependence on the aperture size, where d is the size of the aperture used in the frequency plane. It is found [1,2] that such a figure of merit is maximized when the aperture is close to the Nyquist aperture.

An accurate channel model must account for the light intensity detection due to the CCD. However, most equalization and detection approaches assume a linear superposition channel. Thus, we need a way to approximate a quadratic channel by a pseudo-linear channel model. The discrete spatial magnitude channel [4] provides such a model. In this model, the square root of the CCD output signal is approximated as a linear convolution of the input bit sequence and an effective discrete point spread function (DPSF).

Assuming a square aperture in the frequency plane, the DPSF is given by

$$h_M[k,l] = \int_{-\infty}^{+\infty}\int_{-\infty}^{+\infty}\left(\left\{\mathrm{rect}\left(\frac{x}{\delta x_1},\frac{y}{\delta y_1}\right)\cdot \mathrm{sin\,c}\left(\frac{xd}{\lambda f},\frac{yd}{\lambda f}\right)\right\}\right.$$
$$\left.\cdot\,\mathrm{rect}\left(\frac{x-k\Delta x_2}{\delta x_2},\frac{y-l\Delta y_2}{\delta y_2}\right)\right)\mathrm{d}x\,\mathrm{d}y, \quad (1)$$

where δx_1 and δy_1 are the SLM pixel dimensions, δx_2 and δy_2 are the CCD pixel dimensions, and Δx_2 and Δy_2 are the CCD pixel spacings. Here, $\mathrm{rect}(x,y)$ denotes a function that equals 1 for $|x| < 0.5$ and $|y| < 0.5$ and zero elsewhere, and $\mathrm{sinc}(x,y)$ denotes the function $[\sin(\pi x)/\pi x]\cdot[\sin(\pi y)/\pi y]$.

To illustrate the applicability of this magnitude channel model, we simulated the case with Nyquist aperture, an SLM fill-factor of 100%, a CCD fill-factor of 40%, and a contrast ratio of 3. For this case, pixel magnitudes (i.e. square root of the CCD output) were generated through a complete numerical simulation of the system. The DPSF $h_M[k,l]$ was determined using (1), and modeled pixel values were generated using linear superposition. Figure 2 (left half) shows simulated pixel magnitudes versus modeled pixels. Clearly, the DPSF linear superposition model agrees very well with the simulated noise-free pixel values.

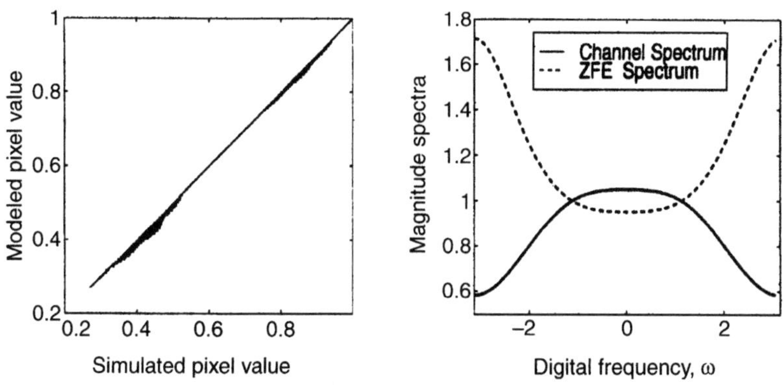

Fig. 2. Pixel values obtained from discrete channel model agree well with simulated pixel values (*left*), indicating the suitability of the model. Magnitude spectra of the discrete channel and the resulting ZFE filter (*right*)

2 Equalization Methods

Equalization can be done, in principle, by optical or electronic means. The insertion loss entailed in optical equalization is usually so prohibitive that it is impractical. Efforts toward equalization have focused mostly on electronic means [5–7]. Equalization is nothing but 2-D filtering of the CCD output array through a carefully designed equalization filter.

The high aggregate data rates usually targeted for holographic data storage systems necessitate parallel architectures for equalization and detection [8]. Another important aspect is that most imaging systems behave differently at the edges compared with their behaviour at the center. This means that different equalizers are needed for different parts of a data page [9]. We can address this spatially varying nature of the ISI by dividing a large data page into many smaller data blocks and designing a separate equalizer for each of these blocks.

A data detector processes the equalized CCD outputs to yield binary information. In the next two subsections we will focus on zero-forcing equalization (ZFE) and linear minimum mean square error (LMMSE) equalization. Both these equalization methods try to remove ISI, and thus simple threshold detection methods that look only at individual equalized pixels can be used. In Sect. 2.3, we will introduce partial response (PR) equalization, which allows a controlled amount of ISI. Since the ISI due to PR equalization does not go to zero, more sophisticated maximum likelihood sequence detection methods that look at more than one equalized pixel at a time must be employed. This cascading of PR equalization followed by ML detection is usually known as the PRML method.

2.1 Zero Forcing Equalization

In ZFE, the idea is to eliminate all ISI. The detector can then be a simple threshold detector. If the equalized pixel value exceeds a preselected threshold, it is decoded as an ON pixel, otherwise as an OFF pixel. Since the ZFE tries to eliminate all ISI, the equalizer transfer function is essentially the inverse of the discrete channel spectrum. Thus ZFE is feasible only if the discrete channel spectrum does not exhibit spectral nulls. Figure 2 (right half) shows a central cross-section of the discrete spatial channel spectrum for the Nyquist aperture case, a 100% fill-factor on SLM and 40% fill-factor on CCD. One can see that there are no spectral nulls, and thus ZFE can be used. The magnitude spectrum of the ZFE filter is also shown in Fig. 2. In Fig. 2, we show cross-sections of 2-D spectra. It is convenient to note that the 2-D DPSF (see (2)) has symmetry and separability. Owing to its separable nature, one can express the 2-D DPSF as an outer product of two 1-D DPSFs.

Thus, designing a ZFE requires knowledge of the frequency response of the discrete channel. This can be obtained in one of two ways. One approach is to derive the channel frequency response from first principles and detailed modeling followed by suitable approximation of the physical system. This is not always easy and it is not always reliable, as the system properties may be somewhat dynamic and unpredictable. A second method is to record reference bit patterns, retrieve the corresponding CCD outputs, and identify the discrete channel from the knowledge of that input–output pair.

Since the ZF equalization is nothing but 2-D filtering, it can be implemented as a convolution of the CCD output array with a 2-D ZF equalization kernel. For complexity reasons, this equalization kernel is usually limited to a 3×3 or 5×5 array. If this equalization kernel (sometimes referred to as the deconvolution mask) can be constrained to be separable, we may be able to achieve some complexity reduction.

2.2 LMMSE Equalization

The principal disadvantage of using ZFE is the noise amplification associated with it. As can be seen from Fig. 2 (*right half*), in the process of attempting to remove ISI completely, a ZFE filter ends up amplifying high spatial frequencies and thus boosting noise. To reduce this SNR penalty, a minimum mean square error (MMSE) filter (also known as the Wiener filter) can be used to strike a balance between ISI and noise. The objective of an MMSE filter is to reduce the overall mean square error (i.e. the distortion in the ideal signal due to noise as well as ISI). Thus, the MMSE filter attempts to minimize the error between the ideal and equalized data values. A particularly convenient MMSE equalizer is the one that processes the CCD outputs linearly. Such a linear MMSE (LMMSE) equalizer can be implemented as a convolution of the CCD outputs with a small sized (e.g. 3×3 or 5×5) deconvolution mask. Let I_{kl} denote the CCD output from the (k, l) pixel and w_{jm} denote the coefficients of the $(2N+1) \times (2N+1)$ equalizer. Then the output of any linear equalizer can be expressed as follows:

$$\hat{d}_{kl} = \sum_{i=-N+1}^{N-1} \sum_{m=-N+1}^{N-1} w_{jm} I_{(k-j)(l-m)}. \tag{2}$$

LMMSE equalizer coefficients w_{jm} are selected to minimize the average squared error between the actual equalized outputs $\hat{d}_{kl}$ and ideal outputs d_{kl}. It can be shown [7] that the resulting LMMSE coefficients w_{jm} must satisfy the following equation, which relates the autocorrelation $R_{II}(m, j)$ of the CCD outputs to the cross-correlation $R_{dI}(m, j)$ between the ideal input and the actual CCD output:

$$R_{dI}(m, j) = w_{mj} * R_{II}(m, j). \tag{3}$$

Here the asterisk symbol denotes 2-D convolution. From (3), we see that we need to estimate the autocorrelation and cross-correlation functions to solve for the LMMSE weights by either frequency-domain methods or linear equation solution methods. Towards this end, we store a known pattern of bits in every data block and use the corresponding CCD outputs to obtain the needed autocorrelation and cross-correlation estimates.

The LMMSE method does not require modeling of the physics of the system. Instead, it relies on observing the output for a known input pattern. As a result, we can design a different equalization kernel for each data block and thus handle the dynamic and space-varying nature of the holographic storage channel. This method was tested [7] with deliberately defocused images from a holographic imaging system. A few observations from that exercise are worth mentioning here. The best BER values were obtained using no more than 3% of the data in a block to determine the LMMSE coefficients. Also, increasing the mask size from 3×3 to 5×5 did not improve the BER results. Finally, using LMMSE resulted in many more data blocks meeting the goal of an average BER of 10^{-3} than was possible with the unequalized case.

2.3 Partial Response (PR) Equalization

The ZFE and LMMSE equalization described before attempt to either eliminate ISI completely or minimize the mean squared error due to ISI and noise. As we employ sub-Nyquist apertures in order to obtain higher storage densities, the extent of the pixel spreading increases, leading to severe ISI. The two equalization methods discussed so far may not be able to alleviate the effects of such severe ISI in the presence of noise. In such cases, more advanced equalization methods such as partial response (PR) equalization coupled with decision feedback (DF) techniques will have to be employed to keep BER within acceptable limits.

In PR equalization, the equalizer does not try to remove all ISI. Instead, a PR equalizer is designed so that a controlled (and hence known) amount of ISI exists between neighboring pixels. This knowledge of the controlled ISI pattern is used in a maximum likelihood (ML) detector that looks at groups of CCD outputs before detecting any single pixel. This combination of partial response equalization followed by maximum likelihood detection is known as the PRML approach. Note that the PR approach we are discussing here deals with the digital post-processing of the CCD output, and is very different from the precoding method introduced by Olson and Esener [10].

The longer the extent of this controlled ISI pattern (also known as partial response) is, the more complicated the task of the ML detector is to unravel the original data pattern. Since the ISI extends in two dimensions, one approach to simplify the ML detector's task is to eliminate ISI in one of the dimensions and apply PRML methods in the other dimension. The former objective can be accomplished through a ZFE or by decision feedback. When decision feedback is used [11], the ISI from already detected rows (or columns) is subtracted off, with the resulting residual ISI being one-dimensional. This 1-D ISI is then reshaped into a controlled ISI pattern through PRML approach. The key assumption here is that the previously detected rows (or columns) were correctly detected and that decision feedback results in the subtraction of correct ISI. In the event that this assumption does not hold, error propagation occurs. The efficacy of this method therefore relies on making accurate decisions in the beginning stages of detection.

We find from our numerical simulations that the most promising PR target for the Nyquist aperture case is of the form $(1 + \alpha D)$, where D denotes a unit pixel shift and α is a parameter to be determined. In general, a PR target is chosen such that the misequalization error is as low as possible. The 2-state ML detector structure (also known as a Viterbi decoder) resulting from such a PR target is computationally attractive. In the next section, we will present some equalization results obtained with ZFE, MMSE and PRML methods.

3 Equalization Results

For linear equalization with the ZFE and MMSE types of filtering, we consider the Nyquist aperture with 100% SLM fill-factor and 40% CCD fill-factor. The amplitude contrast ratio is fixed at 3. The signals are generated from a pseudo-random input data pattern with ISI generated from the 12 most interfering pixels. Rician optical noise as well as Gaussian detector noise are synthesized and added to the signal. We assume in our analysis that the imaging is devoid of any aberrations. The discrete PSF is generated following the magnitude model of (1), and can be approximated by a 5×5 kernel.

Figure 3 (left half) shows the BER obtained as a function of the SNR for the unequalized case as well as with ZF and MMSE equalization. SNR is defined as the ratio of the average ideal signal energy to the average energy in the error (introduced due to ISI and noise). It is clear that both types of equalization lead to improved BER values. We find that ZFE and MMSE differ only marginally in performance. For a target BER of 10^{-3}, we observe an SNR gain of about 3 dB by virtue of equalization. Using the rule of thumb

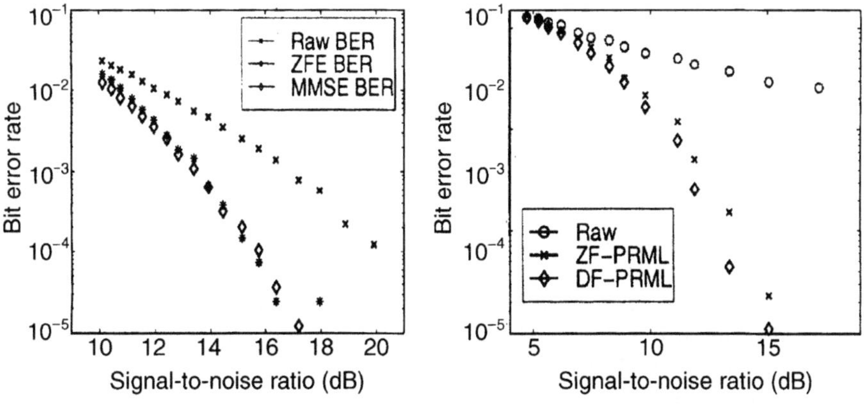

Fig. 3. BER as a function of SNR for the Nyquist aperture case (*left*): $\times$, unequalized case; *, ZFE; as well as $\diamond$, MMSE. BER as a function of SNR for the sub-Nyquist aperture case (*right*): $\circ$, unequalized case; $\times$, ZFPRML; and $\diamond$, DF-PRML

that SNR degrades as $1/N^2$, where N is the number of stored pages, increasing the SNR by 3 dB is approximately equivalent to increasing the number of pages by 40%. For partial response equalization testing, we consider an aperture with dimensions of 0.85 of the Nyquist aperture in one dimension and the Nyquist aperture in the other dimension. The resulting ISI in the channel is approximately 5×7 pixels in extent. In Fig. 3 (right half), we show the BER resulting from ZF-PRML (where the ISI in one direction is eliminated completely using ZFE) as well as DF-PRML (where the ISI from already detected rows is subtracted off) methods for equalization followed by detection. The DF-PRML approach appears to yield slightly better BER, especially at high SNRs. The DF-PRML approach is capable of yielding over 40% density gain with the rectangular, sub-Nyquist aperture.

4 Implementation Issues

In previous sections, we discussed three equalization strategies and showed some illustrative results. But how do we use these equalization methods in a holographic storage system? The answers will depend on the system details. However, we can make some general observations.

Because of its local nature (e.g. convolution with a 3×3 or 5×5 array), equalization lends itself nicely to parallel implementation. Also, a different equalization kernel can be used for each data block, thus allowing for spatially varying ISI. These different kernels can be stored in memory and used as needed. Since the ISI can change not only spatially, but also in time, we need adaptive methods that can adjust the equalization kernel coefficients on a regular basis. Efficient training algorithms (e.g. least mean squares or LMS algorithm) exist for this purpose.

Another implementation issue is the precision with which the equalization operation in (2) needs to be carried out. Simulations (using system parameters suggested before) indicate that using only 4 bits of precision degrades BER significantly, whereas using 8 bits seems to result in no appreciable BER degradation. Thus these equalizers may be implemented using 7 or 8 bits of precision.

5 Summary

In this chapter, we introduced the advantages of equalization. The main role of equalization is to suppress the effects of ISI without worsening the effects of noise too much. We introduced three different equalization strategies: ZFE, which tries to eliminate ISI completely: LMMSE, which tries to minimize the mean squared error using a linear filter; and PRML, which that tries to control the ISI into a simple shape. We also showed results that illustrate the BER improvements due to the use of equalization. These BER improvements can be viewed as effective SNR gains that can in turn be interpreted as allowing increased number of stored data pages. Thus equalization is a key step in any attempt to increase the density of volume holographic storage.

References

1. V. Vadde, B.V.K. Vijaya Kumar, G. W. Burr, H. Coufal, J.A. Hoffnagle, and C.M. Jefferson, "A figure of merit for the optical aperture used in digital volume holographic data storage," *Proc. SPIE Optical Data Storage Meeting '98* **3401**, 194-200 (1998).
2. M. P-Bernal, G.W. Burr, H. Coufal, and M. Quintanilla, "Balancing interpixel crosstalk and detector noise to optimize areal density in holographic storage systems," *Appl. Opt.*, **37**, 5377–5385 (1998).
3. C. Gu, G. Sornat, and J. Hong, "BER statistics of complex amplitude noise in holographic data storage," Opt. Lett., **21**, 1070–1072 (1996).
4. V. Vadde and B.V.K. Vijaya Kumar, "Channel modeling & estimation for intra-page equalization in pixel-matched volume holographic data storage," Appl. Opt., **38**, 4374–4386 (1999).
5. K.M. Chugg, X. Chen and M.A. Neifeld, "Two-dimensional equalization in coherent and incoherent page oriented optical memory," JOSA-A, **16**, 549–562 (1999).
6. X. Chen, K.M. Chugg, and M.A. Neifeld, "Near-optimal page detection for two-dimensional ISI/AWGN channels using concatenated modeling and iterative detection," Proc. ICC 1998, Paper # S27P4, June 1998.
7. M. Keskinoz and B.V.K. Vijaya Kumar, "Application of linear minimum mean squared error equalization for holographic data storage," Appl. Opt., **38**, 4387–4393 (1999).
8. M.A. Neifeld, K.M. Chugg, and B.M. King, "Parallel data detection in page-oriented optical memory," *Opt. Lett.*, **21**, 1481–1483 (1996).
9. V. Vadde and B.V.K. Vijaya Kumar, "Channel estimation and intra-page equalization for digital volume holographic data storage," Proc. SPIE, Optical Data Storage Meeting '97, **3109**, 250–255 (1997).
10. B. Olson and S.C. Esener, "Partial response precoding for parallel-readout optical memories," Opt. Lett., **19**, 661–3 (1994).
11. J.F. Heanue, K. Gurkan, and L. Hesselink, "Signal detection for page-access optical memories with intersymbol interference," Appl. Opt., **35**, 2431–2438 (1996).

Gray-Scale Data Pages for Digital Holographic Data Storage

G.W. Burr and M.A. Neifeld

Many of the attractive features of holographic data storage – such as high capacity, high density, and fast readout – are due to the use of large pixellated data pages, which are recorded into a photosensitive material as volume holograms [1–3]. It is often claimed that each hologram can store as many bits of digital data as there are pixels in the data page. To this end, most of the work performed in the field has used binary encoding, where pixels take one of two distinct states – ON(bright) and OFF (dark), corresponding to binary 1 and 0 – to encode one bit of information. Preceding chapters have shown that the coding introduced to maintain acceptable bit-error-rate (BER) comes with an unavoidable overhead cost, resulting in somewhat less than one bit per pixel. In this chapter, we discuss the use of gray-scale: recording and detecting more than two brightness levels per pixel, leading to *more* than one bit of data per pixel.

1 Motivation for Gray-Scale

If pixels take one of g brightness levels, then each pixel can now convey $\log_2 g$ bits of data. The total amount of stored information per page has increased, so gray-scale encoding would seem to straightforwardly improve both capacity and readout rate. However, gray-scale also divides the system's signal-to-noise ratio (SNR) into $g-1$ parts, one for each transition between brightness levels. Because total SNR depends on the number of holograms, dividing the SNR for gray-scale (while requiring the same error rate) requires reducing the number of holograms that can be stored. In general, finite dynamic range in the holographic storage material causes diffraction efficiency to decrease as $1/M^2$, where M is the number of stored holograms [4]. The presence of a constant noise floor means that total SNR follows diffraction efficiency in this $1/M^2$ dependence. To use more gray levels yet maintain the same SNR *per* transition between brightness levels, the number of holograms must be reduced by the factor $\sqrt{g-1}$. Total storage capacity with gray-scale data pages is the product of two factors that depend on g: the bits per pixel ($\log_2 g$), and the number of holograms, which falls as $1/\sqrt{g-1}$. We show the resulting relationship between capacity and number of gray levels as the thick solid line in Fig. 1. At the optimum value of $g = 5$, the capacity is increased by 15% over binary encoding. Readout rate (user bits/second) is

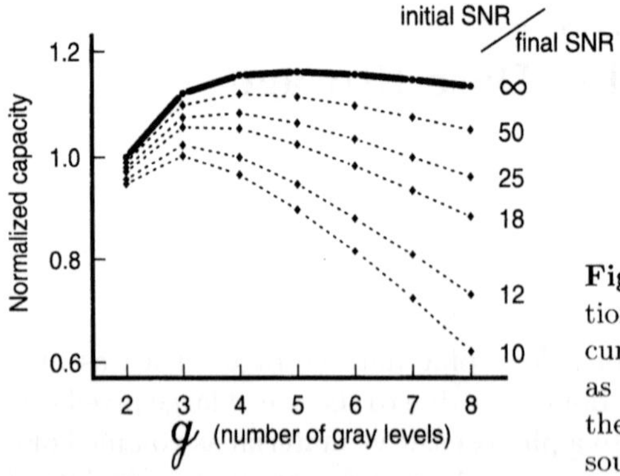

Fig. 1. Normalized capacity as a function of gray level (g). For the solid curve, the SNR for M holograms drops as $1/M^2$. The dashed curves indicate the impact as signal-dependent noise sources become more important [5]

unaffected by the SNR division, and increases over binary encoding by a factor of $\log_2 5 \sim 2.32$.

In addition to a constant noise floor, however, practical holographic storage systems exhibit signal-dependent noise sources. These include any effect in which some of the signal energy from one pixel in the holographic memory ends up reaching a different pixel, either elsewhere on the same data page or perhaps on a different data page. Examples include aberrations, magnification errors, inter-pixel and inter-page cross-talk. Other signal-dependent noise sources include variations in signal strength across the detector array or in time as holograms are recorded and read out. Since all of these noise sources increase with increasing signal, they tend to limit the SNR that can be achieved even when signal levels are far above the constant noise floor (for instance, when only a few, high diffraction efficiency holograms have been stored). More importantly, the presence of these noise sources changes the scaling between SNR and the number of holograms [5,7], favoring a smaller number of gray levels and reducing the gray-scale capacity gain.

In Fig. 1, the influence of the signal-dependent noise sources is shown by the dashed curves. These curves are plotted for various ratios between initial SNR (the SNR with one stored hologram) and final SNR (the SNR at which no further holograms can be added). Initial SNR represents the system performance in the absence of the constant noise floor. Final SNR is dictated by the SNR–to–raw-BER performance of the detection process combined with the raw-BER–to–user-BER performance of the error-correction coding (ECC) [6,7]. Typically, systems must have high-performance optics (to reduce aberrations and maintain focus and magnification) for the initial SNR to exceed the final SNR by more than ~15, even when SNR is measured locally (as is appropriate for any local decoding/decision method). Thus, Fig. 1 indicates that three gray levels ($g = 3$) will most likely be the optimal choice for a practical system.

More elaborate analyses could be performed by considering the raw bit-error-rate (raw BER) seen by the ECC, or by optimizing the level separations or the probability of occurrence of each gray level. However, Fig. 1 serves its purpose of justifying an experimental implementation of gray-level holographic storage, and of emphasizing the importance of high initial SNR (that is, the importance of minimizing signal-dependent noise sources).

We note that gray-scale encoding was briefly analyzed, and reported to be non-advantageous, in a study of the effects of the Rician noise arising from optical scatter [8]. However, very recent work [9] suggests that the scatter noise in holographic storage systems does not follow pure Rician statistics. Instead, because of correlation (or speckle) domains within the scatter noise at the detector array, the variance of the noise is reduced (when compared with Rician statistics). Our discussion to this point has implicitly assumed the background noise to be Gaussian, and the experimental work we show below will be essentially free of scatter noise, so the use of gray-scale in the presence of optical scatter must be considered an open question.

2 Predistortion

Figure 1 shows that signal-dependent noise in the system tends to reduce the capacity improvement offered by gray-scale data pages, with the amount of signal-dependent noise quantified by the initial SNR. A recently developed technique for improving initial SNR is the 'predistortion' technique [10]. Predistortion works by individually manipulating the recording exposure of each pixel on the spatial light modulator (SLM), either through control of exposure time or by relative pixel transmission (analog brightness level on the SLM). Deterministic variations among the ON pixels, such as those created by fixed pattern noise, non-uniformity in the illuminated object beam, and even inter-pixel crosstalk, can be suppressed (increasing the initial SNR). Many of the spatial variations to be removed are present in an image transmitted with low power from the SLM directly to the detector array. Once the particular pattern of non-uniform brightness levels is obtained, the recording exposure for each pixel is simply calculated from the ratio between its current brightness value and the desired pixel brightness [10].

At low areal density, BER improvements of more than 15 orders of magnitude are possible [10]. More importantly, at high density, inter-pixel crosstalk (which is deterministic once the data page is encoded) can be suppressed and BER improved from 10^{-4} to 10^{-12} [10]. Figure 2 shows this experimental result, implemented on the IBM DEMON holographic storage platform [12] with a square aperture of 2.8×2.8 mm^2 placed at the Fourier transform plane of the imaging optics. Another use of the predistortion technique is to increase the contrast between ON and OFF pixel states provided by the SLM. By using interferometric subtraction while recording the hologram, the amount of light received at the OFF detector pixels can be reduced [10].

To create gray-scale data pages, the predistortion technique was modified by simply defining a set of $g-1$ desired pixel brightness levels (instead of

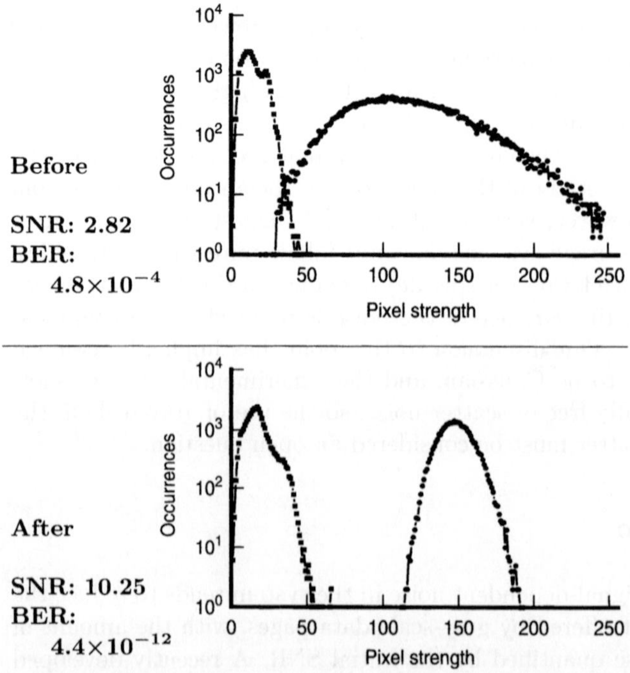

Fig. 2. Brightness histogram for high–areal–density holograms before and after applying the predistortion technique. Before, inter-pixel crosstalk broadens the brightness distributions; after, these deterministic variations are reduced, improving the initial SNR of the system [10]

a single target brightness for the binary pages). The presence of a residual amount of inter-symbol interference necessitated the use of several iterations, using practice holograms, to optimize the individual pixel exposures. (The use of an analog SLM makes it possible to optimize the compensation of each pixel using images instead of stored holograms.) The histogram of a six-level hologram made possible by the predistortion technique is shown in Fig. 3. The SNR of this histogram was 6.39, where the SNR for g levels is defined as

$$SNR = \frac{1}{g-1} \sum_{i=0}^{g-2} \frac{\mu_{i+1} - \mu_i}{\sqrt{\sigma_{i+1}^2 + \sigma_i^2}}, \qquad (1)$$

and μ_i and σ_i^2 are the mean and variance, respectively, of the ith gray level.

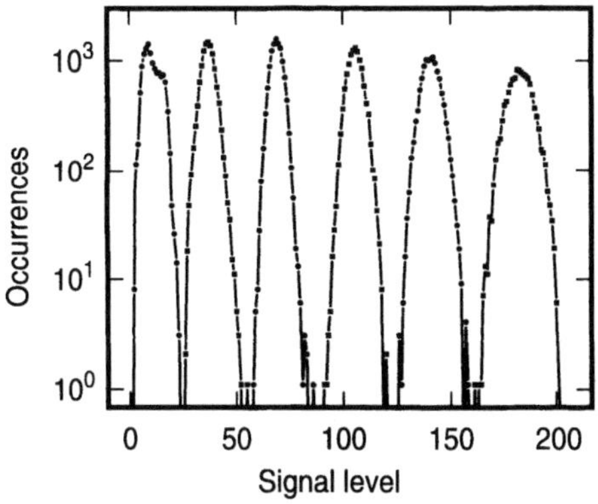

Fig. 3. Histogram of a six-level hologram recorded in the IBM DEMON system using the 'predistortion' technique [5,10]

3 Encoding Digital Data into Gray-Scale Pixels

In order to obtain measurements of the raw BER of gray-level data pages with practical decoding methods, we developed several local-thresholding methods and balanced modulation codes. Three thresholding techniques-termed min–max, multiple parity, and single parity – were implemented. These performed local thresholding over a small block of pixels, and differed only in the approach used to obtain the threshold (and the resulting 'code rate').

The min–max local thresholding method operated on a small block of pixels, sacrificing one pixel from each block in order to guarantee at least one occurrence of both the darkest (zero) and brightest $(g-1)$ level [5]. (The ECC is assumed to cover any errors resulting from the occasional absence of both of these levels.) Upon decoding, the $g-1$ thresholds needed to separate the g brightness levels are derived from the minimum and maximum values within the block, and look-up information about the level separations. We chose to encode over blocks of size 4×4 pixels, to decode over blocks of 8×8 pixels, and to use uniformly-spaced equiprobable gray levels. We anticipate that the performance upon optimizing these choices would be slightly improved. On the other hand, we have not considered how to efficiently encode binary user data into gray-scale pixels, and we expect that this conversion process will reduce the effective code rate (except in the case of four gray levels).

Two other thresholding methods were devised but did not outperform min–max, so we describe them only briefly. In multiple parity, a small section of the data page was assumed to be set aside to carry the number of occurrences of each gray level, for each block on the page. Upon decoding of each block, this information was used along with the total sum of received intensity and look-up information on the level separations to calculate thresholds. In single parity, only the total sum of grey levels was passed as overhead ($n_1 + 2\,n_2 + 3\,n_3 + \ldots$, where n_i is the number of occurrences of the ith grey level). In determining the BER when decoding with either parity method, the overhead information can be assumed to be retrieved without error without seriously affecting the overall BER performance [6].

In addition to these thresholding schemes, several multi-level block-based modulation codes were developed [5]. For $g = 3$, a 15:12 code was implemented (15 bits of user data decoded from 12 detector pixels). The balance constraint was that four of these 12 pixels were always of the '0' level, four were level '1', and four were '2' pixels. Decoding was implemented by sorting the 12 received signal values, assigning the brightest four to level '2,' the next four to level '1,' and the remainder to level '0.'

The complexity involved in even this three-level modulation code led us to implement the remaining gray-level codes by multiple application of binary modulation codes. For instance, a 24:16 decoder for four gray levels was implemented by applying a 6:8 decoder for binary data [12] twice. Each pass divides the histogram into two parts: the first pass separates the class '0'/'1' from the class '2'/'3', while the second pass divides again, resulting in the assignment of pixels to four separate brightness classes. This code is also balanced, so each class has four representatives among the 16 encoded pixels, Note that the 6:8 decoder is used four times on each block of 16 pixels, resulting in $4 \times 6 = 24$ bits of data.

Similarly, for $g = 6$ gray levels, a 48:24 code was created by using the 15:12 decoder (to divide the 24 pixels into three classes: '0/1', '2/3', and '4/5') followed by the 6:8 decoder (to complete the classification into six states). By omitting one of the virtual 6:8 blocks, this was modified into a 42:24 code for $g = 5$. Note that since these codes are built to encode from binary data, there is no loss of efficiency in encoding directly from user data. These modulation codes, and the min–max, multiple parity, and single parity thresholding methods are summarized in Table 1.

Table 1. Code rates (in bits per pixel) for gray-scale codes

Code	Code rate (r_{detect}) for g gray levels $g = 2$	$g = 3$	$g = 4$	$g = 5$	$g = 6$
modulation code	--	1.25	1.5	1.75	2.0
min–max	0.94	1.49	1.88	2.18	2.42
multiple parity	0.84	1.15	1.28	1.33	1.33
single parity	0.84	1.3	1.6	1.86	2.02
$\log_2 g$	1	1.58	2	2.32	2.58

4 Capacity Estimation

The storage capacity with these decoding methods was evaluated using an experimental procedure for capacity estimation [6]. This procedure measures the dependence of raw BER on initial recording exposure, and then uses this data to predict the number of holograms that can be recorded. Holograms were recorded using $LiNbO_3$ (0.02% Fe-doping) in the 90° geometry at 514 nm, using the DEMON tester [12]. Other experimental conditions are described in [6]. After performing the predistortion iteration procedure, three gray-scale holograms were recorded to obtain a reliable measure of raw BER and hologram strength versus recording time. Then the raw BER of a single hologram was measured periodically during erasure. All holograms were angularly separated by > 20 Bragg nulls.

This measured data provides a connection between a given desired raw BER and the recording exposure required to ensure that the holograms remain below this raw BER target even after all subsequent recording exposures. The relationships between exposure time and measured raw BER can then be used to derive a recording schedule [4] which strives to equalize raw BER instead of diffraction efficiency [6]. The number of holograms that can be stored with such a recording schedule is uniquely determined by four measured variables: the initial recording exposure, final recording exposure, erasure duration, and erasure time constant.

The experimental capacity-estimation technique quantifies the relationship between the number of holograms that can be stored, M, and raw BER. Figure 4 shows the results of the capacity estimation procedure for the min–max thresholding algorithm for gray levels from $g = 2$ to $g = 6$. In general, as the raw BER of the system increases, M increases slowly. For instance, with min–max decoding of binary data pages, Fig. 4 shows that each factor of 10 in raw BER allows only 5.5% more holograms to be stored. As we mentioned earlier, the min–max algorithm generally outperformed the other two thresholding techniques, with single parity coming in a close second and multiple parity farther behind. Figure 5 shows how the min–max algorithm compares with the performance of the other decoding strategies for three gray levels ($g = 3$). The parity methods performed best at low values of g, because as g increases their code rates increase more slowly than $\log_2 g$. The modulation codes always outperformed the thresholding methods in the low-raw BER region; however, as we will see in the next section, the performance at a raw BER of $\sim 10^{-3}$ is the critical region.

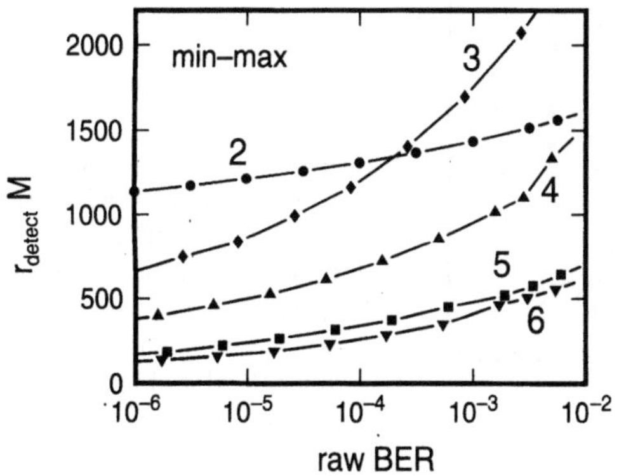

Fig. 4. Number of holograms that can be stored vs. raw BER with the min–max thresholding algorithm, as predicted by the experimental capacity-estimation procedure [5,6]

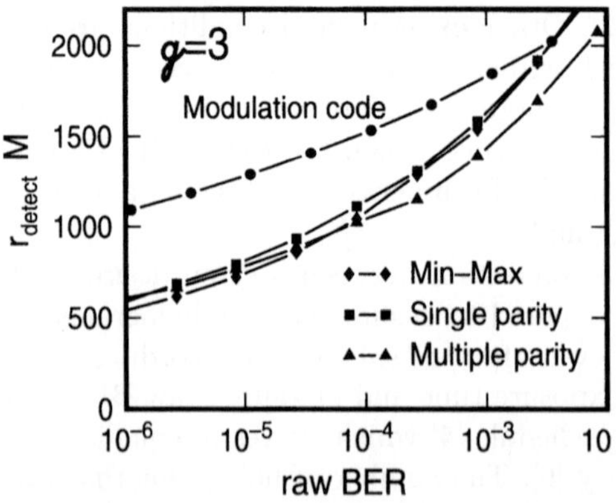

Fig. 5. Number of holograms that can be stored vs. raw BER for three gray levels with various decoding techniques

5 Optimizing the Error–Correction Coding to Obtain User Capacity

In order to maintain a low output user-BER (say, 10^{-12}) as the the raw BER operating point increases, the redundancy of the ECC code must increase. So while the number of holograms increases, the ECC code rate decreases. These two opposing trends create an 'optimal' raw BER, at which the user capacity is maximized [7]. To compare the different decoders fairly, we have taken care to convert all measured error data into a consistent raw BER representative of the BER that would be seen by the ECC decoder. This is necessary, for instance, when a modulation code outputs raw data in blocks (or 'symbols') of 6 bits while the ECC code corrects raw data using symbols of 8 bits.

Figure 6 shows a parametric plot of capacity versus readout rate for the gray-level modulation codes and the min–max algorithm, over the range $g = 2$ to $g = 6$. In order to concentrate on the relative comparison between gray-

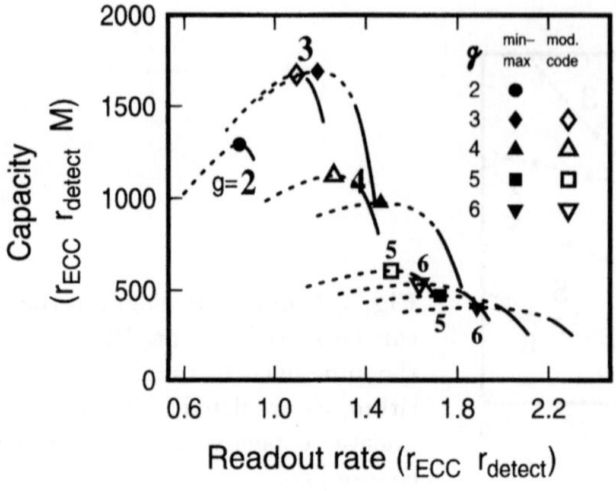

Fig. 6. Capacity vs. readout rate for gray-level data pages using modulation coding and min–max local thresholding. Each curve progresses left to right from stronger to weaker Reed–Solomon ECC codes; the symbol indicates the maximum capacity point

levels and decoding methods, we have not scaled by the number of pixels per page nor by the number of pages per second. Each data point indicates the maximal capacity point; the curve through it indicates the effects of varying the ECC code. The ECC for Fig. 6 utilized t-symbol-correcting Reed–Solomon codes with a block length of 255 eight-bit symbols, using $2t$ symbols of redundancy to achieve a corrected user-BER specification of 10^{-12}. Each curve in Fig. 6 indicates the same range of ECC choices, from $t = 47$ (raw BER $\sim 10^{-2}$) on the left to $t = 4$ (raw BER $\sim 3 \times 10^{-5}$) on the right. Note that distance along the curves is not linear in t. For perspective, we have indicated the point $t = 16$ (raw BER $\sim 10^{-3}$) by changing the curve from dotted to solid. Roughly speaking, the solid line indicates the range of high-speed ECC chips available 'off the shelf'; the dotted line indicates where the complexity of the ECC decoder could have a negative effect on throughput.

6 Summary

We have shown that gray-scale holographic data pages provide an advantage over binary encoding in both capacity and readout rate. A simple SNR partitioning argument shows that when SNR scales as 1 over the number of holograms squared, five gray levels ($\log_2 5$ bits/pixel) would be expected to result in a 15% capacity increase over binary data pages. However, the additional signal-dependent noise sources present in practical systems create an initial (or 'baseline') SNR that reduces both the optimal number of gray levels and the resulting gain in capacity. Under such conditions (in the presence of both signal-dependent noise and a constant noise floor), both the simple theory and our experimental results show that three gray levels produces the highest capacity. However, the SNR model is only accurate in indicating the general trends, making the experimental capacity-estimation procedure [6] invaluable in quantifying the tradeoff between bits-per-pixel and number of stored holograms.

The 'predistortion' technique previously developed for binary page-oriented memories [10] was adapted to record gray-scale data pages with up to six distinct brightness levels, and several new block-based modulation codes for decoding gray-scale data pages were introduced. Experimental results of the capacity-estimation procedure, using data taken with $LiNbO_3$:Fe in the 90° geometry, were shown. Figure 6 shows that three gray levels provides a 30% increase in *both* capacity and readout rate over conventional binary data pages. Implementing three gray levels with a 15-bit-to-12-pixel modulation code is convenient because it expressly encodes from and decodes back to binary data, and requires ECC coding of only moderate complexity. The similar capacity performance obtained with a local thresholding technique, however, indicates that the preference for three gray levels is independent of the particular choice of decoding scheme.

References

1. D. Psaltis and F. Mok. Holographic memories. Sci. Am., **273**(5), 70 (1995).
2. J.H. Hong, I. McMichael, T. Y. Chang, W. Christian, and E. G. Paek. Volume holographic memory systems: techniques and architectures. Opt. Eng., **34**, 2193–2203 (1995).
3. J.F. Heanue, M.C. Bashaw, and L. Hesselink. Volume holographic storage and retrieval of digital data. Science, **265**, 749 (1994).
4. D. Psaltis, D. Brady, and K. Wagner. Adaptive optical networks using photorefractive crystals. Appl. Opt., **27**(9), 1752–1759 (1988).
5. G.W. Burr, G. Barking, H. Coufal, J.A. Hoffnagle, C.M. Jefferson, and M.A. Neifeld. Gray-scale data pages for digital holographic data storage. Opt. Lett., **23**, 1218–1220 (1998).
6. G.W. Burr, W.-C. Chou, M.A. Neifeld, H. Coufal, J.A. Hoffnagle, and C.M. Jefferson. Experimental evaluation of user capacity in holographic data storage systems. Appl. Opt., **37**, 5431–5443 (1998).
7. M.A. Neifeld and J.D. Hayes. Error-correction schemes for volume optical memories. Appl. Opt., **34**(35), 8183–8191 (1995).
8. B.J.F. Heanue, M.C. Bashaw, and L. Hesselink. Channel codes for digital holographic data storage, J. Opt. Soc. Am. **A 12**(11), 2432–2439, 1995.
9. G.W. Burr. High density holographic data storage. In *OSA 1998 Annual Meeting*, October 1998. Paper WAA1.
10. G.W. Burr, H. Coufal, R.K. Grygier, J.A. Hoffnagle, and C.M. Jefferson. Noise reduction of page-oriented data storage by inverse filtering during recording. Opt. Lett, **23**(4), 289–291 (1998).
11. G.W. Burr, J. Ashley, H. Coufal, R.K. Grygier, J.A. Hoffnagle, C.M. Jefferson, and B. Marcus. Modulation coding for pixel-matched holographic data storage. Opt. Lett., **22**(9), 639–641 (1997).

Part V

Demonstration Platforms

System Optimization for Holographic Data Storage Systems

G.W. Burr and M.P. Bernal Artajona

A properly-designed holographic data storage system should preserve the data entrusted to it by a user, and return that same data at some later time. System optimization is the process of maximizing the desirable features of the system (how much data can be stored, how fast can it be returned) while maintaining the mandated fidelity (the output data is really the same as the input data).

The previous chapters have described noise processes that can affect bit error rate (BER) [1–8], and the many coding and signal processing tools [7–21] that can be brought to bear against these noise processes. In this chapter, we briefly summarize these noise sources and the coding and signal processing techniques available. We describe some of the techniques not covered elsewhere in the book, and make some mention of techniques that do not work in combination. The effects of component choices (such as device fill factors and apertures), and the need to use a sufficient portion of the detector's dynamic range, are discussed. Then we describe a capacity-estimation procedure for holographic data storage. This technique makes makes it possible to quantitatively evaluate component, coding, and processing tradeoffs in holographic data storage systems over a range of operating parameters. Finally, the effects of variations in diffraction efficiency on the choice of BER operating point are outlined, including the effect on achievable capacity.

1 Noise

Noise sources were discussed extensively in Part I of this book, and so are treated here only briefly. The basic noise tradeoff in volume holography is between the finite dynamic range of the recording material and the fixed noise floor of the system. For instance, the electronic detection process at the camera tends to contribute the same amount of noise no matter how bright the hologram. However, as the number of holograms superimposed in the same volume increases, the amount of power diffracted into each hologram reconstruction and the resulting signal-to-noise ratio (SNR) decrease. The same problem tends to limit readout rate as well.

Even if all other noise sources are negligible, then there will be a certain hologram strength at which the SNR is inadequate for error-free detection.

The number of detected photons per pixel can be written as

$$n_{\text{photons}} \propto M/\#^2 \, P_{\text{readout}} \, \frac{t_{\text{readout}}}{M^2 \, N_{pixels}}, \quad (1)$$

where $M\#$ is a material/system constant [22], P_{readout}, the power in the readout beam, t_{readout} is the integration time of the camera, N_{pixels}, the number of pixels per hologram, and M, the number of multiplexed holograms. The storage capacity is MN_{pixel}, and the readout rate is $N_{\text{pixel}}/T_{\text{readout}}$. (Storage density is MN_{pixel} divided by the volume or area of each hologram 'stack'.) An increase in either capacity or readout rate leads to a decrease in the number of signal photons [1]. As this signal strength approaches the noise level the BER of the system will rise, and the fidelity of the storage system will not meet the promised specifications. Figure 1, a plot of measured raw BER as a function of the number of detected photons, shows this trend [12]. In this case, the number of equivalent 'noise photons' due to detector noise was ~160.

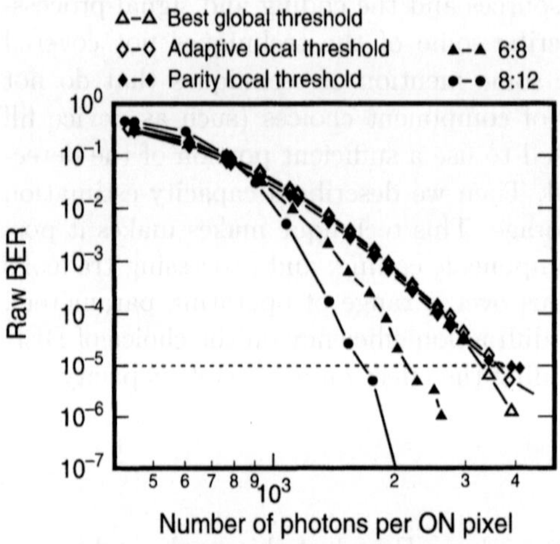

Fig. 1. BER performance vs. signal strength in the presence of a fixed noise floor

While the constant noise floor is usually of primary importance, any additional noise sources will use up part of the SNR budget, increasing the minimum acceptable number of signal photons and reducing the capacity of the system. Noise sources in holographic storage include:

- *Changes in the readout conditions,* including unwanted changes in the reference beam path between the time the hologram is recorded and the time it is reconstructed [4–6].
- *The detector array does not line up with the array of pixels in the reconstructed hologram,* owing to errors in camera registration, rotation, focus, tilt, and the magnification of the image.

- *The detector is receiving undesired light,* either from light scattering off the storage material, from cross-talk from other stored holograms (inter-page cross-talk [2]), or from cross-talk between neighboring pixels of the same hologram (inter-pixel cross-talk [7,8]).
- *There are brightness variations across the detected image,* caused by the SLM, the optical imaging, or the collimation and beam quality of the laser beams themselves. Such variations tend to be deterministic – they don't vary from hologram to hologram.

Given these many noise sources and the need to read back holograms and make bright-vs-dark distinctions with high fidelity, how can one maximize the desirable qualities of the system such as capacity and readout rate? Here are some options:

1. From Equation 1, we can increase capacity or readout rate by increasing P_{readout} (buying a bigger laser) or by increasing $M\#$ (getting a better storage material) [23].
2. Pre-process at the spatial light modulator to either increase signal values [10] or reduce the deterministic variations which are reducing the SNR [15].
3. Post-process at the detector array in order to remove a known point-spread function, with varying degrees of feedback or sequence estimation [7,11,14,19]. Note that post-processing is much more difficult when the point-spread function of the system (the channel response) varies rapidly across the data page. This precludes the use of post-processing in combination with predistortion [15] or dynamic random phase masks [8,24]. Post-processing is also problematic in the presence of 'dead' SLM or CCD pixels, since the convolution can cause each 'dead' pixel to contaminate its neighbors as well, and any remapping of these dead pixels scrambles the known channel response.
4. Use a low-pass modulation code that avoids those pixel combinations prone to inter-pixel cross-talk [8,16,18].
5. Quantize the data with enough resolution to make bright–vs–dark decisions and, if desired, to accurately execute post-processing algorithms [25].
6. Use a decision scheme that produces fewer errors from the same SNR, either with adaptive thresholding [13], or by encoding at the SLM with a balanced modulation code [12].
7. Use interleaving [20,17] and strong error-correction [9] to produce the same target user-BER from a more error-prone stream of raw binary data.
8. Optimize the physical dimensions of the input and output pixel arrays and of the aperture at the hologram, in order to maximize the storage density [18].
9. Arrange the recording exposures so that the BERs of the first- and last-written holograms are equal, reflecting any differences in the noise environment experienced by each [26].

10. Use more than one 'gray' level per pixel, so that each pixel represents more than one bit of information [21].

Points 2, 3, 6, 7, and 10 were discussed in Part IV of this book.

2 Camera Quantization

Before making the experimental measurements in Fig. 1, the gain of the analog-to-digital (A/D) converter was increased so that the incoming signal filled most of the dynamic range. If we had not done this, then the quantization of the detected analog signals might assign ON and OFF pixels of similar brightness – that would otherwise be slightly separated – to the same camera count value, leading to an error. Figure 2 illustrates the effect of this 'quantization noise', showing BER as a function of the quantizing resolution (represented as the $\log_2$ of the number of camera counts separating the means of the ON and OFF pixels). In this simulation, the same two Gaussian noise sources (SNR $\equiv (\mu_1 - \mu_0)/\sqrt{\sigma_1^2 + \sigma_0^2} = 3.5$, $\sigma_1/\sigma_0 = 3$) were applied, and only the resolution of the quantization was varied. Figure 2 shows that for the purposes of decoding performance, one only needs ~3 bits of quantization. It is not important to know the safety margin by which the ON pixels exceed the threshold. However, for the predistortion technique, any of the post-processing techniques, or any decision feedback scheme, we require accurate knowledge of the brightness of pixels. The gain of the A/D converter should then be adjusted so that the brightest ON pixels still fall within the dynamic range of the camera. Maintaining this condition with 8 bits of dynamic range and the brightness distributions used in the simulation above leaves approximately 130 camera count levels (7 bits) betweeen the means.

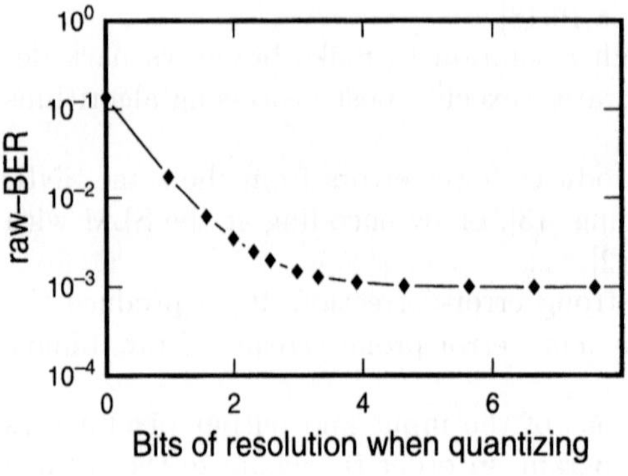

Fig. 2. BER vs. quantization resolution, expressed as $\log_2$ (the number of camera count levels between the means of the ON and OFF pixels)

3 Choice of Fill-Factors and Apertures

Some component choices have an obvious effect on system performance. For instance, the size of the central diffraction order of the SLM scales as

$$D_N \equiv \frac{\lambda f}{\delta}, \tag{2}$$

where λ is wavelength, f is the focal length of the object beam lens, and δ is the spacing between SLM pixels. Essentially, the image will be at the Nyquist sampling condition if a centered square aperture at the Fourier plane has sides of length D_N. If it were any smaller, spatial frequency components that represent data would be cut off. As a result, areal density can be increased by having a short wavelength, short focal length, and large pixels. The limit here is how far the lens design can be pushed to accept a large input field (N_{pix} pixels times δ) yet still maintain a short focal length.

However, maintaining the Nyquist sampling condition does not necessarily guarantee low BER. The aperture is a spatial low–pass filter, so the transmitted image ends up convolved by the system point-spread function (PSF). Since the PSF is the Fourier transform of the aperture, larger apertures mean a narrower PSF and thus less blurring of pixels into their neighbors. Although this reduces the data density per hologram, it leaves more of the SNR budget available for combatting background noise (allowing more stored holograms). In addition to this tradeoff, there is the effect of the fill factors of the SLM and detector array. (The fill factor is the fraction of the space assigned to a pixel that is actually used – modulated in the case of the SLM, or used for collecting photoelectrons in the case of the detector pixel.) Since the shape of the final image of each pixel is a convolution of the original SLM pixel shape and the PSF, it might make sense to have a small SLM fill factor in order to reduce inter-pixel cross-talk yet still allow a small aperture. (An additional effect of a small SLM fill factor is to dramatically reduce the power efficiency of the SLM, since much of the illuminating light not lost at the dead space between active pixel elements is spread into higher diffraction orders.) On the detector array, it might make sense to have very small pixels so that the inter-pixel cross-talk can be effectively removed. This is the Nyquist sampling condition – for the particular square aperture with sides D_N, the contribution of neighboring pixels goes to zero at the exact center of the pixel. So point detectors would be able to avoid all inter-pixel cross-talk. Unfortunately, point detectors don't collect very many photons, so it is impossible to build up enough signal to overcome the fixed background noise. These two opposing trends – small pixels to resist inter-pixel cross-talk, large pixels to collect enough signal – create an optimum fill factor [18]. It turns out that the optimum pair of SLM and CCD fill factors changes for different aperture sizes.

We have studied the interplay between electronic detector noise and inter-pixel cross-talk using a simulation that includes the effects of component fill factors and aperture size [18]. The achievable density is shown graphically by the top curve of Fig. 3, where normalized areal density is plotted as a function of the aperture dimension (itself normalized to the Nyquist aperture). The assumption here is simple global thresholding and a raw BER target of 10^{-4}. We have found that the need to collect photons is more important than suppressing interpixel cross-talk, so using relatively large fill factors (linear SLM f.f. of 90%, linear CCD f.f. of 87%) costs little in terms of density (bottom curve of Fig. 3).

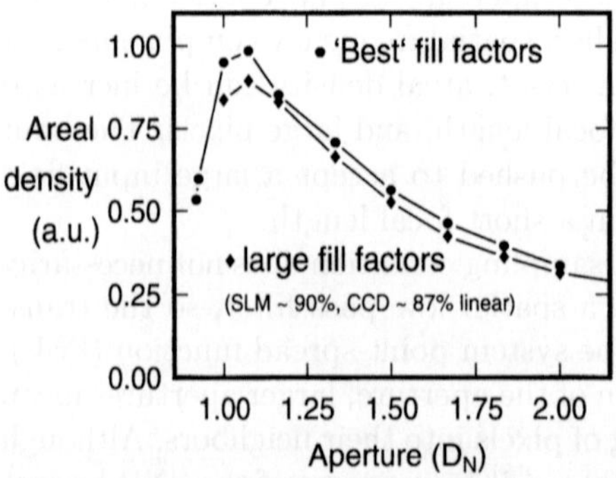

Fig. 3. Normalized areal density vs. aperture size. As the aperture shrinks, density per page increases but SNR decreases due to diffraction-induced inter-pixel cross-talk. For simple thresholding, the tradeoff between the two reaches a maximum for an aperture just larger than D_N, the spatial Nyquist sampling aperture for the system

We have also looked at the case where scatter noise is added to the detector noise and inter-pixel cross-talk [27,28]. The optical noise contribution tends to decrease as the fill factor of the detector pixel decreases, pushing the optimum pixel size lower. As one would expect, adding another noise source tends to reduce the portion of the SNR budget left for writing holograms (that is, for allowing the signal levels to drop). What is surprising is that the statistics of the scatter noise depend on the aperture itself, making larger apertures more favorable because the probability density functions are narrower [28]. As a result, at the Nyquist aperture, when the scatter noise power exceeds ~1% of the received signal power, the raw BER target of 10^{-4} is unachievable (all of the SNR budget is used up by scatter noise and inter-pixel cross-talk). Larger apertures allow the system to tolerate more scatter noise, because the inter-pixel cross-talk takes less of the SNR budget, *and* the scatter noise is averaged over the multiple correlation domains within each detector pixel.

This reflects a common situation in these simulations: as a noise source increases, the capacity or density at a given BER target decreases slowly at first, but then rapidly drops to zero. At this point, the use of modulation coding or the pre-distortion technique (instead of simple thresholding) is doing more than just increasing capacity – it is making it possible for the system to store and retrieve holograms at all under these conditions.

4 Capacity-Estimation Procedure

The recording schedule was originally developed to allow multiple holograms to be recorded to the same diffraction efficiency, η, in read–write holographic materials such as photorefractive crystals [29]. Since the first holograms written must persist through the erasure caused by writing later holograms, diffraction efficiencies can be equalized by carefully scheduling the recording exposures [29]. Initial holograms are written for a longer exposure time and thus grow to a large diffraction efficiency, which is then exactly compensated by the erasure due to the remaining exposures – and the first and last holograms end up with the same diffraction efficiency.

Recently, we developed a variation on this recording schedule, which uses the same mathematical algorithm, but attempts to equalize the raw BER of the first and last holograms. This has the effect of increasing capacity, because the last hologram typically does not experience the same amount of noise as the first hologram. Recording to the same diffraction efficiency is overkill for the last hologram, because this hologram will have a raw BER that is better than required. This extra dynamic range can be used to record additional holograms, creating an increase in total capacity. This is shown in Fig. 4, where the number of holograms that can be stored is plotted against raw BER, for both a conventional, flat–η schedule and the new flat–BER schedule.

The main motivation for this new schedule, however, was not the increase in capacity shown by the measurement in Fig. 4, but just the capability of making such a measurement at all. To measure the number of holograms as a function of raw BER would otherwise be an exhausting series of multiple hologram exposure experiments. Instead, a simple set of repeated measurements of the same hologram gives the relation between exposure time and raw BER. Using the mathematics of the flat-BER recording schedule, this results in a unique value of M – the number of holograms that can be stored – for each target raw–BER.

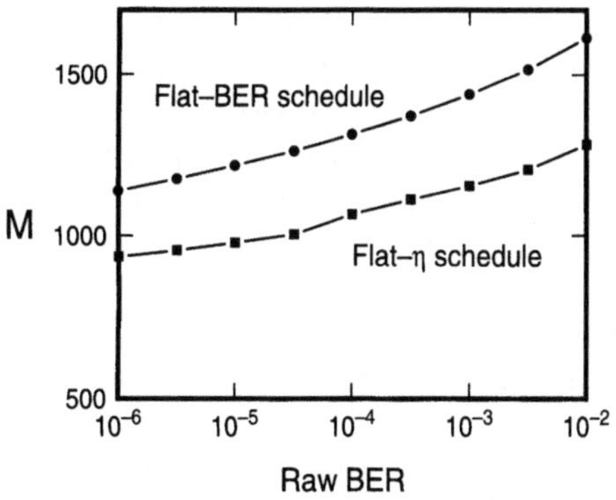

Fig. 4. Number of holograms that can be stored as a function of raw BER. By tolerating a higher raw BER, more holograms can be superimposed. Capacity can also be improved by arranging recording exposures to equalize BER instead of diffraction efficiency [26]

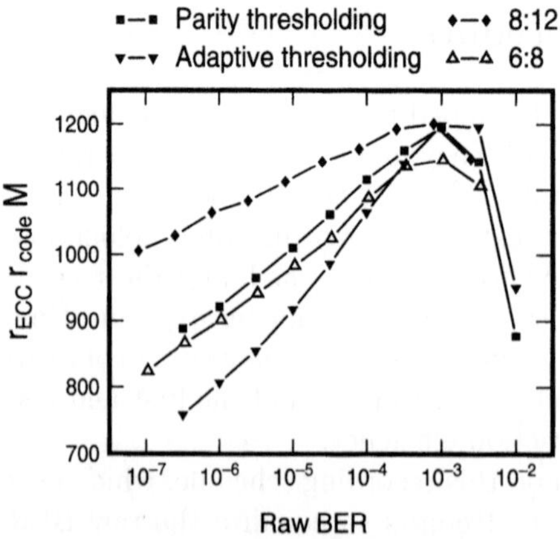

Fig. 5. Total capacity in user bits as a function of the system raw BER. The final user BER is always 10^{-12}, and two modulation codes are compared against two local thresholding techniques. The strong raw BER performance of the codes is counterbalanced by their lower code rate. The presence of the ECC coding makes it possible to find an optimal raw BER [26]

We have used this procedure to measure and optimize the capacity, in user bits, of our DEMON holographic storage system operating at low density (large Fourier transform aperture) [26]. With the M vs raw BER data made possible by the capacity measurement described above, we combine the code rate of the modulation coding and the dependence of the ECC code rate on raw BER. The result of such a study is shown in Fig. 5, where two modulation codes (6:8 and 8:12) are compared against two local thresholding techniques on the basis of total user capacity. Note that the tendency, as raw BER gets larger, for the number of holograms to increase and the ECC code rate to decrease creates an optimal raw BER operating point for the system ($\sim 10^{-3}$ in this case).

5 Choice of ECC Design Point: Effect of Variations in Diffraction Efficiency

We separated the noise sources described earlier into two types: those that are truly random noise (from the detection electronics or optical scatter), and the deterministic variations in signal strength (fixed pattern noise or inter-pixel cross-talk). These sources all tend to make pixel values deviate from the average ON (or OFF) value for the page. One question worth considering is what happens when the mean values themselves are varying from hologram to hologram (because the diffraction efficiency is varying). The dependence of BER on signal power at low density is shown in Fig. 1. If we could count on the diffraction efficiency of all the holograms to be equal, we would multiplex holograms until the page-wide raw BER reached the given target value (say, raw BER of 10^{-3} with ECC coding designed to bring that down to the user BER target of 10^{-12}). In reality, however, the diffraction efficiency of holograms varies from page to page. So, every so often there is a weaker-than-average page that is decoded at a raw BER of 10^{-2}, resulting in a user

BER above the 10^{-12} target, and raising the average user BER. To combat this, we need to increase the average mean of the pages and thus decrease the raw BER target that we try to hit with this average-strength page. Here we estimate how much the average page strength has to increase, and show the resulting cost in user capacity.

We use the dependence of raw–BER on signal strength for the 6:8 code as shown in Fig. 1, and assume that an 8-bit-per-symbol, 255-byte-long Reed–Solomon code is being used. With $t = 14$ bytes of error–correction, this can take a raw BER of 10^{-3} down to a corrected user BER of 10^{-12}. We assume that the user BER roughly follows the raw BER raised to the 14th power (times a correction factor that puts the curve through 10^{-12} user BER for 10^{-3} raw BER). This implies that a change of raw BER by one order of magnitude causes the user BER to swing by 14 orders of magnitude. As a direct result, a small decrease in diffraction efficiency below the average diffraction efficiency results in a huge change in user BER. Assuming that the signal strength follows some probability distribution $p_{\text{signal}}(\eta)$, then the aggregate user BER will be

$$\text{Average user BER} = \int p_{\text{signal}}(\eta) \quad \text{user BER}(\eta)\, \mathrm{d}\eta. \tag{3}$$

As the variation in diffraction efficiency of the pages increases, the average user BER starts to exceed the previously designed-for target because of the weak pages. To counteract the increase in average user BER, we increase the average diffraction efficiency and correspondingly reduce the number of holograms that can be stored. (Since diffraction efficiency scales as 1 over the number of holograms squared, having to increase average diffraction efficiency by a factor f reduces the number of stored pages by $\sqrt{f}$.) This reduction in capacity is shown in Fig. 6, where the horizontal axis shows the ratio of the standard deviation to the mean of the Gaussian distribution of diffraction efficiencies. Note that after a certain point, the average page no longer af-

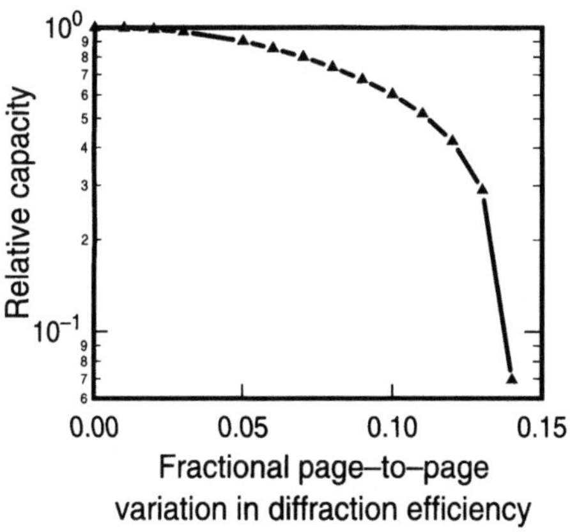

Fig. 6. Loss of density resulting from variation of the diffraction efficiency of the holograms. Pages weaker than the designed-for average-strength page have much higher user BER. The average page must then be made brighter, leading to fewer superimposed holograms

fects the design point – the errors in user data come solely from the weakest data pages. The performance of the system is being maintained by raising the average diffraction efficiency to just strengthen the weakest pages. Alternatively, if the variations are primarily introduced during recording, then the best solution is probably to check each hologram after recording and re-record when necessary.

6 Summary

We have summarized the noise sources that can lead to error in holographic storage systems, and the systems techniques used to suppress these errors while maximizing desirable features such as capacity, density, and readout rate. We have described steps that can be taken (beyond conventional coding and signal processing) to optimize the digital holographic data channel. These include careful choice of components such as the Fourier transform aperture and device fill factors (both SLM and detector), and scheduling of the recording exposures to equalize the raw BER of the holograms. We showed how this recording schedule can be used to quantitatively evaluate otherwise-inaccessible coding, signal processing, and component tradeoffs. Finally, we showed that the detection process is relatively robust when the analog signal values are quantized with low resolution, and that page-to-page variations in the diffraction efficiency of holograms decrease capacity.

References

1. K. Blotekjaer. Limitations on holographic storage capacity of photochromic and photorefractive media. Appl. Opt., **18**, 57–67 (1979).
2. C. Gu, J. Hong, I. McMichael, R. Saxena, and F. Mok. Cross-talk-limited storage capacity of volume holographic memory. J. Opt. Soc. Am. A, **9**(11), 1–6 (1993).
3. J.F. Heanue, M.C. Bashaw, and L. Hesselink. Channel codes for digital holographic data storage. J. Opt. Soc. Am. A, **12**(11), 2432–2439 (1995).
4. R. DeVre, J.F. Heanue, K. Gürkan, and L. Hesselink. Transfer functions based on Bragg detuning effects for image-bearing holograms recorded in photorefractive crystals. J. Opt. Soc. Am. A, **13**(7), 1331–1344 (1996).
5. S. Campbell, S.-H. Lin, X. Yi, and P. Yeh. Absorption effects in photorefractive volume–holographic memory systems. i. beam depletion. J. Opt. Soc. Am. B, **13**(10), 2209–2217 (1996).
6. S. Campbell, S.-H. Lin, X. Yi, and P. Yeh. Absorption effects in photorefractive volume-holographic memory systems. ii. material heating. J. Opt. Soc. Am. B, 13(10), 2218–2228, 1996.
7. J. Heanue, K. Gurkan, and L. Hesselink. Signal detection for page-access optical memories with intersymbol interference. Appl. Opt., **35**, 2431–2438 (1996).
8. J. Hong, I. McMichael, and J. Ma. Influence of phase masks on cross-talk in holographic memory. Opt. Lett., **21**, 1694–1696 (1996).

9. M.A. Neifeld and M. McDonald. Error correction for increasing the usable capacity of photorefractive memories. Opt. Lett., **19**, 1483–1485 (1994).
10. B.H. Olson and S.C. Esener. Partial response precoding for parallel-readout optical memories. Opt. Lett., **19**(9), 661–663 (1994).
11. M.A. Neifeld, K. Chugg, and B. King. Parallel data detection in page-oriented optical memory. Opt. Lett., **21**, 1481–1483 (1996).
12. G.W. Burr, J. Ashley, H. Coufal, R.K. Grygier, J.A. Hoffnagle, C.M. Jefferson, and B. Marcus. Modulation coding for pixel-matched holographic data storage. Opt. Lett., **22**(9), 639–641 (1997).
13. X.A. Shen, A.-D. Nguyen, J.W. Perry, D.L. Huestis, and R. Kachru. Time-domain holographic digital memory. Science, **278**, 96–100 (1997).
14. V. Vadde and B.V.K. Vijaya Kumar. "Channel estimation and intra-page equalization for digital volume holographic data storage," in *Optical Data Storage 1997*, Optical Society of America, 250–255 (1997).
15. G.W. Burr, H. Coufal, R.K. Grygier, J.A. Hoffnagle, and C.M. Jefferson. Noise reduction of page-oriented data storage by inverse filtering during recording. Opt. Lett., **23**(4), 289–291 (1998).
16. J. Ashley and B. Marcus. Two-dimensional lowpass filtering codes for holographic storage. IEEE Trans. Commun., **46**, 724–727 (1998).
17. M. Blaum, J. Bruck, and A. Vardy. Interleaving schemes for multidimensional cluster errors. IEEE Trans. Commun., **44**(2), 730–743 (1998).
18. M.-P. Bernal, G.W. Burr, H. Coufal, and M. Quintanilla. Balancing inter-pixel cross-talk and thermal noise to optimize areal density in holographic storage systems. Appl. Opt., **37**, 5377–5385 (1998).
19. B. King and M.A. Neifeld. Parallel detection algorithm for page-oriented optical memories. Appl. Opt., **37**(26), 6275–6298 (1998).
20. W.-C. Chou and M.A. Neifeld. Interleaving and error correction in volume holographic memory systems. Appl. Opt., **37**(29), 6951–6968 (1998).
21. G.W. Burr, G. Barking, H. Coufal, J.A. Hoffnagle, C.M. Jefferson, and M.A. Neifeld. Gray-scale data pages for digital holographic data storage. Opt. Lett., **23**, 1218–1220 (1998).
22. F.H. Mok, G.W. Burr, and D. Psaltis. System metric for holographic memory systems. Opt. Lett., **21**(12), 896–898 (1996).
23. G.W. Burr and D. Psaltis. Optimization of the oxidation state of $LiNbO_3$ for large scale holographic storage. Opt. Lett., **21**(12), 893–895 (1996).
24. M.-P. Bernal, G.W. Burr, H. Coufal, J.A. Hoffnagle, C.M. Jefferson, R.M. Shelby, and M. Quintanilla. Experimental study of the effects of a six-level phase mask on a digital holographic storage system. Appl. Opt., **37**(11), 2094–2101 (1998).
25. G.W. Burr, J. Ashley, B. Marcus, C.M. Jefferson, J.A. Hoffnagle, and H. Coufal. Optimizing the holographic digital data storage channel. In *Proceedings of SPIE: Advanced Optical Memories and Interfaces to Computer Storage*, volume 3468, 64–75 (1998).
26. G.W. Burr, W.-C. Chou, M.A. Neifeld, H. Coufal, J.A. Hoffnagle, and C.M. Jefferson. Experimental evaluation of user capacity in holographic data storage systems. Appl. Opt., **37**, 5431–5443 (1998).
27. M.-P. Bernal, G.W. Burr, H. Coufal, and M. Quintanilla. Noise in holographic data storage at high areal density. In *CLEO 1998 Technical Digest* (1998). paper CMF3.

28. G.W. Burr. High density holographic data storage. In *OSA 1998 Annual Meeting*, October 1998. Paper WAA1.
29. D. Psaltis, D. Brady, and K. Wagner. Adaptive optical networks using photorefractive crystals. Appl. Opt., **27**(9), 1752–1759 (1988).

Tamarack Optical Head Holographic Storage

S. Redfield

This chapter describes the work at Tamarack Storage Devices, Inc. that lead up the first ever historical effort at integrating the various technologies comprising a holographic storage into a single 5 $\frac{1}{4}$ in footprint drive with a targeted raw capacity of 30 Gbytes

1 Roots

In 1992 Tamarack was formed to make a commercial holographic storage product. It grew out of the work in holographic storage started at MCC (Microcomputer Electronics and Technology Corporation Research Consortium) in 1983. The work at Tamarack culminated in a unique prototype of a holographic storage drive. This drive utilized a novel optical design based on a moving optical head and a disk-shaped medium. The disk had a targeted raw capacity of 1 Gbyte, and the drive held 30 disks.

In the early 1990s it was thought that a 1-Gbyte floppy disk replacement would be a very competitive product. The onset of multimedia in computer applications created a demand for an inexpensive removable medium that had sufficient capacity to hold high-resolution graphics, video clips, and/or music. This holographic medium would have a disk form factor, and the recording material would be photopolymer and so could have a cost of a few dollars. Even though the disk was write-once, since the cost was so low, this did not seem to be a big drawback.

2 Design Evolution

Holographic storage systems traditionally have beam steering opticsm which move both a reference beam and an object beam in tandem to the recording point, or stack site, on the medium. See Fig. 1 for the general layout of a traditional holographic storage system. At the stack site an information-carrying object beam is interfered with a reference beam. This interference pattern (hologram) is recorded in the optical recording medium. Beams and medium are positioned to get to different stacks. The information pattern imposed on the object beam is called a *page*. Angle or phase multiplexing is used to superimpose the holograms for a number of different pages in one stack site. To retrieve a page, a reference beam with the angle or phase code

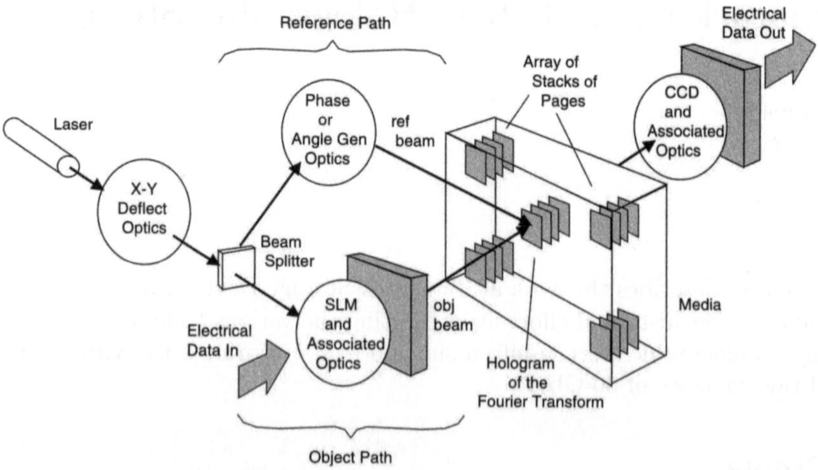

Fig. 1. Generic holographic storage layout

corresponding to the desired page is supplied. This reference beam reconstructs the hologram for that page. This reconstructed recording of a page is captured by a CCD or detector array.

In such a holographic storage system the beam-steering optics stand off from the medium some distance so the surface area on the medium across which the two beams may be moved together can be of a reasonable size. This surface area is called the *access window.* Figure 2 illustrates the concept of an access window.

The design work at Tamarack followed a progression of increasingly more sophisticated breadboards as a deeper understanding of the design problem was achieved. All of the designs that were considered were based upon page-multiplexed stacks of holograms distributed throughout a planar photo-

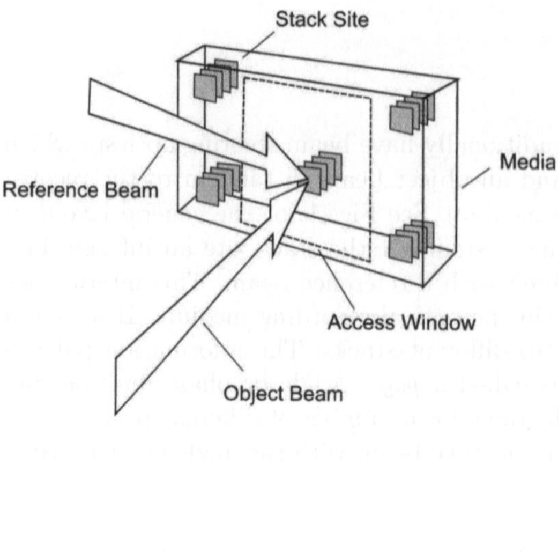

Fig. 2. Access window concept for holographic recording and reading

polymer recording medium that was mechanically moved in some way in order to access the various stacks.

When design work began it was soon concluded that the access window could be one stack wide, a column whose height equaled the height of the medium. This greatly simplified the beam-steering optics. Both beams were moved up down before the beam-splitter. The reference beam would have a phase code or an angle placed on it after the beam splitter.

The medium was initially envisioned as a rectangular tile that held an array of stacks of pages. The access window would cover a vertical column of the stacks, and the tile would be stepped horizontally to reach all the stacks.

There was now a nice idea. Instead of a long sliver for the beam splitter with multiple lenslets for each set of Y stack positions and a mechanism for moving the laser beam to one of the lenslets, why not move a single lenslet beam-splitter combination itself along the laser beam to achieve the Y positioning?

Now that some optics were being moved a radical idea was presented. Why not move all the reference and object beam optics instead of deflecting the beams along different paths in stationary optics? This eliminated the array of balls or grin rods. Also now the optics could be close to the media, as the optics were moving, not the beams, to accomplish the access window. At first consideration it would seem that the optics would be too heavy to be moved very fast. But after some initial design it was found that the optics got so small as a result of being very close to the medium that this was not a problem. Additionally much faster optics (lower f number) could be achieved. The architecture for a moving-head-based optical design is shown in Fig. 3.

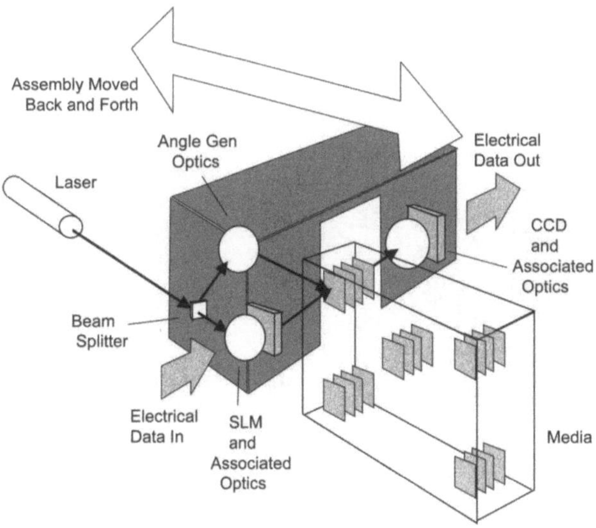

Fig. 3. Moving-head holographic storage system

3 Final System Approach

The final design is a very compact holographic storage drive that would read/write to removable media. It was called *multistore*. The medium is a small circular disk with a raw capacity of 1 Gbyte. The drive contains 30 of these disks in a jukebox arrangement; they can be individually picked and placed at a read station. The access time was comparable to that of a hard disk. An exploded view of the final multistore design is shown in Fig. 4.

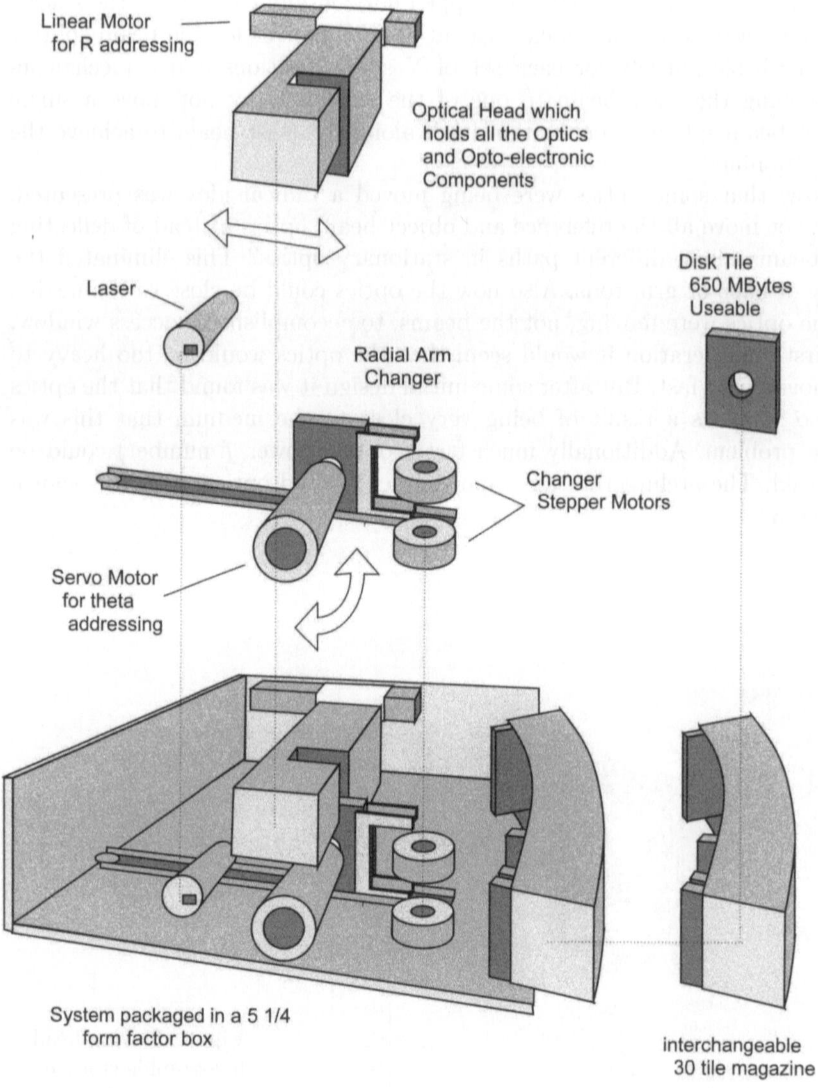

Fig. 4. Multistore

3.1 Requirements

Given the market target of removable folder storage with at least 1 Gbyte of raw capacity, a standard $5\frac{1}{4}$ in form factor, and the same speed as a hard disk, the system then had the requirements shown in Table 1. They are divided into user visible (product) requirements and the resulting internal (hardware) requirements.

3.2 Optical Head

To achieve the necessary compact drive form factor an optical head approach was taken. The holographic optical head (HOH) packaged all the holographic write/read optics. The HOH consists of the spatial light modulator, the CCD detector array, and both reference and object beam optics, all placed in a small package. Both the medium, which is disk-shaped, and the HOH are movable. The storage medium disk rotates to access different azimuthal, or θ, locations, and the HOH is moved linearly across the medium disk in order to access different radial, or R, positions. In this manner any location on the medium disk can then be accessed. The head requires only one input beam, as optics internal to it separate the beam into object and reference paths, perform the data encoding, etc.

Table 1. System requirements

Product level		Hardware level	
Parameter	Target	Parameter	Target
Form factor	$5\frac{1}{4}$ in	Multiplexing	30 pages per stack
Drive capacity	30 medium	Write speed	5 ms
Drive cost	$2 000	Read speed	1 ms
Average access time	10 ms	Signal-to-noise	>10
Peak data bandwidth	65Mbits per s	DRAW time	100 ms
Media type	Removable, Write once	Laser wavelength	685 nm
Media capacity	1 Gbyte raw, 650 Mbyte formatted	Laser power at media	10 mW
Media cost	$2.00	SLM pixels (page size)	256 × 256 (50% "1")
BER (after correction)	<10–12	CCD pixels	512 × 512
I/O interface	SCSI - 2	CCD sensitivity	10^5 photons per pixel
Environment	Office	Object imaging optics	F/2.4

3.3 Media Disk

The final medium for the multistore was a 62 mm disk in a lightproof jacket, which looked very similar to a $5\,\frac{1}{4}$ in floppy disk. The jacket has a covered slot that opens up automatically when the disk is place in the read station. The medium is a photopolymer, which is coated on a PMMA substraite, and which has a mylar cover on the top. The disk has a hub that facilitates its correct placement into a read station. The disk is illustrated in Fig. 5, and a photo of the disk is shown in Fig. 6.

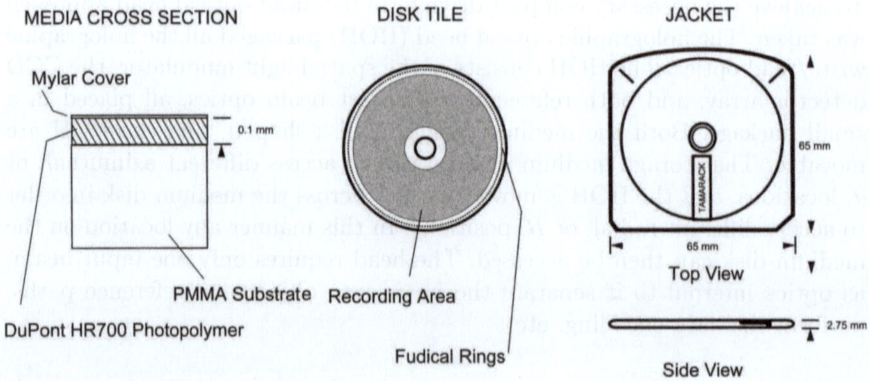

Fig. 5. Holographic storage disk tile

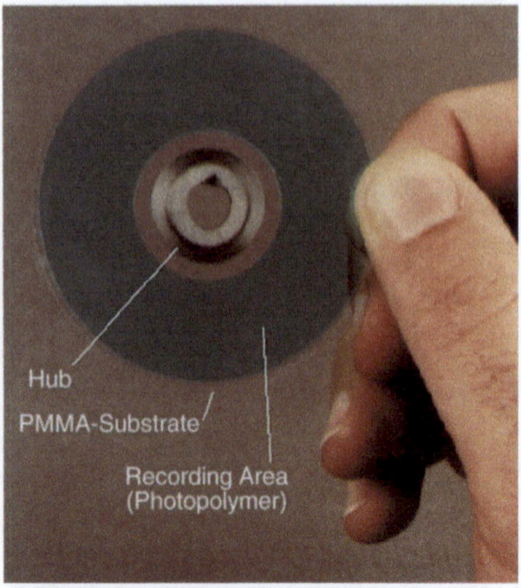

Fig. 6. Photograph of media disk

The recording stacks were arranged on the medium disk radially. There were 2 160 recording stacks throughout the disk. Each stack contains 30 pages of angle-multiplexed data. The area allotted for each stack is 0.857 mm × 1.000 mm, and the disk's outer diameter is 62 mm.

3.4 Data Format

The data is stored as a rectangular array of 256 × 256 black and white spots, white for zero and black for one. The data is pre-encoded so that there are approximately equal numbers of black and white spots. An error correction code is also applied. The readout involves 4 to 1 oversampling and a data recovery process that compensates for any distortion of the spot array. The thresholding to determine if a spot is black or white utilizes local spot neighborhood feedback to compensate for any gradual intensity variation in the readout beam. Finally black-white equalization is decoded and the error correction code is done.

4 Holographic Optical Head

The holographic optical head (HOH) is a very compact assembly using miniature components and extensive optical path folding to achieve its small size. The prototype HOH that has been built weighs approximately 150 g. Figure 7 shows the current HOH optical design configuration. For clarity, the optical beam paths have been unfolded from the manner in which they actually appear in the prototype HOH that has been built.

The HOH contains the components for both the object and reference beam optical systems. The incoming collimated beam is first split into the reference beam path and the object beam path via the penta splitter. A photo of the HOH is shown in Fig. 8.

The overall optical performance of the HOH is summarized in Table 2.

4.1 Reference Path Optical Design

In the reference beam path, we need to deliver a collimated beam with a small cross-section of less than 1 mm^2 and with incidence angles ranging from 10° to 60° with respect to the normal of the recording film. This large angular swing is required to achieve angle multiplexing of multiple pages on the thick photopolymer film. We achieved this goal by means of the small deflector together with an inverted telescope. The deflector, called the *page motor*, has the Galvo design and contains an internal position sensor for closed loop control of its angular position during both recording and readback. The physical angular range of the p-motor is ±6.25° or optically ±12.5°. The inverted telescope causes an angle magnification of 2× to produce the desired ±25° swing while at the same time reducing the beam cross-section to the

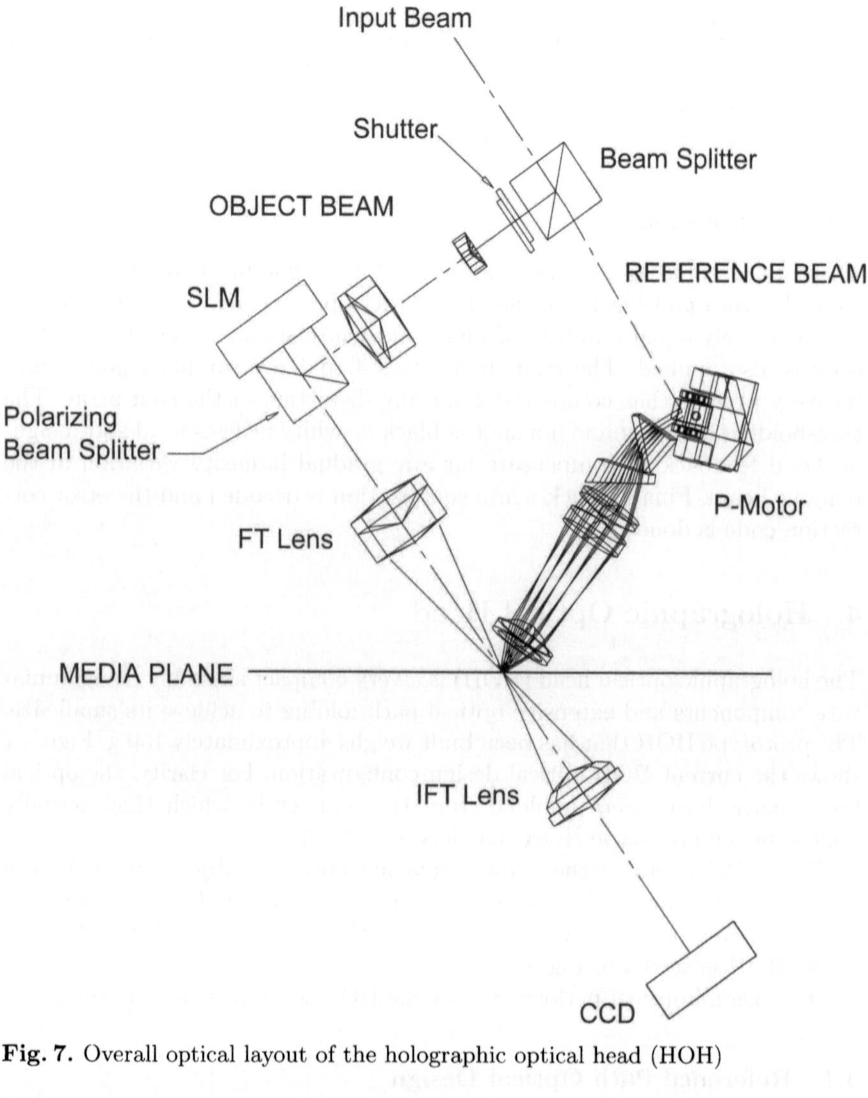

Fig. 7. Overall optical layout of the holographic optical head (HOH)

desired size of $l \times 0.9$ mm^2. A simple inverted telescope design would cause what we call the obliquity problem, the beams at the various incidence angles would have a different cross-section on the medium owing to the cosine effect. This problem is corrected via a pair of prisms in the reference path.

The reference beam optical system consists of the afocal relay lens pair and the page motor for performing the angle-multiplexed recordings at each stack.

The reference beam must sweep out a roughly 50° angular field at the medium surface in order to write the required 30 multiplexed data pages at each stack location. It is typically the case that in most afocal relay systems

Table 2. Optical power budget for the holographic optical head

Reference path	Element (%)	Cumulative (%)	Object path	Element (%)	Cumulative (%)
Penta splitter	50	50	Penta splitter	50.0	50.0
PS/P-motor mirror	98.0	49.0	PS/RWV mirror	98.0	49.0
P-mirror pupil thruput	27.1	13.3	R/W vane	85.0	41.7
P-mirror reflectivity	98.0	13.0	Diverging lens	97.0	40.4
Prism 1	92.0	12.0	Intratelescope Mirror	98.0	39.6
Prism 2	92.0	11.0	Collimator lens	97.0	39.4
Scan lens element 1	96.0	10.6	PBS reflection	98.0	37.6
Scan lens element 2	96.0	10.2	SLM pupil Efficiency	27.1	10.2
Scan/objective mirror	98.0	9.9	SLM reflection	15.0	1.5
Objective	95.0	9.5	PBS transmission	95.0	1.5
			SLM/FTL mirror	98.0	1.4
			Fourier transform Lens	97.0	1.4
			FTL/media mirror	98.0	1.4
Beam ratio		6.98	Laser power into head		30 mW
Splitter transmission		50.0	Ref power		2.84 mW
			Obj power		0.41 mW
Hologram diff. Efficiency		0.15	Power reaching CCD		3.921 μW
Media/IFTL mirror		98.0	Power/pixel		59.8 pW
IFT lens		96.0	Staring time		1 ms
IFTL/CCD mirror		98.0	Photons/s/pixel		2.1E+08 photons
			Photons/pixel		206309 photons

that are used for such scanning applications, the beam cross-section at the record-medium varies significantly with incidence angle at the medium plane. So it is necessary to make special efforts to control the footprint of the reference beam at these various angles. Using an afocal telescope with a pivoting mirror at the entrance pupil as a baseline, and by novel use of anamorphic beam-shaping prisms, Tamarack has designed a reference beam optical system that combines low wavefront error and a uniquely effective method of providing obliquity correction. It provides less than 0.05% change in beam footprint size at the recording plane over a 50° range.

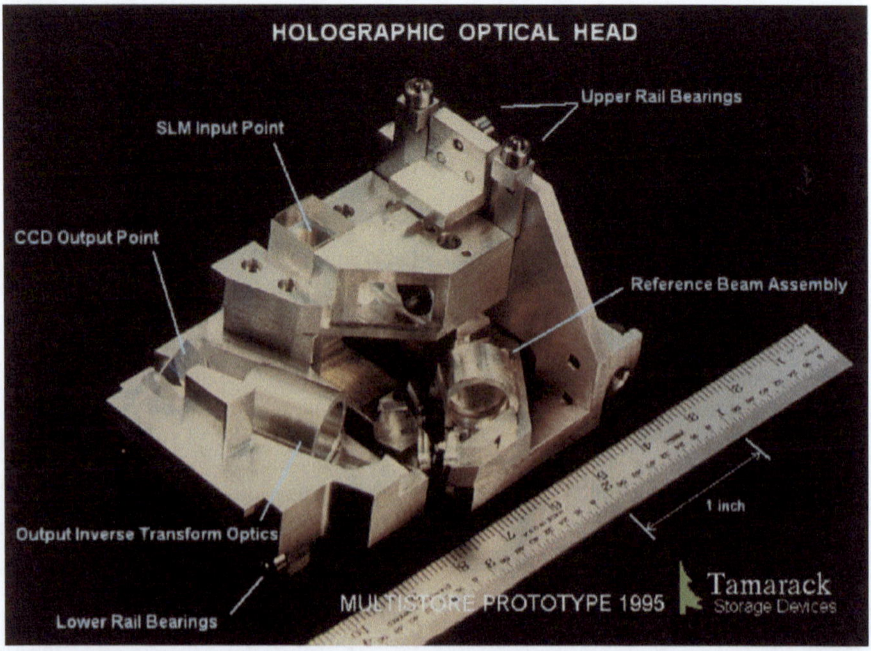

Fig. 8. Photograph of the Holographic Optical Head

4.2 Object Path Optical Design

In the object path the beam first passes through a liquid crystal shutter and then is further expanded to fill the pupil of the SLM, which is mounted on a polarizing cube. The SLM is basically a 2-D array of polarization switches, where the 1s are those pixels whose polarization has been rotated 90° and which propagate downstream, while the 0s retain the original polarization and get reflected back by the cube. The lenses beyond the polarizing cube are the Fourier transform lens and the inverse FF lens. The pair images the data pattern in the SLM onto the CCD with a 1:1 magnification. The recording film intercepts the object beam at the back focal plane of the FT lens to allow the recording of the FT hologram.

A prominent issue in designing the object beam optical system is the $f\#$ of the object beam Fourier transform lens. The mass of the HOH will grow roughly in proportion to the focal lengths of the lenses used. The diameter of the lens components is driven primarily by the SLM size. Thus, low $f\#$s are desired in order to keep the mass of the HOH as low as possible.

In addition, the object beam transform lens $f\#$, along with the SLM's pixel count and the optical wavelength, determine the size and shape of the object beam Fourier transform at the recording plane. The diameter of the transform is proportional to the $f\#$. Thus a small $f\#$ design is desired in order to achieve the smallest stack diameter, and greatest data storage density.

An $f/2.4$ design provided a stack diameter of 800 μm, which permits 2160 stacks to fit onto the 62-mm-diameter photopolymer disk.

A lower transform lens $f\#$, and the correspondingly smaller beam size at the recording medium, can also provide a decrease in the amount of laser energy required to adequately expose the medium, but only up to a point. It was found that for $f/4$ or lower systems, the required exposure energy did not change significantly. This is because the level of scattered light from recording medium imperfections that is gathered by the imaging optics increases as $f\#$ decreases. The extent and magnitude of the random refractive index variations in the photopolymer recording material determine exposure level. Light that is scattered from such random refractive index variations determines the background noise levels at the detector array plane. The signal must assert itself over this background level in order to provide adequate SNR, and to provide an acceptably low error rate during recovery.

5 Mechanical Design

5.1 Page Motor Design

To do the angle encoding of the reference beam a fast acting galvo mirror, called the page motor, was designed and assembled. It fitted within a volume of less than 2 cm^3.

The page motor must provide enough angular scan range for the reference beams for the 30 pages that are multiplexed within each stack to be separated by enough angle to prevent cross-talk between adjacent images. Its angular field of view is $\pm 12.5°$. Together with the 3:1 angular magnification of the reference beam afocal relay lens pair, this provides enough angular field of view. The mirror has a clear aperture of 4 mm by 4 mm, although a smaller portion of the mirror is actually used. The rear side of the mirror is reflective in order to reflect the beam from a laser diode onto a position-sensing diode (PSD). The output from the PSD is used to sense and to control the galvo mirror position in a closed-loop manner.

The mirror rotates $\pm 12.5°$ on miniature jeweled bearings, which have negligible friction. Epoxied on two vertical sides of the mirror are wound copper wire coils. Adjacent to each coil are rare-earth permanent magnets mounted to the housing. The coils are wired in series to cause a force in the same angular direction when direct current of a given polarity is applied.

To fulfill the targeted HDS holographic exposure page rate of 208 Hz, we have defined a mirror settling time of 1 ms. This leaves 3.8 ms for holographic exposure of the recording medium.

5.2 HOH-Media Positioning

To access different stack locations an R-θ positioning assembly was built to move the HOH and the medium disk. The medium disk is rotated (θ) and the HOH is moved along the radial centerline of the medium (R). Discrete stack locations are thereby addressed. The primary components of the R-θ subsystem are the actuators, position sensors, and structural frame. For the targeted HDS holographic exposure page rate of 208 Hz, we have defined a stack access time of 5 ms. An additional requirement is that movement of the medium relative to the HOH be no greater than $\lambda/10$ during the holographic exposure.

The medium disk positioning is achieved with a rotary servomotor. Attached to the motor shaft is a spindle assembly that consists of a housing, bearings and a spindle shaft. The face of the spindle shaft has a disk mounting surface with a feature to prevent rotation of the disk relative to the shaft. This design allows engagement backlash of the disk on the spindle to be less than 10 µm (radial and angular). Disk position is sensed with a Hoetron sensor (as used in optical CD players), which detects a pair of reflective grating markings that are located on the outer perimeter of the disk (5400 lines per revolution), and are in quadrature with one another.

Dual linear servomotors located on either side of the HOH provide positioning of it. The coils of the linear motors are mounted on the HOH, and the permanent magnet/pole pieces are fixed to the frame. Linear rails and rolling element bearings guide the HOH radially. A linear optical encoder provides radial position feedback. The encoder consists of a glass scale with a set of reflective gratings (20 µm pitch) in quadrature and a sensing unit with an LED source and photodiode sensors. The scale attaches directly to the HOH, and the sensing unit is fixed on the frame.

A zero datum between the HOH and the medium disk must be established during system start-up. This is an important part of the position accuracy of the stacks. The datum is established with a homing procedure. During homing, the reference beam is turned on and illuminates the disk as the HOH is slowly moved from the inner to outer radius. A reflective home mark is located along a specific radial location on the disk. When the reference beam reflects from the radial home mark it is sensed by a photodetector located in the HOH, thereby providing the zero datum.

Servo electronics were also developed for control of the R/θ positioners. These are digital controls incorporating a 40 MHz digital signal processor. The analog signals from the R/θ position sensors are digitized by a 10-bit ADC. The digitized signals are entered in the summing junction along with the digital trajectory from the DSP. This creates the error signal to a 12-bit DAC. This signal goes to the power section, which provides current to the R or θ motor.

5.3 Changer

To provide higher capacity the multistore drive incorporates an autochanger. The autochanger is designed to hold 30 disks in a magazine located in the front of the drive. The autochanger can move a single disk at a time to and from the read station. The arm on the autochanger has a mechanism for grasping the disk and simultaneously unlatching and opening a door so that recording can take place. The autochanger is a two-axis device: one axis is a pivoting rail (arm), the other axis is a mechanism (car) that translates along the length of the arm. Both axes are driven by stepper motors via square-tooth rubber belts and pulleys. Each axis has an optical sensor for rotary position sensing. The arm pivots 25° in the horizontal plane. This is to allow access to the user I/O slot on one end of travel to the furthest magazine slot on the other. The car translates along the arm to remove and replace the disk from the magazine or user I/O slot. The combined motion of the arm and car also functions to place the disk on the spindle.

The changer is illustrated in Fig. 9, and a photo is shown in Fig. 10.

The autochanger has a number of optical sensors to keep track of disk position as it is handled. These are located at the user I/O slot to sense when a disk is present and when the autochanger has picked up the disk. Another sensor is located at the end of the arm facing the magazine to sense whether a disk is present in each of the 30 magazine locations. The final sensor tells when the disk is in location to place on the spindle.

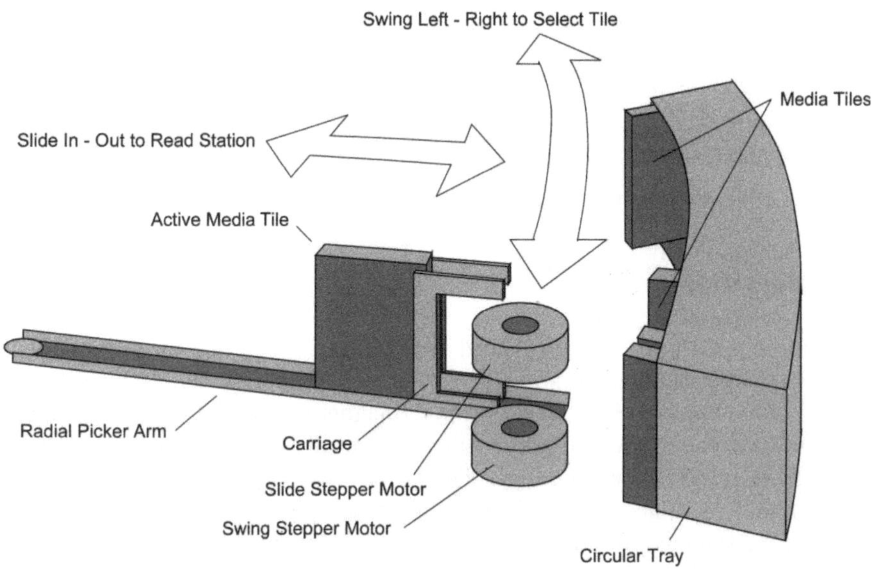

Fig. 9. Disk changer assembly

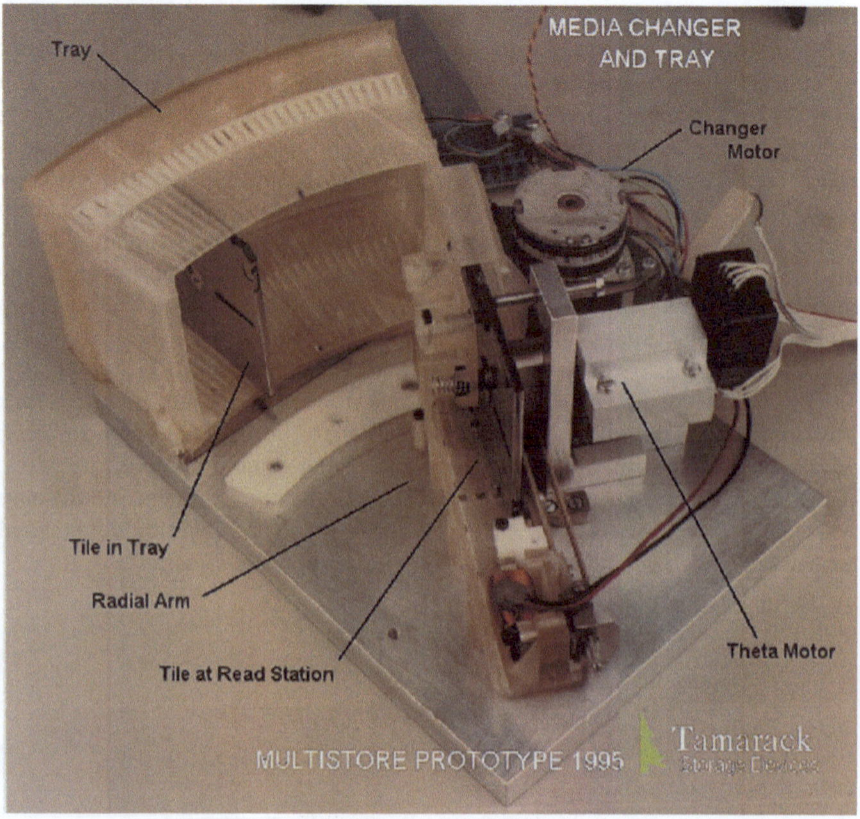

Fig. 10. Photograph of the holographic optical head

6 Summary

Evaluation of the prototype proved the value of the optical head approach. The beam positioning for stack accessing was simplified and could use mechanical technology from conventional magnetic and optical disks. Large lenses were not needed to get a full access window so the optics could be very fast and still not expensive. The is no wasted space because of a need for standoff distance for the beam deflection optics. The circular disk approach for the medium also proved to be an excellent choice.

The biggest drive design concern was capacity. The time to develop the prototype was much longer than planned, and the 1 Gbyte raw capacity of the disk, while attractive when the work was started, is not sufficiently greater than existing conventional technologies to entice a user to the new technology of holographic storage. It was felt that by increasing the thickness of the medium to several millimeters and increasing the diameter to that of a DVD would give it sufficient capacity to again make the drive attractive.

Some adjustments to the prototype would have to be made to accommodate this larger thickness.

Unfortunately as of this date the effort has not resulted in a product because of difficulties that were encountered in test with the planned photopolymer recording material. Photopolymers from a number of different sources has now been tried but none have been completely satisfactory. They have all had shrinkage during recording, not enough sensitivity, and insufficient dynamic range. A promising photopolymer is now being characterized which may solve this problem.

High-Density, High-Performance Data Storage via Volume Holography: The Lucent Technologies Hardware Platform

K. Curtis, W.L. Wilson, M.C. Tackitt, A.J. Hill, and S. Campbell

As illustrated throughout this book, it has long been known that volume holography allows multiple pages of data to be stored in the same volume, thereby allowing densities exceeding the diffraction limit for optical storage. The great attraction is the possibility of large storage capacity, with the individual high-density data pages, (containing $O \sim 10^6$ bits) addressed by changing the angle [1], wavelength [2], or phase code of the reference beams [3], resulting in rapid parallel storage and retrieval of digital data. For this reason holography has long been considered as a emerging technology for developing data storage devices with high density, large capacity, and fast transfer rate.

Although data storage using volume holography has been proposed since the 1970s it has failed to become a commercial technology because of five key reasons:

- *Material:* There has been no viable material for this technology. Photorefractive media have insufficient sensitivity and dynamic range to be of commercial interest.
- *Methods:* Traditional multiplexing strategies proposed result in complex systems, which are difficult to implement. More importantly, geometrical constraints severely limit the maximum density in thin media.
- *Lasers:* Initially proposed laser sources were expensive, complex and extremely unreliable.
- *Detectors:* fast frame rate, large array detectors that are needed for readout have had relatively poor performance for this application, and are generally low-yield, high-cost devices.
- *Data input devices (spatial light modulators, SLM):* Suitable devices developed for the display industry have recently come into existence, but their performance characteristics for storage applications have not been adequately tested.

Our research team has made great strides toward solving many of the problems described above. A list of our key technical accomplishments is summarized below.

1. Using organic, visibly curable, photopolymer-based systems, a new write-once-read-many (WORM) material has been developed. This material system can be fabricated into thick (1 mm) media, with excellent optical quality ($> \lambda/4$).

2. New methods for recording multiplexed holograms have been invented. One innovative technique resolves two of the historic difficulties in implementing holographic storage systems. First, the technique can be implemented with an uncomplicated mechanical geometry, in which individual holograms are addressed by a simple translation of the media. Second, the technique is not dependent upon the Bragg effect: therefore the hologram selectivity is independent of the thickness of the recording medium.
3. A compact, low-cost, reliable, high-power solid-state laser has been developed under contract for storage applications. In addition, commercially available solid-state lasers for the printing and medical markets have advanced significantly.
4. A newly developed CMOS active pixel detector has been adapted for holographic storage applications. This device, which has lower cost and higher performance than CCD detectors, will allow functionality on a pixel level for data processing or buffering. In addition the device inherently supports high data transfer rates.
5. We have modified the new Digital Micromirror (DMD) technology developed by Texas Instruments to resolve the input device need. In principle, DMD technology permits >2000 frame/s data presentation rates, with 1000:1 bit contrast at 1280×1024 pixels per data page. (Modification of the optical window mounting is required to prevent birefringence.)
6. Channel encoding, error correction, and tracking issues have been explored to establish feasibility. A demonstration research prototype has been developed capable of full digital data storage and recovery in polymeric media.

All other technical challenges to holographic storage have been substantially reduced by work described in this book. The implication of these developments for deployment of a technology based on volume holography is discussed below. First, in this section we will review the status of our effort detailing the progress made and directions considered.

1 Materials

Much of the materials development effort has been discussed earlier in this book in the chapter entitled "Photopolymers for Digital Holographic Data Storage." We will simply outline the effort here. Our effort has emphasized photopolymer materials for two dominant reasons. First, the high photosensitivity, high dynamic range, and ease of processing of photopolymer materials for display applications implied that this is a promising material class for adaptation for storage applications [4]. Second, from a system perspective, a WORM (write-once-read-many) drive would provide a simple vehicle to explore feasibility and technology development.

The materials used have been variations on a design strategy in which a cross-linked matrix is formed in-situ, which contains a photoactive species utilized in the recording process. For example, one system consisted of matrix formed by multifunctional acrylate monomers with mono-functional acrylate oligomers that are polymerized during recording [5]. Flat, high optical quality media were fabricated ($< \lambda/4$) using both glass and polycarbonate substrates. The material specifications are also detailed in the chapter referred to above. The materials parameters of greatest importance from a system perspective are dynamic range (quantified using the $M\#$ and scattering floor), the sensitivity, and the optical quality. These key specifications set the system limits and the laser power budget for the storage device in development. Details of materials preparation and media design are reviewed in the referenced chapter. Figure 1 gives an illustration of the optical quality of the prepared medium. The medium is sandwiched between AR-coated glass substrates. The medium typically has $<10^{-6}$ optical scatter and a flatness better than $\lambda/4$ per cm.

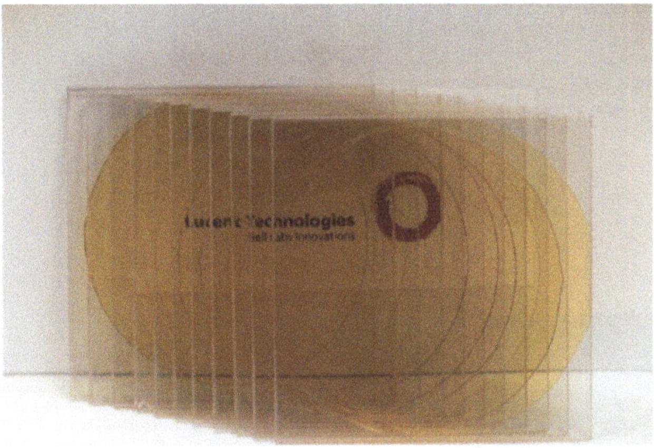

Fig. 1. Examples of the photopolymer used for data storage

2 Multiplexing Methods

The introduction of shift multiplexing, (which was conceived at Caltech [6] and reduced to practice at Bell Laboratories [7]), led to a paradigm shift in holographic storage system design. The idea of accessing holograms by simple translation of the medium in a manner similar to conventional disk drives greatly simplified device architecture concepts. One primary Bell Laboratories innovation was the development of a novel strategy for multiplexing holograms, called *correlation multiplexing* (CM) [8]. In this technique, multiple holograms are stored in an overlapping fashion, accessed by small translations of the medium relative to the incident signal and reference beams. The reference beam is "spatially" modulated with the hologram position information.

This highly complex, *bandwidth-controlled* reference beam relies on the phase, amplitude, and angle differences induced by translating the medium relative to the complex reference for selectivity. The selectivity for a single hologram is independent of the thickness of the storage medium. The stored pattern of a single hologram is recovered when the read-out reference beam is centered on the stored image to an accuracy determined by the correlation function of the reference beam (which in practice can be as short as a few microns). This is in contrast to traditional methods employing the Bragg effect, where the selectivity exists in primarily one dimension and grows with material thickness. Figure 2 illustrates the high shift selectivity of CM. Pictured is (a) a single selectivity map (i.e., hologram diffraction as a function of medium

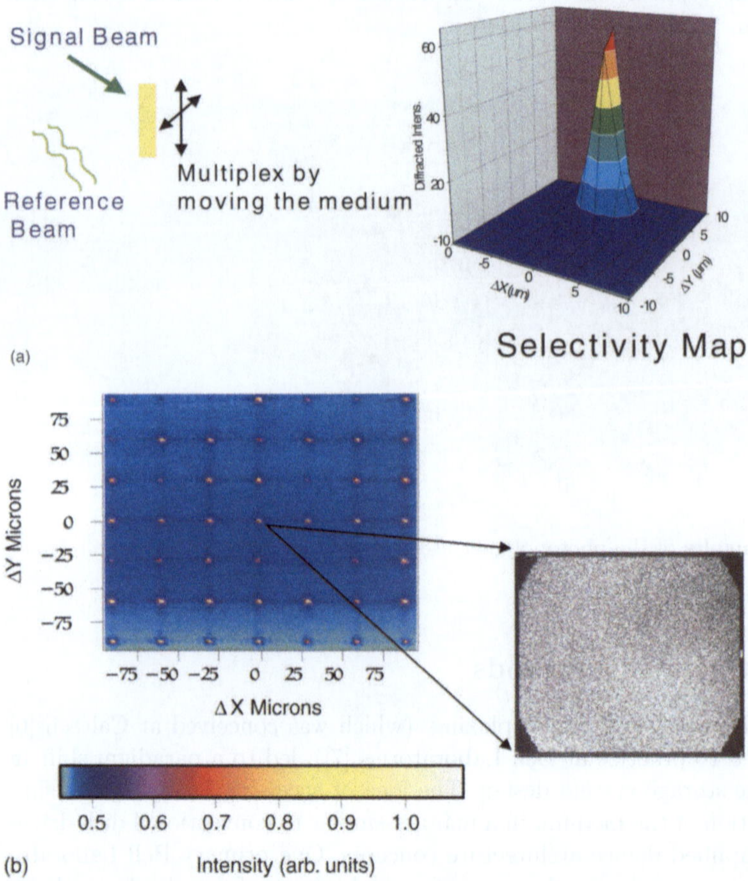

Fig. 2. (**a**) Selectivity map for a single CM-recorded hologram. (**b**) An array of holograms; scan of hologram diffraction efficiency as a function spatial position. At each maximum a greater than 480-kbit data page is stored, as shown. The sample was a 250 μm polymer film

spatial position) and (b) a 7×7 array of stored holograms in a polymeric medium. Shown is a contour plot of the integrated diffracted intensity from the media as a function of position. The diffracted intensity maximums are on 30-μm centers with a full width at half maximum of $\sim$3 μm (each representing a stored data page). It is important to note that CM is unique in that the selectivity obtained is independent of both material thickness and signal bandwidth, and has high SNR. The selectivity of a few other multiplexing methods such as employing fractal sampling grids [9], peristrophic multiplexing [10], or aperture multiplexing [7], are independent of material thickness, but depend on the signal beam bandwidth. Thus the more information each hologram contains, the lower the selectivity. Using our novel techniques, higher densities can be stored in thin media. Using CM, storage densities in excess of 350 channel bits/μm^2 have been demonstrated in 4 mm Fe-doped $LiNbO_3$ at a capacity of the order of 4 Gbits (16 000 holograms).

3 Components

During the last five years several components essential to the development of any viable holographic storage system have been developed. The three key devices are: low-cost, highly coherent laser sources; a high-speed; high-throughput; input device; and finally a high-speed, low-cost output device. Figure 3 shows a block diagram of the necessary components. (Here we have omitted mechanical actuators, concentrating on components unique to holographic applications.) While the figure shows a general configuration read–write device, only the components inside the dotted box are required for a read-only memory (ROM).

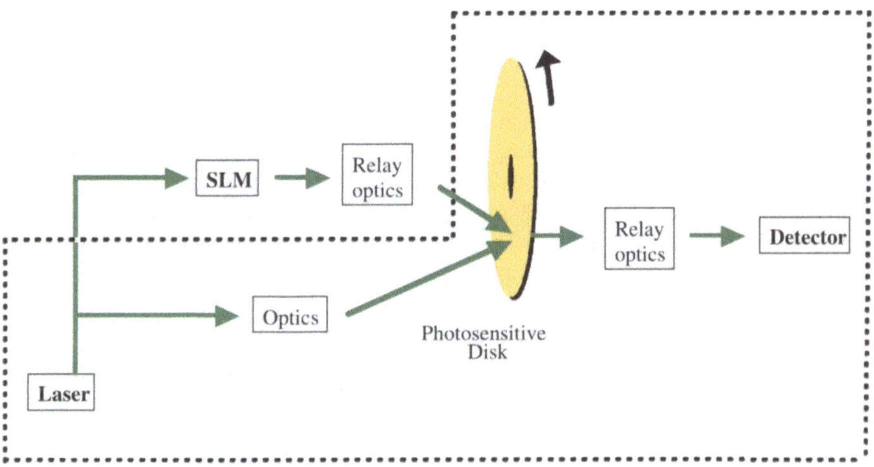

Fig. 3. Block diagram of a holographic storage system. The key components are in bold

The laser source required must be low cost with moderate power (100–200 mW), have a long coherence length (0.1–1 m), and a TEM_{00} beam profile. Diode-pumped solid-state sources allow for flexibility in system design. We utilize doubled Nd:Yag sources (many of these devices are available commercially). A 100 mW, Coherent Radiation model Compass 315 is used in the demonstration system described below. Although technically, lasers of adequate quality have reached maturity, the cost of laser devices is still an issue. Low-cost lasers such as those used in compact disk players, though inexpensive, do not have the required coherence properties. Currently the sources of choice cost in the $2000–20 000 range, and although prices are dropping, the lack of a high-volume application for these devices may limit the rate of this cost decline. One example of a possible source is a monolithic microchip laser [11], which could serve the holographic storage application well. These diode-pumped solid-state devices naturally lase in a single longitudinal TEM_{00} induced by the dimensions of the microresonator. The small volume of active material used may allow high-volume production of these devices. A 100 mW prototype device was developed under contract to demonstrate technical feasibility.

The input device encodes the digital data onto the object beam. This of course needs to be done at a fast frame rate (>100 frames/s) with as high an efficiency as possible to have minimal effect on the laser power budget. Liquid crystal devices are commonly suggested for this application but they have insufficient throughput and frame rate. More viable choices are the Texas Instruments Digital Micromirror Device (DMD) or ferroelectric liquid crystal displays (Displaytech). We have used primarily the DMD as our input device [12]. This device, designed for projection television applications, has 2 kHz frame rates and high throughput (<50% typical). In the demonstration device, the modulator is a 848 × 600 array of 16-μm mirrors on 17-μm cantelevered centers. The individually addressable mirrors steer the input beam either down or out of the optical path, yielding binary "on" and "off" states. Commercial devices exhibited extensive birefringence, introduced as a result of stress in the optical window (Fig. 4a). Though not important for projection applications, birefringence could be fatal in a holographic device. By modifying the way the optical windows were mounted onto the device we eliminated this problem, resulting in an extremely low birefringence device (Fig. 4b). The DMD also has very high contrast. For the first 2.5 Fourier orders of reflected light, the device used had a efficiency of 53% and an 800:1 contrast for digital data. Figure 5 shows a magnified single-bit chessboard pattern illustrating the high contrast of the device.

For the output device speed and cost are crucial. Although high-speed CCD detectors have been developed, the prospect for *low-cost*, high-performance devices is questionable. CMOS active pixel sensors (APS) are a better solution. These devices, made in a conventional silicon process, have a high probability for high-yield, low-cost manufacturing. In addition, these devices

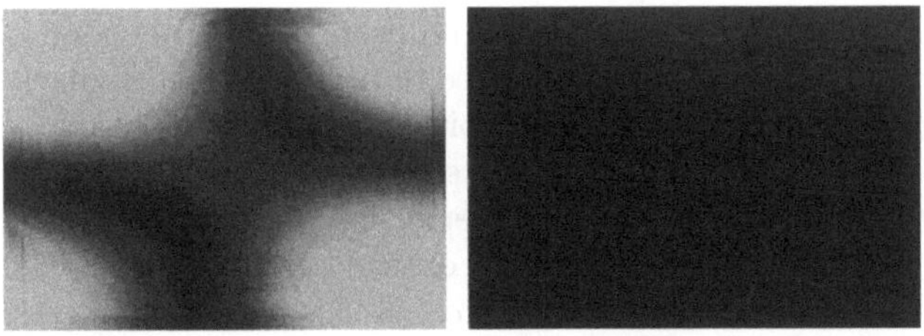

(a) (b)

Fig. 4. Birefringence: (**a**) original window; (**b**) modified window. Both measured interferometricially

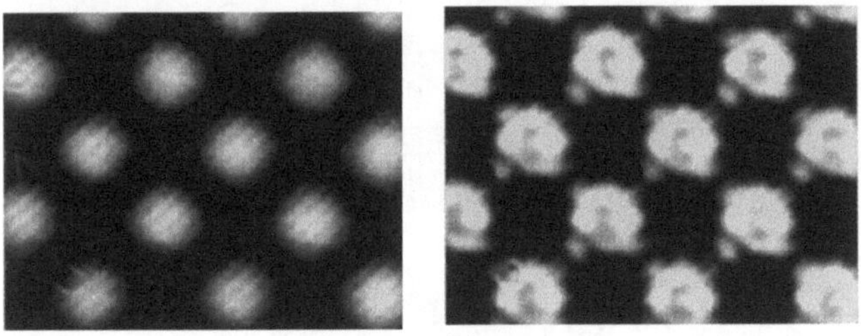

(a) (b)

Fig. 5. ~25 DMD Pixels: (**a**) low-pass, 32% eff; (**b**) 2.5-pass, 53 % eff. Contrast ~800

may allow integration of complex processing functions on the detector. CMOS is an inherently fast (nanosecond switching response), low-power technology allowing great device versatility. APSs typically require only 1% of the power needed to operate comparable CCDs. Generally an APS can operate uncooled at room temperature with resolution in excess of 10 bits at video-rate capture. Furthermore, packaging costs for CCDs are approximately ten times those for APSs because of the heat generated by CCDs. Finally, because CMOS cameras are fabricated in standard CMOS integrated circuit fabrication lines, their production costs are well below those of CCDs, which rely on highly specialized fabrication processes. The APS camera used in our demonstration system was co-designed and fabricated by Photobit, LLC (Pasadena, CA). The detector was designed to match the DMD device mentioned above allowing 1–1 pixel matching. We have also demonstrated oversampled data recovery.

4 Holographic Demonstration System

Our team has developed a demonstration storage system based on CM. Taking advantage of a closely coupled materials and system development effort, a research prototype system capable of automated storage and recovery of pixel matched holograms has been constructed [13]. The system is shown in Fig. 6.

The ~480 kbits/page system is the first of its kind using a photopolymer as medium. Figure 7 shows an example of a recovered data page; the edge markings are used for alignment and feedback. Raw bit error rates of the order of 10^{-5} have been observed for recorded data pages recovered in 280-μm polymer films. Error correction codes have allowed error-free recovery of a wide variety of stored data at high density. Note: all optical components (lenses, beamsplitters, etc.) are standard camera optics or off-the-shelf components; no custom optics of any kind have been used.

Fig. 6. Demonstration system; this 1 ft × 2 ft (300 mm × 600 mm) device is fully functional, and allows for digital data storage and recovery from a polymeric medium

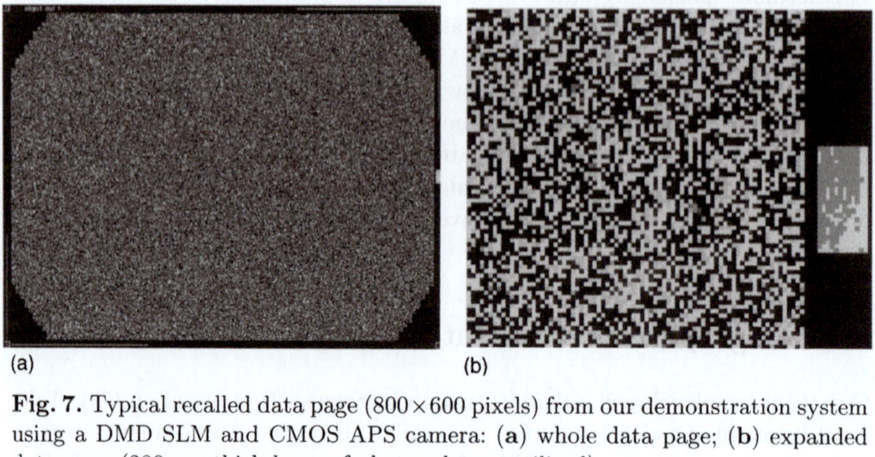

Fig. 7. Typical recalled data page (800×600 pixels) from our demonstration system using a DMD SLM and CMOS APS camera: (**a**) whole data page; (**b**) expanded data page (280-μm-thick layer of photopolymer utilized)

5 System Evolution

Significant progress in characterizing this unique data channel has been achieved, leading to data page normalization and the application of new, powerful, and more compatible error-correction codes. Use of these codes will dramatically reduce system overhead and increase robustness for the next generation device. Currently using shift multiplexing and a new materials formulation [14], we have written ~4000 holograms (~1 Gbit of data), at a user density of ~45 bits/µm^2, corresponding to ~50 Gbytes of user capacity in a $5\frac{1}{4}$ in disk. A plot illustrating the progression of measured density in our shift apparatus is given in Fig. 8. The maximum density achieved has been limited by the material's dynamic range. We now have higher dynamic range (4×) formulations in hand that should support 150 Gbytes of USER capacity in the same format with greater than 30 Mbytes/s transfer rates.

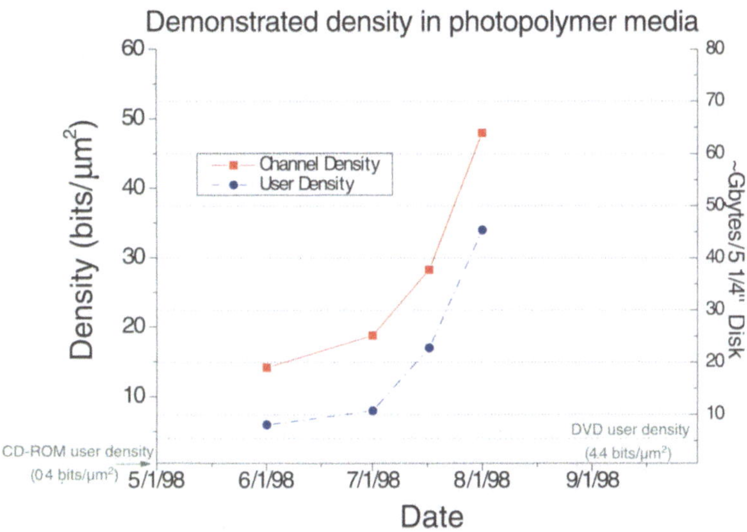

Fig. 8. A plot of the density progress made with our hardware platform; measurements are made using 750-µm media. The plotted density is based on full digital data storage and recovery from a piece of polymeric media

6 Summary

The development of this material class and the evolution of the component technologies have moved the holographic storage closer to viability. The technical feasibility of a high-performance WORM or ROM device has been established. The risks associated with drive engineering are at this point comparable to the industry standard. The most critical risks are still material related: in particular, archival lifetime, shelf life and manufacturability. Preliminary research on these issues for our materials has been promising, but work is ongoing. Finally, the key determining factors in this technology's success will be drive cost and finding the right market for a high-performance WORM system.

Acknowledgments

The authors would like to thank M. Schilling, C. Boyd, L. Dhar, M. Schnoes, A. Harris, A. Hale, H. Katz, and T. Richardson for important contributions to this work.

References

1. D.L. Staebler, W.J. Burke, W. Phillips, and J.J. Amodei, Appl. Phys. Lett. **26**, 182 (1975); F.H. Mok, M.C. Tackitt, and H.M. Stoll, Opt. Lett. **16**, 605 (1991).
2. G.A. Rakuljic, V. Leyva, and A. Yariv, Opt. Lett. **17**, 1471 (1992); F.T. S. Yu, S. Wu, A.W. Mayers, and S. Rajan, Opt. Commun. **81**, 343 (1991).
3. J.E. Ford, Y. Fainman, and S.H. Lee, Opt. Lett. **15**, 1088 (1990); C. Denz, G. Pauliat, and G. Roosen, Opt. Comm. **85**, 171 (1991).
4. W.J. Tomlinson, E.A. Chandross, H.P. Weber, and G.D. Aumiller, Appl. Opt. **2**, 95 (1976).
5. V. Colvin, R. Larson, A.L. Harris, and M. Schilling, J. Appl. Phys. **81**(9), 5913, (1997).
6. D. Psaltis, M. Levine, A. Pu, K. Curtis, and G. Barbastathis, Opt. Lett. **20**, (1995).
7. K. Curtis and W.L. Wilson, AT&T Bell Laboratories, Technical. Mem. #11141-941019-22 (1994).
8. K. Curtis and W.L. Wilson, U.S. Patent, PN:5,719,691, granted (1998).
9. H. Lee, X.G. Gu, and D. Psaltis, Appl. Phys. Lett. **65**, 2191 (1989).
10. K. Curtis, A. Pu, and D. Psaltis, Opt. Lett. **19**, 993 (1994).
11. J.J. Kayhowski and A. Mooradian, Opt. Lett. **14**, 24 (1989).
12. L. Dhar, C. Boyd, S. Campbell, K. Curtis, A. Harris, A. Hill, N. Levinos, M. Schilling, M.C. Tackitt, and W.L. Wilson, *Optical Data Storage*, **8**, OSA Technical Digest Series, pp. 113–115 (1998).
13. S. Campbell, K. Curtis, A. Hill, T.J. Richardson, M.C. Tackitt, and W.L. Wilson, *Optical Data Storage*, **8**, OSA Technical Digest Series, pp. 168–170 (1998).
14. K. Curtis, W.L. Wilson, and M. Tackitt, private communication.

IBM Holographic Digital Data Storage Test Platforms

C.M. Jefferson, G.W. Burr, and J.A. Hoffnagle

IBM has developed three technically advanced test platforms to provide an environment in which media and the technology relating to the storage of digital information using holographic techniques may be studied and developed. These testers are designed to allow experimental access to a wide range of holographic data storage parameters with minimal instrumental contributions to the raw error rate. These testers have been pivotal in advancing the state of the art in holographic digital data storage, and many significant performance benchmarks have been achieved with them.

The first tester to be built was the PRISM (PhotoRefractive Information Storage Materials) tester [1,2]. This tester was designed to allow holographic data storage materials to be studied without concern for artifacts in the data created by deficiencies in the optics or mechanics of the test apparatus. In order to test the technical feasibility of implementing an actual storage device, and to allow the exploration and development of signal processing and error correcting algorithms, the DEMON I (DEMONstration) and DEMON II test systems have been built [3]. These testers are specialized relative to the PRISM tester, in that many of their operational parameters are fixed by design. These testers are designed for the storage and retrieval of multiple pages of arbitrary, pixel-matched digital data, at moderate data rates, and allow modulation and error correction schemes to be implemented "on the fly." DEMON I is relatively unaggressive with respect to data density and data rate, but is capable of being reconfigured for various experimental needs. DEMON II is highly specialized, and not flexible. DEMON II incorporates all the lessons learned in DEMON I and PRISM, and has optics that are both novel and also capable of storage densities compatible with a realistically competitive design point for a product.

1 PRISM Photorefractive Materials Tester

The PRISM tester was conceived as an apparatus that would allow the rigorous evaluation of any holographic storage medium in a pixel-matched, digital data storage context at megapixel data page densities. A broad range of materials, from thin organic films to thick inorganic crystals, was to be studied. It was necessary to allow for operation at multiple wavelengths, ranging from the green to the red. Argon ion (514.5 nm wavelength) and krypton ion (676 nm wavelength) lasers, operated with internal etalons to narrow the linewidth and increase coherence length, were used as the primary experimental light sources. The beams from the lasers were delivered to the optical system with connectorized single-mode polarization-preserving fibers. This allowed the lasers to be interchanged without any realignment of the optics, and also produced a very clean, near-Gaussian beam profile owing to the spatial filtering action of the fiber.

Owing to the diverse nature of the holographic materials to be tested, angles between the object and reference beams of as little as 15° to as much as 165° had to be accommodated. It was necessary to control the intensities of the reference and object beams, and also their ratio over many orders of magnitude. The apparatus was required to be interferometrically stable for timescales ranging up to hundreds of seconds. It was essential that the input data pixels be extremely accurately matched to the pixels of the CCD camera that was to take the data, and the optics had to be capable of reproducing the image of the object data page onto the CCD camera with an error rate small enough to be negligible in comparison with the error rate contributed by the medium under test. All of the functionality of the tester was to be controlled by a computer, as experiments of the type envisioned for the tester required precision control and reproducibility. This combination of requirements had never before been implemented in a holographic storage tester, and was very challenging to satisfy. The result has been highly satisfactory, however, and the PRISM tester has made fundamental contributions to the field of digital data storage by holographic techniques.

The PRISM tester is depicted in Fig. 1. There are four major elements that contribute to its high performance and resolution. The most obvious feature is the large three-legged structure in the center. This tower supports the sample under test, and the precision motion control systems which allow the sample position and orientation to be adjusted with four degrees of freedom. This support structure was necessary because of the innovative method of generating the reference beam, which utilizes the set of rotational stages and mirrors located immediately below the tower.

The reference beam generator, known as the θ-2θ system, is illustrated in Fig. 2. It utilizes a pair of concentric, highly precise rotational stages to reflect a fixed reference beam to a desired angle about an axis centered on the rotational axes of the stages. This reflected beam appears to originate at the center of the stages, and can be swept to any angle. To redirect it so that it is incident on the sample, a retro-reflecting erecting prism is used. This combination of mirror on the axis of rotation, and prism on a separate arm that rotates concentrically with the center mirror, but through twice the angle, allows a reference beam to impinge on the sample from approximately 300 of 360° of azimuthal angle, with an accuracy and reproducibility of about 1 μrad. The PRISM tester is thus uniquely suitable for evaluating samples where unusual reference beam angles are required.

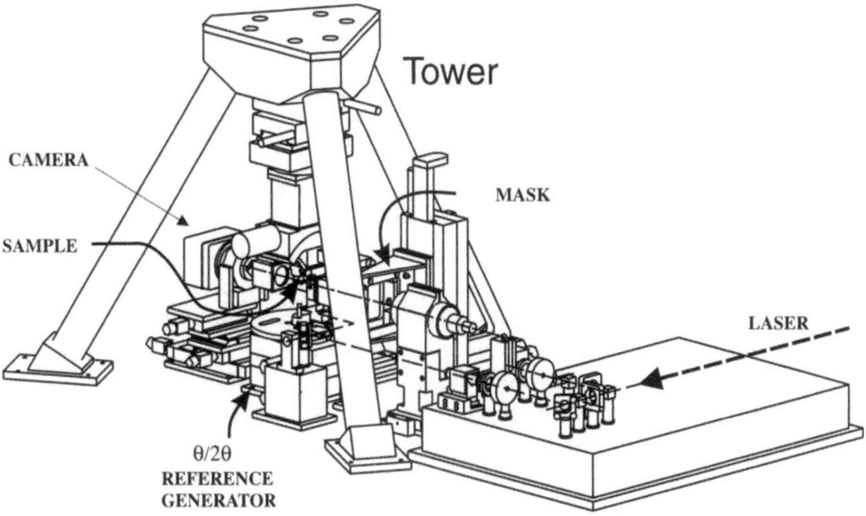

Fig. 1. The PRISM holographic materials test apparatus. The laser beam enters from the right, and is split into the reference and object beam paths. Not shown are the input fiber optic coupler, assorted optical power monitoring and control devices, and the argon or krypton ion lasers used to provide optical energy. The cylindrical object in front of the mask is a beam-expanding telescope used to flatten the intensity distribution of the object beam

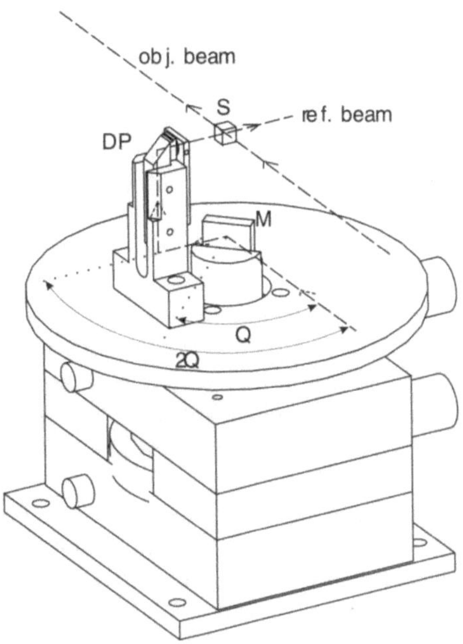

Fig. 2. The $\theta/2\theta$ reference beam generator. This dual, concentric rotational system utilizes a mirror (M) and retro-reflecting Dove prism (DP) to redirect a fixed reference beam into the sample (S) from as many as 300 of 360°, under computer control with microradian precision

Two other features that distinguish the PRISM tester from previous holographic data storage devices are the extremely precise Fourier transform optics and the careful matching of the object beam spatial light modulator pixels with the CCD camera pixels. Although it is commonly believed that the focal plane of a lens contains the Fourier transform of the optical field present on its conjugate focal plane, most real lenses provide only an approximation of this transform. A true Fourier lens has a flat focal plane, and is telecentric in that focal plane. All parallel rays entering the lens come to a focus at the flat focal plane, and are symmetrically oriented around the telecentric chief ray. Thus a point source located at the input focal plane will produce collimated rays in the Fourier focal plane.

These conditions place severe constraints on the design of a true Fourier lens, and make the realization of an inverse Fourier lens with low aberrations very difficult. In addition, since the optical system is afocal, it is not very effective to adjust the magnification by adjusting the relative positions of the object and the image. In PRISM, this was handled by making the image focal plane slightly non-telecentric, and adjusting the positions of the lenses by very small amounts under computer control. This produced an effective magnification change of a few tenths of a percent by trading off a little defocus for a small change in the image field size. The lenses were fabricated very carefully to make the nominal magnification for the system very nearly equal to 1.0000. It has been found that errors in system magnification of only 0.01% are sufficient to cause drastic degradation of the pixel-matched bit error rate.

The objective of the PRISM tester was to provide a platform whose error rate was both extremely low and long-term repeatable. At the time of its design, the state of the art of programmable spatial light modulators (SLMs) was not well advanced, so the PRISM tester utilizes extremely precise, high-contrast photolithography masks as a source of digital data. The masks were fabricated with a pixel pitch exactly equal to that of the CCD camera. These masks were produced with the same tools used to manufacture integrated circuit masks, and are fabricated with chrome deposited on fused silica plates, which are antireflection coated. Pixel spacings of 9 μm, 18 μm, and 36 μm were provided, along with different fill factors. This allows media with poor resolution to be studied, using for example, the 36 μm pitch pixels (mapped 16 camera pixels to 1 SLM pixel), and also allows excellent media to be tested at densities (9 μm pitch) compatible with product design points (1 million pixels or bits in a single image).

The object mask is held in a five-axis position controller, and may be positioned in the field of view of the Fourier optics with 0.1 μm precision and repeatibility. Any of the many masks available on each fused silica substrate may be interchanged under computer control. These masks also include alignment targets of several types, which are absolutely necessary for aligning the optical system.

As previously mentioned, the optical system employed is a Fourier transform geometry, where the object mask is positioned at one focal plane of the input lens, while the sample is positioned at the back focal plane of that lens. The lenses had an effective focal length of 88 mm and were designed for a net magnification of 1.0000. In order to accommodate the various experimental schemes, a long working distance of 75 mm was specified. The front focal plane of the output lens is coincident with the back focal plane of the first lens, and the CCD camera chip lies in the back focal plane of the second lens. This is the so-called 4f optical system. The optical field at the sample (lying at the joint focal planes between the lenses) is the Fourier transform of the field at the input SLM, or data mask. The second lens creates an inverse Fourier transform, and thus the light falling on the CCD camera chip is an "exact" replica of the light at the mask. The PRISM tester optical train is shown in Fig. 3.

One of the areas in which the PRISM tester is supreme is the degree of exactness of the two light fields. In order to obtain very good error rates in the image of the data masks, the alignment, distortion, and magnification of the mask, lenses, and camera chip must be controlled to 0.01%. This is

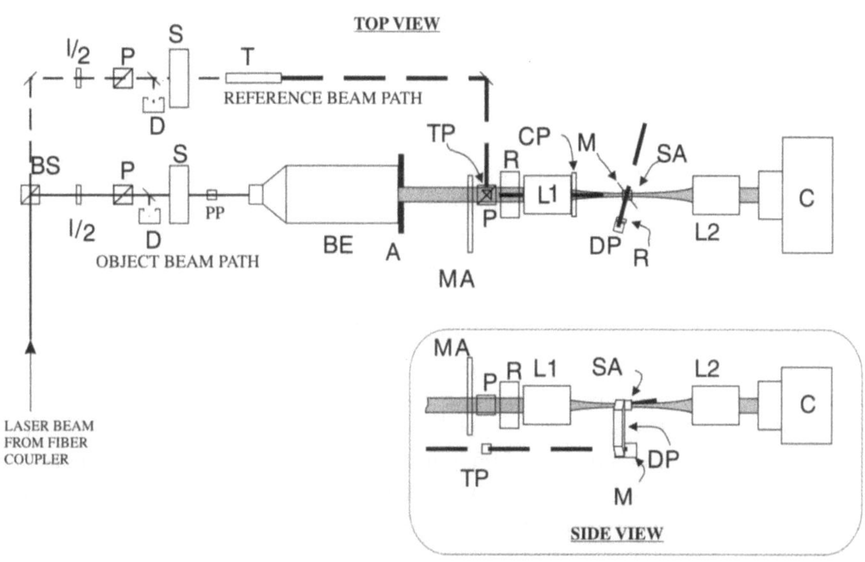

Fig. 3. The optical layout of the PRISM tester. The fiber optic coupler, argon and krypton ion lasers, and some polarization and optical power controlling optics are not shown. P: Polarizing Beamsplitter Cube; BS: Non-Polyrizing Beamsplitter cube; Ω: Half Wave Plate; D: Optical Power Detector; S: Uniblitz Mechanical Shutter; BE: Beam Expanding Telescope; A: Aperture; MA: Data mask; R: Polarization Rotator; L1, L2: Fourier Lenses; TP: Turning Prism; CP: Compensating Plate; DP: Erecting Dove Prism; R: Rotation Stages; T: Telescope; SA: Sample; C: Camera; PP: Periscope Prism

an extremely hard specification to meet, and it may be accomplished only with computer-controlled optimization of the positions of the camera relative to the data mask, and the alignment of the optical system's magnification and focus. Thus all necessary axes of the optics and camera and data mask are controlled to 0.1 μm precision with computer-actuated positioners. Sophisticated algorithms measure deviations of the optical system from perfect alignment, and the positioners are used to adjust the optics for optimum performance.

The sample is located in a position where the reference beam from the θ-2θ beam generator can overlap the Fourier transform of the object beam mask. Thus the sample records and plays back the Fourier transform of the data. This process distributes the information contained in the data mask over the illuminated sample volume. This dispersion of information has the advantage of making the storage relatively immune to errors due to misalignment of the sample or small deficiencies in the sample optical transmission, caused, for example, by dust. It also unfortunately creates a large-amplitude intensity peak along the optical axis in the sample caused by coherent summation of the components of the optical Fourier transform from each pixel with wave vectors parallel to the optical axis. This peak contains no information, and tends to saturate the medium and create optical distortions that degrade the bit error rate. There are several schemes that can reduce the effects of this coherent saturation, with the use of an axicon optical element being the most novel. This optical technique, discovered at IBM Almaden and developed on the PRISM tester, will be described later in this chapter.

The PRISM tester was the first holographic digital data storage tester to implement true pixel-matched imaging with the high-quality optical system necessary to achieve a high-fidelity reproduction of the input data on the CCD camera. This robust optical performance allowed IBM to develop the now standard bit error rate (BER) measuring techniques that have allowed such rapid and substantial progress in both materials and technique. Powerful and novel modulation and error correction codes that allow the storage and retrieval of large amounts of digital data on the DEMON testers could be developed because it was possible to isolate the effects of media deficiencies from errors caused by inadequate tester performance. PRISM thus represents the beginning of the attack on the real difficulties in making digital holographic data storage an industrial reality.

2 DEMON I Holographic Data Storage Engine

DEMON I was an evolutionary step forward from the PRISM tester. While PRISM was designed to handle any conceivable material testing requirement, DEMON I was intended to implement all of the functionality of a holographic digital data storage engine. The reference/object beam geometry was restricted to the 90° geometry, and the range of angles of the reference beam was limited to ±10°. A transmissive, computer-controlled SLM, capable of displaying arbitrary data patterns was used to provide the object beam data. A small, fast CCD camera was used, instead of the large, slow, cooled CCD camera employed in the PRISM tester. Since the DEMON I tester was, to some extent, to be a technology demonstration vehicle, attention was paid to reducing its size relative to that of the enormous PRISM tester, which occupies an entire 4 ft by 16 ft (1.2 mm × 4.9 mm) optical table. The resulting 2 ft by 3 ft (600 mm × 900 mm) footprint (not including the laser) of DEMON I was judged sufficient to illustrate the miniaturizing effect of a specialized design point. The DEMON I tester is illustrated in Fig. 4.

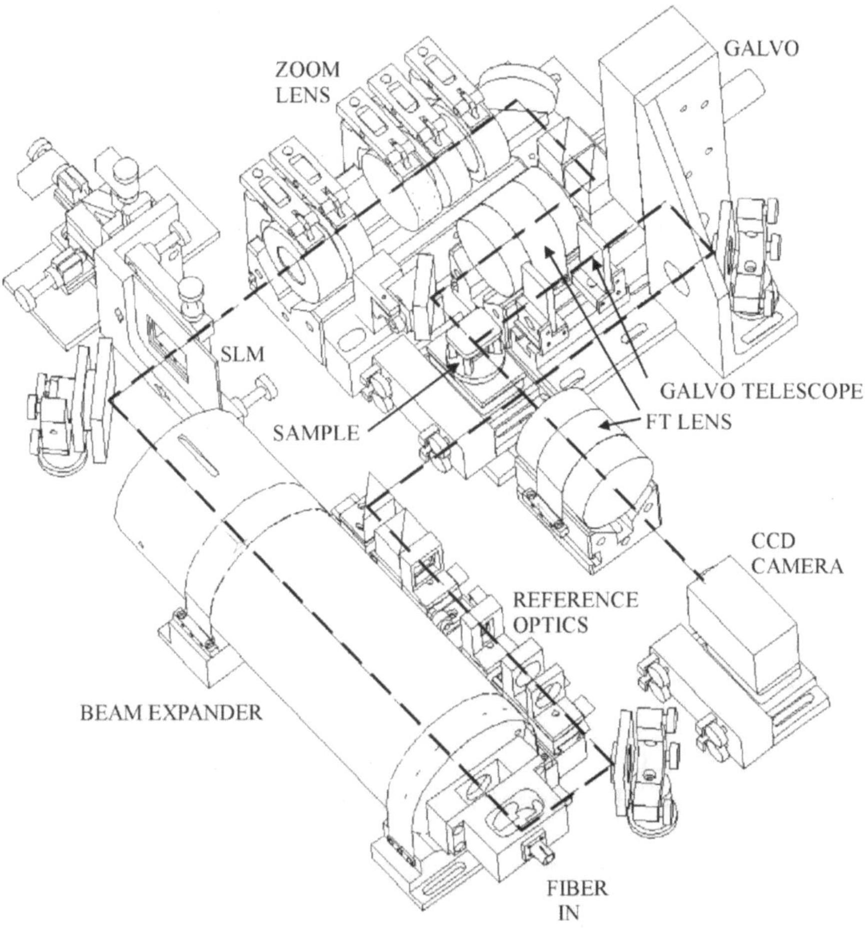

Fig. 4. The DEMON I holographic digital data storage engine. The argon ion laser beam enters through a connectorized optical fiber, and is split immediately into the object and reference beams

The same 88-mm effective focal length Fourier lenses that were used in the PRISM tester were employed in DEMON I. The 42-μm pixel pitch of the Epson SLM used to generate the data was unfortunately different from the 9-μm pixel pitch of the Pulnix CCD camera. A precision custom zoom lens with a variable magnification (or demagnification) ratio of from 0.9 to 3.0 was used to adjust the size of the SLM data field to that of the camera. This zoom lens was designed for distortion over the field of view of less than 0.02%. A single 42-μm SLM pixel was mapped onto four 9-μm camera pixels. Typically the system is aligned so that the light falls mainly on one of the four pixels, and the other three are ignored. Owing to limitations of the optics and camera, only the central 320×240 pixels of the SLM were used. The reference beam angle was varied with a galvo scanner, under computer control. This allowed rapid and accurate control of the hologram multiplexing, and with suitably thick samples several thousand holograms have been recorded and read back in rapid succession with negligible error rate (no net errors), as is described in [3].

DEMON I employs iron-doped nithium niobate as a storage material. Although suffering from deficiencies with regard to sensitivity, data retention time, and dynamic range, this crystal has excellent optical properties, and is easy to use. The laser wavelength used was the green 514.5-nm line from an argon ion laser. The laser beam is delivered to the DEMON tester with a single-mode polarization-preserving optical fiber, which produced a clean Gaussian intensity profile. Optical power delivered to the apparatus prior to the object/reference beamsplitter was as much as 400 mW.

A powerful feature of DEMON I is the computer program used to control it. This elaborate and versatile program allows intricate read/write experiments to be performed that demonstrate the use of holographic data storage as a true computer data storage engine. Large data files can be written and retrieved without errors in a manner analogous to that of a hard disk drive. This capability has allowed the exploration and development of sophisticated and successful data processing and manipulation techniques similar to those used in hard drives. This understanding of the algorithms and protocols of holographic data storage is absolutely vital to making the holographic data storage engine a reality.

3 DEMON II Advanced Holographic Digital Data Storage Engine

DEMON II is a holographic storage engine intended to demonstrate storage densities greater than 100 bits/μm^2. It is conceptually similar to DEMON I, but utilizes very sophisticated optics and a diode-pumped solid-state laser. The layout of DEMON II is shown in Fig. 5. In the configuration illustrated, it is approximately 24 in (600 mm) square, including the laser.

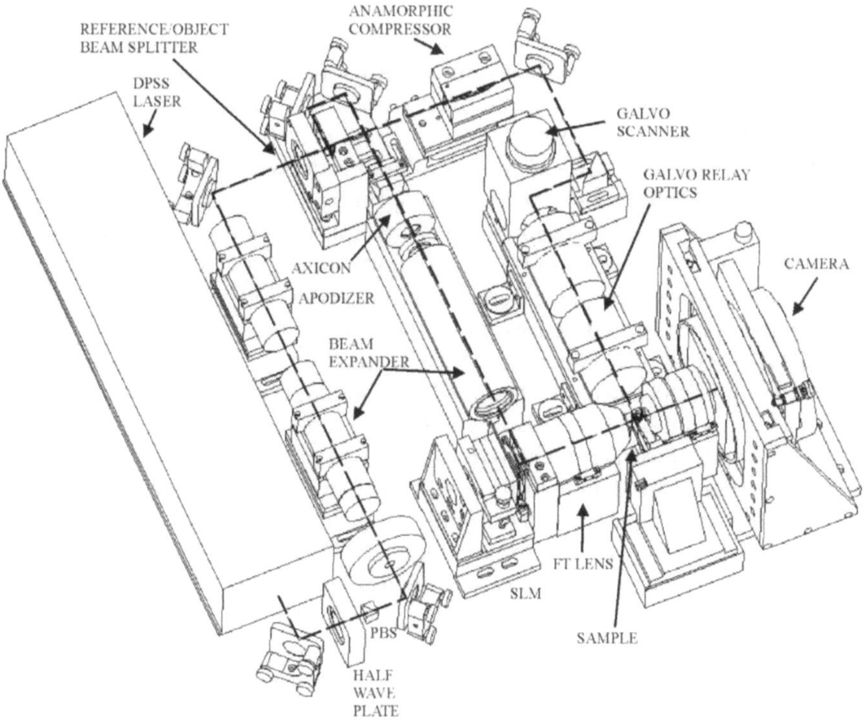

Fig. 5. The layout of the DEMON II holographic digital data storage engine. Utilizing 30-mm-focal-length Fourier transform lenses in the 90° geometry, and a 1-million-pixel spatial light modulator, is expected to this device exceed storage densities of 100 bits/μm^2

The intended storage material is lithium niobate, as in DEMON I. The wavelength of 532 nm is obtained from a diode-pumped doubled YAG laser, with an output power of 400 mW. The data source is a reflective spatial light modulator developed by IBM Yorktown, with 1024×1024 pixels. The Fourier transform lenses have a focal length of 30 mm, approximately three times smaller than that of the lenses in DEMON I. This yields a density in the focal plane approximately nine times higher than the density possible in DEMON I. The output lens has been designed for variable magnification, on the order of $\pm 0.5\%$. This will allow a much finer adjustment of system magnification, a critical parameter for good error rate.

The CCD camera is a 1 megapixel device, capable of operation at 40 frames/s. It is a larger and more sophisticated camera than that used in DEMON I, but less sensitive (although much faster) than that used in PRISM. It is mounted in a full six-axis position controller to facilitate exact alignment relative to the SLM and the optical axis. All alignment operations are carried out via computer, in a manner similar to both PRISM and DEMON I.

The illumination system for the SLM utilizes a novel optical device that turns a Gaussian laser beam into a uniform ("flat-top") beam. This device, described further below, uses custom aspheric optics to deliver about 50% of the incident laser power onto the SLM with a uniformity of a few percent. The reference beam is custom tailored in shape with anamorphic prisms to maximize the power delivered to the volume occupied by the Fourier transform of the SLM data page, and to also optimize the Bragg selectivity of the holograms. The reference beam angle is multiplexed by a galvo scanner as in DEMON I. A set of custom relay lenses allows an angular scan range of $\pm 15°$ with diffraction-limited performance, unlike DEMON I, whose rather crude galvo relay optics caused distortion to the reference beam.

DEMON II is designed for increased speed and very high density, and allows for physical motion of the medium (unlike the fixed medium of DEMON I). It is intended as both a working holographic storage device, on which research into ultrahigh data density storage will be performed, and also as a technology demonstration platform to showcase innovative optical solutions to vexing technological problems such as laser beam shaping. The design points that have been targeted for DEMON II reflect real-world competitive design points for a viable storage device, although DEMON II is not intended to be viewed as an actual prototype of such a storage engine. At the time of writing, DEMON II had achieved a data storage density in excess of 250 Gigabits per square inch, at a user bit error rate less than 1×10^{-12}.

4 Innovative Optics

In the course of development of PRISM, DEMON I, and DEMON II a number of challenging optical design problems have arisen. In several cases, very innovative solutions have been discovered. Two of these that relate directly to holographic data storage will now be described.

4.1 Axicons

As previously noted, the Fourier transform process used to image the spatial light modulator onto the medium has the side-effect of producing a very high-intensity spike in the Fourier plane on the optical axis. This is due to the coherent summation of all the light coming from each pixel of the SLM with a zero wave vector. This high intensity spike carries no information, and although fairly small in spatial extent, can cause catastrophic bleaching or saturation in the medium, with the consequence that the transmitted image (and hologram) of the SLM is severely degraded. It has been known for many years that a potential solution to the problem can be implemented by superimposing a random phase distribution on the pixels of the SLM. This so called "phase mask" has long been touted as the solution to the coherent saturation problem. In work performed by M.-P. Bernal at IBM Almaden, it was shown that although the theoretical redistribution of the intensity in the spike by a phase mask is indeed obtained, the alignment of such a phase mask is critical, and that there are new optical artifacts (dark lines and interference fringe effects) introduced in the image plane of the CCD. These artifacts, and the near-impossibility of aligning the phase mask practically, have made it improbable that it will be the solution to the coherent saturation problem.

M.-P. Bernal, with others, went on to develop alternative optical structures that perform a similar function, namely the spreading of the intensity spike, but which are far less critical of alignment. In particular, it was learned that the introduction of an axicon (conical optic) in the illumination beam of the SLM has the effect of distributing the intensity spike along a ring in the Fourier plane. The diameter of the ring depends on the vertex angle of the conical optic, its index of refraction, and the focal length of the Fourier lens. It is helpful to image the axicon onto the SLM with some relay optics, but the alignment sensitivity is small, and the axicon may be placed at the input focal plane of a Keplerian beam expander, which is used to expand the beam from the laser (usually of the order of a millimeter) to a size large enough to cover the SLM. An axicon used directly behind a transmissive data mask is illustrated in Fig. 6.

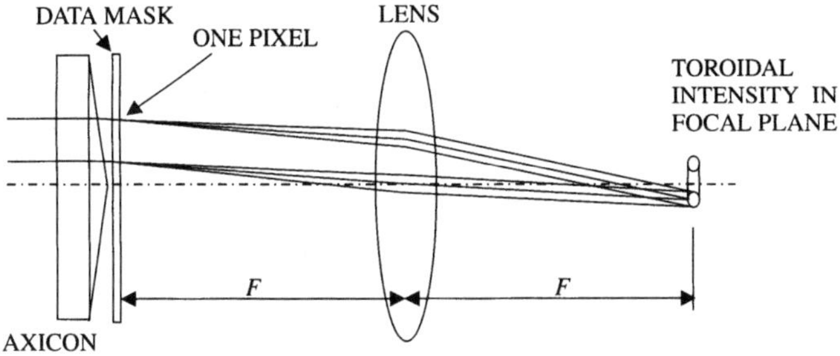

Fig. 6. A convex axicon used to spread the Fourier components of the pixellated data mask into a toroid. This reduces the coherent summation of zero wavevector (dc) waves, which otherwise would create a large spike along the optical axis and saturation and distortion in the holograms

4.2 Aspherical Apodizer

It is an annoying fact that if one illuminates an object with a Gaussian-profile laser beam, the flatness of illumination is about equal to the power in the laser beam over the illuminated area. In other words, if an illumination flatness of 5% is required over a certain area, the laser beam must be expanded to a diameter much larger than the target area, and only 5% of the incident beam will actually be used. This is very wasteful of laser power (which is hard to come by in many instances). It has long been desirable in laser physics to be able to generate a laser beam with a uniform cross-section over a significant percentage of the beam. This is a difficult thing to do, and many ingenious solutions have been proposed. The few that have been implemented generally work only over the first $1/e$ field points of the laser, and commonly suffer from poor flatness and severe diffraction effects. In addition, many solutions, such as diffractive optics, are only flat in one plane in space and both diverge and distort severely as they propagate towards and away from that plane. In addition, there is often severe distortion of the wavefront quality in the apodized beam.

As part of the design of DEMON II, the question of generation of flat-top beams was studied. This was germane not only for DEMON II, but also for ongoing work in deep UV lithography. A new insight was obtained after a review of historical efforts in this field. A two-element telescope with transmissive optical elements was designed that would produce a laser beam flat to 2% over 60% of the incident laser power, and which utilized 99.7% of the incident laser beam power with no apertures. In addition, the roll-off of the intensity profile was carefully crafted to eliminate diffraction effects from the edge of the beam during propagation. The result was an optical device that produces a highly efficient, flat-topped laser beam with the capability of propagating for several meters with little distortion and diffraction-limited wavefront quality. It was also discovered that a single apodizer was usable from the deep UV into the far IR with only a simple focus adjustment.

The Gaussian beam-to-flattop converter utilizes a convex aspheric lens to introduce aberrations into the beam, which redistribute the laser power from an incident Gaussian profile to the desired flat-top profile with a carefully controlled intensity rolloff at the edge. A second aspheric optic recollimates the aberrated beam, restoring the wavefront quality and allowing it to propagate for long distances without spreading.

Fabrication of such aspheric elements has long been very difficult and costly. Recently, new computer-controlled polishing technology has become available, which can make the fabrication of such aspheric surfaces routine. Working closely with the vendor who developed these capabilities has allowed the DEMON II design team to build such an apodizer, and to demonstrate that it works. Currently, the design is being refined, and sophisticated optical testing devices (computer-generated holograms) are being obtained to allow enhancement of the fabrication, and the realization of tighter specifications. This apodizer represents a real step forward in the area of laser illumination control, and many potential applications in a variety of areas have already surfaced.

References

1. M.-P. Bernal, H. Coufal, R.K. Grygier, J.A. Hoffnagle, C.M. Jefferson, R.M. Macfarlane, R.M. Shelby, G.T. Sincerbox, P. Wimmer, and G. Wittman, "A precision tester for studies of holographic optical storage materials and recording physics," Appl. Opt. **35**, 2360–2374 (1996).
2. R.M. Shelby, J.A. Hoffnagle, G.W. Burr, C.M. Jefferson, M.-P. Bernal, H. Coufal, R.K. Grygier, H. Günther, R.M. Macfarlane, and G.T. Sincerbox, "Pixel-matched holographic storage with megabit pages," Opt. Lett. **22**, 1509–1511 (1997).
3. G.W. Burr, J. Ashley, H. Coufal, R.K. Grygier, J.A. Hoffnagle, C.M. Jefferson, and B. Marcus, "Modulation coding for pixel-matched holographic data storage," Opt. Lett. **22**, 639–641 (1997).

Digital Holographic Demonstration Systems by Stanford University and Siros Technologies

L. Hesselink

Its useful capacity, transfer rate and access time measure the performance of a holographic data storage system (HDSS). Data should never be lost, requiring a corrected bit error rate (BER) of 10^{-12} to 10^{-15}. To compete successfully in the large storage marketplace, an HDS drive should be cost-competitive with improved performance over other drives. The exception could be certain niche markets, where unique HDS attributes – all-solid-state implementation with extremely short access times or associative retrieval – are attractive or required.

In the 1970s much fundamental materials and systems work was carried out on HDSS at RCA [1,2], in Russia [3] and Bell Labs [4], and many other industrial and university laboratories worldwide [5]. New and improved components were demonstrated, and incorporated into complex systems. However, there was neither a strong market need, nor a cost-competitive system demonstration to drive large-scale commercial development. Research activities subdued. In the middle 1980s, work at the Microelectronics and Computer Corporation and Stanford University, at Northrop, Caltech and other laboratories restarted a renewed effort in holographic data storage [6]. The primary drivers this time around were new components developed for consumer applications, and a growing commercial need for large-capacity storage devices.

The recording material, however, was not under development for other consumer applications. This led us to the formation of the Photorefractive Information Storage Materials (PRISM) in April 1994, and the Holographic Data Storage Systems (HDSS) NSIC/DARPA/Industry/University consortia in April 1995. The PRISM activities were focused on fundamental materials issues, while new components would be integrated into systems reaching a 100 GB capacity and a raw transfer rate of 1 Gbit/s [7].

Many of the activities of the PRISM and HDSS consortia are described elsewhere. In this chapter we review the demonstration platforms developed by Stanford University and Siros Technologies (formerly Optitek). Siros was the integrator of the PRISM consortium, and Siros and Stanford University jointly developed the HDSS demonstration platform using components and technology developed by HDSS members. For a more in-depth review of holographic data storage systems, the reader is referred to an upcoming invited review paper [8].

System optimization, as usual, is a complex problem involving a large number of tradeoffs. For HDSS in particular, the tradeoff between capacity and transfer rate is different from other storage systems. Fundamentally, holographic data storage is based on multiplexing many holograms in the same volume of the recording medium. For media with a linear response, this implies that the dynamic range for each hologram is roughly equal to the total dynamic range of the medium divided by N, the number of holograms. As the diffraction efficiency of each hologram is proportional to the squared index modulation, readout signal strength drops off as $1/N^2$. The larger the capacity of the device is, the smaller the readout signal strength and the signal-to-noise ratio (SNR) are. In turn, small SNRs cause large raw bit error rates, which above a threshold of 10^{-3} to 10^{-4} cannot be further lowered by error correction schemes. To boost SNR, detector integration times can be increased at the expense of data transfer rates. High-capacity systems are therefore most easily demonstrated for small transfer rates, and vice versa.

1 Optical Architectures

Broadly speaking, holographic data storage materials are divided into two classes: thin (a few hundred microns thick) photosensitive organic media, and thick (a few millimeters to centimeters) inorganic photorefractive crystals. The thin media lend themselves best for transmission-type architectures using a variety of shift- or phase-multiplexing techniques, while thick media are most suitable for angular multiplexing, typically in the 90° geometry. Although phase- and wavelength-multiplexing techniques have been thoroughly investigated, at present the required lasers and phase masks are not suitably developed to achieve large data densities [6].

As an example, consider the case of the 90° geometry shown in Fig. 1. The data-bearing object beam and the reference beam are incident upon a thick photorefractive medium from orthogonal directions. During recording green light is often used, as Fe-$LiNbO_3$ is most sensitive for this wavelength. In conjunction with thermal fixing, a slight wavelength shift is usually required [9] for readout. Alternatively, two-photon recording is accomplished in stoichiometric $LiNbO_3$ using red diode laser light and a blue incoherent gating beam [10]. Nondestructive readout requires only red light.

2 Capacity Versus Transfer Rate Tradeoff

The tradeoff between capacity and transfer rates is estimated by computing the number of photons incident on a detector element during page readout. From this calculation we can estimate the maximum number of superimposed pages readable with a certain SNR:

$$\text{Number of pages} = \left(\frac{M\#^2 \theta QEP \eta_{\mathrm{opt}} \eta_{\mathrm{fix}} t_{\mathrm{read}} 2}{MNp_{\mathrm{min}}} \right)^{0.5} , \quad (1)$$

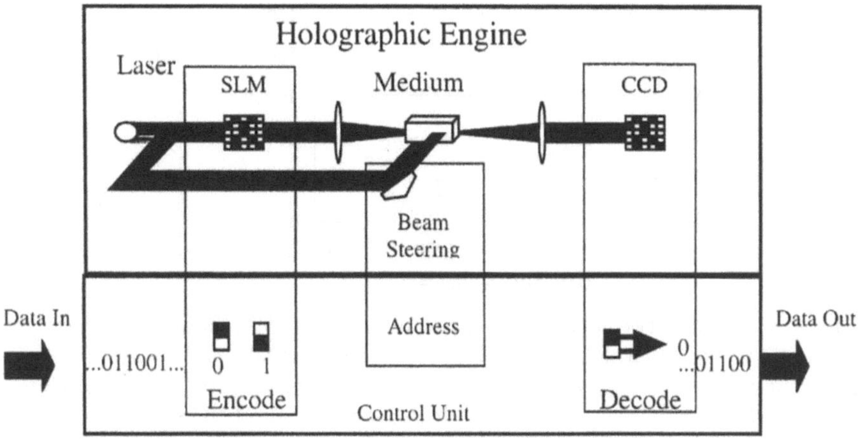

Fig. 1. 90° optical architecture, including the electronic control unit

where $M\#$ is a variable based on system and materials parameters, θ is the number of photons per watt per second, QE is the quantum efficiency of the detector array, η_{opt} is the optical efficiency of the readout system, η_{fix} is the fixing efficiency, t_{read} is the read time of the recalled data, M and N are the number of rows and columns on the data page, p_{min} denotes the number of photoelectrons needed to get a 20 dB SNR, and P is the power of the readout beam. It is assumed that bit patterns are balanced (although this does not have to be the case) by making half the pixels in a page on and half off. From this equation, the tradeoff between capacity (the product of the stored number of pages, the code rate and the number of bits per page) and data transfer rate (the number of bits per page times the code rate divided by the readout time) is determined in a photon-limited system. Optical damage and cross-talk limitations are not considered. In practice, however, optical damage often tends to limit capacity, but can be significantly reduced by recording at elevated temperatures [11]. Equation (1) provides an upper bound.

As an example, assume that $p_{\text{min}} = 10\,000$ photoelectrons to achieve a 20 dB SNR, the page size is $M = N = 1024 \times 1024$, the number of photons per watt per second is $\theta = 3\times10^{18}$, the dynamic range is $M\# = 1$, the camera frame rate is 1000 fps, i.e. $t_{\text{read}} = 1$ ms, the detector quantum efficiency $=$ 0.3 at 0.7 μm, the laser power $I = 5$ W, the fixing efficiency $\eta_{\text{fix}} = 0.5$, and the optical readout efficiency $\eta_{\text{opt}} = 0.5$: then the number of pages per stack is approximately 470, and the number of stacks for 100 Gbytes raw capacity is 2100. High data capacities and transfer rates require substantial spatial multiplexing.

However, capacities of several Gbytes could be achieved with no spatial multiplexing for data transfer rates near a few Mbytes/sec. Using acousto-optic Bragg cells, access times can be made less than 50 μs with no moving parts, described by Rockwell in the chapter entitled "Digital Holographic

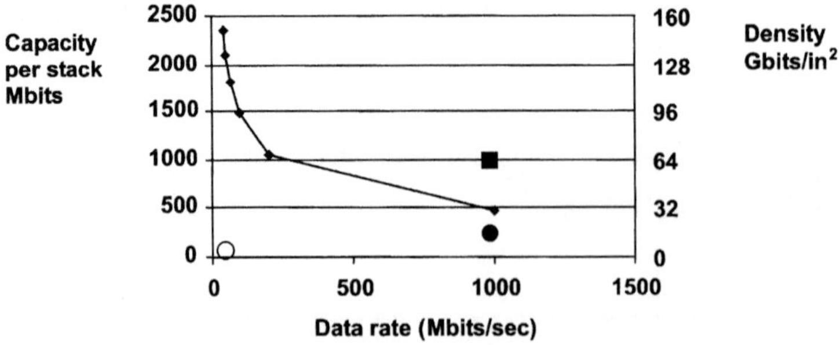

Fig. 2. Tradeoff between capacity and transfer rate for the parameters described in the text. The open dot denotes MO recording at 2 Gbit/in^2 at 40 Mbits/sec. The round filled dot (just above the "1000" marker) represents the Stanford/Siros Gbit/sec demo, while the square dot (above the endpoint of the solid line) denotes the planned final HDSS demonstration performance

Storage with Fast Access." At a transfer rate of 5 Mbytes/s – typical for the current 8× MO drives – data density for HDSS is 50–70 times higher using a laser with two orders of magnitude more power. By varying the materials and systems parameters according to (1), the curve of Fig. 2 can be modified. In particular, improvements in $M\#$ are effective in increasing storage capacity, an area of much current research.

3 Demonstration Platforms

A number of systems have been designed, built and tested at Stanford University and Siros Technologies for a variety of purposes. They were built under sponsorship of the Center for Nonlinear Optical Materials, and the PRISM and HDSS consortia. The primary goals are to gain insight into the underlying physics of the recording mechanisms and the system tradeoffs required to achieve specified performance, and to integrate new components into working demonstration platforms. The most significant demonstrations are tabulated in Fig. 3, and are briefly described in ensuing sections.

4 The Stanford University all Digital System Demonstration (Science, 1994)

In 1994, we demonstrated the first fully digital holographic data storage system [12] by storing digital images and a short soundtrack. The purpose of this experiment was to show that a fully functional system could be implemented with off-the-shelf components. Particular emphasis was placed on using digital signal-processing techniques to overcome noise issues limiting capacity and transfer rates. A new differential coding scheme was introduced that allowed us to significantly increase capacity while adding extra bits for channel

Date	Name	Affiliation	Purpose of demo	Capacity MB	Density Gbit/in²	Read rate Mbit/s
Mar-94	Science	Stanford	Fully functional system	0.2	0.3	0.1
Dec-95	Next Step	Siros	Compressed video	5	3.4	1.7
Mar-96	Nonvolatile	Siros	Long-term storage	10	1.3	1.7
May-96	High-speed I/O	Siros	1000 fps readout	0.4	5.0	45
Sep-96	FPGA	Siros/GTE	Fully electronic readout	11	6.3	1.7
Apr-99	Gbit electronics	Stanford/ Siros	1 Gbit/s electronic readout	1.5	2.5	1000

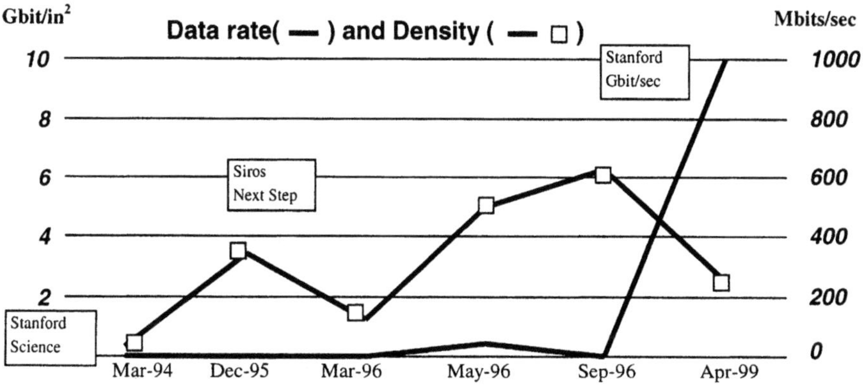

Fig. 3. Siros and Stanford University demonstration platforms in historical order. The laser power for all experiments was 400 mW, except for the April 1999 experiment, which used a 250-mW laser

and ECC coding. It allowed us to evaluate the tradeoffs between storage system capacity, bit error rate, and data transfer rate for several channel and ECC coding schemes.

The capacity is limited for the most part by the attainable diffraction efficiency of the stored holograms. As the number N of holograms stored in a single stack increases, the diffraction efficiency falls as $1/N^2$, mentioned above. As the strength of the diffracted signal decreases, the signal-to-noise ratio (SNR) decreases because the strength of noise due to scatter and CCD noise is independent of N. The total number of holograms that can be stored is thus determined by the minimum acceptable SNR. A holographic data storage system is also limited by page-to-page fluctuations in signal strength, which can occur due to laser instability or, for example, to a non-optimal choice of recording schedule. Intra-page distortion can occur due to imperfections in the optical system or in the SLM or CCD array. These noise sources limit the applicability of direct storage of images in analog form. The majority of previous holographic storage implementations have involved direct storage and retrieval of pictorial information. For example, Mok has demonstrated angular-multiplexed storage of as many as 5000 edge-enhanced analog images in a single crystal [13].

The noise-tolerant nature of digital storage makes it possible to overcome problems associated with the aforementioned noise sources. However, prior to the publication of our Science paper, no fully automated digital holographic data storage system had been implemented, and no comprehensive study of the bit error rate performance of a system operating at reasonable data transfer rates had been carried out. Achievable bit error rates have been projected previously by sampling a small number of digital information bits from a random sampling of 1000 holograms stored in a manually Bragg-tuned system [14]. Using various encoding techniques, we have implemented what to the best of our knowledge is the first fully automated system in which data are written and recalled in digital form. We have used this spatio-rotational multiplexing system to store digital representations of color images and compressed video. Critical features of our implementation that allowed us to overcome previous obstacles include location of the Fourier plane outside the crystal volume, use of cylindrical lenses to implement spatial multiplexing, a novel differential encoding technique used to increase error immunity, use of Hamming error correction codes [15], and the distribution of consecutive information bits over multiple data pages in order to decrease the probability of burst errors. With these techniques, bit error rates of 10^{-6} have been achieved at readout rates of 6.3×10^6 pixels per second. We have used the system to evaluate tradeoffs between bit error rate and storage capacity at a fixed data transfer rate.

A generic diagram of our system is shown in Fig. 4. The storage medium is an Fe-doped $LiNbO_3$ crystal cut such that its c-axis is at 45° to the crystal surfaces. The crystal dimensions are 2 cm × 1 cm × 1 cm, of which only a 0.1 cm^3 portion is used for data storage. The crystal is mounted on a computer-controlled rotation stage, which is in turn mounted on a computer-controlled translation stage, allowing the crystal to be moved in the vertical direction. The SLM is a 480×440 pixel liquid crystal array taken from an InFocus TVT-6000 video projector. It is addressed with an analog video signal produced by a framegrabber board in the computer. The camera is an intensified CCD array. Each stored hologram is read out in 1/30 s. The cylindrical lenses in the reference beam path are used to collimate the reference to an area of approximately 10 mm × 2 mm at the face of the crystal. The combination of the translation stage and the collimation of the reference beam allows holograms to be stored in four different stacks. The signal beam occupies an area of approximately 1.5 mm × 1.5 mm on the front face of the crystal. The Fourier plane is located approximately 3 mm in front of the crystal. A filter is used to select only the central spot of this transform. The filter prevents erasure of previously recorded stacks during writing, and eases alignment tolerances at the CCD plane. By recording in a Fresnel plane, rather than the Fourier plane, the modulation depth is sufficiently uniform to eliminate the need for a diffuser [16]. Because previously recorded holograms were erased in this demonstration as additional holograms are recorded, an appropriate recording schedule must be determined in order to store a large number of pages with equal diffraction efficiency [6].

If all holograms result in equal diffraction efficiencies, a threshold intensity can be chosen to distinguish ON pixels from OFF pixels. Note that we use OFF and ON when referring to SLM or CCD pixels and 0 and 1 when referring to information bits. An incorrect measurement of the time constant can result in a schedule that leads to unequal diffraction efficiencies for each of the stored holograms. A 10% error in the assumed time constant can result in more than a factor of 2 variation in diffraction efficiency. In addition, laser fluctuations or anomalous writing behavior can result in unequal diffraction efficiencies.

In order to circumvent these problems, we used a differential encoding technique, in which the pixel sequence OFF–ON is written to the SLM to represent a 0 and the pixel sequence ON–OFF is written to represent a 1. In addition to being insensitive to page-to-page intensity fluctuations, the differential encoding technique results in a lower probability of error assuming a model of additive white noise that is independent from pixel to pixel. For example, the probability of an error given that a 0 is transmitted in a threshold detection system is given by [17] and [18].

$$p \approx \int_{a_{\mathrm{T}}}^{\infty} \frac{r}{\sigma^2} \exp\left(-\frac{r^2 + a_{\mathrm{L}}^2}{2\sigma^2}\right) I_0\left(\frac{r a_{\mathrm{L}}}{\sigma^2}\right) \mathrm{d}r \tag{2}$$

where a_{H} and a_{L} are the average amplitudes of ON and OFF pixels, respectively, at the CCD array, σ is the standard deviation of the noise, I_0 is the zero order modified Bessel function of the first kind, and the threshold, a_{T}, is set to be the mean of the ON and OFF values.

The probability of an error given that a 0 is transmitted in a differential detection system is given by

$$p \approx \int_{a_{\mathrm{T}}}^{\infty} \frac{x}{\sigma^2} \exp\left(-\frac{x^2 + a_{\mathrm{L}}^2}{2\sigma^2}\right) I_0\left(\frac{x a_{\mathrm{L}}}{\sigma^2}\right) \times \int_{0}^{x} \frac{r}{\sigma^2} \exp\left(-\frac{r^2 + a_{\mathrm{H}}^2}{2\sigma^2}\right) I_0\left(\frac{r a_{\mathrm{H}}}{\sigma^2}\right) \mathrm{d}x\, \mathrm{d}r. \tag{3}$$

These equations show that significant improvements can be made using differential encoding over simple thresholding schemes [12]. Another advantage of the differential encoding technique is that it can be used when single pages reconstruct nonuniformly. This nonuniformity can result from poor overlap of signal and reference beams in the crystal, nonuniform illumination of the SLM, or the introduction of interference fringes due to multiple reflections in the optical system. When a substantial contribution to the variance is due to such large-scale nonuniformities, we expect difference encoding to offer further improvement over simple thresholding techniques.

Ideally, each pixel on the SLM would be used to represent a single bit of data. The liquid crystal array used in these experiments exhibited some interpixel crosstalk. In other words, pixels adjacent to an ON (transmitting) pixel exhibit a significant amount of transmission. Also, the SLM and the CCD array are manufactured with different horizontal to vertical pixel pitch ratios. Thus, using a simple optical system such as ours, it is impossible to image the SLM onto the CCD array such that each SLM pixel is imaged directly onto a CCD pixel. In practice, a multiple lens arrangement could be used to achieve the proper anamorphic imaging; however, this greatly increases system complexity. The pitch mismatch results in a systematic geometrical source of errors if too fine a sampling grid is used. In order to circumvent these problems, we used a block of 8×8 pixels to represent one bit. The differential encoding technique increases the pixel-to-bit ratio to 8×16 pixels per information bit. The data is read out by sampling the CCD output at one pixel for each 8×8 pixel block imaged onto it. The difference in intensity between adjacent samples is used to determine whether the information bit is a 0 or a 1.

We measured the raw bit error rate of our system to be between 10^{-3} and 10^{-4} at video readout rates. In order to improve performance, we implemented a Hamming error-correcting code in which 4 check bits are added to each string of 8 data bits. The Hamming code is capable of correcting a single bit error in the sequence of 12-bits assuming that only one bit error occurs in those 12. In order to reduce the potential of burst errors, each stored data page represented one bit plane in a two-byte sequence. Thus, each bit in the 12 bit sequences was stored in a separate page in order to reduce the probability of multiple errors within a given sequence. With the error-correcting codes, we were able to achieve bit error rates of 10^{-6}, which is adequate for compressed video storage and more than sufficient for uncompressed image storage. We used our system to store both digital color image and low-resolution video data. The total capacity of the system was 1232 pages with 1592 bits per page, resulting in a raw capacity of 245 kbytes. Taking into account the necessary error-correcting bits, the total useful data capacity is 163 kbytes. The total pixel capacity of our system is 2.6×10^8 pixels, with a density of 3×10^9 pixels per cm^3. The transfer rate is 6.3×10^6 pixels per second. The CCD camera outputs one byte per pixel. The total pixel capacity of our system indicates that our storage capacity does not represent a fundamental limit. Rather, the primary limitation in determining the information storage capacity is the necessary oversampling on the SLM. With a higher-quality SLM designed in conjunction with the CCD array, we anticipate that the information capacity and data transfer rates can be enhanced by several orders of magnitude.

5 The Siros First Fully Automated Video Demonstration (1995)

The Stanford demonstration showed that digital encoding techniques could be used to great advantage to improve raw BER, and in combination with error-correcting codes digital images could be retrieved with low error rates. The capacity of this demonstration, however, was rather limited. The next step in the development of digital holographic data storage systems accomplished storage of compressed high-resolution video data, requiring substantially larger data storage capacity, and a good BER of about 10^{-8}. Primarily due to the efforts of John Heanue, Andy Daiber, Ray Snyder, and Jim Colvin at Siros Technologies, the first video movie was stored and retrieved by recording approximately 5 Mbytes of compressed video data. The optical arrangement was similar to the Stanford demonstration, except that the SLM was a Texas Instruments deformable mirror device, with a resolution of 600×480 pixels. This device overcame all problems of the previously used liquid crystal devices.

Both video imagery and sound were recorded in $LiNbO_3$ with the MPEG standard using the system of Fig. 4. Data were encoded using 6:8 channel codes and a Reed–Solomon 15:13 error-correction scheme that corrects one byte error per code word. This system was shown on the Discovery channel in 1995 as part of the Next Step program.

In contrast to the Science demonstration, in this experiment the information was permanently stored using ionic fixing overcoming the earlier erasure problems. After recording of all data stacks, the crystal was removed from the set up and placed in an oven. It was heated to approximately 120° C for

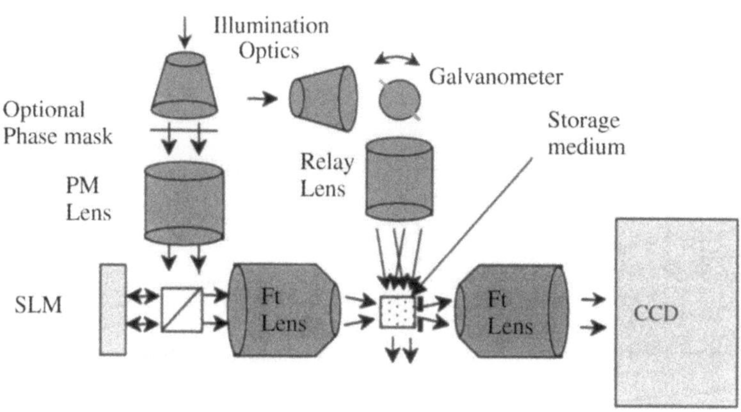

Fig. 4. Generic 90° recording geometry

15 min, and then slowly cooled down to room temperature. As a result of hydrogen defects in the crystal, the thermal fixing efficiency was rather high, approximately 50% (i.e., the diffraction efficiency after thermal fixing was approximately 50% of that before the fixing procedure) [19]. Data retrieval in this experiment was implemented in software.

6 The Siros Fully Automated System with Electronic Readout at Video Rates (PRISM, 1996)

All previous demonstrations were under software control with offline data retrieval. Although it is well recognized that software procedures can often be implemented in electronic hardware, there are many practical issues that make such an implementation far from trivial. At Siros, primarily through the efforts of Bob Okas and Ray Snyder in cooperation with GTE as part of the PRISM consortium effort, a fully electronic readout and control system was implemented in 1996. We believe that this was the first demonstration of a fully automated and electronically controlled system. The block diagram of the electronic layout is shown in Fig. 5. The electronic architecture is based on a VME bus. The optical system architecture was the same as shown in Fig. 4. Temperature fixing in $LiNbO_3$ was implemented for nonvolatile readout, and a total capacity of 5 Mbytes of video data were stored and retrieved at video rates. Channel 6:8 coding, ECC RS 13:15 coding, bit shuffling and data warping were all implemented in electronic hardware, as well as overall system control [20].

A small compact demonstration system using $LiNbO_3$ was built primarily by Mark McDonald at Siros to show that all components could be integrated

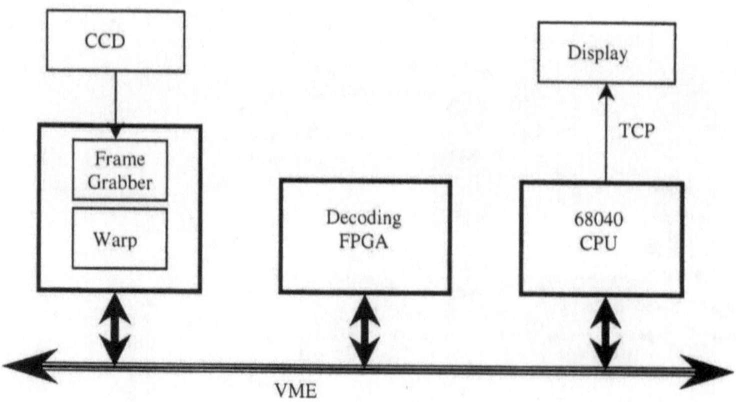

Fig. 5. Electronic block diagram for the Siros real-time video readout system

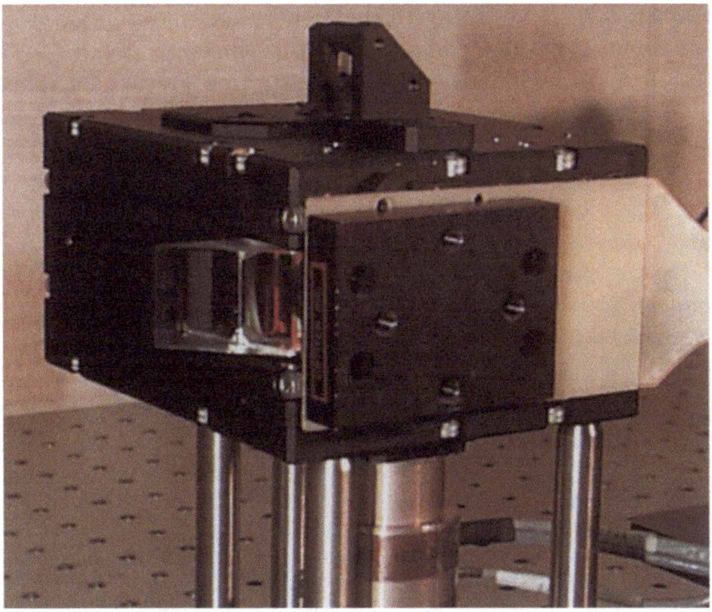

Fig. 6. The Siros small prototype demonstration device incorporating a 800 × 600 DMD, a 1000 × 1000 SLM, and novel optical arrangement. Capacity is limited to less than 10 Gbytes, with a maximum transfer rate of 10 Mbytes/s. The dimensions are 3 × 4 × 5″

into an approximately 5.25 inch form factor, as depicted in Fig. 6. This system includes both recording and readout optics and components.

Associative data retrieval capabilities were added to the Siros 90° demonstration platforms, and content-based searches were demonstrated at 20 Gbit/s burst data rates by Robert McLeod in 1997 [8].

7 The Stanford University and Siros Fully Electronic Data Readout System Achieving 1 Gbit/s (HDSS, 1999)

In previous demonstrations at Stanford and Siros, the total system capacity was limited to tens of Mbytes, with transfer rates on the order of a few Mbytes/s, far from being competitive with current state-of-the-art optical and magnetic drives. The HDSS consortium was formed to develop the components and the systems to demonstrate 1 Gbit/s transfer rate and a capacity of 125 Gbytes on a 5.25 in disk. To achieve this goal required the development of new components and electronics. A reflective 1000 × 1000 element ferroelectric liquid crystal device was designed and built by IBM and Display Tech capable of recording 1000 frames per second. A pixel-matched detector array was developed by Kodak, having 1000 × 1000 pixels, and a frame rate of 1000 fps using 8-bit resolution. The output of this camera is divided into 64 digital channels. A pixel-matching, very low distortion Fresnel lens pair was designed and manufactured by Rochester Photonics to achieve less than 0.3 pixel element distortion over the full one million pixels in the field of view. A novel random phase code reference beam was coaxially placed with the imaging lens. These components were assembled in the optical set up of Fig. 4. In the first high-speed demonstration, a 90° $LiNbO_3$ arrangement was implemented using the electronic architecture of Fig. 7. A 6:8 channel code and a Reed–Solomon 136:156 ECC code for correcting up to ten byte errors in the sequence are used for this demonstration.

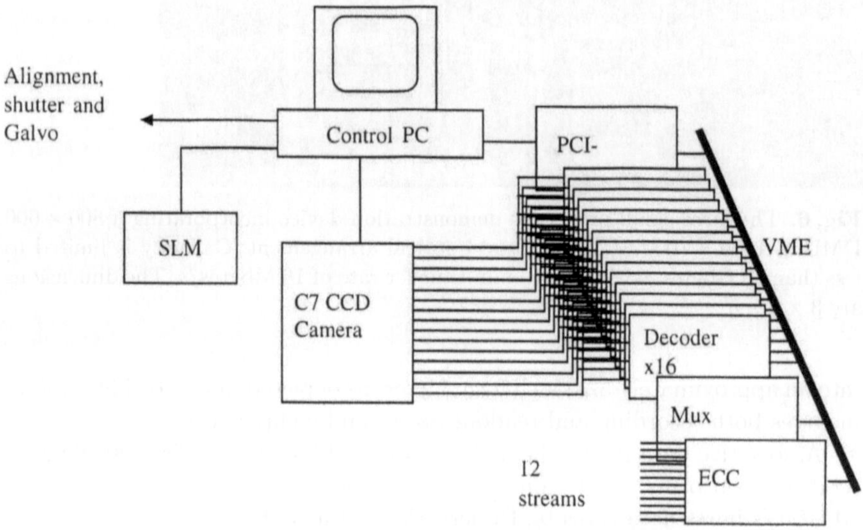

Fig. 7. Electronic block diagram for the 1 Gbit/s electronics

A team at Stanford University and Siros consisting primarily of Robert Okas, Sergei Orlov, Xiaochun Li, Eric Bjornson, Feodor Dimov, Bill Phillips, Ray Snyder, Darren Kwan, and Yuzuru Takashima developed the first implementation of a 1 Gbit/s electronic readout system in August 1999. Six data files were recorded: five JPEG images and one counting sequence. The counting sequence was used so that the data could be easily inspected in numerical format. Each source file was divided into two pieces, giving a total of 12 data sequences, corresponding to the 12 output byte streams.

Each recorded hologram stored data at a 65% efficiency due to 6:8 encoding and RS encoding. Thus each image of 1 048 576 pixels corresponds to 685 600 bits. When this is divided into 12 streams it gives 7140 bytes per image per output stream. Thus our source data files, each of approximately

40 k (split into two 20 k sections) required four images to store. The data were stored as four holograms and retrieved with an integration time of 600 ms, consistent with a 1 ms frame rate. Data passed through the entire system, and the 12 parallel byte streams were collected using the full electronics and checked for accuracy. The raw error rate was 8.7×10^{-5} [8]. Although this system demonstrates ultra high transfer rates, recording sensitivity for $LiNbO_3$ is low. To achieve Gbit/sec recording rates as well and very high capacity storage on a single disk, we developed a rotating disk system based on this photopolymers.

8 The Stanford University and Siros 100-Gbytes Capacity and 1 Gbit/s Readout System Demonstrator (Expected 2000)

For thick media like $LiNbO_3$ in the 90° geometry used in the first high speed demonstrations, capacity is usually limited by media dynamic range and noise rather than by multiplexing. For the final demonstration, a thin photopolymer rotating disk architecture was therefore chosen. For thin media, such as photopolymers, however, this is not the case, and the number of superimposed holograms in a spatial location is largely determined by the limited number of degrees of freedom available for multiplexing. Angular multiplexing in the transmission geometry does not allow sufficient data storage density, and other multiplexing techniques such as shift or perisrophic multiplexing are required. Unfortunately, the displacements necessary for achieving low cross-talk in the shift-multiplexing technique are still too large and require the undesirable stop-and-go architecture for the rotating disk, when using a cw laser. To avoid this problem a very sensitive, relatively thick storage medium is required suitable for pulsed laser recording with nanosecond pulses. Recent developments in the DARPA-funded PRISM consortium have made great steps forward toward achieving such a medium for recording in the green region of the spectrum using a few hundred milliwatts of power. To achieve recording during constant rotation of the disk, a new multiplexing technique had to be implemented based on a phase-modulated reference beam. In such an architecture, performance can be readily analyzed and compared with more conventional technologies. In the system currently under test at Stanford University, densities of over 70 bits/μm^2 are expected for a total system capacity of 125 Gbytes and a transfer rate of 1 Gbit/s using the electronics described above. The optical architecture for this system is shown in Fig. 8. A high spped, high capacity demonstration is started for the fall of 2000.

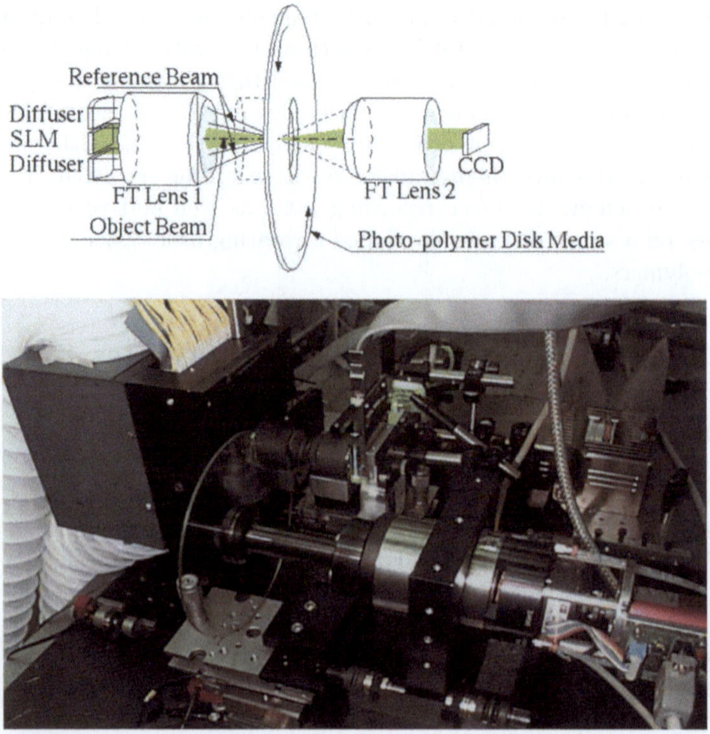

Fig. 8. Optical setup for, and photograph of, the 125-Gbytes capacity and 1-Gbit/s WORM demonstration

Acknowledgments

This work is partially funded by the DARPA/NSIC/Industry/University PRISM and HDSS Consortia. I like to acknowledge all the hard work done by the consortia members, present and past students, postdocs, and visitors to my group at Stanford University. Glenn Sincerbox and Hans Coufal are also acknowledged as co-Principal Investigators on these programs.

References

1. J.J. Amodei and D.L. Staebler, "Holographic pattern fixing in electrooptic crystals," Appl. Phys. Lett., **18**(12), 540-542, 1971.
2. D.L. Staebler, W.J. Burke, W. Phillips, and J.J. Amodei, "Multiple storage and erasure of fixed holograms in Fe-doped LiNbO3," Appl. Phys. Lett., **26**(4), 182–184, 15 Feb. 1975.
3. V.I. Bobrinev, Z.G. Vasil'eva, E.Kh. Gulanyan, and A.L. Mikaelyan, "Multiple rerecording and fixation of holograms in lithium niobate crystals doped with iron," Pis'ma v Zhurnal Eksperimental'noi i Teoreticheskoi Fiziki, **18**(4), 267–269, 1973. Translated in: JETP Lett., **18**(4), 159-160, 1973.

4. D. Von der Linde and A.M. Glass, "Multiphoton process for optical data storage in pyroelectrics", Ferroelectrics, No. 10, 5–8, 1976.
5. F.S. Chen, J.T. LaMacchia, and D.B. Fraser, "Holographic storage in lithium niobate," Appl. Phys. Lett., **13**(7), 223–225, 1968.
6. L. Hesselink and L.M.C. Bashaw, "Optical memories implemented with photorefractive media," Opt. Quantum Electron., **25**(9), S611–S661, 1993.
7. L. Hesselink and G. Sincerbox, Principal and Co-Principal Investigators of the DARPA/NSIC/Industry, University consortia on PRISM and HDSS, 1994, 1995.
8. L. Hesselink and S. Orlov, "Three-dimensional digital holographic optical memories", in preparation, IEEE 2000.
9. E. Bjornson, M. Bashaw, and L. Hesselink, "Digital quasi-phase-matched two-color nonvolatile holographic storage", Appl. Opt., **36**, 3090–3106, 1997.
10. L. Hesselink, S.S. Orlov, A Liu, A. Akella, D. Lande, and Ratnakar R. Neurgaonkar "Photorefractive materials for non-volatile volume holographic data storage", Science, **282**, 1089-1092, 1998.
11. W.J. Burke, D.L. Staehler, W. Phillips, G.A. Alphonse, "Volume Phase Holographic Storage in Ferroelectric Crystals", Optical Eng., **17**(4), 308–316, 1978
12. J.F. Heanue, M.C. Bashaw, and L. Hesselink, "Volume holographic storage and retrieval of digital data", Science, **265**, 749, 1994.
13. F.H. Mok, "Angle-multiplexed storage of 5000 holograms in lithium niobate," Opt. Lett., **18**(11), 915–917, 1991.
14. F. Mok, D. Psaltis, and G. Burr, "Spatially- and angle-multiplexed holographic random access memory," Proc. SPIE, **1773**, 334–45, 1993.
15. R.W. Hamming, *Coding and Information Theory*, Prentice-Hall, Englewood Cliffs, NJ, 1986.
16. R. Brauer, U. Wojak, F. Wyrowski, and O. Bryngdahl, "Digital diffusers for optical holography," Opt. Lett., **16**(18), 1427–1429, 1991.
17. Lee, Wai-Hon, "Effect of film-grain noise on the performance of holographic memory," J. Opt. Soc. Am., **62**(6), 797–801, 1972 .
18. J.W. Goodman, *Statistical Optics*, Wiley, New York, 1985.
19. J. Heanue, M.C. Bashaw, A. Daiber, R. Snyder, and L. Hesselink, "Digital holographic storage system incorporating thermal fixing in lithium niobate," Opt. Lett., **21**(19), 1615–1617, 1996.
20. A. Daiber, R. Snyder, J. Colvin, and R. Okas, "Fully functional digital video holographic storage system", Presentation OSA Annual Meeting, Long Beach, CA, 1997.

Holographic Read-Only Memory

F. Mok, G. Zhou, and D. Psaltis

The most successful use of optical memories so far has been as read-only memories (ROM). A main reason for this success has been the availability of inexpensive methods to mass-produce copies of recorded disks. This has made it possible to publish data (audio, video, databases, computer games) and distribute it widely through normal retail channels. In this chapter, we show results of a holographic read-only memory (HROM) of which digital data on a master disk can be copied onto replicate disks efficiently.

The HROM disk system is composed of three standalone machines: (1) a recorder, (2) a reader, and (3) a replicator (Fig. 1). The recorder records digital, page-formatted data onto a 3-D holographic disk. The replicator reproduces data from a master disk recorded by the recorder onto replicate disks. The reader retrieves stored data from replicate disks.

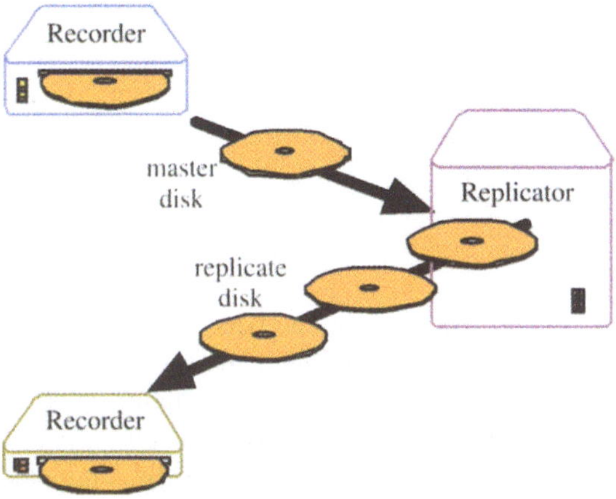

Fig. 1. Schematic diagram of HROM

The HROM system is most suitable for applications that require low-cost distribution of large volumes of data. Currently, the needs of this type of applications are fulfilled by CDs and DVDs. In this chapter, we demonstrate that HROM has significant competitive advantages over CDs and DVDs in recording density and data transfer rate.

1 Specifications

The specifications of the system presented here are given in Table 1. It is important to note that the specifications are for a specific design, and do not represent the limits of the technology. The technology has a great potential for further improvements in many areas including density and data transfer rate. For example, a storage density of 100 bits/μm^2 has already been demonstrated. Recording media with larger M-numbers and higher sensitivity, once developed, can increase data transfer rate for recording and readout, as can more powerful laser sources. Tradeoffs among specifications are generally possible. For instance, data recording rate can be gained at the expense of data readout rate.

Table 1. Specifications of system

Parameters	System*
Disk size	120 mm diameter
Page size	180 000 pixels
Recording density	40 bits/μm^2
Recording rate	9 Mbits/s
Readout rate	1 Gbits/s
Disk replication time	10 seconds
Capacity	340 Gbits

* Numbers here apply to raw data.

2 Recorder

Figure 2 is a conceptual diagram of the recorder for the HROM system. The primary components are a solid-state laser, a spatial light modulator, a head consisting of a multiplexer for changing the reference beam angle, imaging optics, mechanical drives, a CCD, a position-tracking system, and control electronics. Multiple holograms are stored at any one location by changing

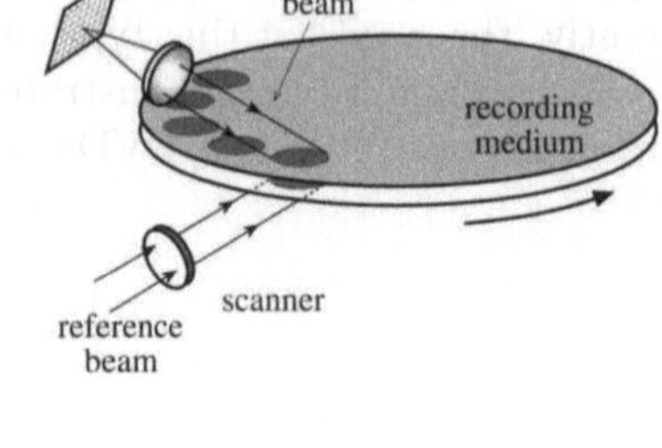

Fig. 2. A conceptual diagram of the recorder

Fig. 3. A photograph of the recorder. The recording medium is a 120-cm disk located near the center of the photo

the angle of the reference beam. Each of the stored holograms can be read out independently for verification. By rotating and translating the disk, the head can record holograms at different locations.

Figure 3 is a photograph of the recorder. The recording medium is a 100-μm thick photopolymer from Dupont, laminated on a glass disk with a 120-cm diameter. A 100-mW doubled Nd:YAG laser (532 nm) provides the optical power for recording. The dynamic range of the medium is $M/5$. The average diffraction efficiency for each hologram is 0.1%. Stepper motors are used to rotate and translate the disk.

F/1 optics focus the object beam to a small spot on the disk. The overlapping region of the object and reference beams occupies an area of 0.32 mm^2. Approximately 180 000 pixels from the 640 × 480 pixels on the SLM are usable. The recording density per hologram, therefore, exceeds 0.5 bit/μm^2. By recording 80 holograms within each location, the recording density reaches 40 bits/μm^2.

For a 100-μm-thick recording medium, the minimal angular separation between successive holograms is close to 1°. Angle multiplexing alone is not sufficient to record 80 holograms. Peristrophic multiplexing is also employed to record the additional holograms needed.

The raw bit error rate (BER) of the recorder shown here is below 10^{-9} for 0 hologram and 10^{-7} for 1 hologram. As the number of recorded holograms increases, the BER increases as well. For a recording density of 40 bits/μm^2, the BER is 10^{-3}. Typically, a BER between 10^{-4} and 10^{-5} for raw data is desired. A number of error correction schemes can be utilized to improve the BER for usable data to less than 10^{-12}, a commercial standard for mass data storage. Results of recording density and BER are shown in Fig. 4.

It can readily be demonstrated that the most significant noise source here is the recording medium. As exposure of the recording medium to optical energy increases, deformation and degradation in optical quality of the medium contribute primarily to increasing the noise level. The level of optical scatter-

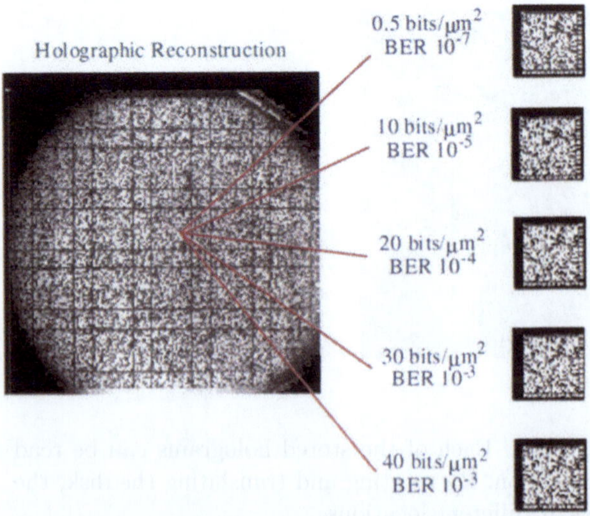

Fig. 4. The dependence of BER on recording density. Here, recording density is increased by recording more holograms

ing of this medium is usually much lower than the signal, and is not believed to cause any major problem.

3 Reader

Figure 5 shows a conceptual diagram of the reader. The reader consists of a solid-state laser, mechanical drives, electronics for tracking, and a readout head comprising a scanner, imaging optics, and a detector array. The reader retrieves data from the holographic disk by reconstructing the recorded holograms. To read out any particular hologram, the mechanical drives rotate and translate the holographic disk to the appropriate location. The readout head positions the scanner to orient the reference beam to the correct angle. The optics image the reconstructed hologram to the detector array. The data are accessed electronically.

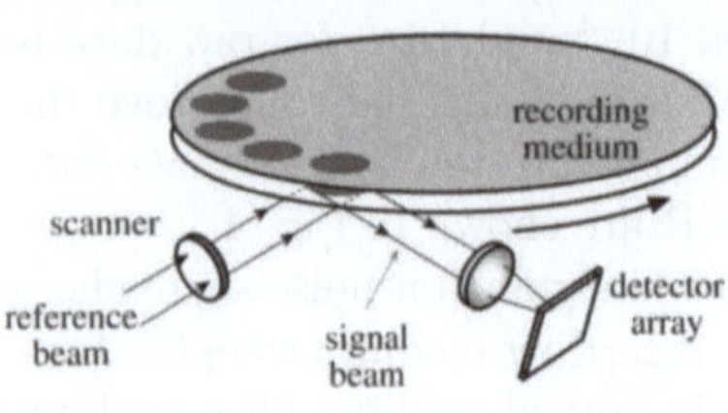

Fig. 5. Conceptual diagram of the reader

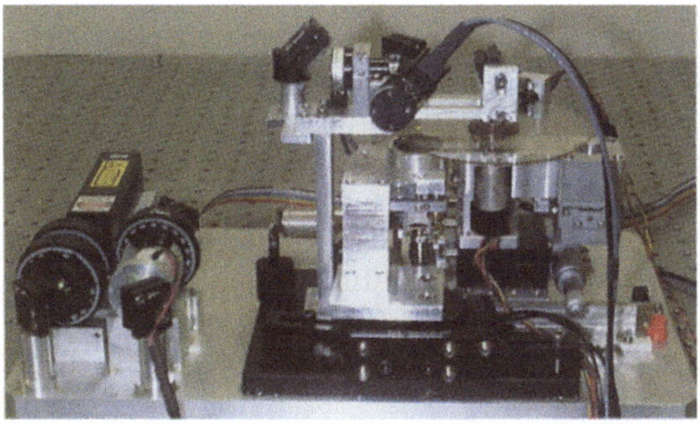

Fig. 6. A picture of the reader. The reader can be packaged to fit within a much smaller volume

The reader shown in Fig. 6 has a 100-mW doubled Nd:YAG laser as its light source. The holographic disk rotates at a rate of 600 rev/min. The response time of the scanner is about 1 ms. The detector array is a CCD with 512×512 pixels. The separation between pixels is 24 μm. Each pixel on the CCD detects exactly one pixel of data from a reconstructed hologram. The maximum distortion within a reconstructed hologram, which occurs at the edges, is approximately 2 μm. Since the size of a reconstructed pixel is about 20 μm, most of the energy of a pixel remains within the confine of a single detector element if there is no alignment error. Compared with medium noise, noise caused by imperfect imaging is negligible. Figure 7 is a picture of a sample detector output.

Fig. 7. The binary image on the left is an image of the optical data that decodes into the picture of a flower on the right

With a cw reference beam, reconstructed holograms sweep across the detector array, making data registration difficult. To essentially freeze reconstructed holograms for readout, the reference beam is pulsed at the right time. By keeping the maximum movements of reconstructed optical pixels to within 10% of the detector pixel pitch, the data registration noise is negligible. This condition is satisfied if the pulse width is no more than 2 μs. The data readout rate, however, is determined by the time it takes to download an entire page from the detector array. That takes less than 150 μs, at a clock rate of 20 MHz and with a 64-bit data bus. The resulting data rate is in excess of 1 Gbit/s.

Owing to rotation and wobbling, six degrees of movement are possible for the recording medium. Translation-type movements of the disk do not cause data registration errors. Translations of Fourier transform (FT) holograms do not translate reconstructed images at the detection plane. Rotation and tilting of FT holograms cause reconstructed images to move laterally. Data registration error caused by the rotation of the disk is minimized by properly timing the firing of the reference beam. Disk tilt is compensated by the adjustment of the orientation of the scanner. Precise firing of the reference beam and tilt compensation requires a real-time error signal for feedback control. High-speed multi-element detector blocks located at the edges of and integrated in the array detector provide the error signal. More details can be found in the chapter “Beam Deflectors and Spatial Light Modulators for Holographic Storage Application.”

4 Replication

To replicate a hologram, a second recording medium is placed close to a master medium. Illumination of the master medium by the appropriate reference beam reconstructs the recorded hologram. The reference beam and the reconstructed object beam create an interference pattern within the second medium that is stored as a replicate hologram (Fig. 8b). The replicate hologram does not have the same phase information as the master hologram. This is unimportant since reconstructing the replicate hologram still produces the same image as that from the master hologram at the detection plane (Fig. 8c).

To replicate multiple multiplexed holograms, reference beams satisfying the Bragg-matching conditions of the holograms provide the illumination. Each reference beam reconstructs and replicates its corresponding multiplexed holograms. The reference beams are mutually incoherent. Consequently, only reference beams and their reconstructed object beams produce stable interference patterns to be recorded. Interference patterns formed otherwise are not stable and hence cannot be recorded. As a result, all the holograms from the master medium within the illumination region are replicated onto the second medium simultaneously.

A replicator with 10 reference beams is shown in Fig. 9. Histograms of a master and replicate holograms of a binary image are shown in Fig. 10. We lost an SNR of 1.3 in the replication process. Yet, no digital errors are detected in the reconstruction of the replicate hologram.

Figure 11 shows a way of generating the plane waves for the replicator. An array of laser sources provides the optical power. There are as many laser sources as multiplexed holograms recorded in each location. For a storage density of 40 bits/μm^2, 80 laser sources are required. Each laser source is collimated into a plane-wave reference beam. The master disk and the replicate disk are placed within the overlapping region of the multiple reference beams. Each reference beam satisfies the Bragg-matching condition for one

REPLICATION

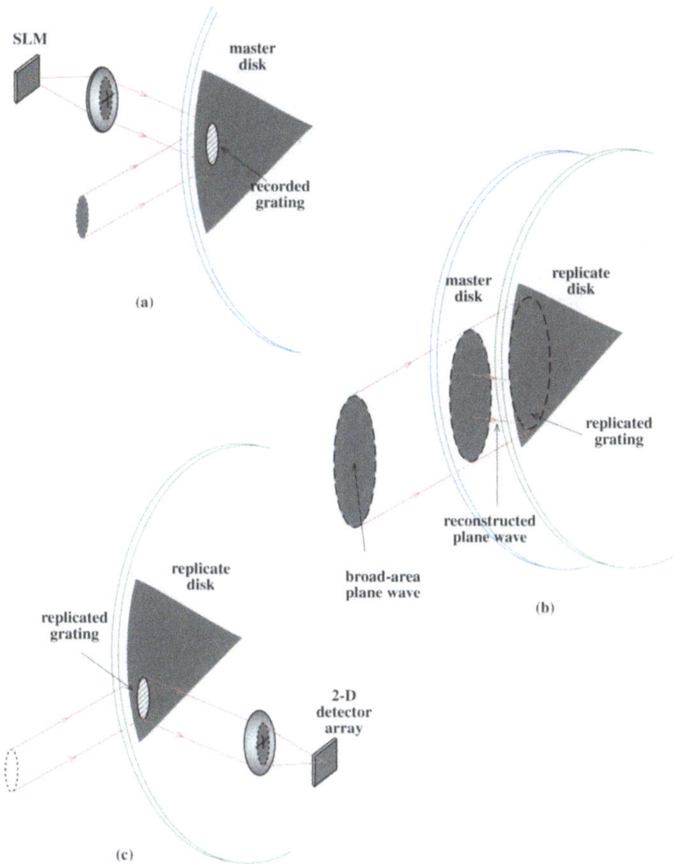

Fig. 8. (**a**) Recording a master hologram. (**b**) Replicating a hologram. (**c**) Reading the replicate hologram

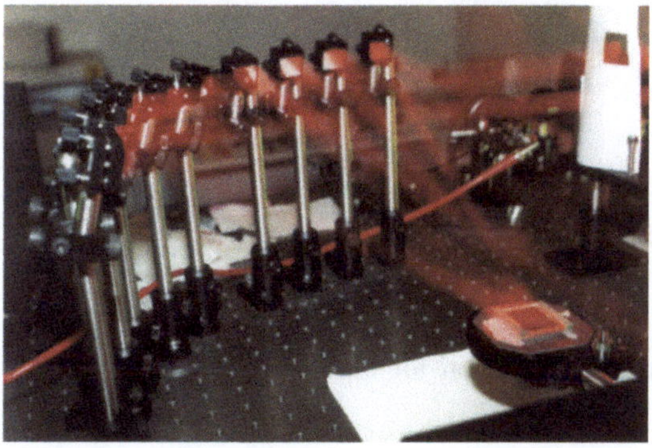

Fig. 9. A 10-reference beam replicator

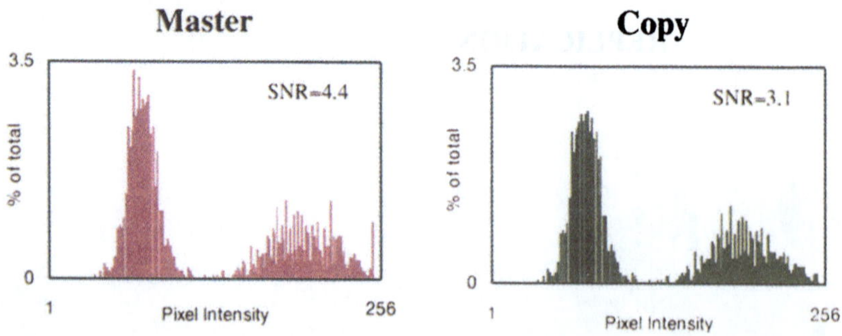

Fig. 10. Histograms of pixel intensity from the reconstruction of a master and its replicate holograms. In each histogram, the lobe on the left represents logic level 0 and the one on the right logic level 1

hologram.[1] By making the reference beams wide enough, holograms in multiple locations can be replicated simultaneously, provided the Bragg-matching conditions remain the same for all locations.

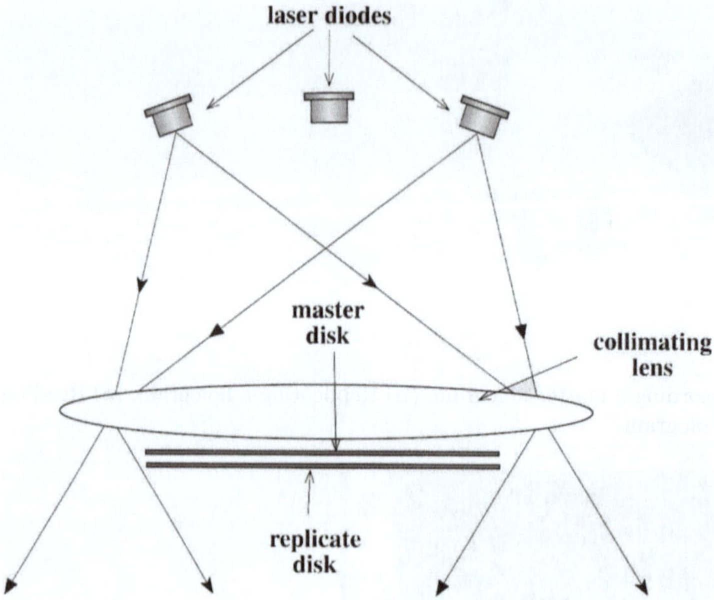

Fig. 11. Only three laser diodes are shown in this schematic of the replicator. In practice 50 diodes will be needed

[1] Shift-multiplexed holograms can be replicated by using noncollimating beams, each satisfying the Bragg-matching condition for one hologram, generated by mutually incoherent light sources.

Holograms recorded using the same reference beam at different locations on a disk cannot all be replicated at the same time by a beam with the same orientation as the reference beam. This is due to the fact that as the disk is rotated, the nominal grating vector of each hologram is rotated as well. Thus, a single reference beam cannot satisfy the Bragg-matching conditions of all the holograms simultaneously. In other words, simultaneously replicating an entire disk cannot be accomplished by the setup shown in Fig. 11.

To replicate an entire disk, the reference beams are expanded to cover a strip along a diameter of the disk. The Bragg-matching conditions for holograms within this strip are independent of their locations. The setup in Fig. 11 can be applied to replicate holograms within this strip. When that is done, the disk is rotated to repeat the replication process. An entire disk is replicated as all locations are covered. Since recording speed is proportional to illumination intensity, the total time taken to replicate a disk using the method described above would be similar to a method that can replicate a disk at once, given the total laser power available is fixed.

Compared with typical holographic recording setups, the replicator is insensitive to vibration. As long as both the master and replicate disks are reasonably secured, any phase changes introduced to the reference beams are automatically transmitted to the reconstructed beams. As a result, the phase differences between corresponding write beams always remain constant.

Digital Holographic Data Storage with Fast Access

J. Ma, T. Chang, S. Choi, and J. Hong

Recent investigations in holographic mass memory systems have produced proof of concept demonstrations that have highlighted their potential for providing unprecedented capacity, data transfer rates and fast random access performance [1–4]. The exploratory nature of most such investigations has been largely confined to benchtop experiments in which the practical constraints of packaging and environmental concerns have been ignored. We have embarked on an effort to demonstrate the holographic mass memory concept by developing a compact prototype system geared for avionics and similar applications, which demand the following features (mostly interdependent factors): (1) solid-state design (no moving parts), (2) fast data-seek time, (3) robustness with respect to environmental factors (temperature, vibration, shock). In this chapter, we report on the development and demonstration of two systems, one with 100 Mbytes and the other with more than 1 Gbyte of storage capacity. Both systems feature solid-state design with the addressing mechanism realized with acousto-optic deflectors that are capable of better than 50 μs data seek time. Since the basic designs for the two systems are similar, we describe only the larger system in detail. The operation of the smaller system has been demonstrated in various environments, including hand-held operation and thermal/mechanical shock, and a photograph of the smaller system is provided as well as actual digital data retrieved from the same system.

1 Introduction

Although data storage has enjoyed enormous success in providing unprecedented performance in several distinct technology options, some important niche roles remain largely unmet. The various data storage options can be categorized somewhat simplistically in terms of two dominant parameters: (1) the storage capacity measured in gigabytes, and (2) the data transfer rate, which represents the access time measured in milliseconds (the time to locate an arbitrary 10 Kbyte block of data), and incorporates both data seek and data transfer rate. Devices that operate on a mechanically actuated principle (e.g. spinning disk, translating head, winding tape) can provide continuous data transfer rates that are sufficiently high for streaming video playback, but typically have limited data-seek performance in locating arbitrary sectors of data. High-performance disk systems, for example, have data seek times on the order of 5 ms. Semiconductor options such as flash memory and DRAM boards configured for high capacity can provide high performance in terms of effective data transfer time, but are in practice limited to capacities smaller than 1 Gbyte. This leaves a technology void corresponding to niche applications that require both high capacity and fast effective data transfer, such as multiple access server systems (e.g. centralized video on demand) and high-performance multimedia databases.

Another important parameter not discussed above is the packaging option, which is highly application dependent. Mobile applications such as avionics typically require a level of environmental robustness that mechanically actuated systems, while extremely cost-effective, cannot provide. Holographic memory implemented in a solid-state architecture is a solution candidate for such demanding applications, which comprise an important set of niche data storage problems. In this chapter, we describe the effort at Rockwell Science Center in developing just such a technology, initially geared toward avionics applications. Since previous chapters have described holographic memory fundamentals in some detail, we will outline the development of our demonstration prototype systems, which are currently operational in our laboratory, in detail without explanation of the underlying fundamental physics (including description of the material $LiNbO_3$ and recording details). The prototype systems that we have developed are: (1) a 100 Mbyte device, which is realized in a compact package that has been tested in some environmentally challenging conditions; and (2) a 10 Gbyte device that is realized on a 1-ft^2 (929 cm^2) platform developed to demonstrate video playback with extremely fast random seek capability. Both systems feature a totally solid-state addressing mechanism based on the use of acousto-optic deflectors, chosen on the basis of ruggedness and scanning speed.

2 System Architecture

Our system is based on a spatio-angular multiplexing technique in which the storage medium is divided into subvolumes, within each of which independent holograms are superimposed using angularly distinct reference or address beams. The system architecture is illustrated in Fig. 1, where both the recording and readout paths are shown, the former indicated by solid lines and the latter by dotted lines. Both address (or reference) and data (or object) beams are derived from a common laser beam, and each is similarly shaped to illuminate the corresonding apertures of the spatial light modulator (SLM) and the acousto-optic deflectors (AODs). The address beam is first deflected vertically by AOD1, which is driven by a digital frequency synthesizer DFS1 at frequency f_v. The deflected light (the dc or undeflected light is removed by spatial filtering) is imaged anamorphically onto AOD2 by the optical subsystem O_1. AOD2 is driven by DFS2 and imparts a horizontal angular deflection, proportional to its driving frequency f_h. The optical subsystem O_2, in concert with O_1 essentially forms a vertically converging and horizontally plane wave at the crystal input aperture in which the vertical displacement of the beam is proportional to f_v and the horizontal incidence angle is proportional to f_h. The details of these optical subsystems are discussed in a later section.

As illustrated in Fig. 1, the storage medium is organized into vertical slabs, with each slab divided into a vertical column of what we call storage subvolumes. While the address beam is directed onto one row of sub-volumes at a time and therefore traverses a number of slabs, the object beam is directed into one particular vertical slab, recording holograms only within that sub-volume. Since the address beam path is deflected by AOD1 and AOD2 and therefore carries a total doppler shift of $f_h + f_v$, the data beam laser beam is first frequency shifted by AOD3 with the sum frequency to impose mutual coherence. The SLM imparts the data spatial modulation onto the data beam and the optical subsystem O_3 forms the Fourier transform of the data beam onto a particular subvolume. The subvolume aperture is designed to be slightly larger than the area covered by the primary order of the diffraction pattern of the data beam so as to allow for its passage through the slab without interference effects due to surface reflections within the slab. The entire data beam path including SLM and O_3 are mounted on a precision translation stage, which allows the data beam spot to illuminate any particular subvolume aperture.

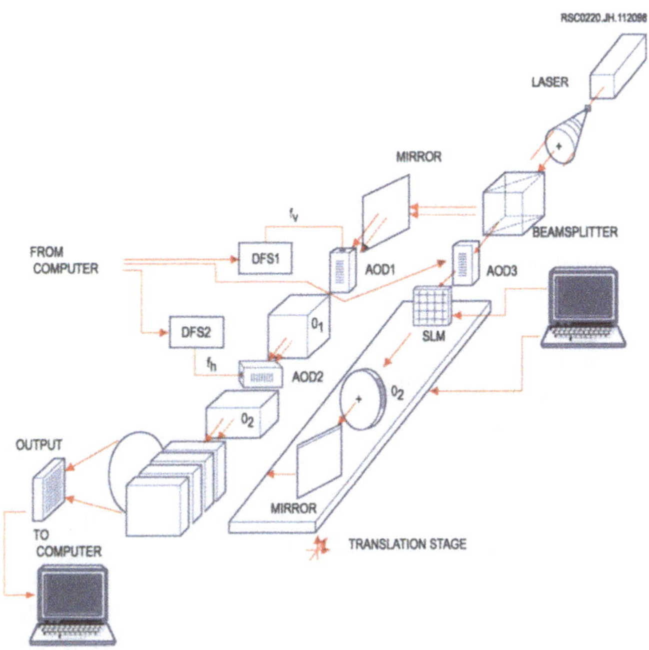

Fig. 1. Spatio-angularly multiplexed holographic memory architecture

During recording, the address beam is positioned to illuminate the bottom row of the slabs, and the data beam bearing the first data page is positioned on the first slab, exposing a hologram in the extreme lower right corner sub-volume. Subsequent pages are exposed, each with different horizontal angle controlled by DFS2. After that sub-volume is completely filled (typically 400 1000 holograms), the data beam is shifted up to the neighboring sub-volume and the address beam likewise is shifted and a new set of holograms are exposed. After the first slab is entirely recorded, it is processed through a fixing procedure (in the case of $LiNbO_3$, the well-known thermal fixing cycle) either in situ or by a well-engineered removable fixture (our current approach) and repositioned in the array. The next slabs are then exposed in similar recording sequences. This particular order of slab recordings was chosen since the reference beam acquires aberrations from the various slabs that it has traversed, affecting the recording conditions for subsequent slabs. The fixing procedure stabilizes any such aberrations so that subsequent slab recordings see constant but possibly aberrated address beam waves.

During readout or data retrieval, the data beam path is turned off and the address beam is positioned onto a selected row and angle, reconstructing a hologram from each subvolume in the row. A linear array of shutters is used (this can be realized using mechanical means or liquid crystal technology) to select one particular subvolume, extinguishing all others. The selected data reconstruction is imaged onto the output detector array with the sensed intensity distribution transformed into an array of electronic pixel values, which are then converted using appropriate post-detection schemes into digital data to be downloaded into the output buffer.

3 System Operation

In what follows, we describe in some detail the design and operation of the Rockwell holographic memory demonstrators: (1) a hand-held prototype, which is capable of storing 100 Mbytes of data, and (2) a larger demonstration unit capable of storing several gigabytes of data.

The optics for the address beam delivery are shown in Fig. 2. Spatial and angular multiplexing were incorporated into the system using acousto-optic deflectors (AODs) controlled by direct digital synthesizers (DDS) for accurate and repeatable beam pointing. The storage medium is divided into a two-dimensional array of cells, each of which stores a set of angularly multiplexed holograms (each hologram is recorded with a unique horizontal angle controlled by AOD-2). Spatial multiplexing AOD-1 is positioned at the front focal point of a telephoto constructed by cylindrical lenses L1, L5, and L8 with a long effective focal length. A small angular deviation of the beam deflected by AOD-1 is converted into a spatial displacement of the beam at the crystal plane. A pair of lenses (L6 and L7) form a 4-f telecentric system that images the angle multiplexing AOD-2 output surface onto the crystal surface

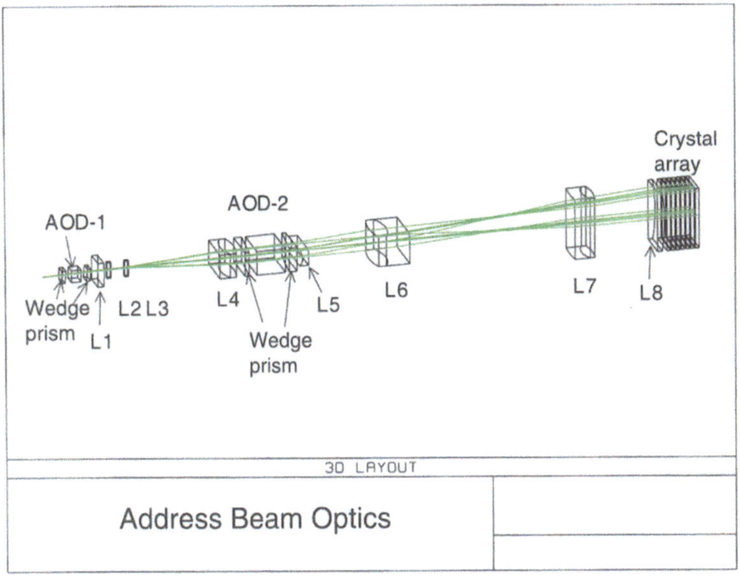

Fig. 2. Address beam optics designed for the multiple Gigabyte system. Fast data access (50 µs) is achieved by using two acousto-optic deflectors (AODs). The first AOD deflects the beam vertically (with respect to the platform surface) to access different layers (shown - two layers are being read). The second AOD is used to access angularly multiplexed holograms within a selected layer by controlling the horizontal deflection angle

with a factor of 3:1 demagnification ratio (from 30 mm of AOD-2 aperture size to 10 mm of crystal size). Angular deflection due to AOD-2 is magnified by a factor of 3 at the crystal plane. With a beam width of 3 mm, a crystal that is 51 mm long yields 17 subcells (layers). An obvious way to increase the storage capacity is to add more crystal slabs, each with dimension 51 mm × 3 mm × 3 mm. There is a limitation on the number of crystal slabs due to the aperture size of the re-imaging lens (shown in Fig. 4) and the absorption of the crystal since the address beam has to pass through multiple slabs. An array of shutters (e.g. ferroelectric liquid crystal devices) is needed to preclude cross-talk among the cells in a given row during readout, since the holograms sharing the same reference beam in a row of cells are reconstructed simultaneously.

The key challenge in such a spatial multiplexing scheme that involves no moving parts is in the design and implementation of the re-imaging optics. With the geometric constraints imposed by the system parameters along with practical considerations of reasonable cost and complexity, we used commercial optical design software to arrive at the solution illustrated in Fig. 3. The optical components were off the shelf, and included 2 doublets from Melles Griot and one 85 mm f/1.4 Nikon camera lens. A totally customized imaging

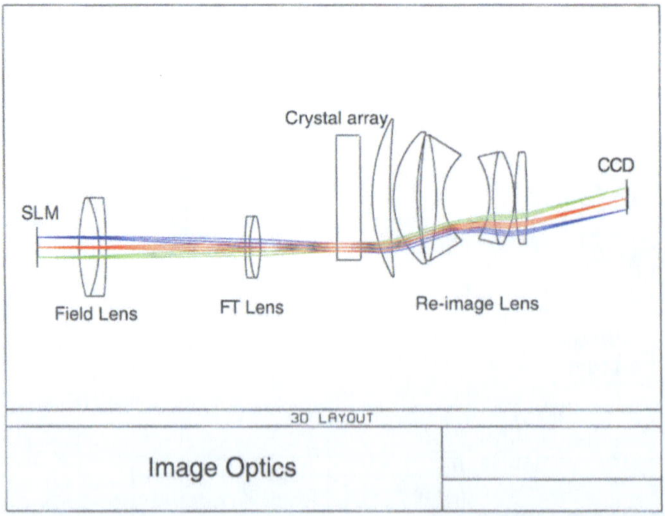

Fig. 3. Image optics design for the multiple gigabyte holographic data storage system

system (design, fabrication and packaging all of the lenses) might improve the performance to some degree but is likely to be much more time-consuming (equivalent to designing a super camera lens) and expensive. We chose to apply a 3×3 pixel over-sampling strategy at the CCD (one SLM pixel matches 3×3 CCD pixels) to compensate for the effects of aberration and distortion due to the optics. The spatial light modulator (SLM) used for imprinting the input data is a Kopin active matrix LCD display with 24 μm pixel pitch (square) and VGA format. The CCD is a Kodak ES1.0 digital camera with 9 μm pixel pitch (square) and 1024×1024 pixels. There are 328×328 pixels in each page, yielding a storage density of 13.4 Mbytes/cell (1000 holograms are superimposed in each cell). A 5 Gbyte system would therefore consist of up to 374 cells requiring 22 crystal slabs.

The impact of the aberration of the optics on the holographic digital data storage and retrieval was measured in terms of the percentage of the reconstructed energy of a single SLM pixel falling on the corresponding set of 3×3 CCD pixels, as exemplified by the cases illustrated in Fig. 4, which shows the dependence of encircled energy (the amount of light from a given SLM pixel that correctly falls on the corresponding 3×3 CCD pixels) as a function of the pixel position at the output detector plane (x–y). Figure 4a shows that for the crystal cell centered on the optical axis, there is no pixel cross-talk since 100% of the reconstructed energy was encircled. However, the situation gets worse as one reads the holograms from off-axis cells. In the worst case scenario shown in Fig. 4b, with the crystal cell in the extreme off-axis position (one of the diagonal corners in the array of cells), a small group (<0.1%) of pixels located on the corner of the page have less than 50% encircled energy.

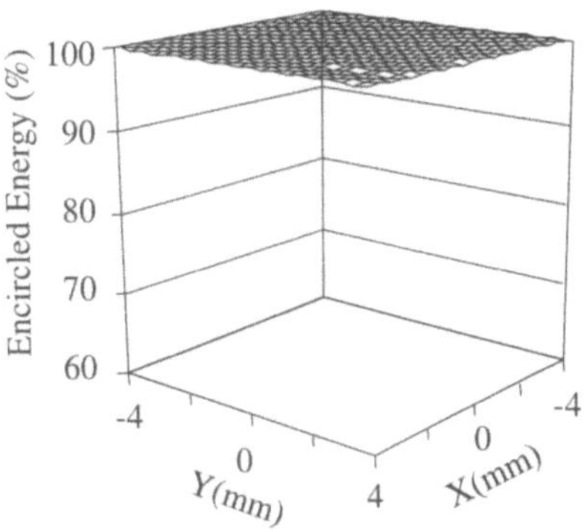

(a) Crystal cell on-axis.

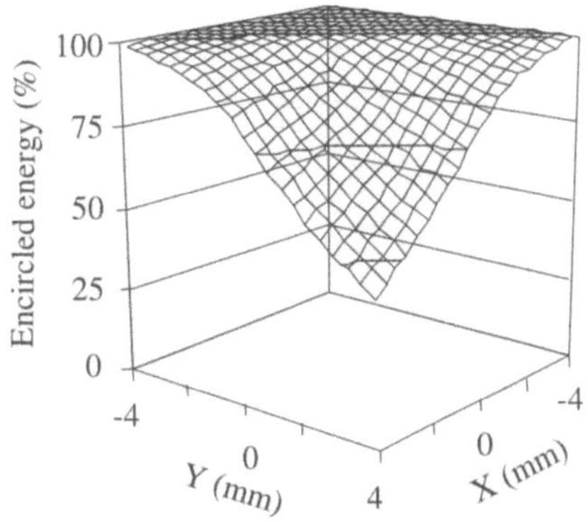

(b) Crystal cell off-axis by Δx=16mm, Δy=16mm.

Fig. 4a,b. Encircled energy distribution across the entire image. Coordinate (X, Y) represents the reconstructed image pixel location on the CCD, where (0,0) represents the on-axis pixel

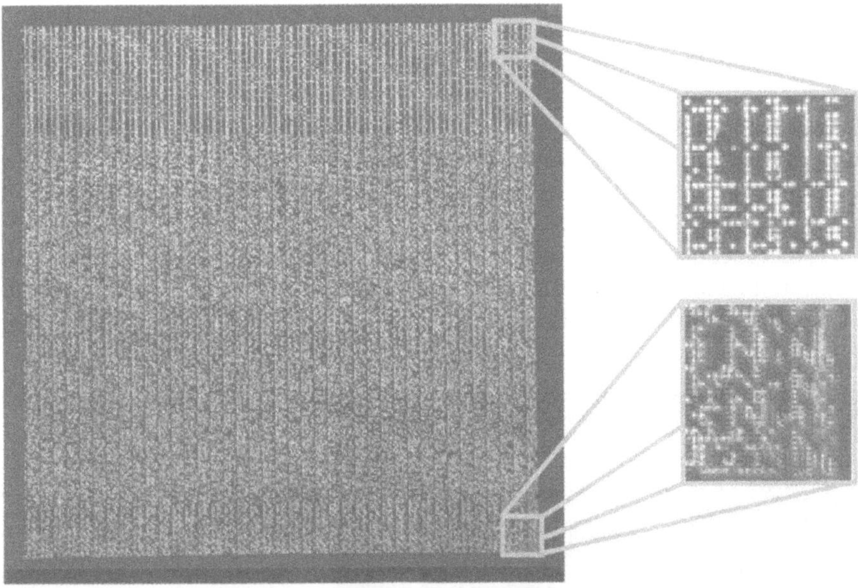

Fig. 5. Image reconstructed from the extreme off-axis crystal cell. A sharp image was obtained except for the right lower corner pixels

We constructed the holographic data storage demonstrator as described, and recorded representative holograms in the various cells in the crystal array. Figure 5 shows a picture of the worst-case hologram reconstruction from the extreme corner cell. The two right corners are enlarged to illustrate the difference in imaging quality at these locations. The digital data stored was

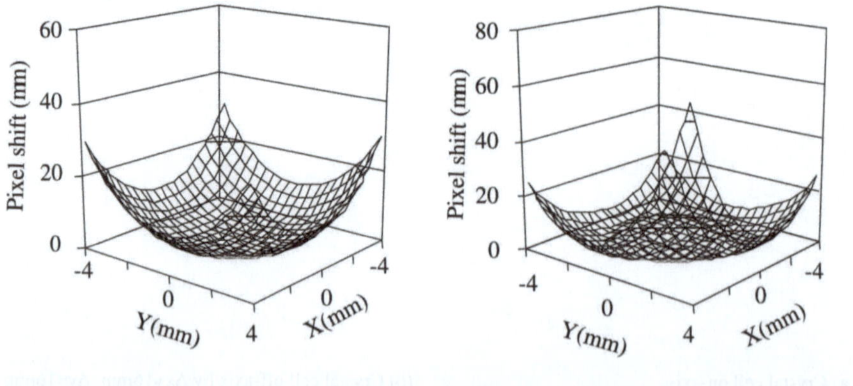

(a) Crystal cell on-axis. (b) Crystal cell off-axis by Δx=16mm, Δy=16mm.

Fig. 6. Distortion distribution across the entire image. Coordinate (X, Y) represents the image reconstructed image pixel location on the CCD, where (0,0) represents the on-axis pixel

a pseudo-random binary sequence. The original can be compared with the reconstructed data (the analog pixel values are binarized using an adaptive thresholding algorithm) to produce an error map, which shows that most of the errors were concentrated at the lower right corner. The raw bit error rate (BER) of that page was measured to be 1×10^{-3}. By sacrificing a small percentage (<0.1%) of capacity at this corner, a BER of 10^{-4} was obtained.

The effect of optical distortion on digital holographic data storage was also measured as a function of the cell location across the entire page. Figure 6a shows the distortion distribution of an image on the CCD measured as the

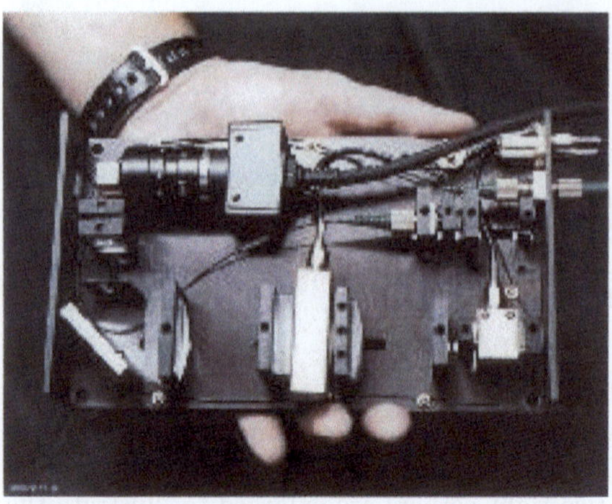

Fig. 7. A picture of a packaged prototype demonstrator of a read-only memory with 100 Mbytes capacity and 50 μs access time

absolute value of pixel shift with the crystal cell centered on the optical axis. A maximum distortion of 27 μm was found on the four corners. A distortion map in the worst-case scenario with holograms read out from a crystal cell in the extreme off-axis corner is presented in Fig. 6b, showing a maximum distortion of 65 μm located at the corner where the largest off-axis of the pixels occurs. With the application of over-sampling at the CCD, the contribution of distortion to BRE can be minimized. We have developed an adaptive distortion compensation technique with which two complementary checker boards were recorded in each cell that measures the vector of pixel shift for every pixel through the entire page. The distortion map was then used to locate the CCD pixels of the corresponding bits that were read. Using this technique, we were able to obtain a raw BER of 10^{-4} consistently.

A prototype demonstrator of a read only memory (ROM) with a package size of 8.2 in × 4.7 in × 1.6 in (208 × 119 × 41 mm) and a capacity in the order of 100 Mbytes was constructed and implemented (shown in Fig. 7): this contains two AODs, an Fe-doped $LiNbO_3$ crystal, a CCD detector, and all the optical components needed to form a reference beam. A docking unit, which includes a spatial light modulator and the necessary optical components for constructing an object beam, was used to record the holograms into the ROM device. Figure 8 (see next page) shows examples of actual retrieved color photos from the holographic memory without error correction.

Acknowledgments

This work has been supported in part by DARPA through the HDSS (Holographic Data Storage Systems) consortium. The authors thank Drs. Ian McMichael, Bill Christian, and Monte Khoshnevisan for their valuable help.

References

1. J.F. Heanue, M.C. Bashaw, and L. Hesselink, "Volume holographic storage and retrieval of digital data," Science, **265**(5173), 749–752, (1994).
2. J.H. Hong, I. McMichael, T.Y. Chang, W. Christian, and E. Paek, "Volume holographic memory systems – techniques and architectures," Opt. Eng., **34**(8), 2193–2203 (1995).
3. D. Psaltis and F. Mok, "Holographic memories," Sci. Am., **273**(5), 70–76 (1995).
4. M.-P. Bernal, H. Coufal, R.K. Grygier, J.A. Hoffnagle, C.M. Jefferson, R.M. Macfarlane, R.M. Shelby, G.T. Sincerbox, P. Wimmer, and G. Wittmann. Appl. Opt., **35**, 2360–2374 (1996).

Fig. 8. Examples of actual retrieved digital color photos from the packaged holographic ROM unit without error correction

A Demonstration Platform for Phase-Coded Multiplexing

C. Denz, K.-O. Müller, F. Visinka, and T. Tschudi

Among the techniques that have been envisaged for page-oriented data storage in volume holographic memories, orthogonal phase-coded multiplexing [1,2] offers several advantages over angular and wavelength multiplexing having the same storage capacity. A holographic data storage system based on phase-coded multiplexing operates with a fixed wavelength and a fixed geometry, avoiding mechanically moving parts. At the same time it provides short readout times and high energy efficiency. The signal-to-noise ratio of phase-coded multiplexing for a given number of holograms is more than two orders of magnitude higher than for angular and wavelength multiplexing [3,4]. Moreover, it allows one to perform arithmetic operations such as addition, subtraction or inversion of different data pages directly during readout of the memory [5–8]. Therefore, using phase-coded optical memories, high-capacity data storage and high-speed data processing become possible simultaneously. Additionally, orthogonal phase-coded multiplexing can be combined with a random-phase key in order to realize an encrypted holographic data storage system [9].

For these reasons, we realized a demonstration platform for phase-coded multiplexing at the Institute of Applied Physics at the Darmstadt University of Technology. The aim of our work is to demonstrate experimentally the potential of phase-coded multiplexing compared with other multiplexing techniques. The memory is designed to store up to 480 data pages in a single interaction region of a photorefractive $LiNbO_3$ crystal. Our demonstrator will not compete with other existing memory systems-but-clarify the possibilities of these addressing methods for volume holographic memories.

1 Phase-Coded Multiplexing

Figure 1 shows the configuration principle of phase-coded multiplexing. Each image is stored with a fixed set of N reference beams incident on the crystal with a discrete angular spectrum. The incident angles of the references obey the Bragg condition. In contrast to angular multiplexing, all reference beams interfere simultaneously with the signal beam in the storage crystal, each having a relative phase shift of 0 or π. Therefore, it is not the direction of each reference that is varied, but its phase. Each stored data page can be retrieved independently with the appropriate phase address that was used for storage.

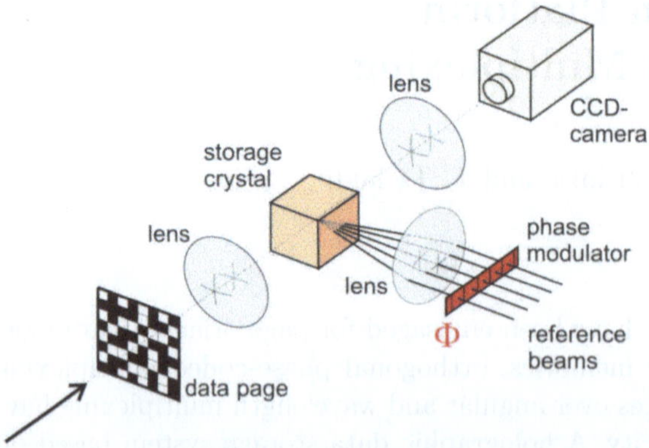

Fig. 1. Principle of phase-coded multiplexing

During recall, reconstructions of undesired images interfere destructively, resulting in zero intensity. This principle of destructive interference as a means of selective recall is the reason for the significantly higher signal-to-noise ratio of this method compared with other multiplexing techniques [4].

1.1 Phase Code Generation

The set of adjustable phases $(\varphi_{i1}, ..., \varphi_{iN})$ represents the phase address ϕ_i of the ith data page. To obtain the reconstruction of the desired jth image without cross-talk, the phases of the reference beams have to fulfill the condition [1,10]

$$\sum_{k=1}^{N} \chi_{jk} \cdot \chi_{ik}^{*} = \sum_{k=1}^{N} \mathrm{e}^{\mathrm{i}(\varphi_{jk}-\varphi_{ik})} = \begin{cases} 0 \text{ for } (i \neq j) \\ N \text{ for } (i = j) \end{cases} . \tag{1}$$

Equation (1) can be solved algebraically by unitary matrices χ fulfilling $\chi \cdot \chi^{*} = 1$, where 1 is the identity matrix. The phase codes for each data page are given by the N orthogonal columns of the matrix χ determining the maximum number of data pages that can be stored independently. For phase-coded multiplexing, the components χ_{ij} of the matrix χ are pure phase factors and can be chosen equal to 0 or π, which can be easily implemented with phase-only spatial light modulators.

For the storage of N holograms, N orthogonal phase codes have to be used. The generation of the phase codes can be achieved by the Walsh–Hadamard matrix $H^{(n)}$, which has only the two possible values: -1 and 1. The construction algorithm for the H-matrix is

$$H^{(1)} = \begin{pmatrix} 1 & 1 \\ 1 & -1 \end{pmatrix}, \quad H^{(n+1)} = \begin{pmatrix} H^{(n)} & H^{(n)} \\ H^{(n)} & -H^{(n)} \end{pmatrix}, \tag{2}$$

where the value 1 corresponds to a phase shift of 0 and the value -1 corresponds to a phase shift of π. The dimension N of these matrices is restricted to $N = 2^n$ $(n = 1, 2, 3, \dots)$ if pure Walsh–Hadamard matrices are used. Using Paley- or Williamson-type matrices, binary phase codes of the order $N = 4 \cdot m$ $(m = 1, 2, 3, \dots)$ can be constructed [11]. These construction rules allow one to adapt the number of phase codes to the available number of reference waves or pixels of the phase modulation device.

1.2 Arithmetic Image Operations

In addition to high-capacity data storage, phase-coded multiplexing offers the potential of performing arithmetic operations such as addition or subtraction of two data pages, linear combinations of several pages, and inversion of a single page during readout of the memory [6,8]. Therefore, data processing such as the comparison of two stored images or logical operations (OR, XOR, NOT) with digital data pages can be realized in a phase-coded memory.

To perform linear operations on the stored data pages, it is necessary to decrease the selectivity of the phase codes. For this purpose, only a subset of reference beams has to be used for the readout of the memory, giving the same phase code configuration for certain images. For addition of data pages during recall, the remaining references have to be chosen in such a way that the phase codes are equal. Subtraction of data pages can be achieved in a similar way, using a reference subset that has a phase difference of π between the phase code configurations. Furthermore, image inversion can also be realized during readout of the memory by subtracting a page from a plane wave stored in the photorefractive medium. Of course, linear combinations of images combining addition and subtraction are also possible by choosing the appropriate decrease in selectivity. These operations can be realized experimentally with an additional amplitude spatial light modulator by addressing the dynamic memory with only a subset of the original phase-coded beams used for storage of the appropriate images.

2 Design and Implementation of the Demonstrator

Figure 2 shows the schematic architecture of the phase-coded memory in a typical 90° geometry. The spatial light modulator (SLM) in the data arm is an electronically addressable liquid crystal display with active TFT technology. Actually, the usable area of the display is determined by the circular aperture of the optical components, resulting in a page size of 320×320 pixels, which corresponds to a capacity of about 100 kbits per page. To increase the information content of each hologram, several gray levels could be assigned to each pixel. Lens L3 performs a two-dimensional Fourier transform of the input data page, and projects the Fourier spectrum into the storage medium.

In the reference arm, addressing of data pages is realized by a phase-modulating device (PM) in combination with a computer-generated hologram

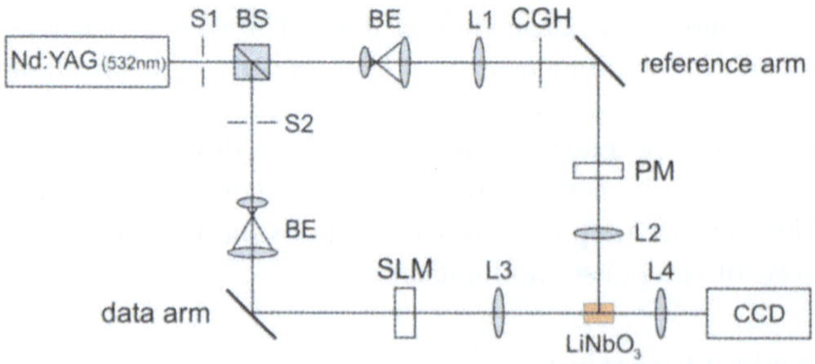

Fig. 2. Storage demonstrator using phase-coded multiplexing

(CGH) serving as a beam separation unit. It converts the laser beam into an array of 480 separate beams with equal amplitudes in the Fourier plane of lens L1. The purpose of this hologram is to ensure an equal intensity distribution of the reference beam into the 480 references, at the same time minimizing intensity losses. To achieve this aim, the hologram is realized as a kinoform detour phase type, where the concept of the detour phase is combined with highly efficient carrier gratings [12]. All 480 beams can be phase-modulated separately by the electronically driven liquid crystal phase modulator (PM), is positioned in the back focal plane of lens L1. It performs binary phase modulation to encode the reference beam array with the orthogonal phase codes. The PM is a twisted nematic liquid crystal stripe array consisting of 480 stripes with a pitch of 80 μm. This device allows one to modulate the phase of each beam simultaneously and independently without modulation of the amplitude. By means of the driving computer, the orthogonal phase codes are generated and transferred to the phase modulator. Lens L2 focuses the phase-modulated array into the storage medium.

In our experiment, an iron-doped $LiNbO_3$ crystal with a size of 1 cm^3 is used as storage medium. The c-axis of the crystal is in the plane of the data and reference beam at 45° with respect to the vertical faces of the crystal. This 90° configuration allows perpendicular incidence of signal and reference beam on the storage medium, ensuring maximum Bragg selectivity.

During the data recording process, the exposure times are controlled by a computer, which steers the shutter S1 appropriately. For recall of the stored data pages, shutter S2 will be closed. Only the reference beams illuminate the crystal encoded with the appropriate phase code. The Fourier spectrum is then read out without cross-talk, owing to destructive interference of the undesired contributions. After an inverse Fourier transform by lens system L4, the retrieved data pages are detected on a CCD camera with a resolution of 1280 × 1024 pixels.

The recall time of the system is determined by the frame rate of the phase modulator. In our experiment, we are using a liquid crystal stripe modulation device capable of switching a whole phase code in 10 ms, which results in a readout rate of 100 Hz. However, the data transfer rate is actually limited by the CCD camera with a frame rate of 8 Hz for the full resolution of 1280×1024 pixels.

The complete experiment – storage as well as recall – is controlled by a single computer, which steers the shutters, supplies the input data pages on the SLM, and generates the phase codes for the PM.

3 Experimental Results

This section describes the experimental results achieved with our phase-coded demonstration platform. Since the first results in 1991 [1], a storage capacity of 64 pages has not been exceeded using deterministic phase-coded holographic storage, although several preliminary function demonstrations have been presented [5,13]. Because these were proof-of-principle experiments with limited capabilities, it was difficult to show that this technique is competitive with angular multiplexing for holographic data storage. In 1998, we could present the realization of a fast and fully automated volume holographic memory based on phase-coded multiplexing that is capable of storing 88 data pages [14]. Recently we increased this number to as many as 180 multiplexed data pages [15].

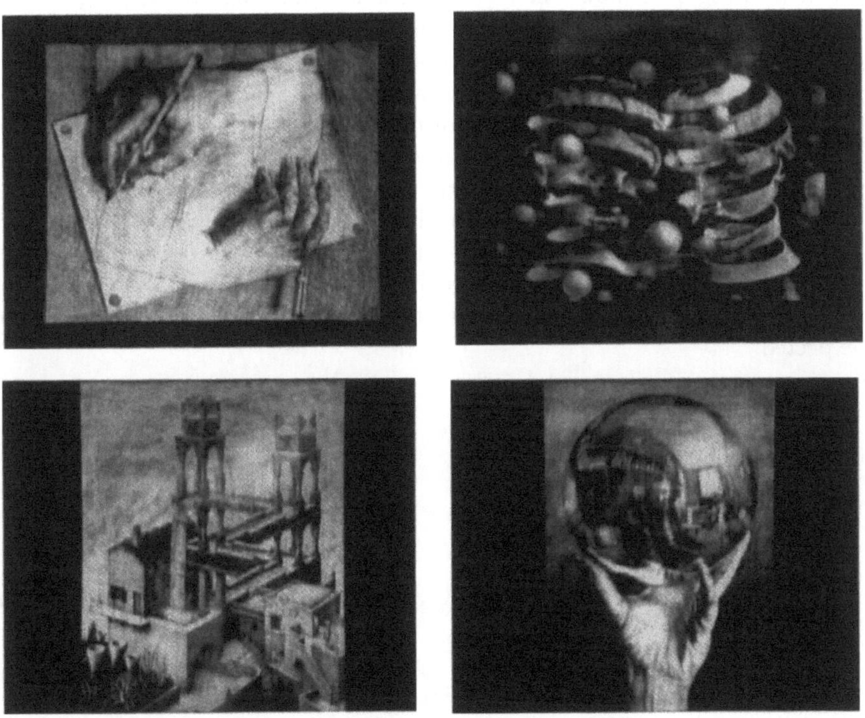

Fig. 3. Analog data storage: images by the artist M.C. Escher

Fig. 4. Digital data pages stored with phase-coded multiplexing

To demonstrate the performance of our system we stored several analog data pages. The reconstructions of some examples are shown in Fig. 3. Most promising applications of holographic memories up to now are in the area of digital data storage. In order to approach the requirements of digital data storage in our phase-coded memory, we investigated the transfer of binary pixels of a single data page. For this purpose, each pixel of the spatial light modulator or a combination of them represents a digital bit. The method of oversampling SLM pixels allows one to reduce the bit-error-rate, but at the same time it reduces capacity and transfer rates in proportion to the oversampling rate. The pixel-to-bit ratio is limited by the distortions of our imaging system. Using highly corrected, adapted optical systems, it is possible to reduce the number of pixels per bit in order to achieve higher storage capacities. Figure 4 shows the reconstructions of four digital data pages with an oversampling of 5×5 pixels per data bit.

A complete encoding reconstruction cycle of a digitally encoded image page was realized for a pixel-to-bit ratio of 5×5 pixels representing one single bit. Using the actual aperture of our system, this results in a page size of 4096 bits. A 256-colour image was used for an initial testing of the conversion of digital data back into the image, using 96 pages for the storage of the image. For the image shown in Fig. 5, we achieved a raw BER in the order of 10^{-3}. Note that actually no modulation or error correction coding was applied for storing the data. Applying novel techniques of noise reduction as well as modulation coding, a reduction of several orders in BER can be expected.

Fig. 5. Digital data storage: the VW Beetle original (*left*) and reconstruction (*right*) using phase-coded multiplexing

3.1 Arithmetic Image Operations

To demonstrate the readout of linear combinations of stored data pages, we stored three images with phase-coded multiplexing. Figure 6 shows the original reconstructions of the three images (upper row), the addition of image A

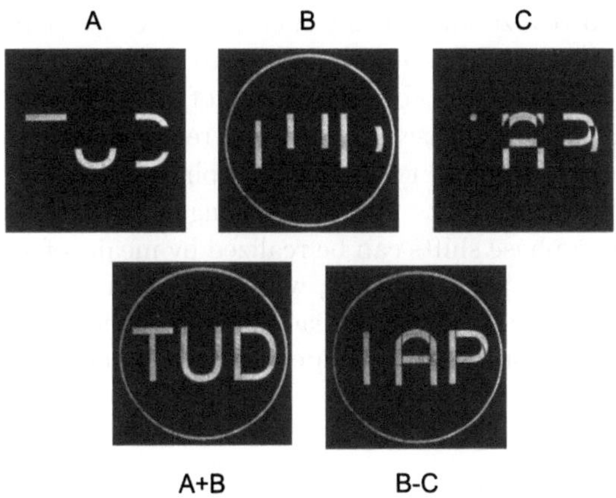

Fig. 6. Example for linear combinations of stored data pages

and B (lower left), and the subtraction of image B and C (lower right). The addition of two images is detected as constructive interference and subtraction as destructive interference. Because the result of this image processing is detected with an intensity-sensitive CCD camera device, this operation is reduced to give only the absolute value of the subtraction. However, the subtraction process itself conserves the sign of the subtraction owing to its coherent nature. Image inversion can be achieved in our memory by storing a supplementary plane wave in the memory and performing subtraction on this page in combination with the page to be inverted. An example of analog image inversion is shown in Fig. 7. Applying these operations to binary or digital data pages realizes logical operations such s XOR (subtraction) and OR (addition). Moreover, the subtraction of a digital data page from a plane

Fig. 7. Example for image inversion

wave is a NOT operation. Therefore, information processing and image operations are possible directly during readout of a phase-coded holographic memory.

3.2 Data Encryption

Phase-coded multiplexing allows us to combine orthogonal phase codes with a random phase key in order to realize an efficient encryption device. This method to prevent unauthorized users from accessing the stored data takes advantage of the freedom in choosing the absolute phase of the reference beams wavefront. By adding a random phase shift to each reference beam, the references still satisfy the requirements for orthogonal-phase-code multiplexing as long as the random phase shifts remain unchanged for writing and reading. Appropriate random phase shifts can be realized by means of an additional random-phase plate in the reference arm, with correlation length smaller than or equal to the reference beam size. Readout of the memory is not possible unless the correct random phase is added to each plane wave component.

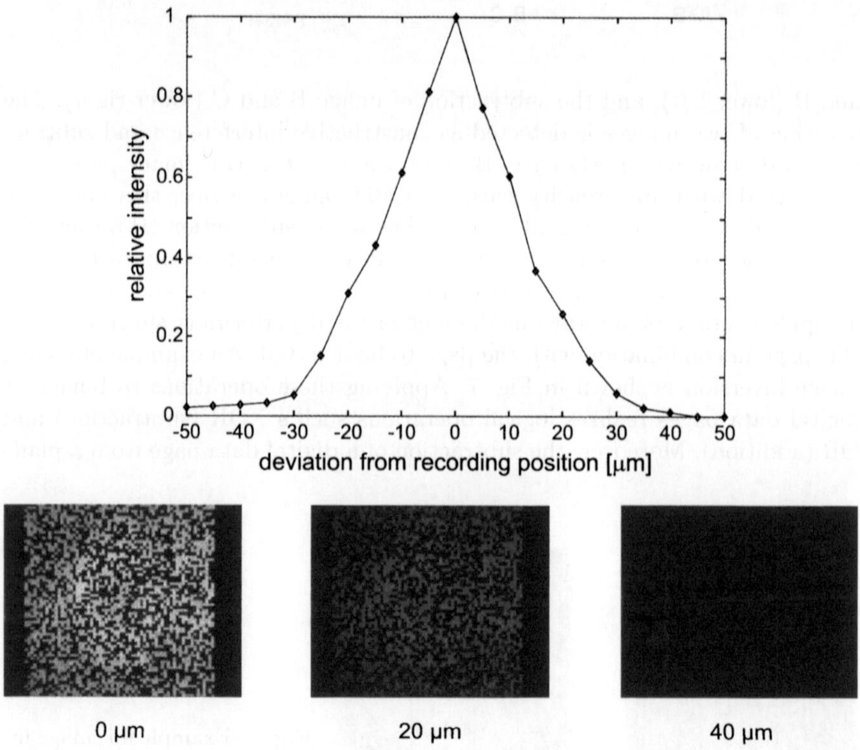

Fig. 8. Data encryption by means of a random-phase key

To demonstrate the encryption of stored data pages, we implemented a random-phase key with a correlation length of 32 μm directly in front of the phase modulator in the reference arm, and stored several digital data pages with deterministic phase-coded multiplexing. The phase key remained in a fixed position for the entire recording process. All images are still multiplexed by deterministic phase codes, but the phase key encrypts all data pages simultaneously. Figure 8 shows the image reconstruction in dependence of the position of the random-phase key. For a key shift of more than the correlation length the data can no longer be retrieved. Without knowledge of the correct phase key the phase distribution has to be guessed to access the data inside the memory. If we assume only one random phase shift per reference beam and only a limited number k of phase steps, we can roughly estimate the probability P of guessing the correct phase key to be

$$P(k) = k \cdot \frac{(N/k)!^k}{N!}, \tag{3}$$

where N is the number of reference beams. A number of $N = 180$ reference beams and assuming only $k = 5$ phase steps already results in a probability of guessing the correct phase key of $P = 10^{-121}$. Using $N = 480$ reference beams with $k = 10$ phase steps in the phase key, the probability becomes $P = 10^{-496}$. This estimation demonstrates the high potential of phase-coded memories for data encryption.

4 Summary

We presented a fully automated volume holographic storage, demonstrator based on phase-coded multiplexing for analog and digital data storage and reported the phase-coded storage of 180 holograms. Compared with the more commonly used technique of angular multiplexing, this technique allows recording and retrieving of data pages without introducing moving parts or frequency-shifting elements into the setup, being at the same time energy efficient and fast. The device can also be used for image processing, being able to realize arithmetic operations with the stored data pages during readout of the memory. Moreover, phase-coded multiplexing allows the encryption of the stored data by means of an additional random-phase key. The current limitation of phase-coded multiplexing is the lack of commercially available reliable phase modulators, but the next years will certainly bring considerable progress to commercial high-quality phase modulation devices. The performance of our demonstrator with all the natural advantages of phase-coded multiplexing shows very promising features for future applications in the field of volume holographic data storage.

References

1. C. Denz, G. Pauliat, G. Roosen, and T. Tschudi, "Volume hologram multiplexing using a deterministic phase encoding method," Opt. Commun. **85**, p. 171, 1991.
2. Y. Taketomi, J.E. Ford, H. Sasaki, J. Ma, Y. Fainman, and S.H. Lee, "Incremental recording for photorefractive hologram multiplexing," Opt. Lett. **16**, p. 1774, 1991.
3. M.C. Bashaw, A. Aharoni, J.F. Walkup, and L. Hesselink, "Cross-talk considerations for angular and phase-encoded multiplexing in volume holography," J. Opt. Soc. Am. B **11**, p. 1820, 1994.
4. K. Curtis and D. Psaltis, "Cross talk in phase-coded holographic memories," J. Opt. Soc. Am. A **10**, p. 2547, 1993.
5. J. Lembcke, C. Denz, T.H. Barnes, and T. Tschudi, "Multiple image storage using phase encoding - latest results," *Digest of the Topical Meeting on Photorefractive Materials, Effects and Devices*, Theophania, Kiev, p. 574, 1993.
6. J.F. Heanue, M.C. Bashaw, and L. Hesselink, "Recall of linear combinations of stored data pages based on phase-code multiplexing in volume holography," Opt. Lett. **19**, p. 1079, 1994.
7. C. Denz, G. Pauliat, G. Roosen, and T. Tschudi, "Potentialities and limitations of hologram multiplexing by using the phase-encoding technique," Appl. Opt. **31**, p. 5700, 1992.
8. C. Denz, T. Dellwig, J. Lembcke, and T. Tschudi, "Parallel optical image addition and subtraction in a dynamic photorefractive memory by phase-code multiplexing," Opt. Lett. **21**, p. 278, 1996.
9. J.F. Heanue, M.C. Bashaw, and L. Hesselink, "Encrypted holographic data storge based on orthogonal phase-code multiplexing," Appl. Opt. **34**, p. 6012, 1995.
10. J. Lembcke, C. Denz, and T. Tschudi, "General formalism for angular and phase-encoding multiplexing in holographic image storage," Opt. Mat. **4**, p. 428, 1995.
11. X. Yang, Y. Xu, and Z. Wen, "Generation of Hadamard matrices for phase-code-multiplexed holographic memories," Opt. Lett. **21**, p. 1067, 1996.
12. S. Sinzinger and V. Arrizón, "High efficiency detour-phase holograms," Opt. Lett. **22**, p. 928, 1997.
13. C. Alves, G. Pauliat, and G. Roosen, "Dynamic phase-encoding storage of 64 images in a $BaTiO_3$ photorefractive crystal," Opt. Lett. **19**, p. 1894, 1994.
14. K.-O. Müller, C. Denz, T. Rauch, T. Heimann, and T. Tschudi, "High capacity holographic data storage based on phase-coded multiplexing," Optical Memory & Neural Networks **7**, p. 1, 1998.
15. C. Denz, K.-O. Müller, T. Heimann, and T. Tschudi, "Volume holographic storage demonstrator based on phase-coded multiplexing," IEEE J. of Sel. Top. in Quant. Elec. **4**, p. 832, 1998.

Volume Holographic Optical Correlators

P.A. Mitkas and G.W. Burr

Volume holograms have received considerable recent interest for high-capacity data storage [1–3] because of the ability to multiplex many holograms within the same volume. Volume holographic memories may also provide an attractive solution to the massive capacity and transfer rate requirements of multimedia applications [4]. The page-oriented data format of holographic memories is a natural match for the format of certain database objects such as tables of records, images, and video frames. The massive parallelism provides both high capacity and high data rates, which can alleviate the I/O bottleneck of multimedia databases [5].

In a conventional all-electronic implementation, searching a database is often accomplished by sequentially retrieving data records from secondary memory and comparing their attributes with a set of selection criteria. As we describe below, when Fourier holograms are used to record pages of data in a volume holographic memory, *parallel* associative searching of the entire memory space for matches to a user query is possible [6–8]. Such a holographic associative processor, operating as a special-purpose functional unit connected to an electronic host, can rapidly perform otherwise time- and data-intensive selection operations, producing lean data sets that are rich in content. The associative search generates the addresses of the records that contain the desired data with a single optical input, making search time independent of database size. The electronic computer can then proceed with complex data processing tasks on a much smaller subset of the database, reducing the I/O traffic in the system many times [8–10].

The key to this optical database machine is parallel optical correlation. As we show below, optical correlation is a byproduct of the spatial Fourier-transform properties of lenses [11], while the parallelism is provided by volume holography [1].

1 Optical Correlation

Optical information processing systems are often built around the ability of a lens to perform a two-dimensional (2-D) Fourier transform in spatial coordinates. Such systems typically use the 4-F configuration: two lenses separated by the sum of their focal lengths, with 2-D input and output planes located one focal length in front of and behind the lens pair. Simply inserting an aperture in the center focal plane shared by the two lenses implements spatial filtering [12]. Alternatively, a transparency (either a computer-generated hologram [13], a phase-only filter [14], or a synthetic discriminant function [15]) can be inserted for arbitrary filtering operations. Multiplying the 2-D Fourier transforms of two different functions at this filter plane and then Fourier transforming with the back lens implements the 2-D convolution between them.

In the Van der Lugt correlator [12] shown in Fig. 1, a hologram stores the Fourier transform of the first function so that the second function can be presented at some later time at the same input plane (typically on a pixellated spatial light modulator (SLM)). If the input function is $f(x, y)$, its Fourier transform $F(u, v)$ in spatial frequency coordinates appears at the filter plane. (u and v are related to the real spatial coordinates $\mathbf{r}'$ through the lens focal length f_L and λ [11].) The hologram stored between $F(u, v)$ and a plane-wave reference $e^{-j\mathbf{k}_r \mathbf{r}'}$ (proportional to the intensity of their interference pattern) can be written as

$$| F(u, v) + e^{-j\mathbf{k}_r \mathbf{r}'} |^2 . \tag{1}$$

When this hologram is illuminated by the Fourier transform $G(u, v)$ of a second function, several diffracted components are generated, including

$$G(u, v)\, F^*(u, v)\, e^{-j\mathbf{k}_r \mathbf{r}'} \quad \text{and} \quad G(u, v)\, F(u, v)\, e^{j\mathbf{k}_r \mathbf{r}'} . \tag{2}$$

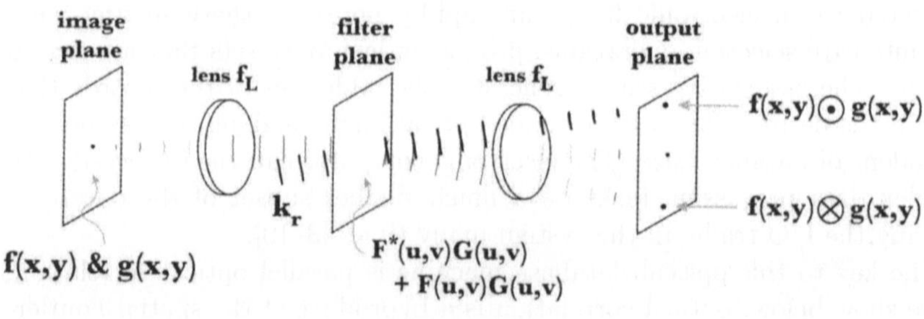

Fig. 1. 4-F system, composed of two identical lenses separated by the sum of their focal lengths. Because each lens performs a spatial Fourier transform, the system performs both the correlation (⊙) and the convolution (⊗) between the pair of 2-D optically-input functions

The convolution theorem shows that, after passing through the second lens, these two terms result in the cross-correlation between f and g,

$$f(x, y) \odot g(x, y) \equiv \iint f(x', y')\, g^*(x + x', y + y')\, dx'\, dy' \tag{3}$$

$$= \mathcal{FT}\{F(u, v)\, G^*(u, v)\} , \tag{4}$$

as well as the convolution of f and g, $f(x,y) \otimes g(x,y) = \mathcal{FT}\{F(u,v)\,G(u,v)\}$, where $\mathcal{FT}$ represents the 2-D Fourier transform operation. The VanderLugt arrangement has the advantage that the desired outputs leave the filter plane on the spatial carrier, $e^{\pm j\mathbf{k}_r \mathbf{r}'}$, corresponding to the angle of the reference beam used to store the hologram. In the output plane, the convolution and correlation terms are then separated from the transmitted image of the illuminating second function.

2 Volume Holographic Correlators

Instead of the single planar hologram, a set of multiplexed volume holograms can be used as the filter in the Van der Lugt correlator described above. After recording multiple holograms between the data-bearing object beam and different reference beams, the stored holograms are then illuminated with a new object beam to reconstruct the reference beams. Because the readout of each hologram multiplies two 2-D Fourier transforms (the transform of the input page multiplied against the transform of the original stored data page), multiple 2-D correlations are performed. If the reference beams were originally separated in angle (as in angle or peristrophic multiplexing), the second Fourier transform lens focuses each reference beam to a distinct correlation peak in the output plane. The use of angle-multiplexed volume holograms allows the optical correlator to compare an input against many templates in parallel. Alternatively, the volume hologram can be scanned under the object beam to reconstruct shift-multiplexed reference beams, or the wavelength of the object beam scanned to sequentially correlate against wavelength-multiplexed holograms.

As described in the chapter entitled "Volume Holographic Multiplexing Methods," the diffraction efficiency of the reconstructed hologram depends on the similarity between the readout beam and the original recording beam. Small angle changes of the readout beam in the horizontal plane (the plane of interaction formed by the reference and object beams) result in Bragg mismatch and no diffracted output. For a volume holographic optical correlator, this Bragg-mismatch eliminates the convolution term of the Van der Lugt correlator, and limits the correlation output to a narrow vertical slice of the 2-D correlation function, $f(x,y) \odot g(x,y)$ [16].

Such volume holographic correlators have been used for multi-channel target recognition [17], to implement optical neural networks [18] for problems such as face recognition [19], and to implement rapid digital database searches [6,8]. For optical neural networks and target recognition, the input page typically contains a complex analog image. In contrast, for a digital database searching the input page contains structured pixel blocks that represent the digital database fields [6,8]. The holographic system is then a content-addressable data storage device, with searches through the stored database performed in parallel. For large and complex databases, the search time can be significantly lower than for a conventional software-based, sequential search method [8].

3 Volume Holographic Database System Architecture

A volume holographic database system (VHDS), developed at Colorado State University, has been used to characterize associative recall of binary and gray-scale data in a range of multimedia applications including database management, interactive video, image compression, and video indexing.

As shown in Fig. 2, the core of VHDS is an angularly multiplexed, 90° geometry [20], volume holographic memory with a 1-cm^3 $Fe{:}LiNbO_3$ crystal and a 150 mW 532 nm Nd:YAG laser. The spatial light modulator (SLM) that provides the data input has been either a 640×480 resolution transmissive liquid crystal display or an 800×600 reflective digital micromirror display. The system can operate in three different modes: the *recording mode*, the *addressed recall* mode, and the *associative recall* mode.

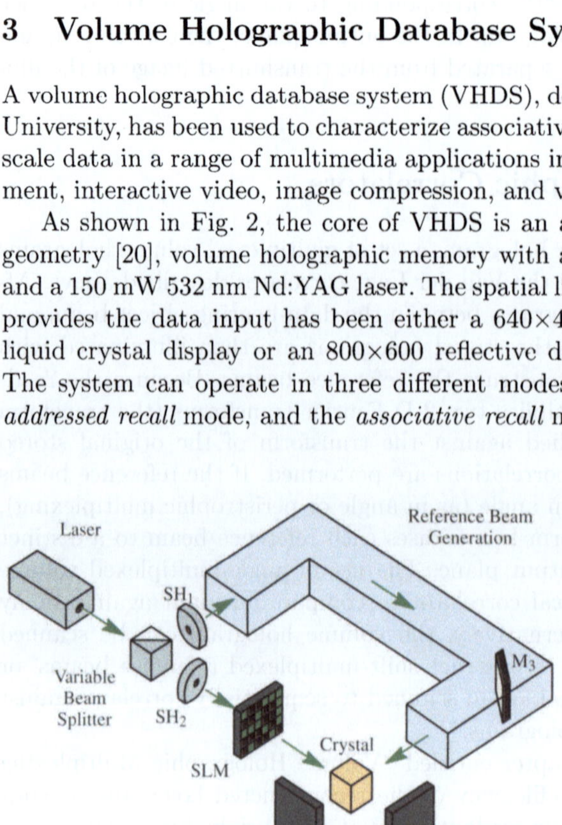

Fig. 2. Functional diagram of VHDS with a transmissive spatial light modulator

In the *recording mode*, data are stored in the memory in the form of 2-D binary or gray-scale pages. During *addressed recall*, a data page can be retrieved onto CCD_1 by illuminating the crystal with the reference beam at the angle that was used to record the desired data page.

Associative recall is the process of searching the entire contents of the memory, in parallel, with a search argument loaded in the SLM. The search argument can be any portion of a page from a maximum of 100% to a minimum percentage beyond which associative recall is not possible. The set of N holographically reconstructed and focused reference beams (the "reference beam profile") is captured by CCD_2 and is converted to an array P of N values. Each value $P(i)$ corresponds to the strength of the correlation between

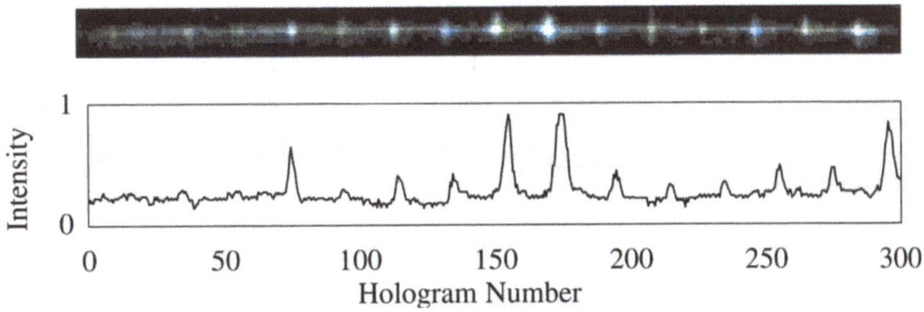

Fig. 3. A reference beam profile obtained from VHDS. Both the image captured by the CCD and the corresponding intensity profile are shown

the search argument and the contents of page i. The reference beam profile, such as the one shown in Fig. 3, is then processed to select the high-intensity peaks that indicate search hits. The location of these peaks can be mapped directly to the addresses of qualifying pages.

3.1 Associative Recall with Binary Data

To evaluate the ability of VHDS to perform searches in a binary database, records were stored from a membership database containing the following fields: Name, Address, Company/Affiliation, and Phone Numbers. According to a multiblock sparse encoding scheme with embedded parity bits (for error correction in addressed recall mode) [21], alphanumeric characters were mapped to columns of 15 bits, with only 2 of the 15 allowed to be ON at a time. A parity column was appended to every group of seven consecutive characters. A logical bit is represented in the SLM as a rectangular block of pixels, all ON or OFF, depending on the bit value. Typical experiments stored between 100 and 800 holograms in the VHDS, usually with one record per page.

An example of an associative search showing multiple hits is shown in Fig. 4. Table 1 summarizes the results of searches performed with arguments of various sizes on 530 records from the membership database, each with 210 characters. The minimum number of 8×8 pixel characters required to obtain a discernible signal at CCD_2 was 8 in this experiment, which corresponds to light passing through less than 0.4% of the pixels of the SLM.

VHDS is capable of finding exact and partial matches, single and multiple matches, or no matches. It can easily perform conjunctions (e.g. name=John AND city=Pasadena). Disjunctions (e.g. city=Pasadena OR city=Rochester) can be performed by searching the memory with a number of arguments and then combining the results. More elaborate searches can also be implemented with a few iterations of the basic VHDS associative recall operation. Thus we have the building block for a fast and functional database search engine.

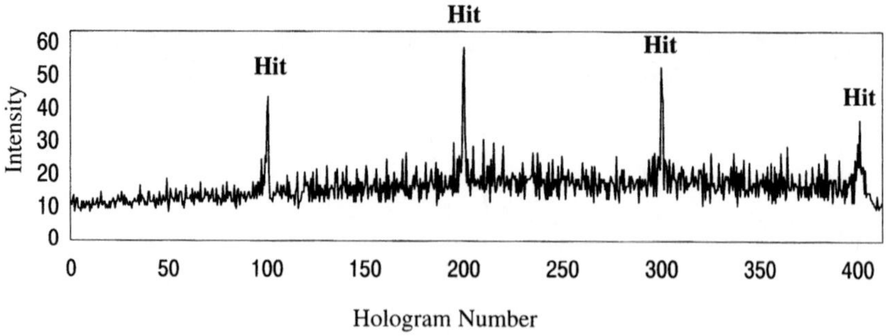

Fig. 4. A reference beam profile obtained for associative search of 400 holograms (4 repetitions of 100 records). The digital search argument was 40% of page #100

Table 1. Associative recall results from a database management experiment

	Argument size (characters)					
	8	10–19	20–29	30–39	40–49	50–52
Correct hits	33	21	13	62	42	21
False hits	4	3	2	15	4	7
Misses	1	1	1	1	0	0

3.2 Associative Recall with Image Data

In addition to binary data, gray-scale image data can be stored in VHDS at VGA or SVGA resolution. Associative recall on image data can be used to determine pattern similarities or exact matches among analog images such as faces, target imagery, or even video frames. In fact, because image data exhibit a wider spectrum of Fourier frequencies than digital data, the intensity peaks in the reference beam profile are generally higher. However, brighter images tend to produce higher peaks whether they match the search argument or not, generating more false hits than binary data pages.

One multimedia application demonstrated on VHDS is video indexing: the process of identifying the range of video frames in which a certain pattern (i.e., object, target, face, or signature) appears. Owing to the multitude of parameters involved, the amount of data that must be processed, and the lack of standard techniques, video indexing has been an extremely computationally intensive operation and, therefore, seldom available using conventional technology. In contrast, the analog correlation capabilities of VHDS are well-suited to this task.

In a recent experiment [22], a video sequence was recorded in VHDS comprising 126 frames from the trailer for the last *Star Trek* movie. When a portion of one of these frames (#46) was used as a search argument, the reference beam profile shown in Fig. 5 was obtained. The series of strong correlation peaks around hologram #46 identify the sequence of similar video

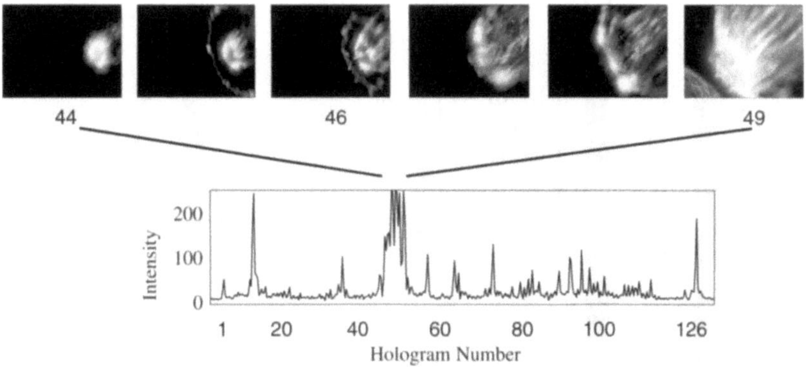

Fig. 5. Results of a video indexing experiment

frames, demonstrating that video indexing is indeed possible. Note that it is not necessary to store the whole video sequence in the memory – a representative sample (e.g. 1 frame/s) should suffice.

4 Fuzzy Volume Holographic Search Engine

The previous section demonstrated both exact searching ("hit or no-hit") of digital data, and similarity searching of analog images. Recently, a novel data encoding method was developed at IBM Almaden that allows similarity or fuzzy [23] searching of digital database records. This technique encodes similar data values in similar pixel block patterns [8]. As shown in Fig. 6a, b, data values are encoded by the position of a block of ON pixels within a vertical track, creating a 'slider'. When searching for data values near the stored value x, the partial overlap between the input and stored sliders produces a correlation peak whose brightness indicates the similarity between the input value and x. This fuzzy search encoding was experimentally demonstrated [8] using the IBM Almaden DEMON holographic storage demonstration platform [24]. Figure 6c shows a search of a single fuzzy-encoded data field as the input data value approaches and then exceeds the stored value. The amplitude response (the square root of measured power) resembles a triangle function [25,8]. The vertical extent of the slider blocks on the stored and input pages controls the detected signal power and the search range, and the search slider can be changed on a search-by-search basis. For further functionality, two slider blocks can be used to implement an OR search (i.e. records with values <16 or >240), or two tracks for increased resolution (one track for 0–256, another for the data value modulo 16). Hybrid combinations of exact and fuzzy coding allow a tradeoff between search flexibility and the most efficient use of SLM pixels.

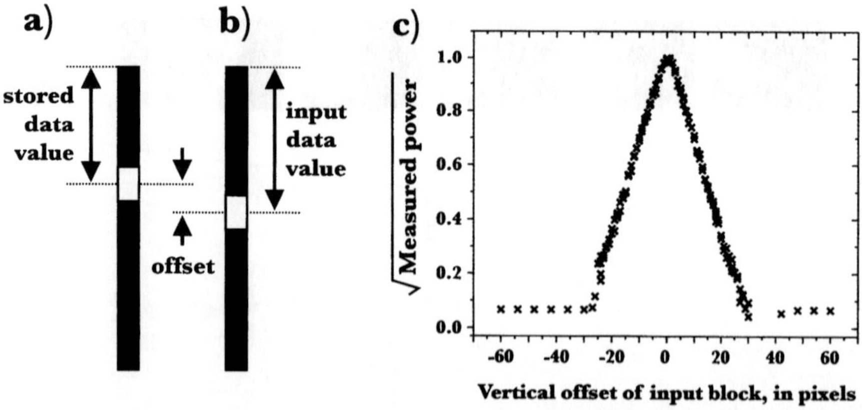

Fig. 6. Data encoding for fuzzy searching. (**a**) A stored data value is encoded by the location of a block of pixels within a 'slider' track on the SLM. (**b**) The input query is encoded similarly. (**c**) The measured optical power indicates the similarity between the data values

4.1 All-Optical Search-and-Retrieve Demonstration

To demonstrate high-fidelity parallel searching of a holographic content-addressable memory, a small multimedia database was stored in the IBM Almaden DEMON system [8]. Each hologram represented one record from an IBM Query-by-Image-Content (QBIC) database [26]. In the QBIC system, searches are performed across feature vectors previously extracted from the images, rather than on the images themselves. Each record included several alphanumeric fields (such as image description and image number) encoded for exact searches, and 64 fuzzy sliders containing the color histogram information (percentage of each given color within the associated image). A separate portion of the SLM page, pixel-matched onto a CCD detector for conventional address-based holographic readout, was encoded with the binary GIF data for the image thumbnail (using a 6:8 modulation code [24] and 7:11 Hamming error-correction code). Using an Ar^+ laser at 514.5 nm, 100 holograms were recorded in an $8 \times 15 \times 15$ mm^3 $LiNbO_3$ crystal in the 90° geometry [24] using a $\pm 2.5°$ span of reference angles. These reference angles were mapped onto a second CCD detector, and a 4 pixels wide by 7 pixels tall region on the detector was assigned empirically to each. After recording, the portion of the SLM not containing search data was blocked. A single calibration factor was measured for each correlation bin using an input page with all pixels OFF (see Sect. 5).

Each search, initiated by a user query, ran under computer control, including display of the appropriate patterns, detection of the correlation peaks (averaging multiple measurements to reduce detector noise), identification of the eight highest correlation scores, mapping of correlation bins to reference beam angle, address-based recall of these eight holograms, decoding of the pixel-matched data page, and finally display of the GIF thumbnail images on the computer monitor. The total optical readout required only 0.25 s. To find images based on color similarity, the 64 sliders were used to input the color histogram information for the upper left image in Fig. 7a; the holographic search process then output the other three images as the closest matching images. Figure 7b quantifies the search fidelity. As expected, the lefthand image correlated strongly with its own feature vector, but the system was also able to correctly identify the images with the highest cross-correlation.

These fuzzy sliders could also be used to select images by color distribution. Figures 7c, d correspond to a search for images containing 20% white and 20% light gray. Although several images were ranked slightly higher than they deserved (circle), the system performance was impressive considering the background 'dark' signal was twice as large as the correlation signal. Note that an area of 300 pixels corresponds to about 0.1% of the 640×480 SLM. In Figs. 7e, f, the alphanumeric description field was used to search for the keyword 'shore.' Note that because many characters are involved, both the expected and measured scores are large. However, similar results were obtained for exact search arguments as small as a single character.

5 Evaluation of Associative Recall

The fidelity of the associative recall operation (making correct distinctions between bright and dark correlation peaks) is determined by the quality of the reference beam profile. The simplest technique for determining a hit is to compare the correlation scores in P with a hard threshold, Th. To check fidelity, however, one must decide which correlation peaks should have resulted in a hit. Since a search may be looking for or, alternatively, trying to avoid partial matches of data pages, the desired result of the search must be described as a thresholding of the 2-D inner product $[E(i) \equiv \iint S(x', y')\, M_i^*(x+x', y+y')\, \mathrm{d}x'\, \mathrm{d}y']$ between the ith memory page, M_i, and the search data, S. For each stored page i for which $E(i)$ exceeds some desired θ, then a *hit* occurs when $P(i)$ exceeds Th. If $P(i) < Th$, then a *miss* has occurred. For example, if one wishes to find those pages containing even a minute fraction of S ("Is $S \in M_i$?"), then θ would be close to 0. In this context, a *false hit* occurs when $E(i) < \theta$ and $P(i) \geq Th$, while a *don't care* occurs when $E(i) < \theta$ and $P(i) < Th$.

Let if the total number of hits, false hits, misses, and don't cares generated by a particular search be denoted as h, f, m, and dc, respectively, maintaining the total record count, $N = m + f + h + dc$. Then, the expected number of correct hits can be expressed as $n_{\text{hit}} = m + h$. The fidelity of the associative recall will be higher as m and f approach zero. Generally, a small number of false hits can be tolerated and removed during postprocessing or by subsequent electronic processors. Misses, however, can be detrimental and should be eliminated. One way to reduce misses is provided when the overall

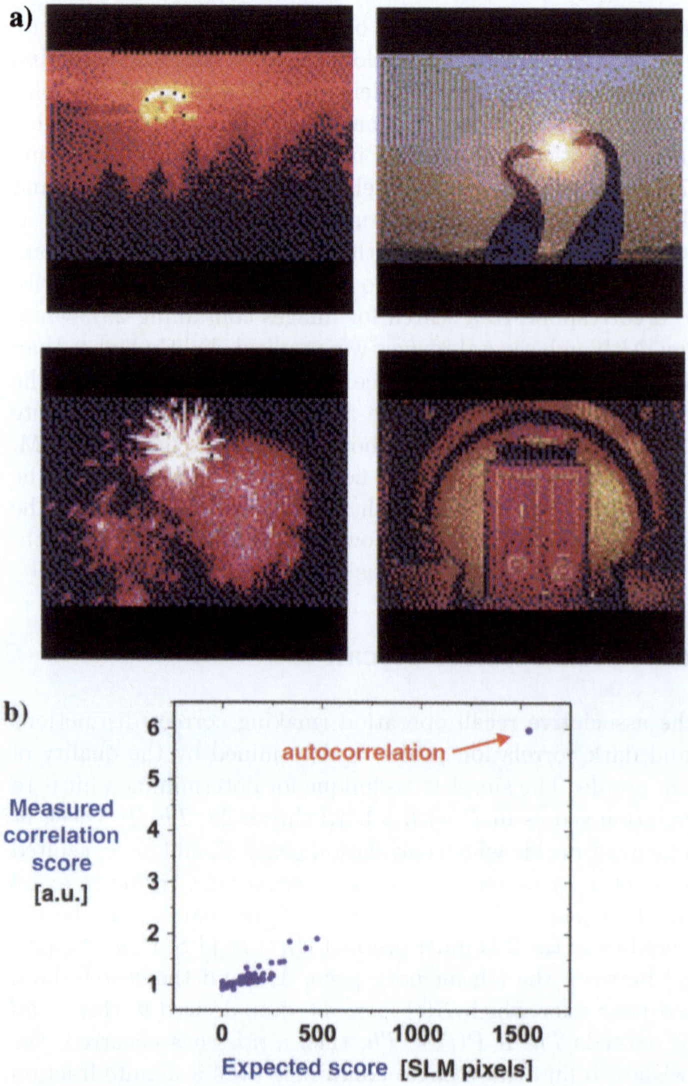

Fig. 7a–f. Three experimental search results from an all–holographic search–and retrieve engine, operating on a database of 100 feature vectors from the IBM Query-by-Image-Content (QBIC) image database [26]. (**a**), (**c**), and (**e**) show the associated images, while (**b**), (**d**), and (**f**) show measured correlation score (detected signal scaled by calibration value) as a function of the expected response (number of matching SLM pixels). (**a**) and (**b**): The search query was the color feature vector for the left most image

c)

d)

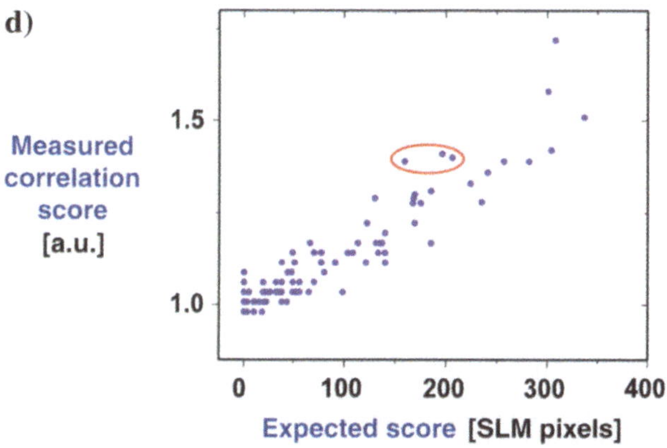

Fig. 7a–f. (Continued). (**c**) and (**d**): Fuzzy search for 20% white and 20% light grey

e)

f)

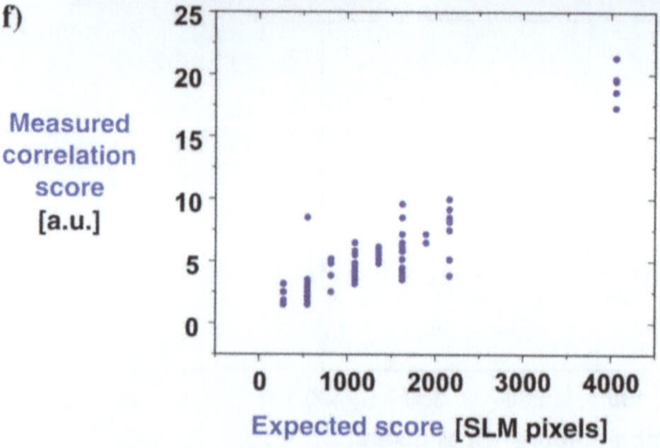

Fig. 7a–f. (Continued). (**e**) and (**f**): Exact search for the keyword 'shore', encoded into 5 characters with 3 sub-fields per character

system task is to find the "best" q matches. If the optical subsystem passes on *more* than q matches to the subsequent electronic processor, misses are greatly reduced without losing the speed-up provided by the parallel optical search.

Two functions can be defined to express the performance of data search engine. *Recall* is the ability of the system to retrieve relevant holograms, and is defined as $R = h/(h+m)$. An associative search that results in a high value for recall is considered an exhaustive search. *Precision* is the ability of the system to reject non-relevant holograms and is defined as $Pr = h/(h+f)$. An associative search that results in a high value for precision represents a specific search. Precision and recall are accuracy related functions, and ideally should have a value of 1.

The two major factors that determine the success of an associative search are selection of the right threshold value, and the signal-to-noise-ratio (SNR) of the reference beam profile. Threshold selection is not always an easy task, but is simplified by our knowledge of the size of the search argument, and thus the amount of optical power illuminating the crystal. The threshold Th can be calculated from the expected peak correlation value ($E_{\max} = S \odot S$, when a page matches the search data exactly), the targeted degree-of-match (θ), and our knowledge about how the different contributions to the diffracted correlation peak tend to combine [16].

Calculation of the appropriate SNR is complicated by the possibility of partial matches. In the simple case where a data page is expected to either match the entire search argument or be effectively disjoint, the SNR might be written as $\mathrm{SNR} = (\mu_{\mathrm{hit}} - \mu_{\mathrm{miss}})/\sqrt{\sigma_{\mathrm{hit}}^2 + \sigma_{\mathrm{miss}}^2}$. Here $\mu_{\mathrm{hit}}, \mu_{\mathrm{miss}}$ and $\sigma_{\mathrm{hit}}, \sigma_{\mathrm{miss}}$ are the mean and standard deviation of the peak intensities and the rest of the reference beam profile, respectively. Several system factors can affect the signal strength, including the number of pages recorded in the memory, the page size, the type of data, and the data encoding scheme [27]. In general, there will be two types of noise processes: background noise from electronic detection or optical scatter, and hologram–to–hologram variations.

To investigate the dependence of the correlation signal on the size and location of the SLM pixel blocks, a diffraction model was developed at IBM Almaden by Sebastian Kobras [25,16]. This numerical simulation solves for the diffracted output of a 90° geometry hologram. To reduce complexity, the model assumes that the diffraction only varies significantly along the vertical dimension. Figure 8 shows the comparison between the model and experimental measurements taken in the IBM Almaden DEMON system [25,16], for the correlation peak resulting from an auto-correlation of a simple two-block pattern. The model and experiments also agree on the dependence of signal strength on horizontal and vertical block size, and on the unexpected interactions between multiple blocks as the spacing in SLM pixels between them changes [16]. These strange effects arise from placement of the storage material away from the true Fourier transform plane of the system, a

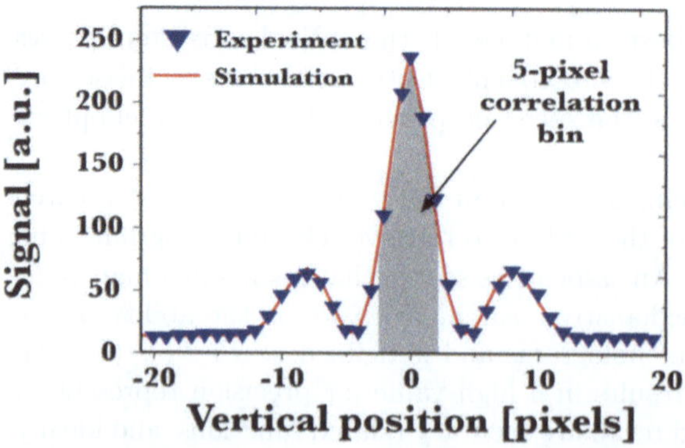

Fig. 8. Measured and simulated optical correlation peak from auto-correlating a two-block pattern, as a function of vertical position within the correlation plane

choice that is often necessary to avoid material saturation at the strongly focussing 'dc peak' of the object beam [25,16]. In addition, the diffraction model predicts that SLM contrast ratio will have an extremely important impact, creating a background 'dark' signal at the correlation plane [25,16].

If the desired signal is too dim, then the correlation peaks will tend to fade into either the background noise or the residual 'dark' signal. This is especially troublesome when the search arguments contain few characters, or when fewer pixels are used per character (in order to have more characters per record) [25,8]. However, dim correlation peaks can be augmented by modifying the search argument to allow more light to pass through the spatial light modulator (SLM) [28]. This modification must not affect the results of the search in any way other than producing an increase in the average signal brightness in the reference beam profile.

To do this, we can either use a grey level of intensity I_{d} over a large area of the SLM, where $0 \leq I_{\mathrm{d}} \leq I_{\max}$ [28], or set a smaller area of the SLM to $I_{\max}$. In either case, it is important to ensure that each stored page gets brighter by exactly the same amount [28]. For example, in the block coding scheme described in Section 3.1, illuminating the columns of parity bits has been shown to increase the intensity peaks above the noise level [28]. In the case where partial responses are expected, however, care must be taken to choose the amount of light introduced carefully, in order to avoid altering the scaling between partial and full responses caused by the interferometric nature of the optical diffraction process [16].

Despite this signal amplification, hologram-to-hologram variations in diffraction efficiency can still create misses and false hits. These variations can be suppressed by postprocessing the reference beam profile captured by the photodetector with a set of calibration values, $P_{\mathrm{B}}(i)$. These calibration values are measured by inputting an argument into the system that is supposed to produce an identical response from every hologram. Examples include the all-ON or all-OFF pages [25,8], a uniform gray background [28], or a small gauging block purposely placed on every stored page [25]. The postprocessing can be implemented by simply scaling the detected optical signal,

$$i\text{th adjusted score} = i\text{th raw score} \times \frac{\bar{P_{\mathrm{B}}}}{P_{\mathrm{B}}(i)}, \tag{5}$$

where $\bar{P_{\mathrm{B}}}$ is the average calibration factor. Alternatively, more complicated schemes [25] can be used to compensate for the interactions between SLM pixel blocks described above.

6 Conclusions

With the wide array of possible applications demonstrated, and the novel coding techniques and signal-to-noise enhancements being introduced, this is an exciting time for volume holographic content-addressable data storage. Optical correlation is emerging as an attractive method for rapidly searching vast databases with complex queries, providing faster performance than software methods through massive parallelism. Areas of current investigation include finding and demonstrating new application areas keyed to the strengths of optical correlation, implementing system architectures that support many thousands of simultaneously searched records, and quantifying (and optimizing) capacity-reliability tradeoffs.

References

1. D. Psaltis and F. Mok. Holographic memories. Sci. Am., **273**(5), 70–76, 1995.
2. J.F. Heanue, M.C. Bashaw, and L. Hesselink. Volume holographic storage and retrieval of digital data. Science, **265**, 749–752, 1994.
3. J.H. Hong, I. McMichael, T.Y. Chang, W. Christian, and E.G. Paek. Volume holographic memory systems: techniques and architectures. Opt. Eng., **34**, 2193–2203, 1995.
4. D.A. Adjeroh and K.C. Nwosu. Multimedia database management-requirements and issues. IEEE Multimedia, 24–33, 1997.
5. A.J. Daiber, R. Snyder, J. Colvin, B. Okas, and L. Hesselink. Fully functional digital video holographic storage system. In *OSA Annual Meeting, Long Beach, California*, Paper ThR3, 1997.
6. B.J. Goertzen and P.A. Mitkas. Volume holographic storage for large relational databases. Opt. Eng., **35**(7), 1847–1853, 1995.
7. L.J. Irakliotis, G. Betzos, and P.A. Mitkas. Optical associative processors. In A. Krikelis and C.C. Weems, editors, Associative Processing and Processors, 155–179. IEEE Computer Society Press, 1997.
8. G.W. Burr, S. Kobras, H. Hanssen, and H. Coufal. Content-addressable data storage using volume holograms. Appl. Opt., **38**(32), 6779–6784, 1999.
9. P.B. Berra, A. Ghafoor, P.A. Mitkas, S.J. Marcinkowski, and M. Guizani. The impact of optics on data and knowledge base systems. IEEE Trans. Knowledge and Data Eng., **1**, 111–132, 1989.

10. L.J. Irakliotis, C.W. Wilmsen, and P.A. Mitkas. The optical memory/electronic computer interface as a parallel processing architecture. J. Parallel and Distributed Computing, **41**, 67–77, 1997.
11. J.W. Goodman. *Introduction to Fourier Optics.* McGraw–Hill, 2nd edition, 1996.
12. A. Van der Lugt. Signal detection by complex spatial filtering. IEEE Trans. Inf. Theory, **IT–10**, 139–145, 1964.
13. B.R. Brown and A.W. Lohmann. Complex spatial filtering with binary masks. Appl. Opt., **5**(6), 967–969, 1966.
14. J.L. Horner and P.D. Gianino. Phase-only matched filtering. Appl. Opt., **23**(6), 812–816, 1984.
15. B.V.K. Vijaya Kumar. Tutorial survey of composite filter designs for optical correlators. Appl. Opt., **31**(23), 4773–4801, 1992.
16. S. Kobras, G.W. Burr, H. Coufal, and G. Abstreiter. Opt. correlation of digital data using volume holograms: diffraction analysis. Unpublished.
17. S.H. Lee, editor. *Optical Information Processing – Fundamentals.* Springer–Verlag, 1981.
18. E.G. Paek and D. Psaltis. Holographic implementation of a neural network model. J. Opt. Soc. Am. A, **3**(13), 32, 1986.
19. H.-Y.S. Li, Y. Qiao, and D. Psaltis. Optical network for real time face recognition. Appl. Opt., **32**(26), 5026–5035, 1993.
20. G.W. Burr, F.H. Mok, and D. Psaltis. Angle and space multiplexed holographic storage using the 90° geometry. Opt. Commun., **117**, 49–55, 1995.
21. B.J. Goertzen and P.A. Mitkas. An error correcting code for volume holographic storage of a relational database. Opt. Lett., **20**, 1655–1657, 1995.
22. G.A. Betzos, K.G. Richling, and P.A. Mitkas. Optical associative processing for multimedia database applications. In Proceedings of the 4th International Workshop on Multimedia Database Management Systems, 190–197. IEEE Computer Society Press, August 1998.
23. L.A. Zadeh. Fuzzy sets. Inf. Control, **8**, 338–353, 1965.
24. G.W. Burr, J. Ashley, H. Coufal, R.K. Grygier, J.A. Hoffnagle, C.M. Jefferson, and B. Marcus. Modulation coding for pixel–matched holographic data storage. Opt. Lett., **22**(9), 639–641, 1997.
25. S. Kobras. Associative recall of digital data in volume holographic storage systems. Master's thesis, Technische Universität München, May 1998.
26. M. Flickner, H. Sawhney, W. Niblack, J. Ashley, B. Qian Huang Dom, M. Gorkani, J. Hafner, D. Lee, D. Petkovic, D. Steele, and P. Yanker. Query by image and video content: the QBIC system. IEEE Computer, **28**(9), 23–32, 1995.
27. P.A. Mitkas, G.A. Betzos, S. Mailis, and N.A. Vainos. Characterization of associative recall in a volume holographic database system for multimedia applications. In Proceedings of the SPIE, volume 3388, 198–208, April 1998.
28. G.A. Betzos, A. Laisne, and P.A. Mitkas. Improved associative recall of binary data in volume holographic memories. Opt. Comm. **171**(1–3), 37–44, 1999.

Part VI

Competing Technologies

The Continuing Evolution of Magnetic Hard Disk Drives

E. Grochowski

Since the first magnetic hard disk drive was introduced in 1956, this technology has become a cornerstone in every information processing system used today, and it is considered as the dominant storage technology for many years to come. This role is continually growing, fueled by new advances in head, media, electronics and mechanical technologies, which have resulted in storage devices that demonstrate even higher data densities. Increased drive capacity, higher performance and progressively lower price per megabyte of storage are attributes that have created a vast popularity among computer users. This remarkable story has changed the standards for evaluating future, nonmagnetic storage that could potentially replace magnetic hard disk drives, and this story is continually improving. It is interesting to analyze the evolution of magnetic HDDs, the technology innovations that have resulted in this success, and what new technologies are on the horizon that could further enhance the application of future designs.

The fundamental architecture of HDDs has remained intact since 1956, the concept of a direct access storage device (DASD) in which data can be directly written or read when a sensor element scans the surface of a rotating disk as shown in Fig. 1. Data is written along concentric rings, or tracks, at today's densities of nearly 40 000 tracks per inch. A precision electrical/mechanical servo system, using an actuator motor, positions the head or sensor element over each track and maintains this position as data is being read in a track-following mode. Along each track, magnetized bits of information are written at today's linear densities of more than 450 000 bits per inch of track. A precise combination of head geometries, aerodynamic spacing between the flying head and rotating disk, and magnetic properties of this disk determine linear or bit density. While a high degree of precision is necessary in each of the components of a magnetic HDD, the basic design is simple, straightforward to manufacture with high yields, and easily applied as a storage device: facts that must have attracted the attention of the early inventors and users alike.

The principal evolutionary direction in HDD technology is to increase the storage capacity of this device by increasing the data density on the disk surface, i.e. the areal density, and to contain as well as to utilize the increased internal data rate to maximize performance.

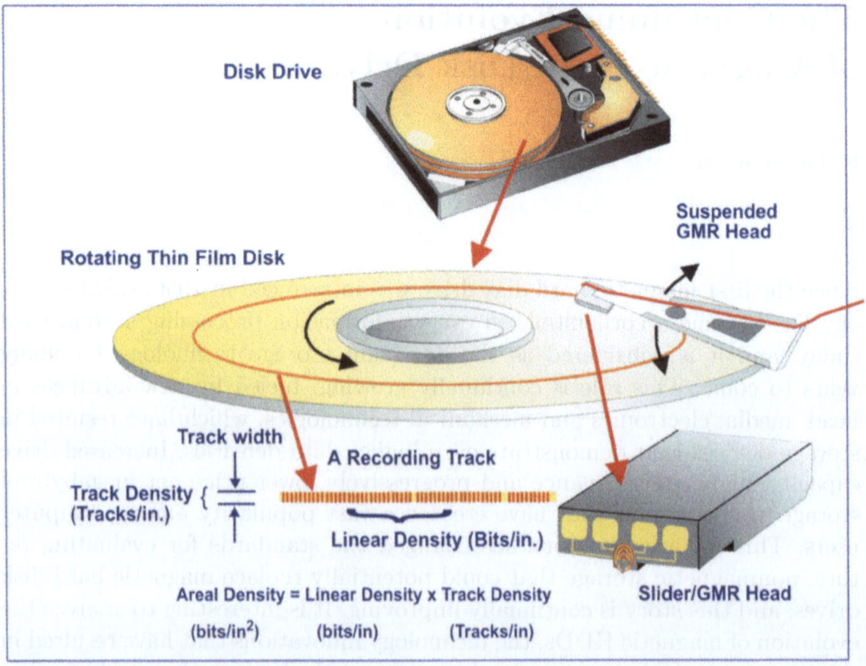

Fig. 1. Magnetic hard disk drive basics

1 Areal Density

The data density of bits per unit surface, areal density, is computed as the product of track density and linear density, and indicates the true technology power of any storage device. Figure 2 is a historical indication of HDD areal density growth since the first device, RAMAC, in 1956. In 1991 this growth rate accelerated to a compound growth rate of 60% with the application of magnetoresistive (MR) read heads, to 100% with the application of giant magnetoresistive heads (GMR). Today's areal densities are greater than 17 Gbits/in^2, with an indication that areal densities much greater than this are very possible in the near future. This optimism is based on a series of laboratory demonstrations culminating with nearly 50 Gbits/in^2 attained by several investigators. The intent of these magnetic recording demonstrations in a lab environment is to display the feasibility of using advanced heads, media, and electronics to attain higher areal densities for future products, and definitely indicates a possible technical trend for magnetic recording. Giant magnetoresistive read heads were introduced in 1997, and it is predicted that this new head design could further accelerate the areal density curve to 100%, i.e. doubling density every year. It is obvious that these magnetic recording products will demonstrate areal densities beyond 20 Gbits/in^2 in the

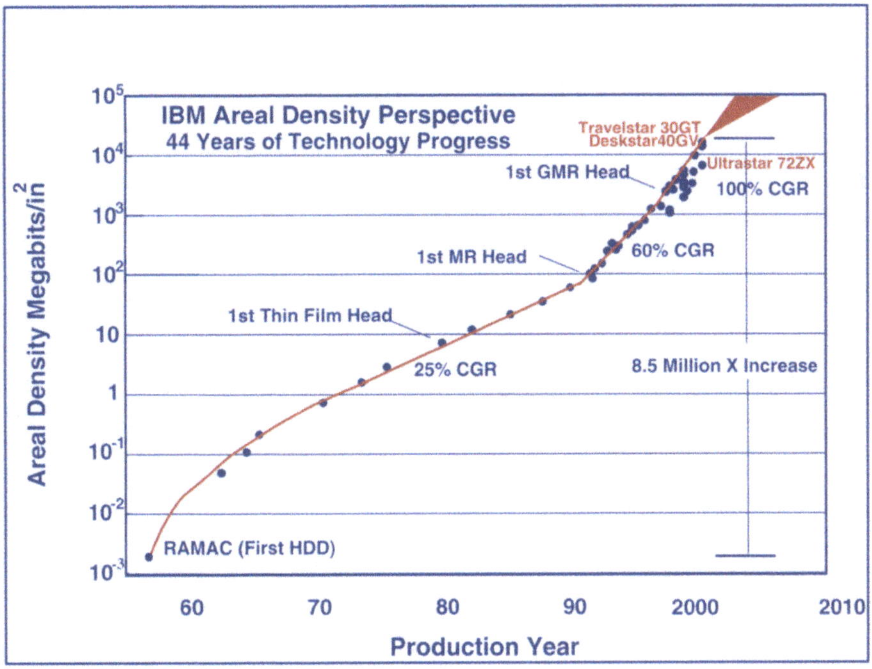

Fig. 2. IBM areal density perspective

near future, and will continue to maintain their usefulness for many years to come.

Figure 3 shows the evolution of magnetic heads to support this growth in areal density, and magnetic read/write heads are considered a key component enabling this progress. The latest read head technology to be employed to achieve densities beyond 17 Gbits/in^2 is the giant magnetoresistive or spin valve head, and this sensor technology has also been used in the laboratory to attain an areal density demonstration of nearly 50 Gbits/in^2. GMR heads show enhanced sensitivity and high changes in resistance at high areal densities, and are a logical follow-on technology to MR read heads as the technology evolution continues. This evolution involves a reduction in sensor length, for reading increased track density, as well as a reduction in GMR film thickness, which allows reading of higher bit densities. This evolutionary approach to scaling critical dimensions has been used to increase areal density with past previous head technologies – thin film inductive, MR and GMR – and will continue with the application of a new head design beyond GMR that could employ an advanced reading process such as junction tunneling. This evolutionary increase in areal density and its corresponding increase in HDD capacity is expected to continue well into the next decade, and in fact does not indicate an insurmountable limitation to the extension of magnetic recording anytime in the foreseeable future.

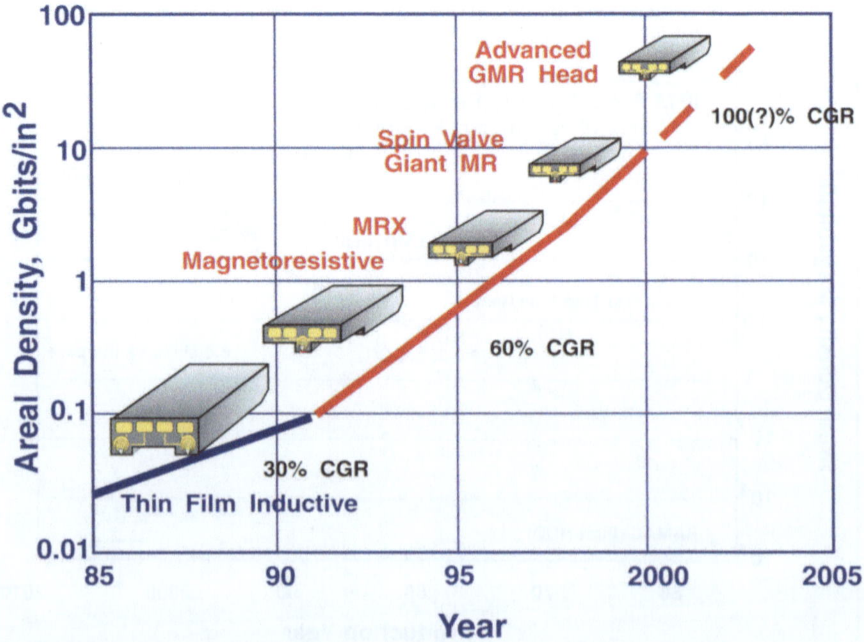

Fig. 3. Magnetic head evolution

2 Magnetic Recording Head Physics GMR

The basic magnetic recording head design used in today's advanced HDD designs contains an inductive write element and a GMR element for reading in a integrated film structure. GMR heads exhibit a significant voltage change and signal amplitude at areal densities exceeding 5 Gbits/in^2. In reading, GMR sensors, which comprise two films separated by a very thin conducting spacer, yield a voltage change in response to the medium's magnetic field, as shown in Fig. 4. This voltage change at constant current I is given by

$$V_{\mathrm{GMR}} = I\{R + \Delta R \sin\theta\},$$

and

$$\sin\theta \sim H_{\mathrm{disk}},$$

where R is the resistance change of the GMR stripe, θ is the angle between the GMR film's magnetization and a horizontal disk plane, and H_{disk} is the magnetic field originating from the disk surface as a consequence of the recorded magnetized bits.

GMR sensors exploit the fundamental quantum nature of electrons, which have spin orientations in either the up or down directions, and allow conduction (low resistance) only when electrons have their spin direction parallel to

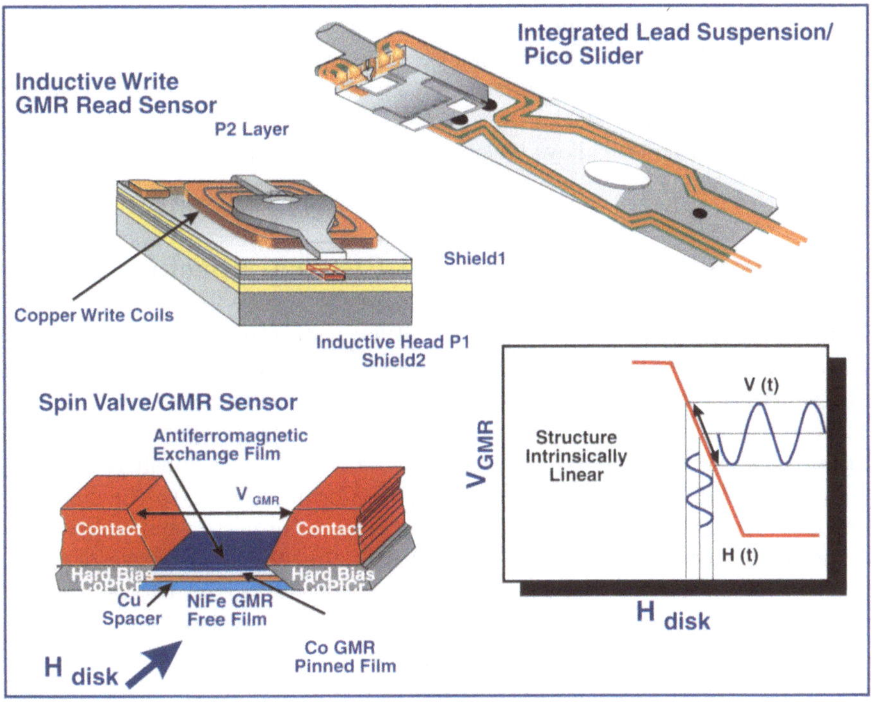

Fig. 4. GMR heads biasing

the GMR films' magnetic direction and can move freely through the structure. One film in the GMR assembly, the pinned film, has a fixed magnetic orientation through contact with an antiferromagnetic exchange film; the second film has a variable magnetic orientation as influenced by the disk's magnetic field. When both variable and pinned films have the same, parallel magnetic orientation, conduction electrons travel easily throughout both films, crossing the thin conducting spacer. When the disk's magnetic field causes a rotation of the free film's magnetization, now differing from the pinned film's direction, conducting electrons are prevented from traveling through the entire structure, colliding frequently in either film. Magnetization swings within the free film (usually an alloy of NiFe) are very large; they produce signal amplitude changes easily detected by a precision electronic data channel with a very low error rate, and are ideally suited for higher areal densities. GMR heads are in production today and are being applied to disk drives across all computer platforms: server, desktop and mobile. There is flexibility to continue the rapid areal density increase with scaled GMR heads to even smaller film thicknesses and lithographic dimensions.

Magnetic heads have the ability to erase previously written data while writing new information, and this single-pass write operation is a significant advantage in HDD technology. The inductive write element has evolved to smaller gaps and thinner pole dimensions to maintain the ability to magnetize progressively smaller regions of the disk. Increasing areal density translates into creating magnetic reversals in progressively smaller medium volumes, requiring an increase in the disk coercive field, H_c, and in most cases this requires new magnetic alloys beyond the CoPtCrB quaternary materials now in use.

3 Magnetic Disk Design and Physical Spacing

The magnetic medium, or disk, is a multilayer structure consisting of a substrate, traditionally an aluminum/magnesium alloy but now evolving into a high-strength glass for modulus, shock resistance, and smoothness properties. With increasing areal density the heads must fly progressively closer to the disk's surface, and therefore an ultrasmooth surface finish is required. The overlying, deposited magnetic films conform precisely to the substrate topography, and glass demonstrates a high degree of perfection, which yields defect-free disk surfaces. Figure 5 indicates the rate of decrease of physical spacings for 3.5 in server platform HDD products, the Ultrastar series, and the 2.5 in mobile platform HDD products, the Travelstar series. Today's head/disk spacings are less than 20 nm, with a projection of 10 nm in the

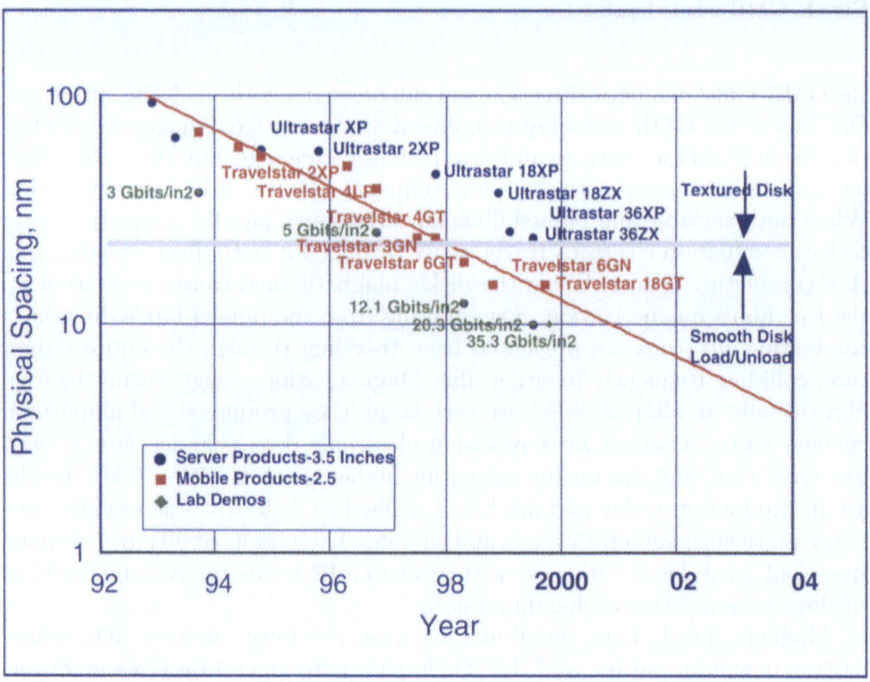

Fig. 5. Physical spacing and disk surface evolution

very near future. The trend in increased areal density and lower spacings can be expressed as

$$PW_{50} = 2\left[(g/2)^2 + (a+d)^2\right]^{1/2},$$

where

$$a = K\left[(M_r\delta/H_c)\{d(d+\delta)\}^{1/2}\right]^{1/2},$$

PW_{50} is the signal amplitude pulse width at a 50% value, g is the inductive write head gap, d is the head/disk spacing, K is a proportionality constant with a value near 0.87, M_r is the remanent magnetization of the medium's magnetic film, δ is the thickness of the magnetic film, and H_c is its coercivity. The smallest values of the quantity a can be considered as the narrowest possible transition allowing recorded bits to be at a minimum spacing, resulting in a maximum in linear density.

Ultra-smooth disk surfaces to allow for physical spacings less than 20 nm avoid the problem of head/disk contact and stiction through the use of a ramp load/unload mechanism that retracts the flying slider/head assembly during power-off times. This concept, a evolutionary approach from the use of textured or ID zone textured disks, enables the continual decrease in physical spacings to 10 nm and less, which will be required for future areal density increases to maintain or exceed the existing 60% CGR.

Figure 6 shows the fundamental disk multilayer structure used in today's magnetic disks. Future trends will involve a progressive thinning of the magnetic film, an increase in disk coercivity, and a gradual decrease in lube and carbon overcoat thickness. This allows the actual magnetic spacing to more closely approach the physical spacing or flying height. There does not appear to be any significant limitation to the trend of reducing physical spacing until the ultimate contact recording regime is attained.

4 The Mechanical HDD Design and Form Factor Evolution

Figure 7 describes the reductions in HDD form factor over the previous 10–20 years without a reduction in storage capacity, all based on rapid increases in areal density. HDD form factors have decreased from the 24-in diameter disks of the 1956 RAMAC, through 14 in for the 3380 series, to the 3.5-in (95-mm diameter) form factor that constitutes the largest percentage of disk drives used today. HDDs of 3.5 in find application to both the high-capacity, high-performance server market and the cost-performance desktop market. The mobile computer market utilizes the 2.5-in form factor, 65-mm disks for low-power performance at high capacity. Today this family of drives maintain the highest areal density lowest physical spacings, and exercise a leadership position for the introduction of advanced technologies into HDDs.

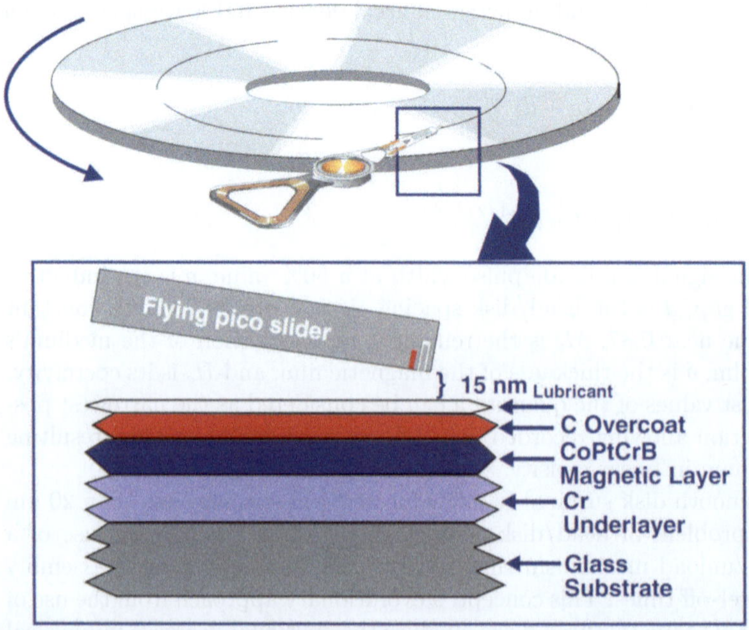

Fig. 6. Thin-film disk technology

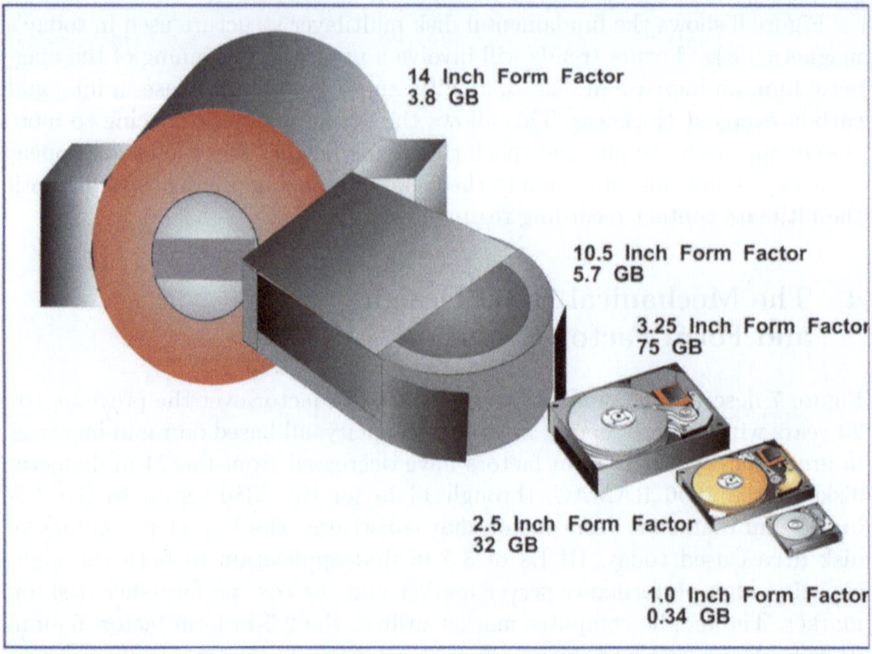

Fig. 7. HDD form factor evolution

Since internal or media data rates are moderated as disk diameter is reduced by

$$\text{Media Data Rate} \sim \text{RPM} \times \text{kbits/in} \times \text{disk diameter}$$

it is probable to anticipate application of the 65-mm-diameter-disk to the high-rotation-rate, high-performance server market in the near future. This application reduces dependence on progressively higher-bandwidth silicon technology for the HDD data channel electronics.

Application of HDDs to consumer, noninformation-processing markets is a logical extension of the technology, based on low prices per megabytes and expanded capacity. HDDs are expected to impact on the entertainment industry in such forms as home TV set-top boxes. A 1-in form factor HDD has been introduced, containing GMR heads, a load/unload mechanism, and glass substrates. This design is expected to compete directly with semiconductor flash memory cards for mobile applications, such as digital cameras, hand-held communications devices, and palmtop computers. If a 1-in HDD were to be included inside a cellular phone, for example, it would be possible to access the telephone directories for the entire United States.

Figure 8 shows projected capacity and performance evolution for magnetic HDDs applied to all major platforms. Actual progress as indicated could very well exceed the trend shown in this figure. Volume of shipped products determines price, and HDDs are approaching 200 million units per year. Application to all computer platforms as well as emerging consumer-based markets has the effect of ensuring this large unit volume, so that magnetic HDD devices will remain the dominant storage technology for many years into the future.

Today's voice-coil-driven actuator designs are required to locate progressively smaller tracks, as track densities exceed 50 000 tpi, during a seek operation, and to maintain this track position during reading/writing in the track-following mode. The addition of a second, fine, actuator to the structure could improve the ability of the actuator in both seeking and track-following modes. This micro actuator design could consist of mechanical piezoelectric mounting bars between the actuator and suspended slider, i.e. a dual-stage actuator. Voltage applied to these bars results in a small pivoting of the slider/suspension to correct for track position. Ultimately this piezoelectric or electrostatic element could be positioned near the slider to achieve near-0.1-μm track widths and a servo bandwidth of 5 kHz or more. In both designs it should be possible to read as many as four tracks without mechanical motion of the actuator, and this would improve adjacent track seek times to less than tenths of milliseconds.

5 Price

Figure 9 indicates that price per megabyte for magnetic HDDs is decreasing by about 40% per year, and this is a direct result of the 60% CGR for areal

	1999	2000	2002	2004
Server 3.5 Inch HDD 41.3 mm high 	36 GB 10000 RPM 5.4 ms Tseek	73 GB 10000 RPM 4.9 ms Tseek	73 GB 15000 RPM 4.5 ms Tseek	
Entry-Server 3.5 Inch HDD 25.4 mm high 	37 GB 7200 RPM 8.5 ms Tseek	50 GB 7200 RPM 8.5 ms Tseek	72 GB 10000 RPM 7.0 ms Tseek	150 GB 15000 RPM 4.0 ms Tseek
Mobile 2.5 Inch HDD 9.512.5/17 mm high 	25 GB 5400 RPM 12 ms Tseek	35 GB 5400 RPM 10 ms Tseek	50 GB 5400 RPM 9.0 ms Tseek	100 GB 7200 RPM 7.0 ms Tseek
Consumer 1.0 Inch HDD 5.0 mm high 	0.34 GB 4500 RPM 15 ms Tseek	0.6 GB 4500 RPM 15 ms Tseek	1.3 GB 4500 RPM 15 ms Tseek	3.2 GB 5400 RPM 12 ms Tseek

Fig. 8. HDD design projections

density:

$$1 + CGR_{\text{price/megabyte}} = 1/\{1 + CGR_{\text{areal density}}\}.$$

This expression assumes only a modest decline in unit price with time as compared with areal density, and describes the behavior of magnetic hard disk drives as well as DRAM silicon technology, as shown in the figure. The trends for both price/megabyte lines are diverging, which has a far-reaching implication concerning the replacement of HDDs by a silicon technology, even such a nonvolatile example as flash cards. The difference in price per megabyte between the two trend lines is at least a factor of 20. On the basis of continual improvements in areal density, HDD price per megabyte would be expected to decline below \$0.01 soon after the year 2000. Eventually, assuming further density increases, future price declines beyond this would be a certainty as the decade progresses.

The trend towards fewer components per HDD, i.e. heads and disks, is also a consequence of higher areal density at a fixed or nearly fixed capacity. This trend would significantly affect unit cost, so that a lower price per megabyte would result. In fact, the activity directed toward low-cost drives for the PC market reflects this trend, and could only be possible at higher areal density to meet capacity requirements. In general, the trend to simpler HDD designs is also driven by reducing power requirements at high spindle

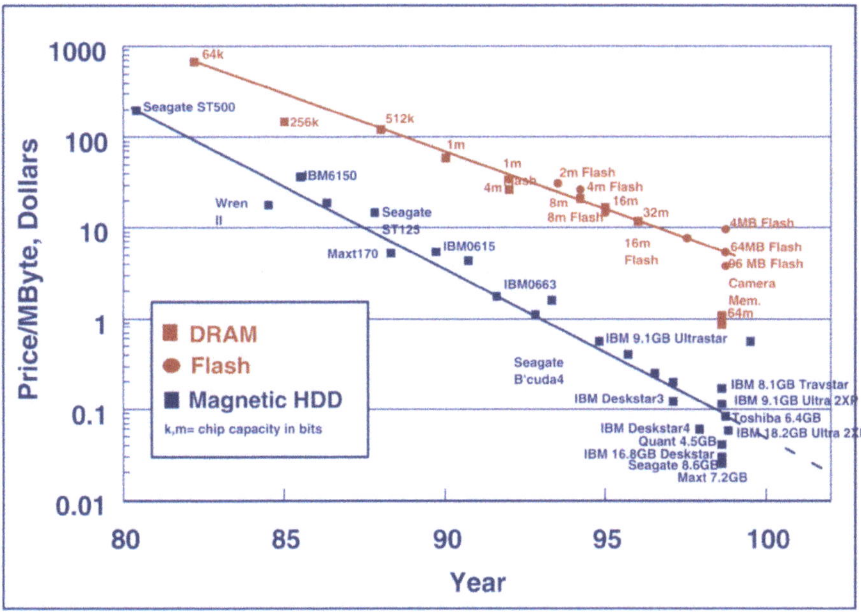

Fig. 9. Average price of storage

motor speeds greater that 15 000 rpm, registration demands at tpi values greater than 40 000 – this determines nonrepeatable run-out (NRRO) and track misregistration (TMR) – and finally the anticipated peaking of the demands for storage capacity in some specific applications.

6 Performance and Coding

Figure 10 shows that internal or media data rate is increasing with time with a CGR of 37%, based on increases in linear density, HDD rpm, and to some extent the application of zoned bit recording, which minimizes the decrease in linear density from ID to OD of the disk, allowing for a variable data rate. Today's high-performance server HDD products, at 10 000 rpm, have internal data rates greater than 60 Mbytes/s, and with further increases in linear bit density, spindle motor speeds of 15 000 rpm or faster, internal data rates can be expected to exceed 60 Mbytes/s within the next few years, and to exceed 100 Mbytes/s early in the next decade. The performance of electronic data channels, approaching 500 MHz today, is equivalent to the fastest microprocessor, and will continue to require the very latest CMOS processing at lithographic line widths of 0.25 mm or less. Figure 10 also indicates the internal data rate approach to 1 GHz, which should occur within the next five years. Internal data rate to a certain extent determines interface bus bandwidth, the rate at which the HDD communicates with the computer through

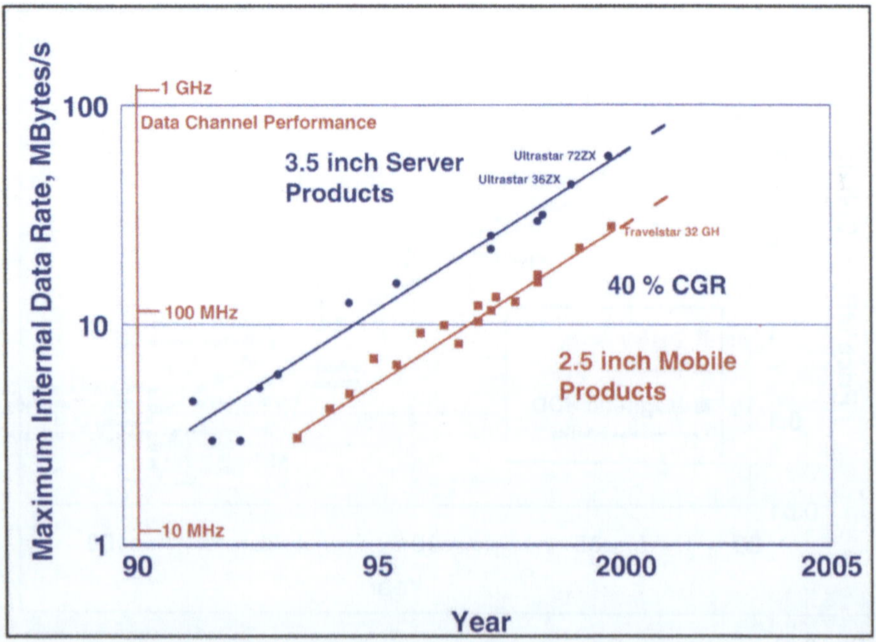

Fig. 10. Internal (media) data rate trend for IBM products

the controller circuits. For example, in SCSI devices, interface bandwidth can be considered to be four times the HDD internal data rate, so that in general faster media rates allow faster interface rates for storage in future information processing systems.

Average access time is determined by latency, which decreases inversely with rpm, and seek time, which is a function of the actuator motor capability and the data bandwidth:

$$t_{\text{seek}} \sim (\text{actuator inertia}/\text{actuator power})^{1/3} \times (\text{data bandwidth})^{2/3}.$$

Today's efficient actuator motors and low-mass assemblies have resulted in seek times below 6 ms, and it is possible to reach 3–4 ms with low-mass materials combined with new motor designs. At spindle speeds exceeding 15 000 rpm this would result in a 2 ms latency and as low as a 5 ms average access time.

Today's spindle motors use precision ball bearings in an assembly buried within the hub of a disk stack. While these bearings exhibit low startup torque and low bearing loss, damping of the spindle motor vibrational modes is not acceptable at higher speeds. As an alternative, hydrodynamic spindle motor designs offer promise to reduce NRRO, acoustic noise, non-operational shock resistance, good motor speed control and improved damping. There continues to be a higher bearing loss based on shearing forces, and a higher starting torque is associated with hydrodynamic spindles, which must be overcome with improved deigns. The use of ramp load/unload has been observed to offset some of the starting torque, particularly in a multi-disk configuration. Hydrodynamic spindles are considered as a future objective in HDD designs with high spindle speeds.

ECC is absolutely critical in keeping user bit error rates sufficiently low while maintaining adequate sensitivity for data detection at higher data density and data rate. This is valid for magnetic recording as well as all other nonmagnetic recording techniques. ECC methods for each stored sector employ three-way interleave for sectors of 512 bytes using Reed–Solomon code. ECC techniques must have the power to correct burst errors that effect multiple bytes. Future HDD products at high areal density will stress today's Reed–Solomon codes to the upper bounds. This will require new, more powerful ECC codes, which inevitably involve a higher degree of complexity. Candidates for such codes might be Reed–Solomon codes with larger symbol size as 10 bit-bytes rather than an 8 bit-bytes, algebraic geometry codes, or turbo codes. The latter function by utilizing the soft information output from the Viterbi detector. In addition to ECC codes, modulation of codes could be implemented to impose constraints on recorded data sequences, necessary to facilitate timing recovery. The sequence of modulation code evolution from the use of peak detect to PRML to EPRML has involved increasing rates as:

$$2/3 \longrightarrow 8/9 \longrightarrow 16/17 \stackrel{?}{\longrightarrow} 32/33.$$

The higher rate codes will be necessary to address weaker constraints, and will be enabled by improvements in signal processing such as improved timing recovery algorithms.

Evolution to a larger data block size is also a future trend. Today's areal density require about 5% overhead for ECC with a 512-byte block size. Future higher areal densities will require ECC overheads of 20% at this same 512-byte block size, but only 5% if a 4096-byte block is used as a data format.

Increases in areal density are prompting a new generation of high-data-rate electronics, from the data channel, arm electronics amplifier module, to the controller electronics, and these will be transformed by progressively higher levels of electronics integration, thereby increasing circuit count per chip. This will result in further price reductions, lower power dissipation, and very much improved performance.

7 Super Paramagnetism and "Limits" for Magnetic Recording

It has been proposed that increasing areal density, which decreases the bit area as shown in Fig. 11, will eventually result in reduced thermal stability of the bit magnetization, which determines its signal-carrying characteristics.

This behavior is based on the magnetic decay time relationship as shown in the figure, and is dependent on the energy barrier to decay per volume, the bit volume and recording temperature. At constant K_u, reducing the volume per bit reduces the total energy barrier for demagnetization, allowing a magnetization reversal as shown in the energy band diagram. At today's areal densities, magnetic media have $K_u V/kT$ large enough to yield decay times very much greater than 10 years. At areal densities greater than 40 Gbits/in^2 it is projected that this decay time is reduced to a value that could be significant for the stability of magnetic recording.

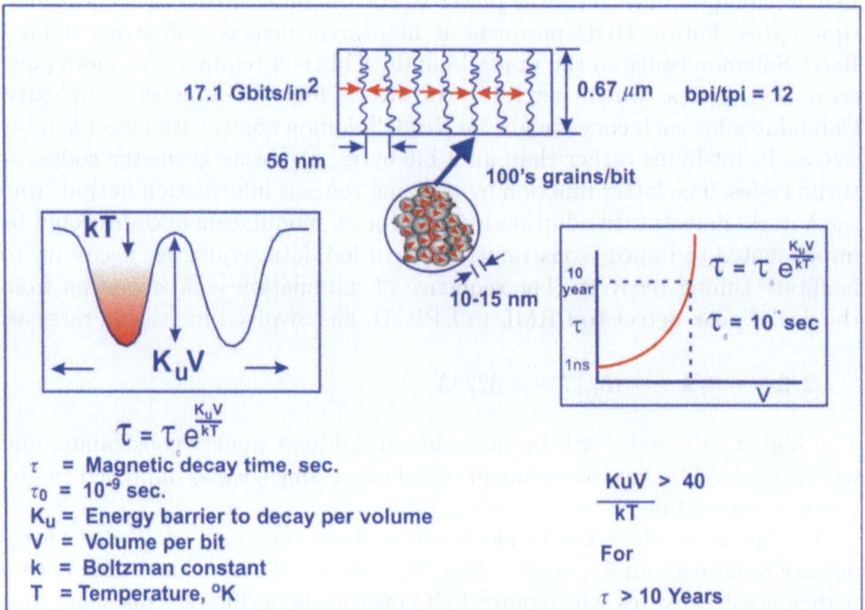

Fig. 11. Thermal stability for magnetic recording

The following are techniques for minimizing superparamagnetic effects, and could extend magnetic recording to densities far beyond 40 Gbits/in^2. The rectangular magnetized regions that constitute bits of data are formed by conglomerates of metallic grains of the medium, and the boundaries of these regions are where magnetizations oppose each other. This extends for the width of the track, and is usually irregular based on the grain shapes. By changing the aspect ratio of track width to bit length one could modify the peripheral configuration of each bit, reduce the contact region between opposing magnetizations with respect to the entire bit volume, and effectively maintain V at progressively higher areal densities.

Figure 12 indicates the progression of probable techniques that could continue to increase areal densities. Conventional scaling of critical dimensions for both heads and media are expected to yield above 20 Gbits/in^2. Selection of new disk materials which would increase with increased H_c and K_u combined with enhanced coding would be expected to result in 40 Gbits/in^2, while lowering the bit aspect ratio approaching 4:1 combined with new $4\pi M$ high-moment materials for write head yoke and pole tip structures, including FeTaN and FeNiCo, could extend this to 100 Gbits/in^2 or beyond. These new technologies are considered very probable within a period of five years, and will extend magnetic HDD products to these higher data densities and capacities. Beyond this, newer, more exotic approaches are believed to be required, including patterned media. In this advanced media structure, individual bits are isolated from each other by a magnetically inert matrix, thereby reducing the demagnetizing effects of adjacent bits. Patterned media are not considered an extension of today's media technology, but an entirely new structure involving disk processes as photolithography, microfinishing and metal alloy deposition, which have not been practiced on a manufacturing scale to date. A 500-Gbits/in^2 areal density could involve discrete magnetic patterns only hundredths of a μm on a side. In addition, alternative techniques including continuously exchanged media for increased stability, totally new magnetic writing methods beyond conventional inductive write heads, and finally perpendicular recording, could be used. These advanced techniques are mentioned to indicate that there are many possibilities for extending magnetic HDD products to new levels of recording densities and performances.

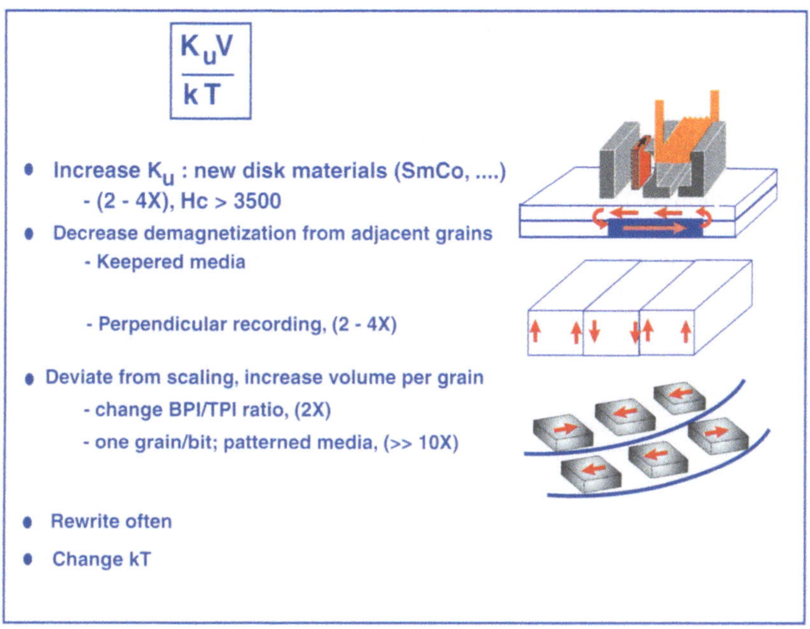

Fig. 12. Magnetic recording future

8 Conclusion

The importance of magnetic HDD storage to the information processing industry has prompted a rapid increase in the technological innovations applied over the past 44 years, resulting in a compact, low-cost, high-performance product that is universally used in quantities approaching 200 million units per year. The stream of introductions of HDD innovations is expected to continue throughout the next decade; some of them are based on such simple evolutionary concepts as scaling of dimensions of critical parameters, while others involve more complex concepts in materials and geometries. There is sufficient interest in and importance of HDD advances in the information processing industry to assure that many of these approaches will be successful. In addition, new applications for HDD products, such as an emerging consumer market that goes beyond information processing in today's viewpoint, promise to continue and expand market growth and further accelerate technological progress.

References

1. K.G. Ashar, *Magnetic Disk Drive Technology*, IEEE Press, New York, 1997.
2. M.L. Williams and E. Grochowski, "Design of magnetic inductive write heads for high density storage", Datatech **2**, 1999.
3. E. Grochowski, "Emerging trends in data storage on magnetic hard drives", Datatech **1**, 11–16, 1998.
4. C. Tsang et al., "12 Gb/in^2 recording demonstration with spin valve read heads and conventional narrow pole-tip write heads", to be published in IEEE Trans. Magn., 1999.
5. S.H. Charap et al., "Thermal stability of recorded information at high densities", IEEE Trans. Magn., **33**(1), 978–983, 1997.
6. J.M. Harker et al., "A quarter century of disk file innovations", IBM J. Res. Develop., **25**(5), 677–689, 1981.
7. E. Grochowski and D.A. Thompson, "Outlook for maintaining areal density growth in magnetic recording", IEEE Trans. Magn., **30**(6), 3797–3800, 1994.
8. S.X. Chen et al., "Future high speed spindle and components for hard disk drives", Insite, 26–28, 1999.

Optical Disk Storage Roadmap

B.H. Schechtman

Optical disk technology is the only optical data storage technology with a notable market. After more than a decade of persistent struggle, optical disk storage has established itself as a significant, albeit not dominant, segment of the computer data storage industry. Worldwide revenue for optical disk drives in 1998 was $6.1 billion, which compares with $34.2 billion for magnetic hard disk drives [1]. The success and growth of optical storage in the last few years has been strongly fueled by the emergence and acceptance of the CD-ROM device and media as the de facto standard for software distribution on mass market desktop computer systems. CD-ROM drives accounted for 92% of the optical drive units shipped in 1998, although this represented only 75% of the overall optical drive revenue [1]. For software distribution, the attractive features of CD-ROM are that it is a removable storage medium with very low cost and very high capacity, which can be easily mass-replicated. The ubiquitous presence of CD-ROM drives in desktop computer systems has in turn provided a foundation for the recent growth of recordable and rewritable devices in the CD form factor. This strong, established base will have an important influence on the future development of optical disk storage.

The dominant presence of CD-based devices notwithstanding, the industry continues to manufacture and ship a variety of drives in many other form factors, with media sizes ranging from the 2.5-in MD-Data format to large 14-in disks for very high-capacity jukebox libraries. In addition to a diversity of physical sizes, these products involve several different implementations of underlying technology for the recording of data, primarily dye or phase-change (PC) based for write-once-read-many (WORM) systems, and phase-change or magneto-optical (MO) for rewritable systems. Historically, the MO approach has had broader utilization for rewritable systems, but this has changed recently with the success in the marketplace of CD-RW products, which are based on PC recording.

As we look ahead in terms of the newest developments that will impact optical disk storage, those that seem particularly significant are the progress in GaN-based short-wavelength lasers [2], the interest in near-field optics to extend areal density [3–5], and the introduction of several new approaches in MO recording to improve the optical detection of extremely small marks [6,7]. These and other developments will be discussed in Section 2.

Much of the material described here was developed as part of the 1997 Optical Disk Storage Roadmap study [8] conducted by the National Storage

Industry Consortium (NSIC) and the Optoelectronics Industry Development Association (OIDA). Since optical storage is a rapidly advancing field with a sizable international research investment, additional material has been included here to represent the latest developments.

1 Product Categories

Optical disk products encompass a wide range of applications, ranging from the familiar CD-audio, CD-ROM and video disk players, used for playback of prerecorded digital audio, computer software and movies respectively, through large-capacity WORM drives for archival storage of data on computers, to rewritable MO and PC drives, which are used alone in workstations or in jukeboxes that store libraries of data for large computer systems. Space does not permit giving an explanation of the design and operating principles of these products, but such material is available elsewhere [9].

Prior to the strong emergence of the 120-mm CD-based optical storage products, the main data storage product categories that came to market were large-format (mostly 12-in or 14-in media) jukeboxes and smaller standalone drives (with 3.5-in or 5.25-in media). The large-format devices were generally offered as unique designs proprietary to each manufacturer, with little regard for standards and no provision for interchange of media among different systems. This approach afforded the developers quite a bit of flexibility in their technical approaches, and innovations in areas such as media formulation, media packaging, and optical head design could be easily introduced. Over time, the large-format devices have been losing their economic competitiveness, and the smaller form factors, primarily 5.25 in but also CD-based, have begun to dominate the automated jukebox library applications.

In the past 10 years, rewritable devices and media based on MO technology have been developed to ISO-based standards to permit and encourage media interchange among different manufacturers' systems. Although for standalone applications the MO drives have remained expensive compared with magnetic disk drives of comparable capacity e.g. a mainstream 5.2-Gbyte MO drive has a price of \$1500 whereas an 8.4-Gbyte magnetic drive sells for a tenth of the price – this standardization has led to increased customer confidence in MO technology, and modest growth in units shipped over recent years. Another significant development for the 3.5-in and 5.25-in MO products over the last five years has been the coordination, among many of the manufacturing companies, to define a sequence of twofold capacity improvements for introduction approximately every two years, based on a consistent set of technology innovations for each of the product generations. For example, for 5.25-in products the 1X generation had a (two-sided) capacity of 650 Mbytes and recently the 8X generation has been introduced with a (two-sided) capacity of 5.2 Gbytes. This coordination among the manufacturers has been facilitated by the Optical Storage Technology Association (OSTA)

[10]. The roadmap published by NSIC in its Optical Disk Storage Roadmap [8] differs a bit in detail from the OSTA roadmap. The NSIC roadmap was based on the consensus, developed among a group of 50 worldwide technical experts, on when the required technology innovations would reach the market, and is reproduced here in Figs. 1 and 2 for the 5.25-in and 3.5-in form factors respectively. The technology elements represented in these figures will be discussed in the next section.

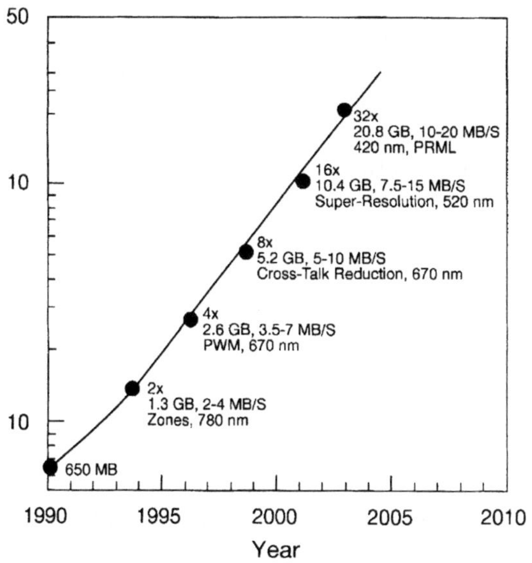

Fig. 1. 5.25 in – ISO standard MO drive capacity

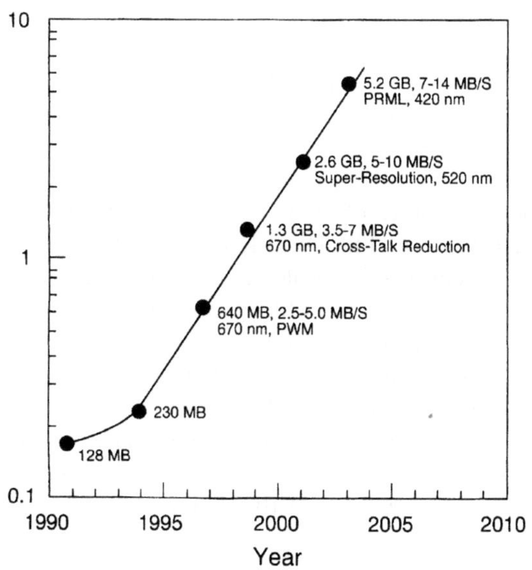

Fig. 2. 3.5 in-MO drive capacity

In addition to the MO-based products, both WORM and rewritable small form factor drives based on PC recording technology have been marketed. While the WORM products have earned some market share, until recently the rewritable versions have had less acceptance than the MO products. In general, the PC-based products have offered lower performance than the comparable generation of MO products, and for some applications the finite specification on the maximum number of rewrite cycles offered with PC media (up to 100 000 cycles in recent products) has been a drawback.

In the last few years, recordable WORM and rewritable versions of CD optical storage, denoted as CD-R and CD-RW respectively, have significantly increased their presence in the marketplace. The CD-R products are based on ablative recording in dye layer media, and CD-RW products are based on PC formulations of chalcogenide alloys. Initially, the market acceptance of CD-R was poor because there were system limitations on the ability to separately record files in different writing sessions, and the recorded disks could not always be read on the CD-ROM drives that had become so ubiquitous. These problems have since been remedied, and combined with a dramatic drop in the price of CD-R media to below $2 per disk (representing <0.3 cents/Mbyte), 2.2 million units of CD-R drives shipped in 1998 [11]. That number is quite significant when compared with the total of 1.8 million WORM and rewritable optical drives shipped in the 3.5-in and 5.25-in sizes during the same year [1]. In 1998, the CD-RW products for the first time surpassed the CD-R products, with 5.0 million CD-RW units shipped for the year [11], despite higher drive cost and significantly higher (by about a factor of 5) media cost.

At the time of the NSIC roadmap study (1996–97) it was expected that by now DVD technology would be well on its way to supplanting the above base of CD-derived products. This has turned out to not be the case. Although read-only versions of DVD have been shipping in modest volumes for video and computer ROM applications, with 6.1 million units in 1998 [11], the expected follow-on products for recordable (DVD-R) and rewritable (DVD-RAM) functions have not developed significant market penetration. Both families of products have been introduced, but there has been slow customer acceptance for several reasons. Both products initially have significantly less capacity than the 4.7 Gbytes of DVD-ROM, with DVD-R at 3.9 Gbytes and DVD-RAM at 2.6 Gbytes. The DVD-R products have been introduced as expensive specialty recorders for professionals, rather than as devices for consumers. As such, this design is unlikely to achieve the low cost necessary for mass market competition. The rewritable product arena has been confounded by the announcement of several other competing approaches to DVD-RAM, including DVD+RW and DVD-R/W, each with its own distinct technology subtleties and none to be compatible with any of the other approaches. All these approaches plan to achieve 4.7 Gbytes single-layer capacity in their second generation to match DVD-ROM. For the moment, the momentum of rewritable DVD products is stalled while the customers wait for the dust to settle.

To add further confusion to the array of offerings for high-capacity removable storage media, a multi-company effort called Advanced Storage Magneto-Optic (ASMO) is seeking to develop advanced MO-based products in the 120-mm form factor with single-layer capacity of 5–6 Gbytes [12].

2 Technology Status and Outlook

The technology elements annotated in Figs. 1 and 2 represent many of the factors that are critical to the continued advance of optical disk storage. The increasing capacities and data rates in the sequences of products shown have in large part come from a reduction in the diffraction-limited spot size that can be focused at the recording medium. This spot size varies directly as the wavelength of the light source used, and inversely as the numerical aperture (NA) of the read–write head objective lens.

Early products were limited to use of infrared wavelengths of 830–780 nm, as these were the only diode laser devices then available with sufficient reliability (typically 10^4 hours) at the power output levels (typically 30 mW) necessary to record at acceptable data rates. Present 4X and 8X generation MO drives employ visible red wavelengths of 685–650 nm. The push to even shorter wavelengths in the 650–635 nm region has been driven by the recent development of the DVD family of products. Continued advances toward even shorter wavelengths depend on development of new materials families for diode lasers. There has been significant laboratory progress in green diode lasers (490 nm) based on ZnSe [13], and even more dramatic progress in blue lasers (420 nm) based on GaN [2]. In addition, considerable work has been done by NSIC and by several companies to develop blue lasers based on the use of nonlinear materials to achieve second-harmonic generation (SHG) from a high-power infrared diode laser [14–16]. Such SHG devices, at least at power levels of a few mW suitable for read heads, have been expected to appear commercially in the last two years, but the initial devices are very expensive, and have not become a significant factor in the optical storage market. The projection, from the NSIC roadmap study [8], of future availability of short wavelength lasers is shown in Fig. 3. There has been greater than expected progress in GaN laser development since Fig. 3 was created, and the availability of low-power blue lasers is beginning to occur in 1999.

In addition to the significant activity to develop shorter-wavelength lasers, recent devices have been designed with increased NA to further reduce the working spot size. The CD devices have operated at NA = 0.45, whereas the 8× MO products use NA = 0.55 and the DVD devices use NA = 0.60. Although these improvements have provided a direct benefit to optical storage capacity, they also impose more restrictive tolerances on important operating parameters of the drive, such as the focus servo and the disk tilt.

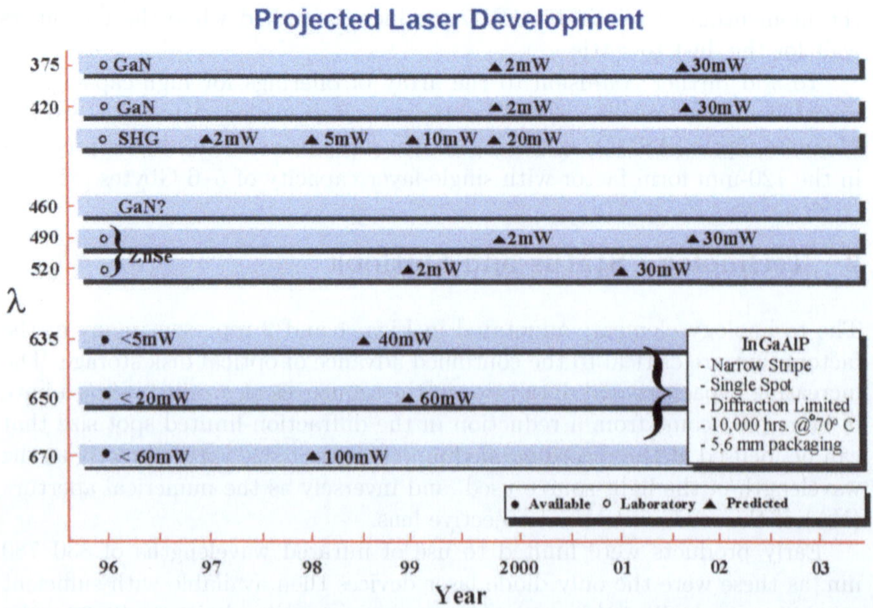

Fig. 3. Projected laser development

Perhaps the most innovative directions in optical disk storage to have emerged in the past two years involve two configurations of the heads and media that depart significantly from mainstream designs of the past decade. Both these configurations involve miniaturizing the optical head and "flying" it close to the disk, in the fashion that has been practiced for decades in magnetic disk drives. Both approaches use MO media in a surface-recording mode, and will forgo the conventional approach of burying the medium active layers behind a thick plastic substrate. If one or both of these designs proves commercially successful, this will represent a major shift in the direction of optical disk storage and in its competitive positioning with respect to magnetic disk drives and tape storage.

One such approach, based on near-field recording using a solid immersion lens or SIL [3,4], has been moved toward commercialization by TeraStor [5] in the form depicted in Fig. 4. The crucial aspect of this technology is the achievement of smaller than diffraction-limited spot size by the use of total internal reflection in the SIL, which is a high-index glass element. The spot size advantage comes at a costly penalty, however, since the coupling of the light from the SIL to the media is via the exponentially decaying evanescent wave. This requires a head–medium separation smaller than the wavelength, putting this technology in the regime of hard disk drives with respect to engineering and controlling difficult materials issues at the interface. It remains to be seen whether reliable products with removable optical media can be produced with head–medium separations <100 nm.

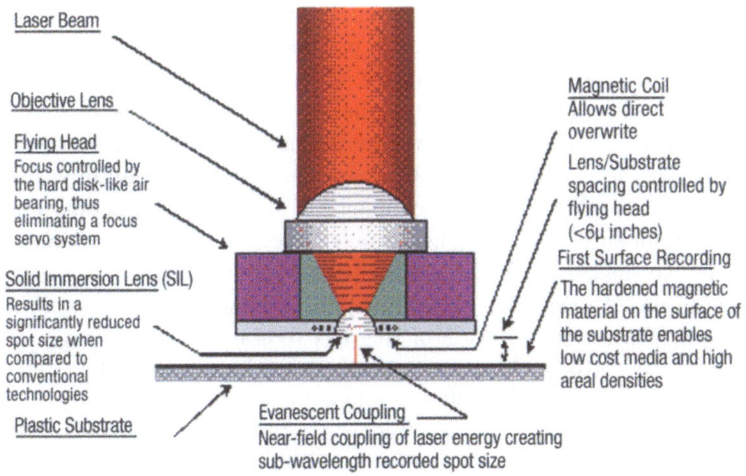

Fig. 4. TeraStor approach to near-field optical storage

The second major departure from mainstream optical drive designs has been pursued by Quinta. In this approach, parallelism is introduced by the use of a switching module and optical fiber delivery system to permit a single laser and detector to access many surfaces in a closely spaced stack of optical disks [17]. Mirror deflectors based on micro-electromechanical devices are incorporated into the flying heads to support the tracking servo function. Although this design does not appear to operate in the near-field regime to attain reduced spot size, it represents the first attempt to develop an optical drive with many heads and platters. This provides a means to achieve very high volumetric density, as has been common practice in magnetic hard drives. A key unknown with this approach is whether the large number of heads per drive can be manufactured at sufficiently low cost for the drives to be competitive in terms of dollars per megabyte.

Concurrent with the major new technology directions described above, many other advances have been developed to improve mainstream optical storage devices and media. The higher-performance MO drives have, over time, made numerous improvements to their recording channel, including the adoption of zoned recording to maximize use of disk real estate, the incorporation of pulse width modulation (PWM) to supplant pulse position modulation (PPM) at a significant linear density advantage, and the introduction of cross-talk cancellation techniques between adjacent tracks. MO drives have also increased the block size of the stored data format from 512 bytes to 2048 bytes, which provided a 26% capacity increase. The ASMO group plans to further increase the block size to 32 KB [12]. In addition, both ASMO and future versions of the MO products represented in Figs. 1 and 2 plan to introduce advanced digital channels, such as partial response maximum likelihood (PRML) detection schemes, to provide additional capacity increases. There are also developments aimed at increasing the capacity of optical channels by encoding multiple bits per mark, first for ROM devices [18] and eventually for rewritable technology [19].

As mentioned earlier, if one or more of the rewritable DVD product formats succeeds in the marketplace, this will represent a shift from MO toward PC media technology for mainstream optical storage devices. A counteracting force, however, is the ability of MO technology to invoke novel media configurations to significantly enhance the storage function. An early example of this was the introduction of MO media with a complex stack of magnetic layers to provide direct overwrite capability [20], in contrast to earlier MO media that required an extra rotation of the disk to erase before writing.

As we look to the future, there are several recent innovations that have been reported for MO media that offer promise for additional capacity improvements. Particularly interesting among these are the so-called magnetic super resolution (MSR) and magnetically amplified magneto-optic system (MAMMOS) techniques. These approaches were developed in recognition of the fact that it is relatively easier to write a very small mark with an oversize optical spot than it is to detect the mark with the same spot. The reason for this is that the writing process is a thermal process, which can be engineered to be a highly nonlinear function of the incident beam intensity. Spots can be written, therefore, with diameters much smaller than the fullwidth at half maximum (FWHM) of the incident beam, by effectively writing only where the peak of the incident light intensity is sufficient to exceed the necessary thermal threshold. Reading, however, is a strictly optical process in which the signal is typically a linear function of the incident intensity. In this case, the full profile of the beam is effective, and if the mark to be detected is much smaller than the beam's FWHM, unwanted background detection results in degradation of signal-to-noise ratio. Both the MSR and MAMMOS techniques address this problem by introducing a thermally driven step into the reading process.

The MSR technique [6], as depicted in Fig. 5, introduces a magnetic read layer that, upon laser heating, reduces its coercivity and replicates the data of the storage layer beneath it. The surrounding areas of the read layer, which have not reached sufficiently high temperatures, form a magnetic mask that does not permit the beam to see the unwanted background areas of the storage

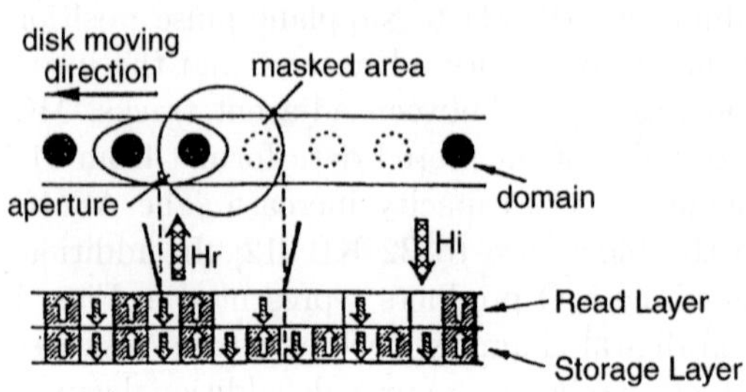

Fig. 5. Principle of magnetic super resolution

layer surrounding the desired small data mark. Several implementations of the MSR technique have been reported from numerous laboratories, and it is planned for incorporation into the 16X generations of 5.25-in and 3.5-in MO products, as well as the 120-mm ASMO products.

In a more recent development, the MAMMOS technique [7] introduces a magnetic amplifying layer above the storage layer, as shown in Fig. 6. Upon its heating with the reading beam, the amplifying layer replicates and enlarges the small magnetic domain representing the data mark in the storage layer to a size comparable to the beam diameter. This method appears to offer a significant increase in signal-to-noise ratio, and has been described as potentially providing an improvement in MO-based optical storage that is akin to the improvement that magnetoresistive heads provided for magnetic storage. Although product plans incorporating MAMMOS have not yet been announced, impressive laboratory results have been reported, including detection of mark sizes as small as 80 nm [7].

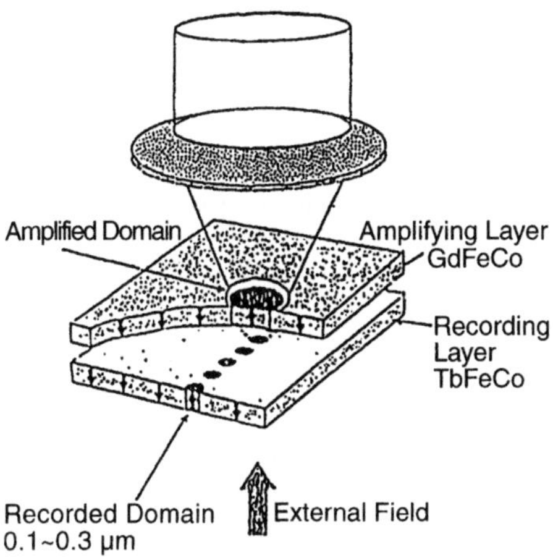

Fig. 6. Principle of magnetically amplified magneto-optics

An additional high-resolution reading technique called domain wall displacement detection (DWDD) has been reported very recently [21]. Initial studies with DWDD, which utilizes the laser-induced temperature gradient to expand the data domains, indicate that quite reasonable carrier-to-noise levels can be obtained for mark sizes down to 100 nm.

One other direction worth noting is the possibility that optical storage will exploit the potential to access two or more spatially separated storage layers by adjusting the focal distance of a far-field objective. This method is already available commercially in the 9-Gbyte versions of DVD video software, but multilayer storage will be much more difficult to implement in recordable and rewritable devices because of the tight margins required to control the thermal and optical properties of each recording layer and its surrounding dielectric layers. Nonetheless, encouraging laboratory results of two-layer MO recording have been reported [22], and further development of these techniques would seem worthwhile.

3 Summary

The rapidly increasing use of digital information for both consumer entertainment and business records, including very large data files incorporating sound, graphic images and video, is driving an accelerating demand for digital storage capacity. Optical disk storage devices represent an important and growing part of the storage industry's ability to address this demand, especially through devices that offer removable storage media that can provide virtually unlimited near-line or online capacities.

The optical storage market has gained momentum and volume shipments through the proliferation of CD-based 120-mm size products, but many of the traditional products in other form factors continue to exist, and each has its market niche and plans for continuing to improve its underlying technology. There are many developments occurring that will offer opportunities for further capacity and performance advances in optical disk storage. Those that appear especially noteworthy are the emergence of blue wavelength lasers based on GaN, the introduction of near-field and multi-head drive configurations, and the development of novel techniques to detect extremely small marks in magneto-optical media.

References

1. Private communication, J. Porter, Disk/Trend, Inc.
2. S. Nakamura, M. Senoh, S. Nagahama, N. Iwasa, T. Yamada, T. Matsushita, H. Kiyoku, and Y. Sugimoto, "InGaN-based multi-quantum-well structure laser diodes," Jpn. J. Appl. Phys. **35**, L74 (1996).
3. B.D. Terris, H.J. Mamin, and D. Rugar, "Near-field optical data storage," Appl. Phys. Lett. **68**, 141 (1996).
4. I. Ichimura, S. Hayashi, and G.S. Kino, "High-density optical recording using a solid immersion lens," Appl. Optics **36**, 4339 (1997).
5. G. Knight, "Beyond the superparamagnetic limit I: near-field recording," Data Storage **5**, 23 (1998).
6. M. Kaneko, K. Aratani, and M. Ohta, "Multilayered magneto-optical disks for magnetically induced superresolution," Jpn. J. Appl. Phys. **31**, 568 (1992).
7. H. Awano, S. Ohnuki, H. Shirai, and N. Ohta, "Magnetic AMplifying Magneto-Optical System (MAMMOS)," in *Optical Data Storage '97*, H. Birecki and J. Kwiecien, Editors, Proc. SPIE **3109**, 83 (1997).
8. *Optical Disk Storage Roadmap*, published by National Storage Industry Consortium, June 1997.
9. M. Mansuripur and G. Sincerbox, "Principles and techniques of optical data storage," Proc. IEEE **85**, 1780 (1997).
10. The OSTA roadmaps may be accessed at www.osta.org

11. Private communication, W. Schlichting, International Data Corporation.
12. D. Dalton, "Advanced Storage Magneto-Optic," in *Optical Data Storage*, Vol. 8, OSA Technical Digest Series (Optical Society of America, Washington DC, 1998).
13. N. Eguchi and A. Ishibashi, "An optical pickup using a blue-green laser diode to read a high-density disk," in *Optical Data Storage '97*, H. Birecki and J. Kwiecien, Editors, Proc. SPIE **3109**, 128 (1997).
14. W.P. Risk, S.D. Lau, R. Fontana, L. Lane, and Ch. Nadler, "Type-II second-harmonic generation and sum-frequency mixing in uniform KTiOPO4 channel waveguides," Appl. Phys. Lett. **63**, 1301 (1993).
15. S. Siala, J. Webjorn, S. DeMars, V. Wong, A. Schoenfelder, R. Waarts, and R. Lang, "Compact visible laser source based on frequency-doubled α-DFB laser diode," in *Conference on Lasers and Electro-Optics*, **6**, 1998 OSA Technical Digest Series (Optical Society of America, Washington DC, 1998), p. 381.
16. Y. Kitaoka, K. Yamamoto, K. Mizuuchi, K. Narumi, and M. Kato, "Blue second-harmonic generation waveguide device and its application to high-density optical disk," in *Conference on Lasers and Electro-optics*, Vol. 6, 1998 OSA Technical Digest Series (Optical Society of America, Washington DC, 1998), p. 270.
17. J.E. Davis, "High performance, high capacity optically assisted winchester (OAW(tm)) technology for data storage," in *Optical Data Storage*, **8**, OSA Technical Digest Series (Optical Society of America, Washington DC, 1998), pp. 28-29.
18. S. Spielman, B.V. Johnson, G.A. McDermott, M.P. O'Neil, C. Pietrzyk, T. Shafaat, D.K. Warland, and T.L. Wong, "Using pit-depth modulation to increase capacity and data transfer rate in optical disks," in *Optical Data Storage '97*, H. Birecki and J. Kwiecien, Editors, Proc. SPIE **3109**, 98 (1997).
19. M.P. O'Neill and T.L. Wong, "Multi-level data storage system using phase-change optical disks," in *Optical Data Storage 2000*, to be published.
20. T. Hosokawa, A. Okamuro, K. Miyata, H. Akasaka, and A. Sakemoto, "Production of LIM-DOW media," Proc. MORIS, J. Magn. Soc. Jpn. **20**, Suppl. No. S1, 205 (1996).
21. T. Shiratori, E. Fujii, Y. Miyaoka, and Y. Hazumi, "High-density magneto-optical recording with domain wall displacement detection," Proc. MORIS, J. Magn. Soc. Jpn. **22**, Suppl. No. S2, 47 (1998).
22. K.A. Rubin, H.J. Rosen, T. McDaniel, and W. Tang, "Volumetric magneto-optic storage on multiple recording surfaces," in *Joint International Symposium on Optical Memory and Optical Data Storage*, Vol. 12, 1996 OSA Technical Digest Series (Optical Society of America, Washington DC, 1996), p. 10.

Alternative Storage Techniques

G.R. Ashton and W.C. Mitchell

The previous chapters of this book have detailed progress in the research and development of holographic data storage. This effort has been undertaken with the hope of eventually producing a commercially successful data storage device based on holography. The goal may never be realized. Improvements in integrated circuit, magnetic disk, optical disk, and magnetic tape storage may be sufficient for the near term. Or perhaps alternative data storage technologies will prove to be more successful in the markets that holographic storage is attempting to claim. The purpose of this section is to examine some of these alternative techniques.

1 Three-Dimensional Optical Recording

Optical systems are finding ways to store information in the third dimension to increase storage density and data access times without holography. DVD and the pit depth modulation employed by Calimetrics are just two systems among several that store data in more than two dimensions.

1.1 Electron Trapping Optical Memory

Another method of obtaining multiple data bits per spot is to use electron-trapping material as the storage medium [1]. To store data in these materials one wavelength of light is used to excite electrons from the valence band to the conduction band of the material. These electrons get trapped between the conduction and valence bands until released with illumination from a second wavelength light source. Releasing electrons from the traps produces light from the material at yet a third wavelength. This third wavelength of light is sensed and used to retrieve the data bits [2]. Intensity modulation can be used to store multiple bits per illuminated spot on the storage medium.

1.2 Liquid Crystal Optical Disk

Three-dimensional optical storage can be achieved by stacking separate information layers. Unlike magnetic recording, optical recording allows the possibility of accessing multiple layers by a beam penetrating the stack. The problem is to extract information independently from each layer with minimum interference. Rubin et al. of IBM demonstrated reading and writing of information from a multiple-layer stack using depth of focus to discriminate between layers [3,4]. The commercial DVD disk format now includes a dual-layer stack with focal depth discrimination. However, the number of layers accessible by focal depth discrimination alone is limited by the relatively large separation required between layers (tens of microns) and by the loss of light from the partial reflector at each layer.

Reveo is investigating the use of cholesteric liquid crystals to increase the number of independently readable layers. The liquid crystal layers can be designed to reflect only a specific polarization in a relatively narrow bandwidth, while passing the rest of the beam unperturbed. Reveo suggests that the combination of polarization discrimination multiplexing, wavelength division multiplexing and focal depth discrimination may permit 500 independent layers to be accessed [5–7].

1.3 Surface-Enhanced Raman Optical Data Storage

It has been observed that molecular interaction with a rough metal surface can enhance Raman scattering by eight orders of magnitude, leading to the possibility of detection of a few molecules. Furthermore, the amplitude and frequency dependence of the surface-enhanced Raman scattering can be easily modified by changing the surface interaction on a molecular level, e.g. by local heating with a focused laser beam. Vo-Dinh of Oak Ridge National Labs has been attempting to exploit these characteristics for optical recording [8–10]. Vo-Dinh has investigated the use of surface-enhanced Raman scattering media with near-field fiber-addressing techniques, where the optical power reaching the substrate is low. Alternatively, he suggests that surface-enhanced Raman scattering could reduce cross-talk in multilayer recording if the layers were constructed to have different Raman frequencies. Implementation of surface-enhanced Raman scattering has been limited because increased sensitivity does not by itself provide increased storage density; a very localized write process is needed to convert sensitivity to storage density.

1.4 Optical Tape Technology

In a sense a tape pack is a method of storing data in the third dimension. The main advantage of tape is its large surface area compared with a disk medium owing to the use of the third dimension. Tape media have traditionally been used to hold high volumes of data that can be accessed in a serial bit stream and often at very high speed. Optical media have been contemplated for use in tape form for many years in write-once dye polymer, re-writable magneto-optical, and phase change forms. A commercially available optical tape recorder was also available until recently [11]. Current work in optical tape has been directed toward increasing data density on the tape surface by optical tracking and increasing the data rate by using multiple read and write channels and rapid tape movement [12]. Current goals are to produce more than one terabyte of data per rewritable cartridge with data rates at more than 100 Mbytes/s [13].

2 New Storage Forms

2.1 Persistent Spectral Hole Burning

Data can be stored in materials that have absorption spectra broadened by variations in the local environment at the absorption sites [14]. Changes in narrow-frequency portions of the absorption spectra can be caused by incident light at the appropriate frequency. These "holes," which are "burned" in the absorption spectra, remain for long times and can be detected as stored data bits: hence the name "persistent spectral hole burning" data storage. Major limitations of this technology are the need for a frequency-tunable light source, and the fact that the data storage materials require operation at cryogenic temperatures. Possible advantages of the technology are high data rates and high data densities. Eight gigabits per square inch at 20 Mbits/s data rates have been demonstrated in the laboratory [15,16]. A present target for this technology is a device for operation as a large random access memory with 10 ms latency. Estimated capacities are 50 Gbytes at a cost of about $2000 per gigabyte [17].

2.2 Two-Photon Three-Dimensional Recording

A two-photon process occurs when two photons simultaneously interact with a molecule to provide the energy for a transition that is inaccessible to a single photon. The signature of a two-photon process is the dependence of transition rate on the product of beam intensities. If two-photon processes are used for both writing and reading, then true 3-D recording can be achieved by using orthogonal beams; only the intersection of the beams, which could be a diffraction-limited three-dimensional spot, is activated. Both two-photon write and two-photon read (at longer wavelength so as not to overwrite) have been demonstrated in materials such as spyrobenzopyrans [18].

Call/recall is researching and developing 3-D optical recording systems based on two-photon processes. They recently reported a page-addressed memory with each 80-μm-thick page addressed by an anamorphically focused sheet of 532-nm radiation, and with a 100×100 bit array on each page formed by 1064-nm radiation through a chrome mask. The written data was read by illuminating the page with a 543-nm beam and monitoring the induced fluorescence with a CCD array. They believe that they can improve this to store 1000 pages with 10^7 bits per plane [19,20].

2.3 Charged Particle Beam Technology

Archival document storage in human-readable format as well as high-density CD-ROM-like disks are being produced by focused ion beams [21,22]. The ion beams are used to ion implant silicon wafers. The implanted ions reduce the etch rate of the silicon, and are thus used to create raised areas on the silicon substrates. Nickel is then electroformed on the etched silicon. If the nickel original is produced with tiny bitmapped images of the original document it becomes the archival original. Alternatively the nickel master can be used to form copies much as is done for CD-ROMs. Archival originals are read using scanning electron beam microscopy. It is envisioned that the CD-ROM-like data disks will be read using atomic force microscopy or scanning interferometric apertureless microscopy [23]. While this technology does not attempt to take advantage of a third dimension for data storage, it takes the reduction of bit size to the extreme at 50 nm/bit. Current capabilities include 165 Gbytes per 120-mm disk with 15 ms access time, write rates of 20–50 Mbytes/s and read rates of 6–10 Mbytes/s [24].

2.4 Optical Storage Card

Holographic processes are not the only means for reading data in two-dimensional data pages or patches. Work is presently being done to develop small plastic data cards capable of storing 128 Mbytes in 5000 data patches of 32 Kbytes per patch [25]. The plastic data cards would be produced at low cost similar to CD-ROMs by replication of the 450-µm-diameter data patches and plastic lenses. The illumination source will be an organic light-emitting diode array that has one light source for each of the 5000 data patches. This arrangement means that there are no moving parts for the reader, resulting in low power consumption and fast access times. Current goals for the system are a cost of $2–3 per 128 Mbyte data card, a data transfer rate of 1.6 Mbytes/s and power consumption of about 200 mW only during the 10 ms access period per patch.

2.5 Scanning Probe Storage

Atomic force microscopy (AFM) has been proposed as a technique for rewritable data storage [26]. The extremely small bit size of this technique is what results in its high data storage capacity without a third dimension. Small pits or bumps representing the stored data are produced in a surface by electron beam writing in photopolymer and etching, or by heating the tip of an AFM probe as it moves across the surface [27]. Reading is accomplished by sensing the force on the AFM tip using a laser or by a sensor incorporated into the cantilevers of the tip. Concerns with this technology are data rates and tip wear. Present laboratory systems produce 50-nm data bits and read at 100 kbits/s [28].

3 Conclusions

This section has presented a brief look at a diverse group of alternative data storage techniques. With history as a guide, the probability is small that any of the technologies presented in this chapter will be incorporated in large-volume consumer products. However, the examination of all technological alternatives is necessary to anticipate and plan for technological progress [29]. Indications of the changes to come in the data storage industry are revealed by two observations. First, of the techniques considered here, atomic force microscopy and charged particle beam technologies are the only non-optical techniques. Second, none of the techniques considered stores data magnetically. If these observations hold true, sometime in the distant future optical storage will prevail, and magnetic technology will no longer dominate data storage.

References

1. D. Carlin, "Electron trapping promises superdense optical storage." Data Storage Magazine, May/June, 41, 1996.
2. A.D.McAulay and J. Wang, "Methods for addressing electron-trapping optical memory material." Proc. SPIE, **2754**, 210–215, 1996.
3. K. Rubin, H. Rosen, T. Strand, W. Imaino, and W. Tang, In: *Optical Data Storage Topical Meeting*, Tech Dig. Series **10**, 104, 1994.
4. H. Rosen, K. Rubin, T. Sincerbox, T. Strand, and J. Zavlislan, "Multiple data surface optical storage system." US patent 5,202,875, 1993.
5. S. Faris and B. Fan, "Liquid polymers store a google of gigabytes." Data Storage Magazine, October, 21–26, 1998.
6. W. Schlichting, S. Faris, B. Fan, J. Haag, Z. Lu, L. Li, T. Milster, and H. Luo, "Recording and readout of a cholesteric liquid-crystal-based multilayer disk." In: *1996 Int Symp on Optical Memory and Optical Data Storage*, Paper OWB3, 1996.
7. S. Faris, "Optical mass storage system and memory cell incorporated therein." US patent 5,353,247, 1994.
8. T. Vo-Dinh and D. Stokes, "Serods: a new medium for high-density optical data storage." In: *Optical Data Storage '98 Conference*, Proc. SPIE, **3401**, 284–290, 1998.
9. T. Vo-Dinh,"Surface-enhanced Raman optical data storage (SERODS): principle of a new optical data storage system." In: *Optical Data Storage '94 Conference*, Proc. SPIE, **2338**, 148–155, 1994.
10. T. Vo-Dinh, "Surface-enhanced Raman optical data storage system." US patent 4,999,810, 1991.
11. D. Gelbart, "An optical tape recorder using linear scanning." Proc. SPIE, **1316**, 65, 1990.
12. R. Stahl, "Will optical tape find a niche in automated data libraries?" Data Storage Magazine, March, 49, 1996.
13. W.S. Oakley, "A Novel Digital Optical Tape Recorder." Proc. SPIE, **2604**, 256–262, 1995.

14. W.E. Moerner, *Persistent Spectral Hole Burning: Science and Applications.* Springer, Berlin, 1988.
15. L. Hai, T. Wang, and T.W. Mossberg, "Demonstration of 8 Gbits/in^2 areal storage density using swept-carrier frequency-selective optical memory." Opt. Lett., **20**, 1658, 1995.
16. T.W. Mossberg, "Seminar on optical dynamic RAM," Department of Physics, University of Oregon. http://decryptor.uoregon.edu/ mosswww/Home.html.
17. T. Mossberg, "Spectral holographic memory promises high areal densities." Data Storage Magazine, April, 49, 1996.
18. S. Hunter, F. Kiamilev, S. Esener, D. Parthenopoulos, and P. Rentzepis, "Potential of two-photon based optical memories for high performance computing." Appl. Opt., **29**(14), 2058–2066, 1990.
19. I. Cokgor, F. McCormick, A. Dvornikov, M. Wang, K. Coblentz, S. Esener, and P. Rentzepis, "Multi-layer recording using 2-photon absorption and the numerical simulation of the recording process." In: *Optical Data Storage '97 Conference*, Proc. SPIE, **3109**, 182–186, 1997.
20. A. Dvornikov, I. Cokgor, M. Wang, F. McCormick, K. Coblentz, S. Esener, and P. Rentzepis, "Materials and systems for two photon 3-D ROM devices." IEEE Trans. Components, Packaging and Manufacturing Technol. A, **20**(2), 203–212, 1997.
21. Bartlett et al., "Will charged-particle systems redefine high-density storage?" Data Storage Magazine, November/December 37, 1997.
22. B.C. Lamartine, R.A. Stutz, and J.B. Alexander, "Long, long-term storage." IEEE Potentials, **16**(5), 17–19, 1997/98.
23. F. Zenhausern, Y. Martin, and H.K. Wickramasinghe, Science, **269**, 1083, 1995.
24. J. Bishop, "Ultra high density analog and digital storage." *THIC Meeting in Albuquerque NM*, 21–22 April 1998.
25. Ioptics Internet site at http://www2.ioptics.com/default2.htm.
26. A. Sato and Y. Tsukamoto, "Nanometer-scale recording and erasing with the scanning tunneling microscope." Nature, **363**, June, 431, 1993.
27. B.W. Chui, T.D. Stowe, Y.S. Ju, K.E. Goodson, T.W. Kenny, H.J. Mamin, B.D. Terris, R.P. Ried, and D. Rugar, "Low-stiffness silicon cantilevers with integrated heaters and piezoresistive sensors for high-density AFM thermomechanical data storage." IEEE J. Microelectromechanical Syst., **7**(1), 69–78, 1998.
28. B. Terris, H.J. Mamin, R. Ried, and D. Rugar, "AFM-based storage: route to ultra-high areal densities?" Data Storage Magazine, August, 21–26, 1998.
29. P. Asthana, "Jumping the technology S-curve." IEEE Spectrum, June, 49–54, 1996.

Index

4-F configuration, 30, 429
90°-geometry, 28, 57

absorption, 105
access, 9
– time, 458
acid generator molecule, 173
acousto-optic deflector, 243, 251, 409, 410
actuator motors, 458
adaptive threshold, 284
algorithms: Berlekamp–Massey or Euclidean, 298
angle Bragg selectivity, 37
angle multiplexing, 7, 13, 38, 63, 242
– in-plane, 25
– out-of-plane, 25
aperture, 310
areal density, 448
arithmetic image operations, 419, 421, 425
Arrhenius-type dependence on temperature, 128
ASMO group, 467, 469
aspherical apodizer, 379
associative
– data, 393
– recall, 433
– retrieval, 9
searching, 429
atomic force microscopy, 478
axicons, 265, 378
azo group, 217

$Ba_{0.77}Ca_{0.23}TiO_3$ (BCT), 119
balanced block code, 285
$BaTiO_3$, 119
beam conditioning, 259
beam deflector, 241, 253
– speed, 250
beam steering, 343
$Bi_{12}TiO_{20}$ (BTO), 119
binders, 174
bipolaron, 139, 152
bit error rate, 16, 91, 101, 108, 206, 283, 309, 320, 331, 374, 383, 401
blanced strip codes, 287
block code, 283, 297
Born approximation, 31, 33
Bragg
– angular detuning, 181
– matching, 37, 404
– methods, 24, 51
– mismatch, 67, 79
– multiplexing methods, 54
– peaks, 202
– selectivity, 63, 246
Bässler formalism, 162

Call/Recall, 477
capacity, 117, 294, 337, 384
cationic ring-opening mechanism (CROP), 172
CD-audio, 464
CD-R, 466
CD-ROM, 463, 464
CD-RW, 466
channel
– encoding, 360
– modeling, 310
charge transport, 114
charge-coupled devices (CCDs), 271, 274, 309, 344, 347, 370, 388, 400
– fast, 375
charge-injection devices, 274
chromophores, 212

complementary metal-oxide-semiconductor (CMOS), 271, 274, 360
– active pixel sensors, 364
coding, 283
coercivity, 453
coherent noise, 96
coherent scattering, 294
combinations of multiplexing methods, 28
communication channel, 295
compensated grating, 128
component segregation, 180
congruent composition (CLN), 151
constant-weight codes, 285
constraint, 283
convolutional codes, 297
correlation
– detection, 285
– multiplexing (CM), 361
correlators, 429
cross-talk noise, 63, 66

dark
– current, 277
– decay, 107, 117, 156
– storage time, 117
data
– encryption, 426
– transfer, 409
decay during readout: fixing, 107
decision feedback (DF), 314
defocusing, 260
deformable mirror device, 391
DEMON I, 369, 374
DEMON II, 369, 376
detection scheme, 283
detector arrays, 271
deterministic phase encoding, 8, 25, 45, 419
differential coding scheme, 386
differential encoding, 246, 285
diffraction efficiency, 24, 36
digital micromirror, 360, 364
digital mirror, 244
digital signal processing techniques, 386
diode laser devices, 467
diode-pumped frequency-doubled Nd:YAG, 231
diode-pumped frequency-doubled YAG laser, 377
diode-pumped solid-state lasers, 232
diode-pumped solid-state sources, 364
direct
– access storage device, 447
– digital synthesizers, 412
direct-gap semiconductor, 234
disk shaped media, 343
distributed
– Bragg reflector (DBR), 235
– feedback, 236
DMD display, 248
domain wall displacement detection, 471
doped stoichiometric lithium niobate, 144
DVD disk, 476
DVD technology, 466
DVD-R, 466
DVD-RAM, 466
dynamic range, 104, 145, 155, 164, 273

ECC, 309, 338, 387, 459
electrical fixing, 122, 136, 149
electro-optic
– coefficient, 159
– chromophores, 160
– polymer, 159
electron trapping optical memory, 475
encoder, 283
encrypted holographic, 419
equalization, 309, 311
error correction, 293, 297, 360
Ewald sphere, 42
extrinsic defects, 114, 144

Fabry–Pérot lasers, 234
fast CCDs, 375
ferroelectric domain reversal, 107, 136
ferroelectric liquid crystal, 244
fiducial patterns, 284
figure of merit for photorefractivity, 159
fill-factors, 276, 310, 335
fixing, 127
– by spontaneous polarization modulation, 136
flat-top beams, 380
flying heads, 447, 469

form factor, 453
Fourier
– holograms, 429
– lenses, 372, 376
– plane, 30, 260
– transform, 5, 259, 372, 378, 429
– transform lenses, 377
fractal multiplexing, 24, 51, 52
Fresnel
– spectrum, 260
– zone, 30

GaAs, 119
gated recording, 108, 138
gating ratio, 143, 146, 154
Gaussian
– detector noise, 315
– functions, 97
Gaussian-profile, 379
giant magnetoresistive or spin valve head, 449
global threshold detection, 284
gray-scale capacity gain, 319
guest/host approach, 159

Hadamard function, 47
Hamming
– distance, 297
– error-correcting code, 390
Harris-Intertype wide-band recorder, 13
holographic data storage system (HDSS), 383
high-low fixing, 134
histograms, 94, 206
historical development, 3, 10
Hitachi holographic video disk, 14
hologram fidelity, 108
holographic disks, 15, 55, 241
holographic optical disk, 241
Holographic Random Access Memory (HRAM), 57
holographic storage in the Soviet Union, 14
Holoplex memory device for fingerprint verification, 17
Holoscan, 14
HOST tester, 91

IBM DEMON, 18, 321
IBM Holographic Optical Storage Tester (HOST), 11, 102
image plane, 29
incoherent noise, 96
inductive write element, 450
InGaP DFB, 231
inorganic photorefractive materials, 113
inter-page cross-talk, 293, 333
inter-pixel
– cross-talk, 321, 333
– interference, 288, 293
inter-symbol interference, 310, 312
interference heliography, 3
interleaving, 293, 300
intrinsic
– scattering noise, 76
– defects, 114, 139
– light scattering, 102
inverse Fourier lens, 372
ionic
– charge pattern, 128
– conduction, 130
– fixing, 128, 391
iron-doped lithium niobate, 135, 376
iron impurities, 156
isomerization, 210

jukeboxes, 464

k-space, 42
K-space, 74
Kanji character generation system, 15
$KNbO_3$, 119
krypton ion laser, 370
$KTa_{0.52}Nb_{0.48}O_3$ (KTN), 119, 138

laser sources, 231
latency, 458
lens aberrations, 293
lifetime of fixed ionic gratings, 131
light
– absorption, 181
– scattering, 120
linear minimum mean squared error (LMMSE) equalization, 312, 313
liquid crystal
– optical disk, 475
– scanner, 243
$LiTaO_3$, 119, 138

lithium niobate ($LiNbO_3$), 106, 119, 138, 149, 377, 388, 412
local threshold, 284
Lommel functions, 40
low-pass codes, 288

M number, 24, 104, 117, 135, 155, 232, 333, 385
magnetically amplified magneto-optic system, 470
magnetic
– hard disk drives, 447
– heads, 449
– super resolution, 470
magnetoresistive read heads, 448
magneto-optical recording, 463
MAMMOS, 470
manufacturing defects, 293
mass transport, 178
master oscillator, 237
MSR, 470
material stability, 165
matrix precursors, 201
mechanical deflectors, 243, 251
media requirements, 101
mesogenic
– group, 218
– side chain, 212
minimum mean square error filter, 313
mirror deflectors, 469
modulation
– codes, 283, 324
– encoding, 297
monomers, 171
moving optical head, 343
multifunctional polymer, 159
multiple parity, 323
multiplexing, 5, 7, 21, 22, 191, 361
multistore, 346

Nd:YAG laser, 401
noise, 273, 276, 331
– gratings, 79, 86, 175
nonsystematic, 294
nondestructive readout, 121
nonvolatile storage, 127
Nyquist, 310, 335

optical
– anisotropy, 214
– fiber delivery, 469
– quality, 101
– storage card, 478
– system misalignment, 293
– tape, 476
optically gated recording, 138
Optitek, 383
organic
– molecules, 171
– photopolymers, 106
– photorefractive (PR) materials, 159
orientational
– birefringence, 160
– photorefractivity, 160, 163

page motor, 349, 353
parity thresholding, 286
partial response
– (PR) equalization, 314
– maximum likelihood (PRML), 314
performance limits, 118
peristrophic multiplexing, 8, 25, 53, 242
persistent spectral hole burning, 477
phase code generation, 420
phase mask, 378
phase-change, 463
phase-coded multiplexing, 8, 25, 45, 419
phase-modulated reference beam, 395
phase-modulating device, 421
photoaddressable polymers, 209
photochemistry of azobenzene, 210
photodiode arrays, 274
photogeneration, 161
photoinduced noise gratings, 79
photoinitiated polymerization, 171
photon shot noise, 276, 294
photon-gated recording, 149
photopolymer recording, 345
photopolymers, 171, 199, 242, 359, 360, 395, 401
photorefractive
– crystals, 79, 104, 106
– $LiNbO_3$, 127, 419
– materials, 127, 369
– media, 149

- polymers, 159
- storage, 76
photosensitizer molecule, 173
photovoltaic
- damage, 293
- effect, 116, 130
pixel matched bit error rate, 372
pixel-to-pixel matching, 18
Pockels
- coefficient, 159
- effect, 76
point-spread function, 310, 335
polarization-preserving fibers, 370
poling field, 163
polymerization, 200
power amplifier, 237
pre-exposure processing, 184
predistortion, 321
PRISM, 16, 91, 369, 383
pseudo-blanced strip codes, 287
pseudo-random phase masks, 263

quantum efficiency, 275

radical generator, 173
RAMAC, 448
random
- access, 409
- error sources, 294
random-phase
- key, 419
- masks, 261
Rayleigh scattering, 76
RCA holographic memory, 12
read-only memory (ROM), 241, 399
readout rate, 278
recording
- channel, 91
- density, 244
- rate, 250
- schedule, 325
redundant storage, 6
Reed–Solomon codes, 297, 327, 339, 394, 459
reflection geometry, 29
refractive index changes, 177
remanent magnetization, 452
removable media, 346
replication, 404
response time, 117
Rician noise, 315, 321
roadmap, 463
Rockwell read-only demonstrator, 17
rotary servomotor, 354

second-harmonic generation, 467
seek time, 458
semiconductor lasers, 234
sensitivity, 103, 117, 138, 142, 153
servo system, 253
shift Bragg selectivity, 41
shift multiplexing, 8, 38, 361
- in-plane, 25
- out-of-plane, 27
shrinkage, 178, 185, 209
signal processing, 279, 283
signal-to-noise ratio (SNR), 66, 246, 314, 331, 419, 441
single parity, 323
sliding block decoder, 286
small polaron, 139, 152
- levels, 139
solid-state design, 409
solid-state laser, 360, 400
sorting detection, 284
space bandwidth product, 50, 273
space charge pattern, 128
space-charge field, 114, 115, 128
sparse code, 285
spatial light modulators (SLM), 4, 21, 241, 259, 310, 347, 372, 375, 377, 388, 400, 410
- characteristics, 246
spatial low-pass filter, 335
spatial multiplexing, 27
spatial segregation, 179
spatio-angular multiplexing, 410
speckle noise, 294
spectral sensitivity, 164
spectroscopy, 152
speed, 166
spindle motors, 458
spyrobenzopyrans, 477
$Sr_{0.61}Ba_{0.39}Nb_2O_6$ (SBN), 119, 137
standing wave, 3
stoichiometric lithium niobate (SLN), 138, 150, 151
storage

– capacity, 72
– channel, 295
– density, 23
strip code, 286
strip constraint, 287
super paramagnetism, 460
surface-enhanced Raman scattering, 476
surface storage density, 23
system optimization, 331
systematic errors, 293

Tamarack, 343
– multistore, 16
temperature fixing, 392
thermal fixing, 122, 127, 149
– cycle, 412
– uses, 107
thermal noise, 294
thermally assisted ionic fixing, 128
Thompson CSF read–write memory, 13
three-dimensional holographic disks, 50, 399
thresholding method, 323
transfer rate, 393
transit time, 250
transmission geometry, 28
transport, 161
twisted-nematic liquid crystal, 244
two-color holography, 149
two-color recording, 108
two-photon holographic, 138
two-photon recording, 122, 138, 146, 477

undoped stoichiometric lithium niobate, 139

van der Lugt geometry, 30
van der Lugt correlator, 431
video disc players, 464
viscosity, 184
Viterbi decoder, 315
voice-coil-driven actuator, 455
volatility, 105
volume diffraction, 31

Walsh–Hadamard codes, 47
wavelength Bragg selectivity, 45
wavelength multiplexing, 8, 25, 63, 231
Wiener filter, 313
WORM
write-once read-many (WORM), 149, 171, 175, 359, 360, 463

zero forcing equalization, 312

Springer Series in
OPTICAL SCIENCES

New editions of volumes prior to volume 60

1 **Solid-State Laser Engineering**
By W. Koechner, 5th revised and updated ed. 1999, 472 figs., 55 tabs., XII, 746 pages

14 **Laser Crystals**
Their Physics and Properties
By A. A. Kaminskii, 2nd ed. 1990, 89 figs., 56 tabs., XVI, 456 pages

15 **X-Ray Spectroscopy**
An Introduction
By B. K. Agarwal, 2nd ed. 1991, 239 figs., XV, 419 pages

36 **Transmission Electron Microscopy**
Physics of Image Formation and Microanalysis
By L. Reimer, 4th ed. 1997, 273 figs. XVI, 584 pages

45 **Scanning Electron Microscopy**
Physics of Image Formation and Microanalysis
By L. Reimer, 2nd completely revised and updated ed. 1998,
260 figs., XIV, 527 pages

Published titles since volume 60

60 **Holographic Interferometry in Experimental Mechanics**
By Yu. I. Ostrovsky, V. P. Shchepinov, V. V. Yakovlev, 1991, 167 figs., IX, 248 pages

61 **Millimetre and Submillimetre Wavelength Lasers**
A Handbook of cw Measurements
By N. G. Douglas, 1989, 15 figs., IX, 278 pages

62 **Photoacoustic and Photothermal Phenomena II**
Proceedings of the 6th International Topical Meeting, Baltimore, Maryland,
July 31 - August 3, 1989
By J. C. Murphy, J. W. Maclachlan Spicer, L. C. Aamodt, B. S. H. Royce (Eds.),
1990, 389 figs., 23 tabs., XXI, 545 pages

63 **Electron Energy Loss Spectrometers**
The Technology of High Performance
By H. Ibach, 1991, 103 figs., VIII, 178 pages

64 **Handbook of Nonlinear Optical Crystals**
By V. G. Dmitriev, G. G. Gurzadyan, D. N. Nikogosyan,
3rd revised ed. 1999, 39 figs., XVIII, 413 pages

65 **High-Power Dye Lasers**
By F. J. Duarte (Ed.), 1991, 93 figs., XIII, 252 pages

66 **Silver-Halide Recording Materials**
for Holography and Their Processing
By H. I. Bjelkhagen, 2nd ed. 1995, 64 figs., XX, 440 pages

67 **X-Ray Microscopy III**
Proceedings of the Third International Conference, London, September 3-7, 1990
By A. G. Michette, G. R. Morrison, C. J. Buckley (Eds.), 1992, 359 figs., XVI, 491 pages

68 **Holographic Interferometry**
Principles and Methods
By P. K. Rastogi (Ed.), 1994, 178 figs., 3 in color, XIII, 328 pages

69 **Photoacoustic and Photothermal Phenomena III**
Proceedings of the 7th International Topical Meeting, Doorwerth, The Netherlands,
August 26-30, 1991
By D. Bicanic (Ed.), 1992, 501 figs., XXVIII, 731 pages

Springer Series in

OPTICAL SCIENCES

70 **Electron Holography**
By A. Tonomura, 2nd, enlarged ed. 1999, 127 figs., XII, 162 pages

71 **Energy-Filtering Transmission Electron Microscopy**
By L. Reimer (Ed.), 1995, 199 figs., XIV, 424 pages

72 **Nonlinear Optical Effects and Materials**
By P. Günter (Ed.), 2000, 174 figs., 43 tabs., XIV, 540 pages

73 **Evanescent Waves**
From Newtonian Optics to Atomic Optics
By F. de Fornel, 2000, 277 figs., XVIII, 268 pages

74 **International Trends in Optics and Photonics**
ICO IV
By T. Asakura (Ed.), 1999, 190 figs., 14 tabs., XX, 426 pages

75 **Advanced Optical Imaging Theory**
By M. Gu, 2000, 93 figs., XII, 214 pages

76 **Holographic Data Storage**
By H.J. Coufal, D. Psaltis, G.T. Sincerbox (Eds.), 2000
228 figs., 64 in color, 12 tabs., XXVI, 486 pages

77 **Solid-State Lasers for Material Processing**
Fundamental Relations and Technical Realizations
By R. Iffländer, 2000, 230 figs., 73 tabs., XVI, 350 pages

78 **Holography**
The First 50 Years
By J.-M. Fournier (Ed.), 2000, 266 figs., XII, 460 pages

79 **Mathematical Methods of Quantum Optics**
By R.R. Puri, 2000, 13 figs., XIV, 285 pages

Location: http://www.springer.de/phys/

You are one **click** *away*
from a **world of physics** *information!*

Come and visit **Springer's**

Physics Online Library

Books

- Search the Springer website catalogue
- Subscribe to our free alerting service for new books
- Look through the book series profiles

You want to order? **Email to: orders@springer.de**

Journals

- Get abstracts, ToC's free of charge to everyone
- Use our powerful search engine LINK Search
- Subscribe to our free alerting service LINK *Alert*
- Read full-text articles (available only to subscribers of the paper version of a journal)

You want to subscribe? **Email to: subscriptions@springer.de**

Electronic Media

- Get more information on our software and CD-ROMs

You have a question on an electronic product? **Email to: helpdesk-em@springer.de**

Bookmark now:

http://www.springer.de/phys/

Springer

Springer · Customer Service
Haberstr. 7 · D-69126 Heidelberg, Germany
Tel: +49 6221 345 200 · Fax: +49 6221 300186
d&p · 64378/MNT/SF · Gha.

MIX
Papier aus verantwortungsvollen Quellen
Paper from responsible sources
FSC® C105338

If you have any concerns about our products,
you can contact us on
ProductSafety@springernature.com

In case Publisher is established outside the EU,
the EU authorized representative is:
Springer Nature Customer Service Center GmbH
Europaplatz 3, 69115 Heidelberg, Germany

Printed by Libri Plureos GmbH
in Hamburg, Germany